Available in the same series:

VULTEE AIRCRAFT 1932-1947	ISBN 0-913322-02-4
MARTIN AIRCRAFT 1909-1960	ISBN 0-913322-03-2
FAIRCHILD AIRCRAFT 1926-1987	ISBN 0-913322-04-0
NORTH AMERICAN AIRCRAFT 1934-1998, Volume 1	ISBN 0-913322-05-9

AF727409

NORTH AMERICAN AIRCRAFT 1934-1999
Volume 2

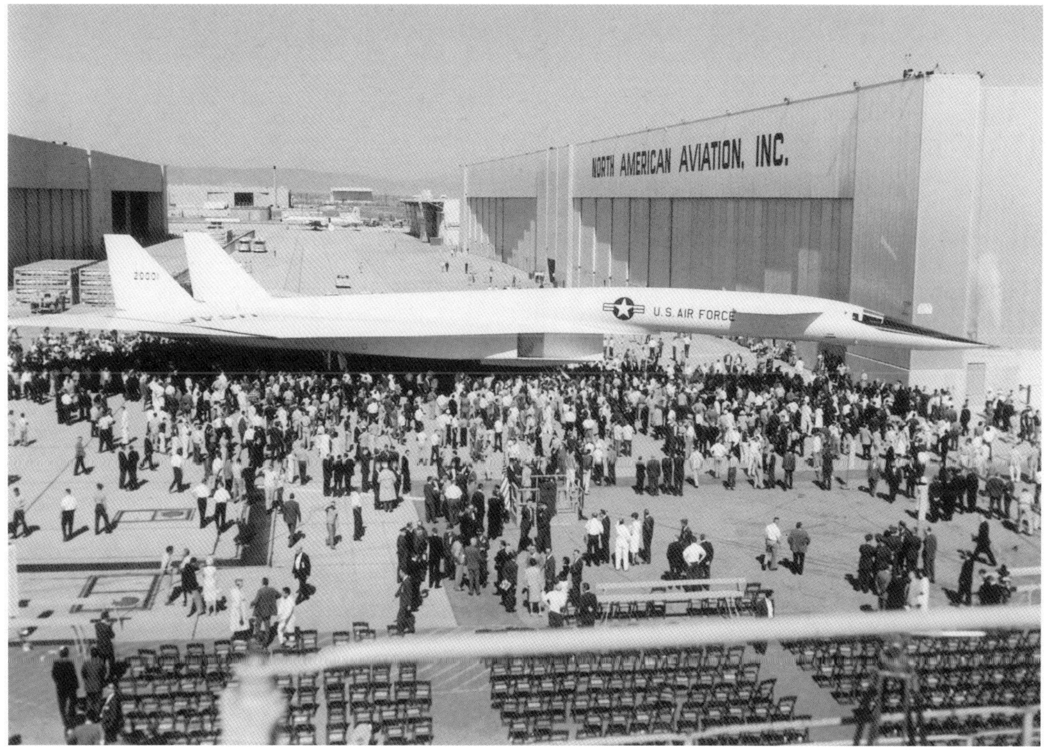

By the mid-1960s most of North American Aviation's new U.S. Air Force projects were generated at the company's Palmdale facility, adjacent to Edwards AFB, 50 miles north of Los Angeles. Here, on May 11, 1964, the first XB-70 was rolled out of its hangar to the amazement of employees, Air Force officers and members of the press.

Title pages: Side-by-side at the Los Angeles factory, the fifth F-100A-1, 52-5760, and a brand-new YF-107A, 55-5120.

North American

AIRCRAFT 1934-1999 Volume 2

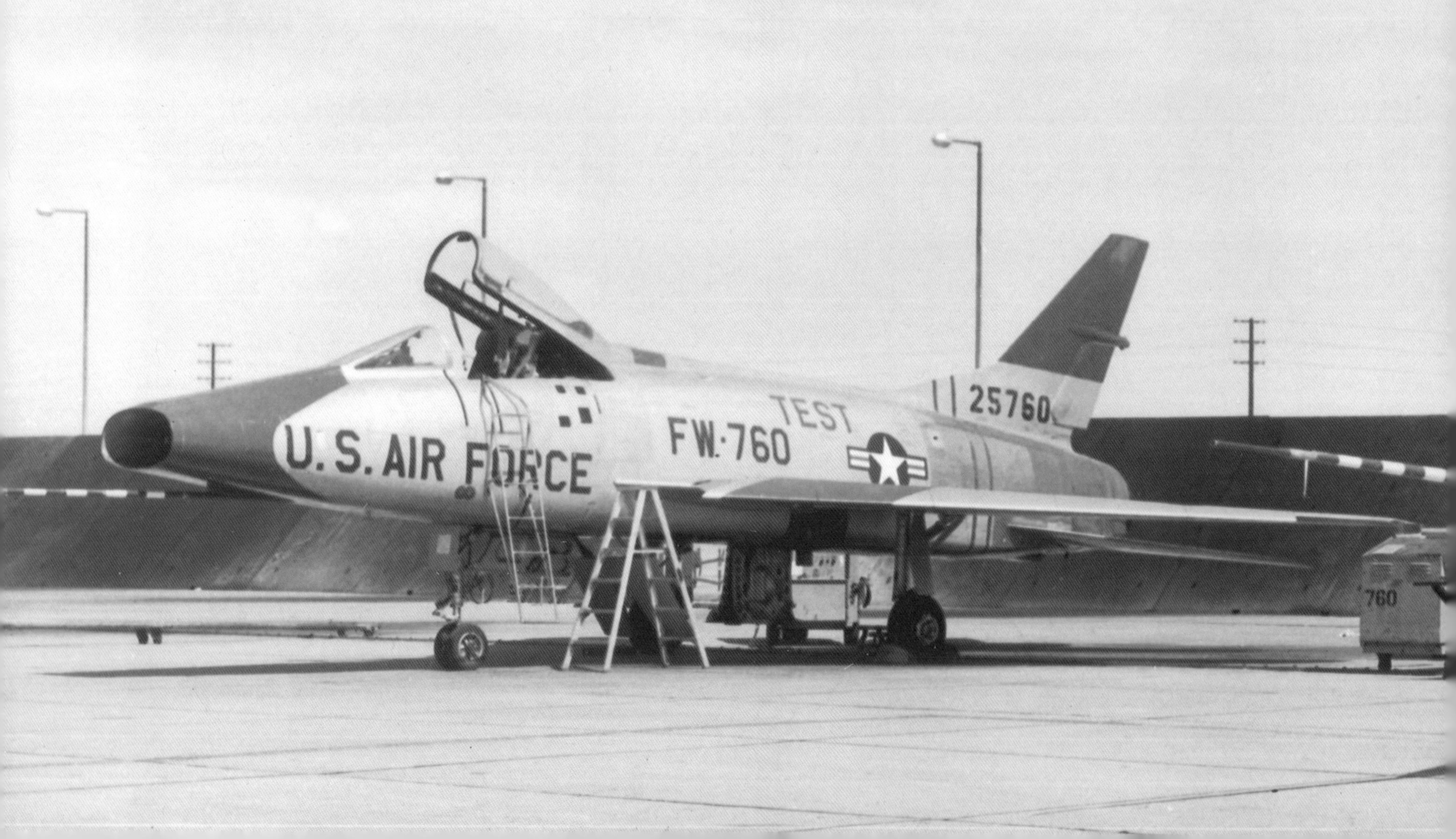

KEVIN THOMPSON

Narkiewicz // Thompson

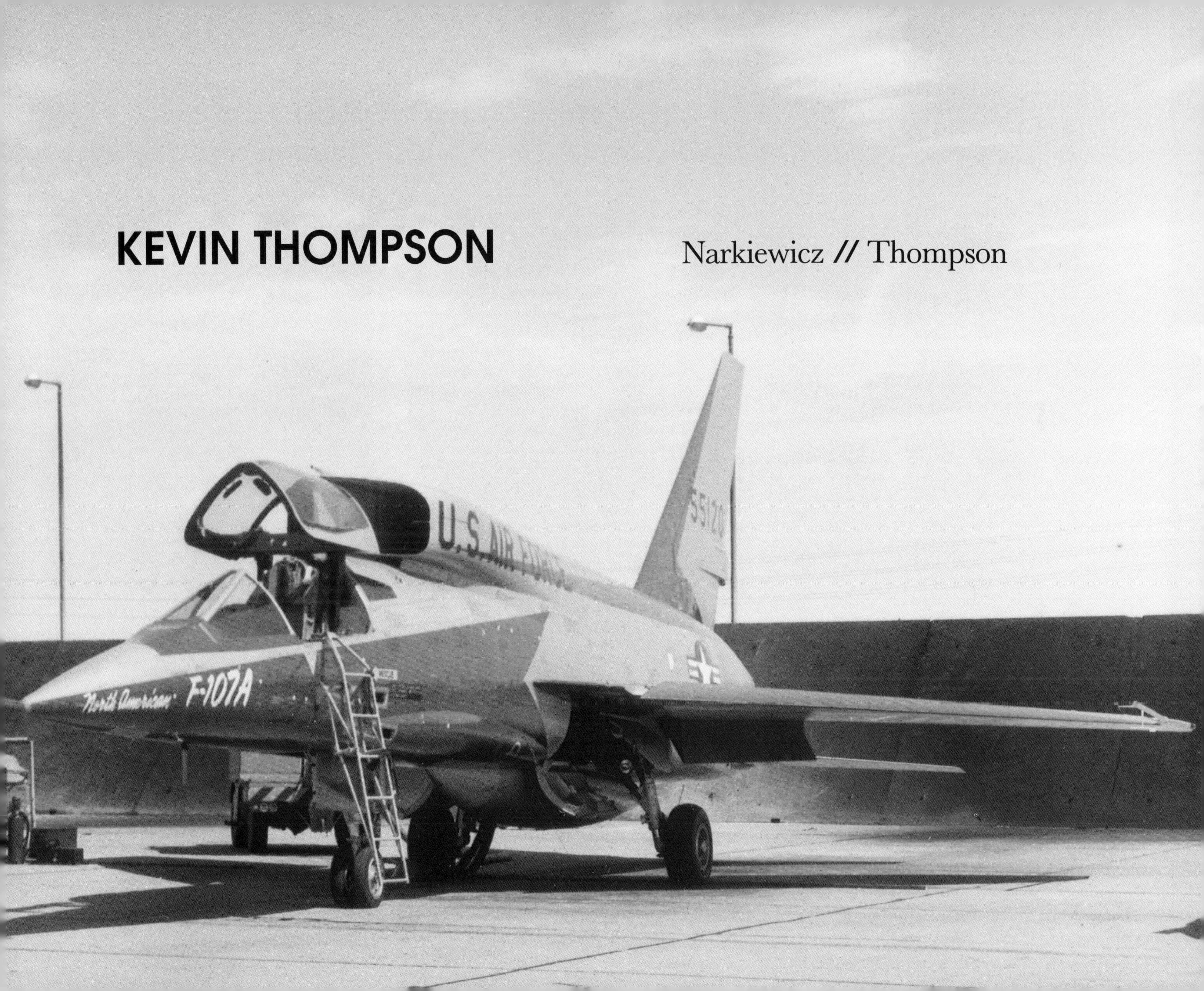

NORTH AMERICAN AIRCRAFT 1934-1999 Volume 2

by Kevin Thompson

Beverly S. Narkiewicz, Editor
Jonathan Thompson, Publisher

Library of Congress Catalog Card Number 98-60468
Copyright 1999 by Jonathan Thompson/Publisher

All rights reserved. No part of this book may be reproduced or transmitted in any form or by any means, electronic or mechanical, including photocopying, recording or by any information storage or retrieval system, without permission in writing from the publisher.

For Darlene

Narkiewicz//Thompson
1331 South Birch Street
Santa Ana, California 92707

(714) 542-7145 / FAX 542-1914

Printed in the United States of America
by Greens Inc. Printers & Lithographers

ISBN 0-913322-06-7

Acknowledgements

IN THE COURSE OF RESEARCHING and compiling the materials for this book, as well as for *Fairchild Aircraft 1926-1987* and *North American Aircraft 1934-1998*, Volume 1, the publisher has depended very heavily on the superb archives and generous cooperation of aviation historian Gerald H. Balzer. In addition to his great store of information, Gerry provided more than 200 photographs for this volume. We absolutely could not have produced the book without his help.

Norm Avery, the author of Volume 1, has continued to provide assistance on the whole NAA project, while Kevin Thompson has ably written the the text for the postwar North American and Rockwell models. Great help, especially on the Columbus projects, has come from Frank Compton, the undoubted authority on the subject of the Navy aircraft, as well as from Hal Andrews. The assistance of Dave Menard of the Air Force Museum has been essential on all aspects of USAF aircraft, especially the F-100. Lee Atwood continued his active involvement with NAA history until his death earlier this year.

Highly valued are the assistance and contributions of Peter M. Bowers, Commander Brian Dempsey USN Ret., Robert F. Dorr, Lowell Ford, Fred Freeman, George Gehrkens, Dan Hagedorn, Alan W. Hall, Lieutenant Jack Keegan USN Ret., Tony Landis, Mike Lombardi of The Boeing Company Historical Archives; Richard Marmo; Mike Mathews and Chris Wamsley of Boeing North American Aircraft Files; Brett Morse,

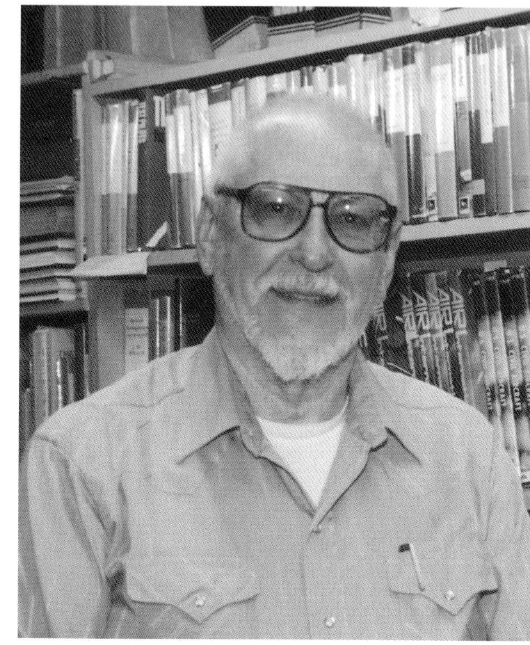

Gerald H. Balzer

Brian Nicklas, Stan Piet, Jim Tuttle, and of course the excellent historian, Ray Wagner of the San Diego Aerospace Museum. We also want to thank Joyce Baker and Raymond Puffer of the Air Force Flight Test Center, Edwards AFB; and Walt Bohl, Byron Calomiris, John Dzurica Jr., Harry Gann, Al Hansen and Tim Williams of the American Aviation Historical Society, an organization that continues to provide a solid foundation for aviation research.

Jonathan Thompson
Santa Ana, California, July 1999

Lee Atwood, 1904-1999

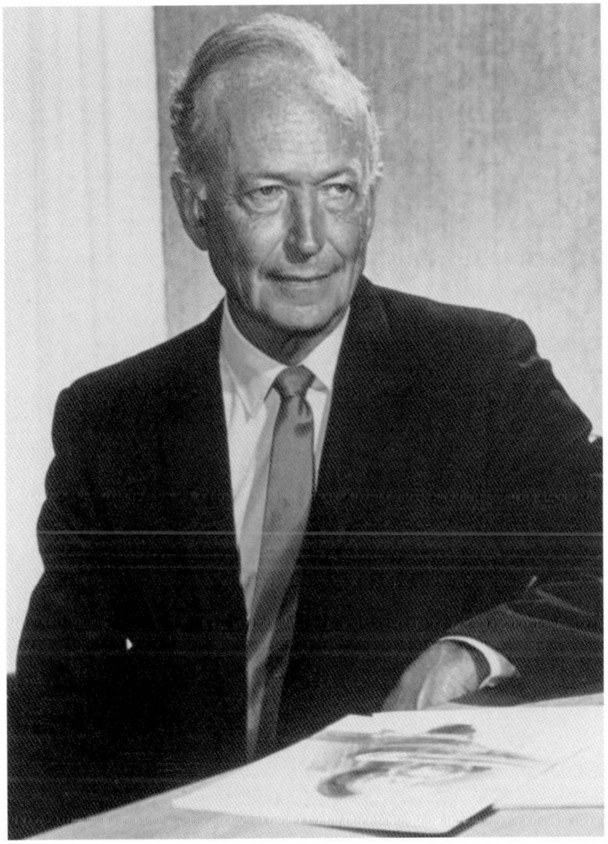

CERTAINLY ONE OF THE GREAT industrialists of the Twentieth Century, John Leland Atwood worked closely with James Howard Kindelberger in shaping North American Aviation, Inc., from 1934 through the 1950s. Succeeding "Dutch" as chairman of the board in 1962, the quietly dedicated Atwood ran the company with great expertise until his retirement from Rockwell International in 1978. He continued to nourish NAA history, particularly in clarifying the details of P-51 development, until his death in March 1999.

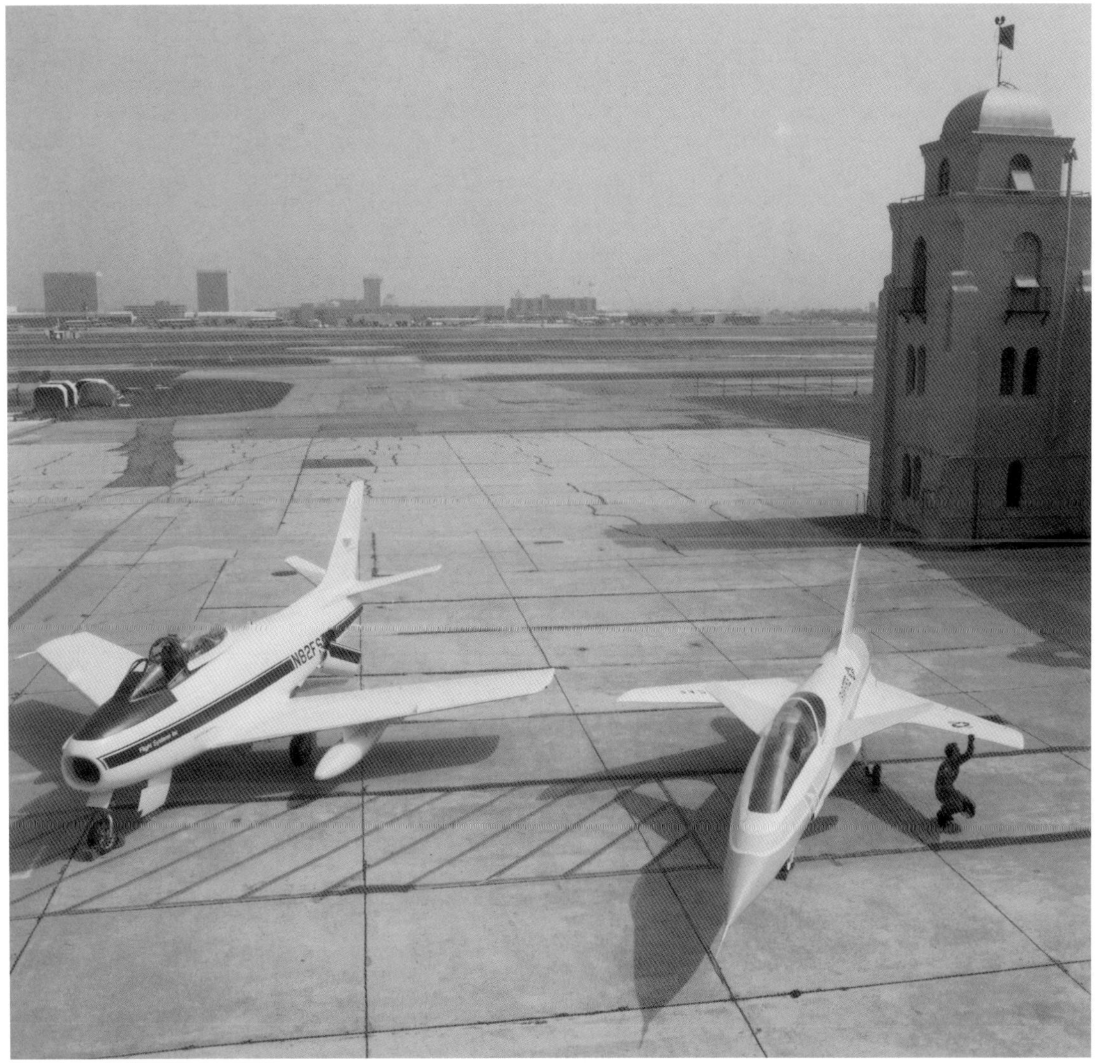

Posed next to the old 1930s Los Angeles airport tower, adjoining the original site of North American Aviation, Inc., in Inglewood, California, are an F-86 Sabre, N82FS, operated by Flight Systems Inc., and a mock-up of the USAF/DARPA Rockwell FSW Sabrebat Forward-Swept Wing Technology Demonstrator, later displayed at the Paris Air Show in 1979.

North American NA-142 XSN2J-1 NA-159>NA-307 XBT/T-28 / T-28A/B/C/D Trojan/ Nomad / NA-284 YAT-28E Turbo Trojan

EARLY IN 1945, two and a half years before starting preliminary design on the XBT (T-28) proposal, North American had begun design activity in support of the U.S. Navy's New Training Plan. This called for the introduction of two new trainers, primary and basic/advanced, which would cover an integrated program from ab initio to finale.

Fairchild won the primary competition and was given a contract to produce the XNQ-1, while North American received a go-ahead to build the XSN2J-1. The Navy issued a letter of intent to NAA on September 1, 1945, for two prototypes and one static test article.

Production plans for both the XNQ-1 and XSN2J-1 were soon shelved because of funding problems in Congress. The USAAF version of the XNQ-1, the T-31, was also dropped following a primary trainer fly-off at Wright Field, and Beechcraft's T-34 was selected for production. Congress then diverted funds earmarked for the Navy's integrated training plan to the T-6G modernization program.

When measured against the SNJ (AT-6), the XSN2J-1 was indeed a "next generation," high-performance trainer. The structure was optimized to meet the criteria for catapult take-offs and high-rate-of-sink, arrested landings, although there was only minor accommodation for quick access to equipment and systems com-

As its Navy designation indicated, the XSN2J-1 was a direct successor to the AT-6/SNJ Texan. Although retaining the earlier model's general layout and dimensions, it was significantly more powerful.

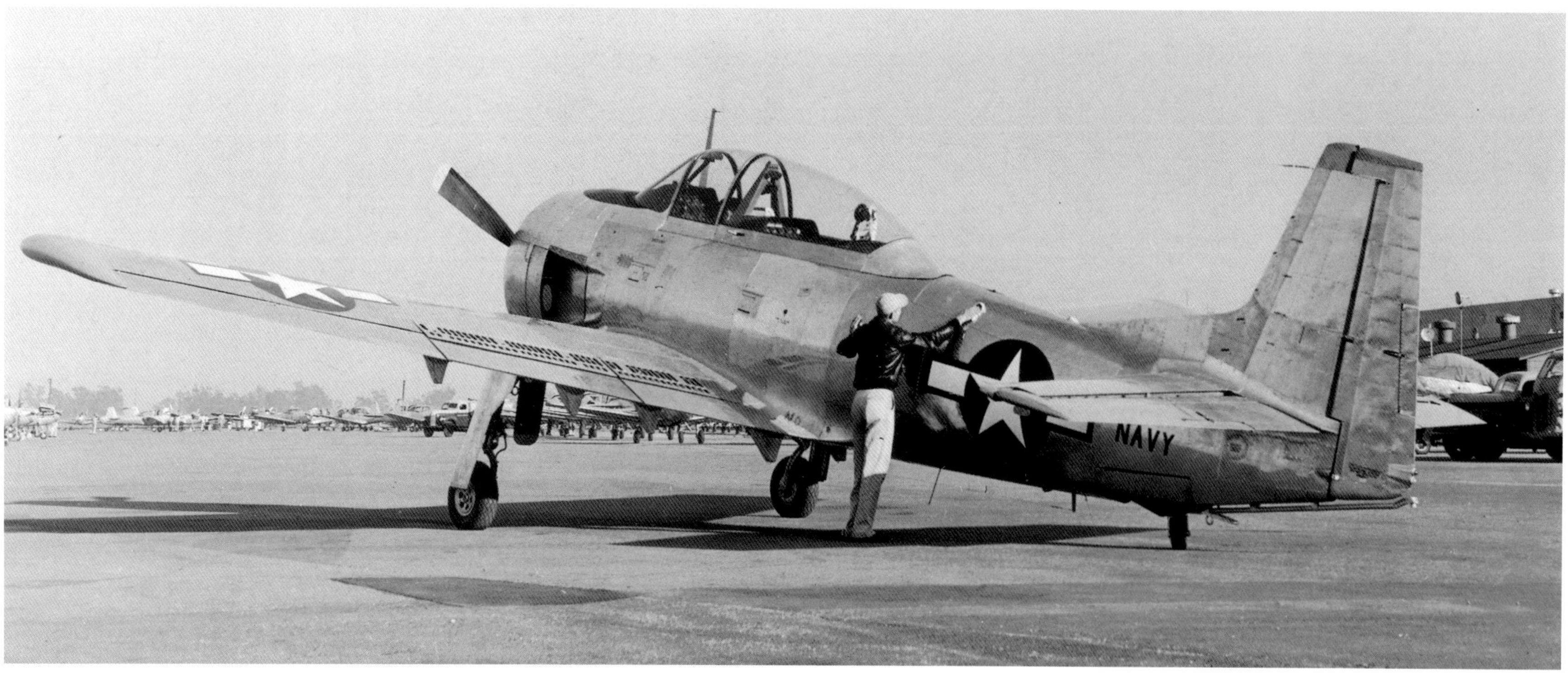

ponents. NAA engineering test pilot George Welch made the first flight on February 10, 1947.

In May, Howard Evans, NAA-LAD (Los Angeles) chief designer, received an advance copy of the preliminary requirements for the USAAF XBT trainer, intended to replace the aging AT-6 and serve as a transition aircraft for the new jet-age Air Force. At this point, the principal engineers for the preliminary design were selected: Frank Compton, project leader; Hugh Elkin, power plant installation; George Douglas, structural design; Al Arslan, stress analysis; and an aerodynamics team that included Mac Blair, Roy Jackson and Jack Bonner.

Initially the XSN2J-1 design was the baseline, but it quickly became evident that it was incompatible with the XBT requirements. The nosewheel configuration, the structural compromises necessary for improved maintenance access, and

Specifications: North American XSN2J-1

Description: U.S. Navy primary/intermediate trainer. Single-engine, low-wing monoplane of all-metal construction with retractable landing gear. Crew 2.

Engine: 1100-hp Wright R-1820-78 Cyclone nine-cylinder radial air-cooled.

Dimensions: Span 42 ft 11.25 in, length 33 ft 10 in, height 14 ft 7in, wing area 302.6 sq ft.

Weights: 5948 lb empty, 8406 lb max. take-off.

Armament: Two .50-caliber machine guns, provisions for bombs and rockets.

Performance: Maximum speed 308 mph, climb 2000 ft/min, service ceiling 32,200 ft, range 1600 miles.

Production: North American XSN2J-1

NA-142 XSN2J-1 BuAer 121449/121450 2

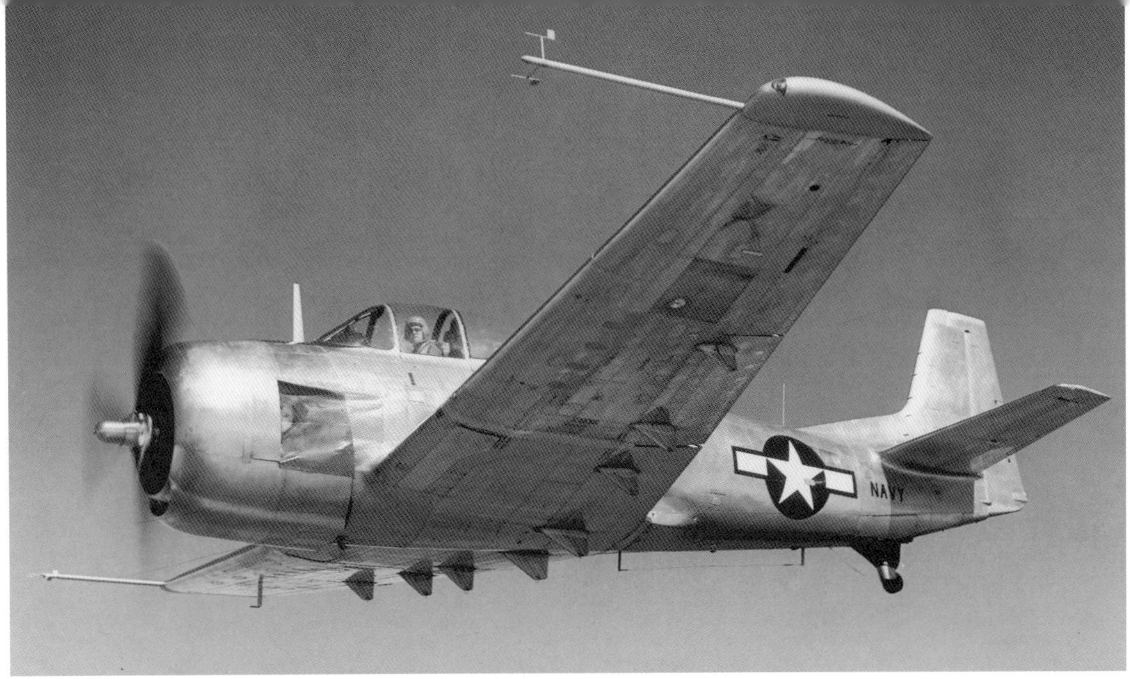

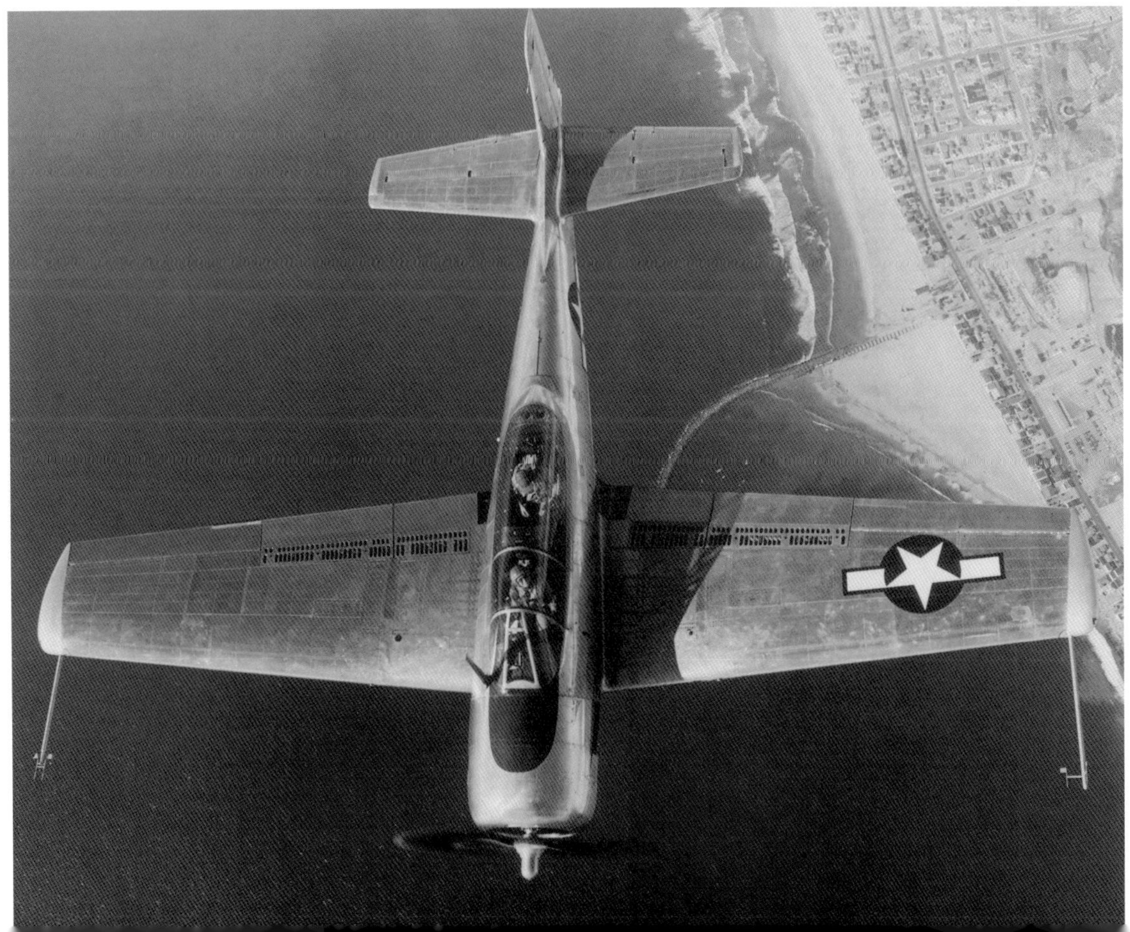

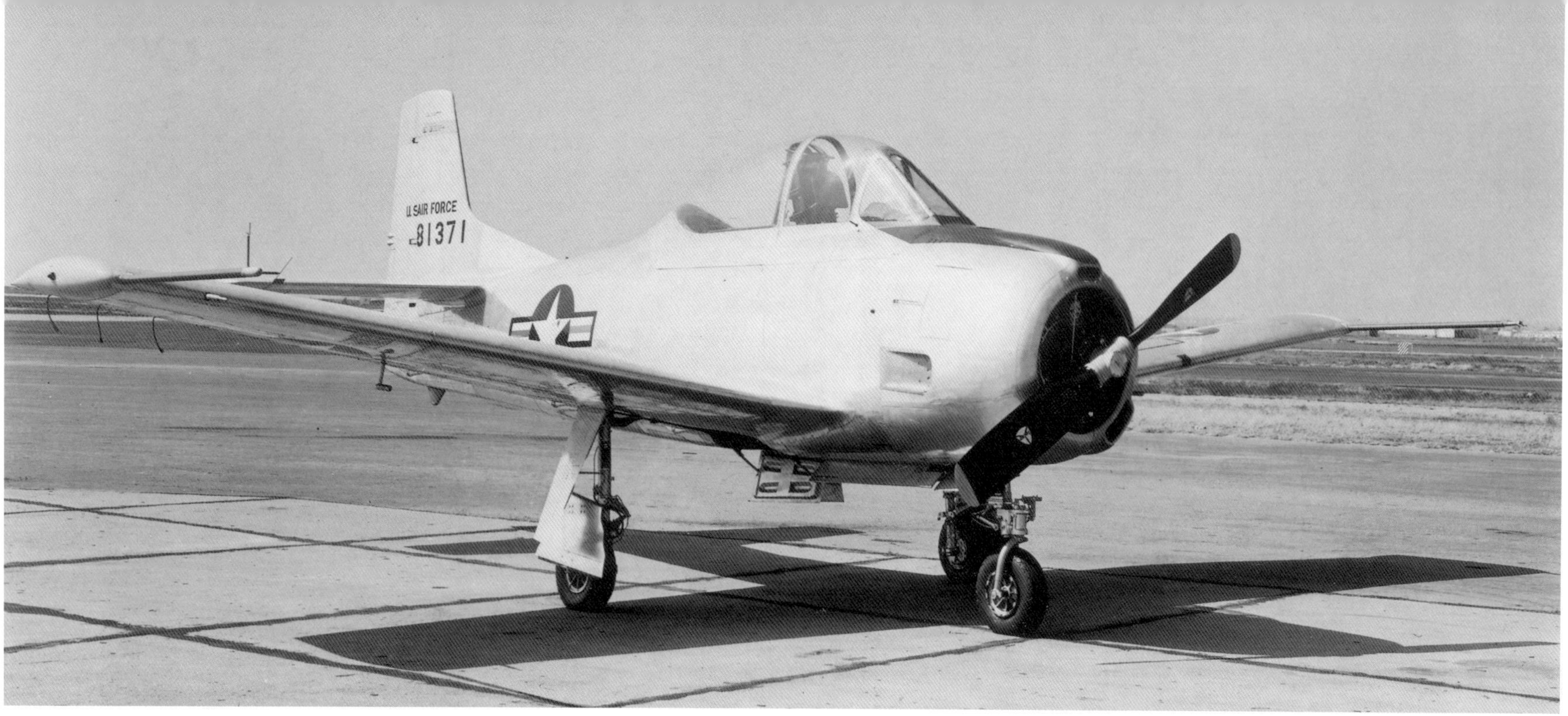

The first of two XT-28s, AF48-1371, showed a general similarity to the XSN2J-1, but was in reality a different design, initially less powerful and employing tricycle landing gear.

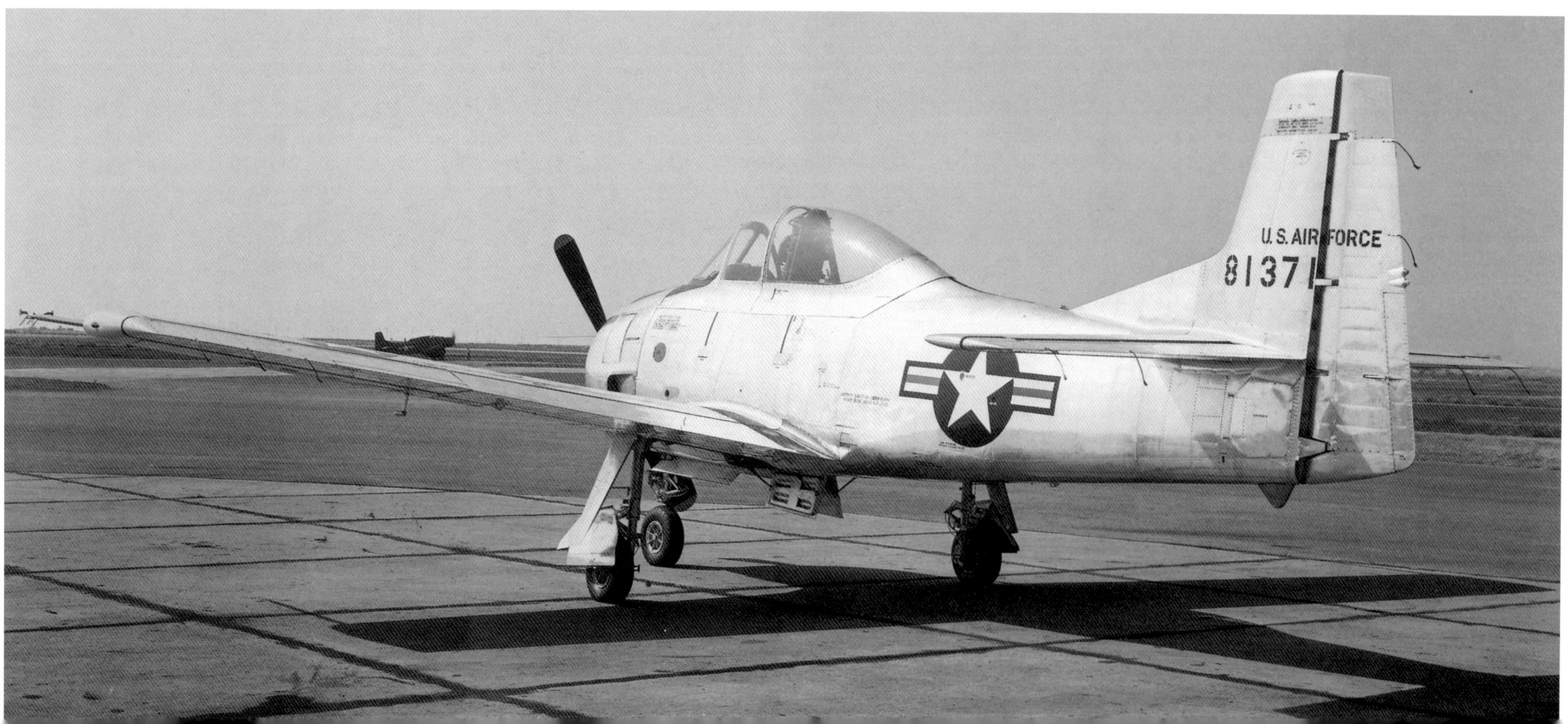

the differences between the USAAF's handbook of instructions for airplane design and the Navy's SWD24F design handbook required a fresh start for the XBT. Further, to meet performance guarantees with the 800-horsepower Wright R-1300-1A Cyclone engine, the wing and tailplane were designed around the NACA 64 series of low-drag airfoil sections. The only usable parts of the XSN2J-1 design were the canopy operating mechanism, the rudder pedal hangers and a few odds and ends. The XBT concept did have a strong family resemblance to other NAA aircraft, including the XSN2J-1, but even Fairchild's XNQ-1 was a dead ringer for the bigger "J bird."

With Douglas Aircraft known to be a strong competitor, North American needed to provide an outstanding design feature that would improve its chances of winning the XBT contract. In a meeting with the preliminary design staff, NAA's chief engineer Raymond Rice and Howard Evans decided that the company should emphasize its "superior maintainability," even though the structural compromises would increase the weight and reduce the performance. The design team was confident, however, that it could meet the performance guarantees specified by the customer.

Quick-opening bi-fold doors behind and under the rear cockpit afforded unrestricted access to instruments, electrical and hydraulic components, radios, oxygen equipment, aircraft controls and wiring. Up front through the nose gear well, access was provided to the engine accessory compartment by doors in the firewall, and to the engine cylinders and intake and exhaust elements by quick-opening, self-supporting cowling segments. Facing aft in the nose well, the control and support system components and wiring could be readily inspected and maintained.

On April 5, 1948, six months after the USAAF had become the independent United States Air Force, NAA received a letter of intent authorizing the construction of two XBT prototypes, redesignated as XT-28s, and one static test article. Dave Hoffman was assigned to the program as project engineer. The Air Force's post-award critique advised North American that the overriding factor in the selection was the emphasis on maintainability. In contrast, Douglas's proposal was quite complex, featuring an R-1300 buried in the fuselage and a shaft running forward to a nose-mounted propeller; an advantage was supposed to be the possibility of converting to turbojet power but the USAF Materiel Command was not impressed.

The XT-28 had a wingspan just over 40 feet, comparable with the T-6/SNJ and Vultee BT-13/SNV at 42 feet. The cockpit was the conventional tandem arrangement with a two-piece sliding canopy. Tricycle landing gear was employed, with a hydraulically-operated steerable nose wheel standard on USAF aircraft produced in the 1950s. This was deleted on Navy T-28B/C models because maneuverability and deck handling required unrestricted rotation of the nose gear in carrier operations. An underfuselage speed brake, located just aft of the main landing gear, was later deleted on production T-28As.

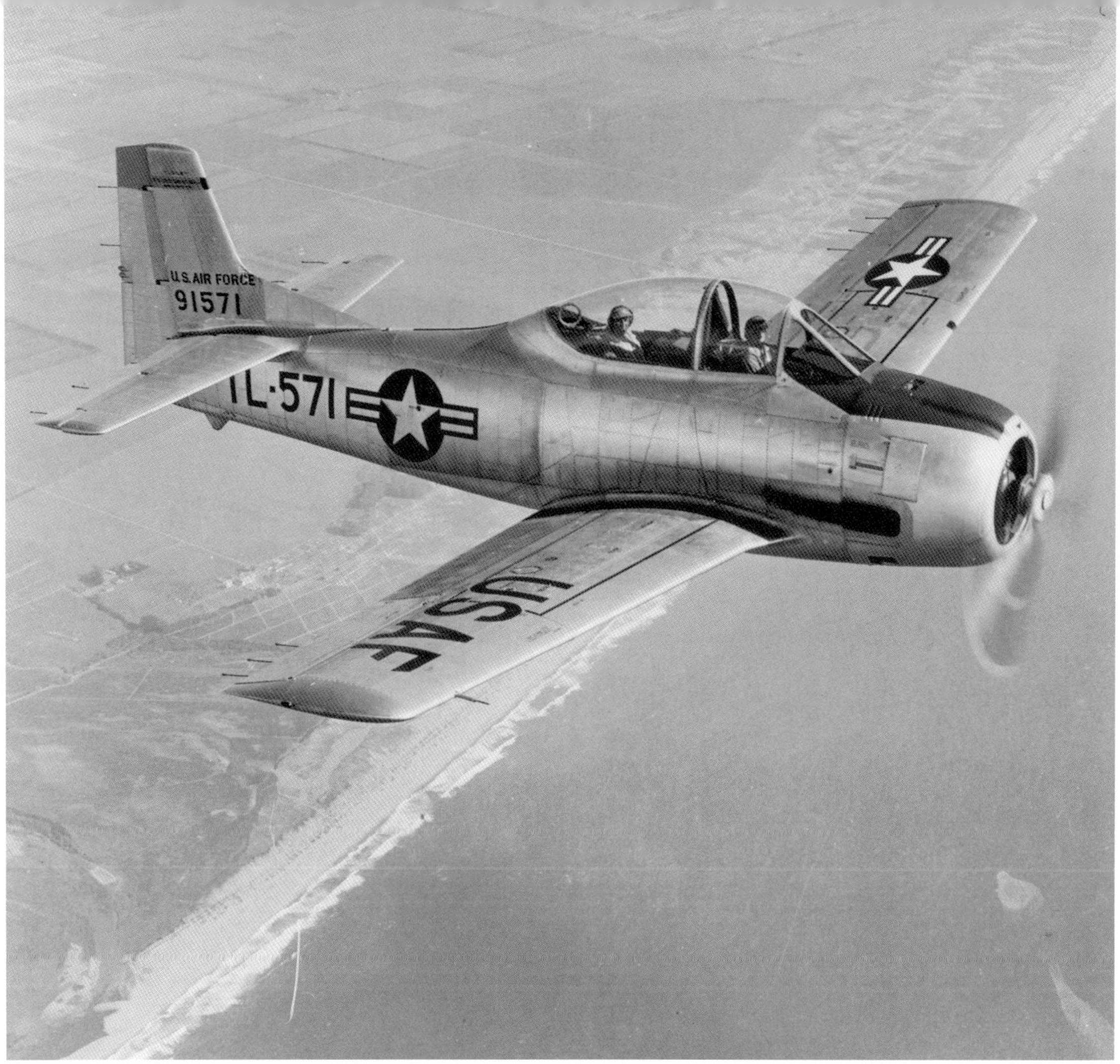

The first production model of the Trojan was the T-28A, of which 1193 were built for the Air Force. Above is AF49-1571.

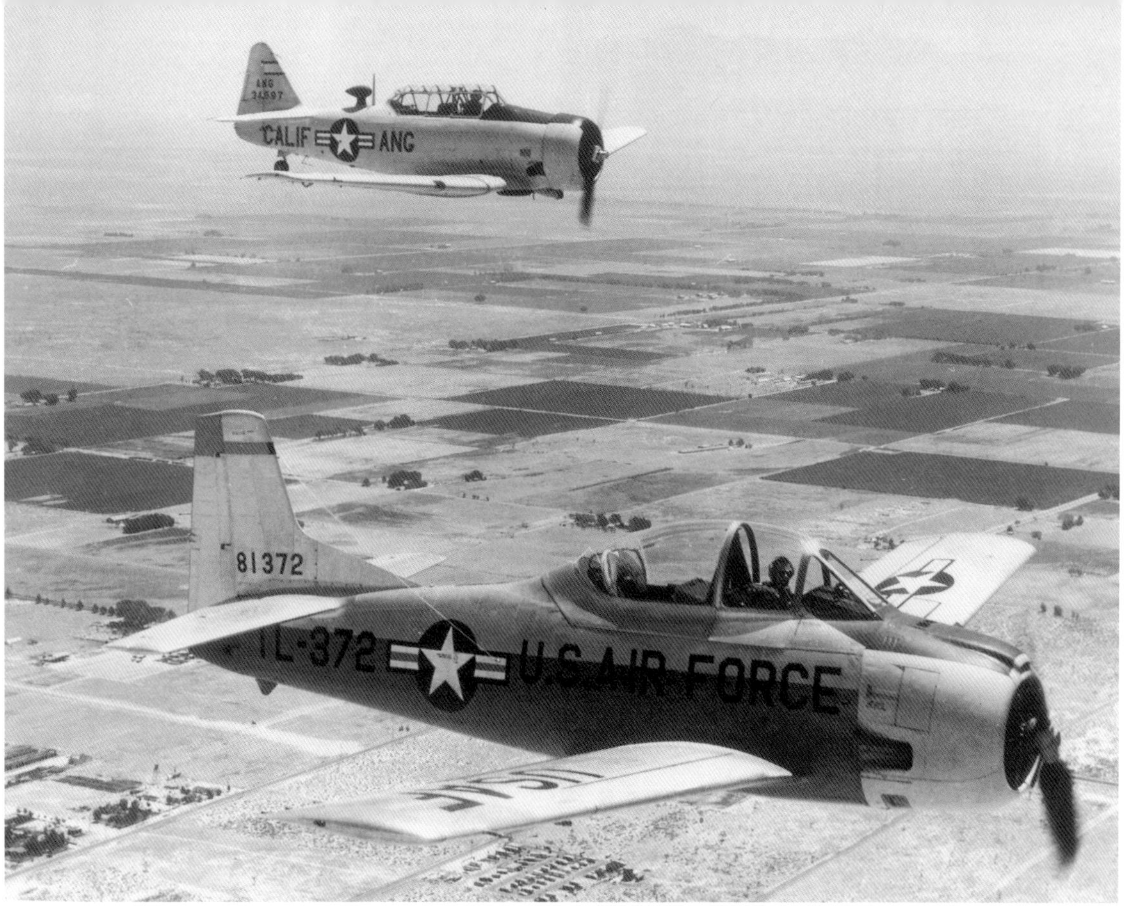

In the early T-28s the R-1300-1A engine, in effect a halved twin-row R-2600, suffered failures from undamped crankshaft vibrations until a damping system was included on all Training Command T-28As. The engine drove a 10-foot diameter AeroProducts two-blade propeller.

The first flying prototype XT-28, AF48-1371, took off at Inglewood on September 24, 1949, with engineering test pilot Skip Ziegler at the controls. The early flights showed excellent performance, including a top speed of 292 mph and a range of 800 nautical miles. The second prototype, AF48-1372, joined the flight test program shortly thereafter and the promising results led to an Air Force order for 266 production aircraft designated T-28A and subsequently named Trojan.

The first production T-28As entered service with the USAF Air Training Command in April 1950. With communication/navigational equipment installed, the T-28A weighed approximately 365 pounds over specification; compared to previous training aircraft, it realized an improve-

Left, the old and the new: a California Air National Guard T-6G keeping company with the second XT-28, AF48-1372. Below, a T-28A, AF51-3508, from the third production run.

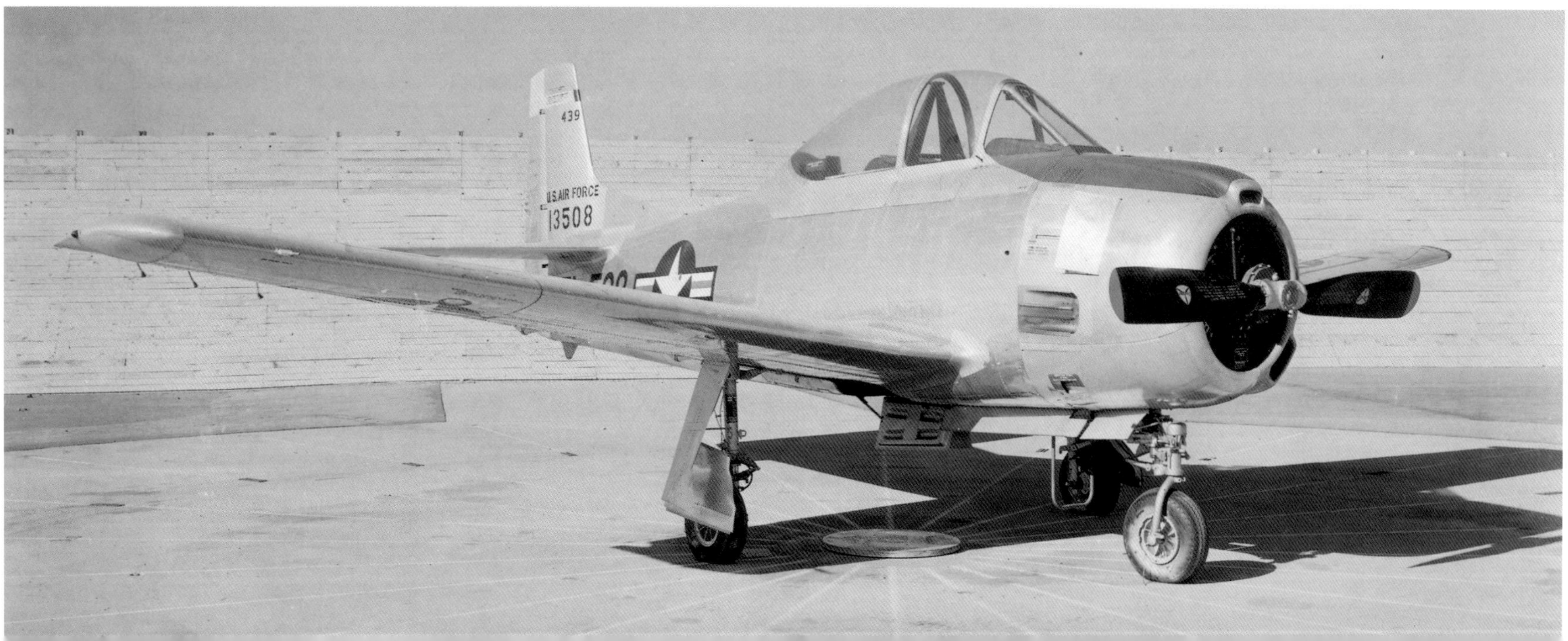

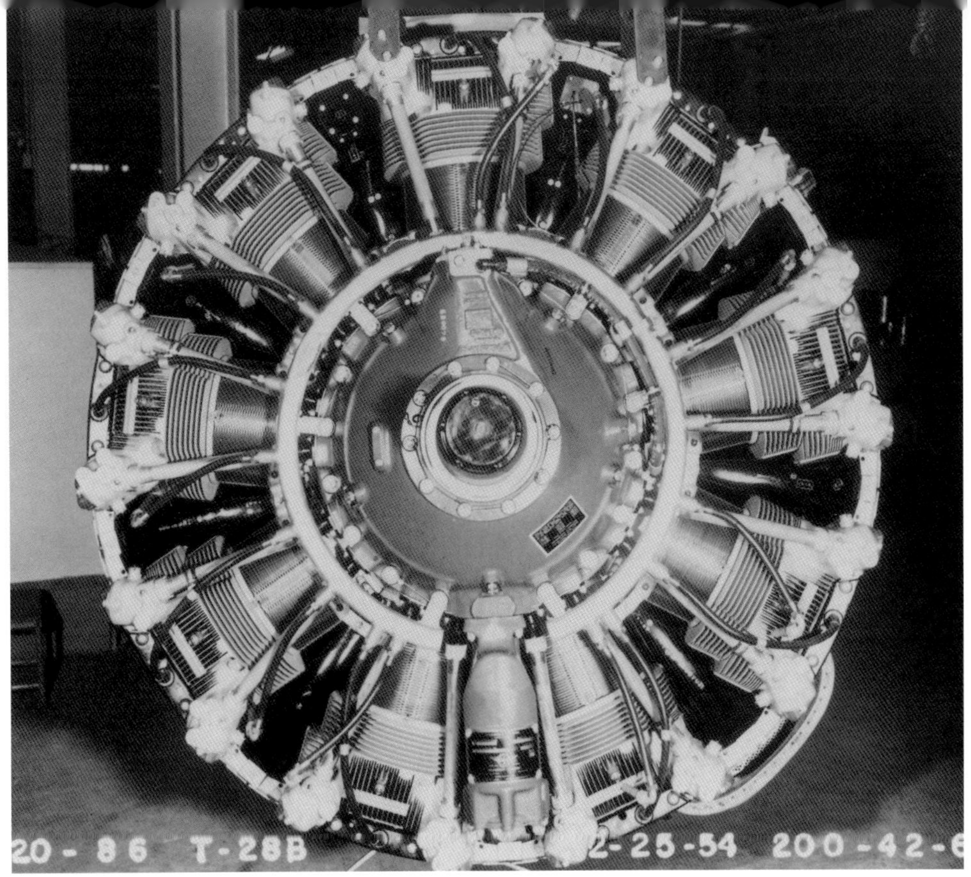

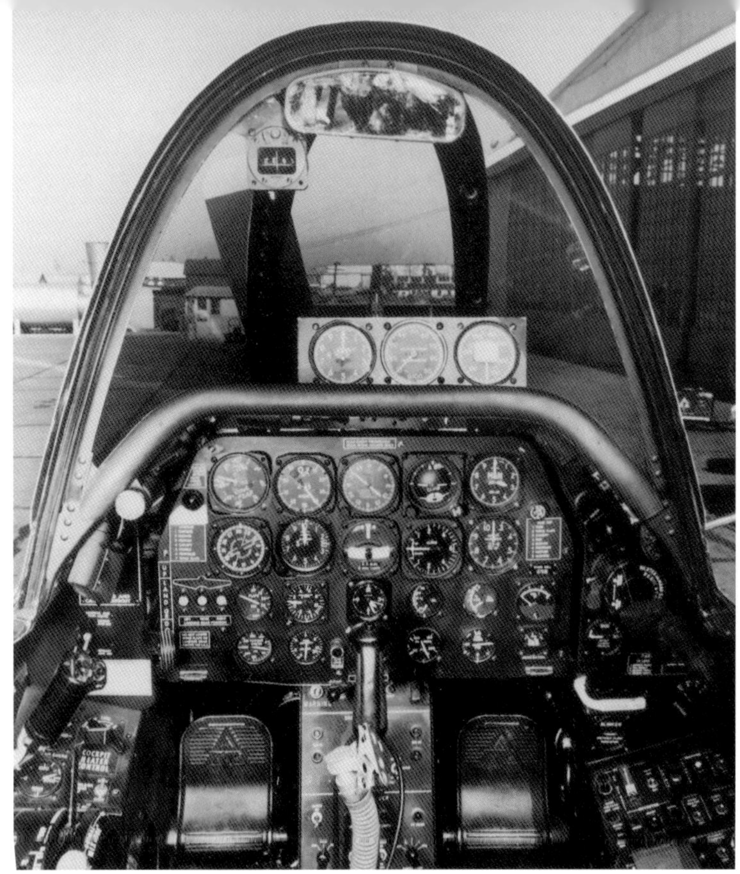

Above, the Navy T-28B's 1425-horsepower Wright R-1820 Cyclone radial. Right, the T-28B front cockpit. Below, the Navy's Trojan was easy to identify with its hefty 3-blade propeller.

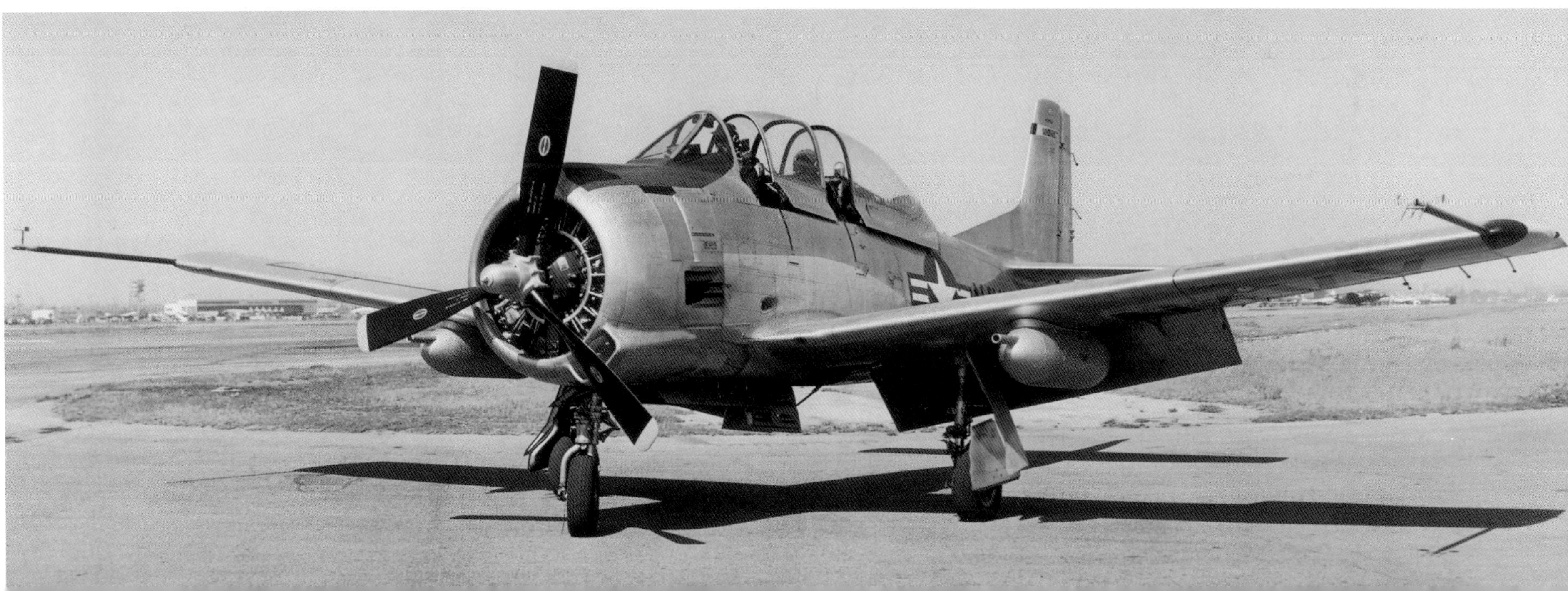

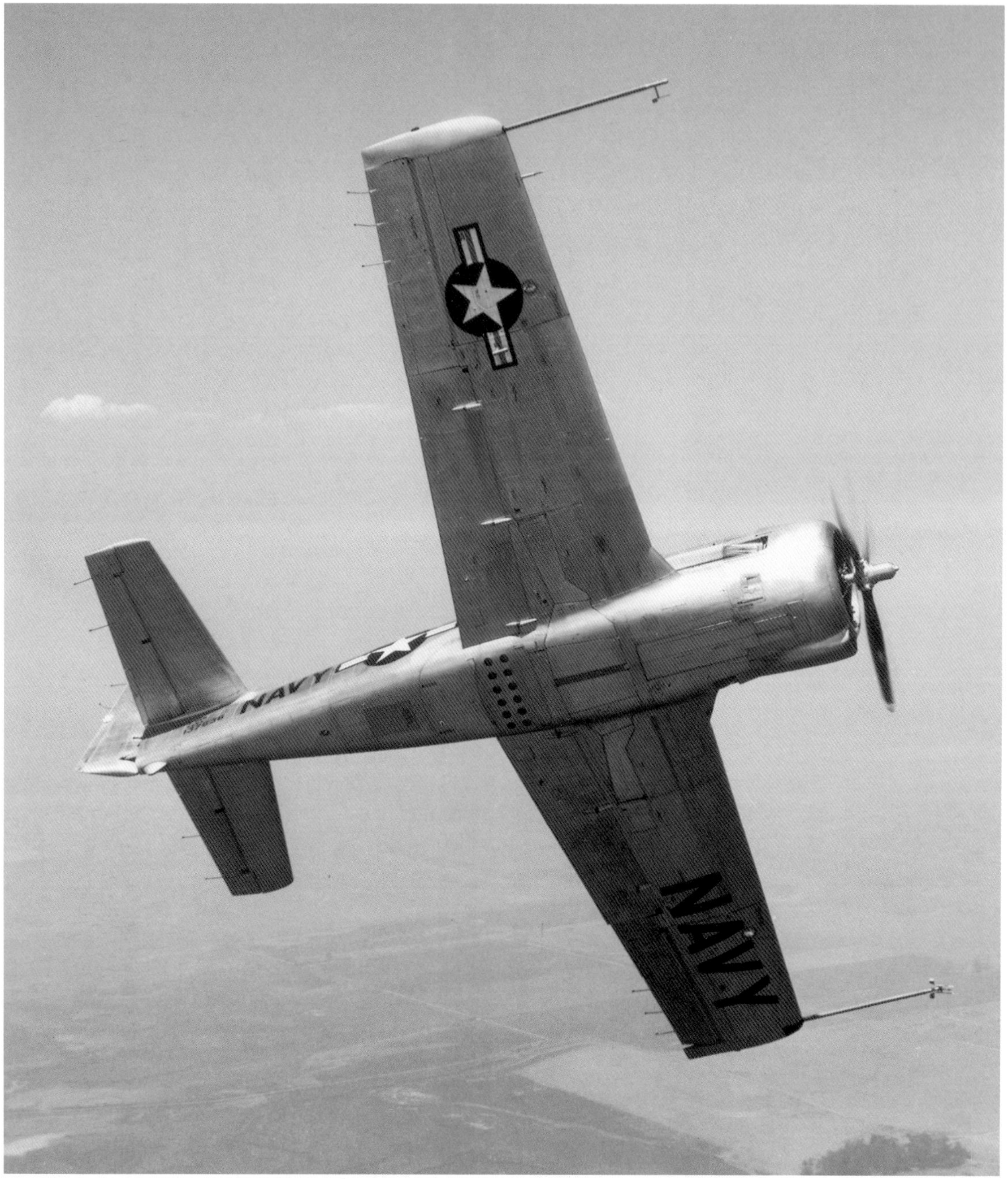

An underside view of the first T-28B, BuAer 137636. The perforated panel behind the gear doors is a retractable air brake.

ment in power loading of only 10 per cent, but this was deemed adequate for the primary/basic training curriculum. (The power loading of the 450-horsepower BT-13A was 9.1, the 600-horsepower T-6G's was 8.7 and the T-28A's 7.9.) The Trojan was well received by flight instructors and ground crews alike. It exhibited excellent handling qualities, with exceptional spin departure resistence and stall recovery, markedly improved over the venerable Texan.

Improvements made to the T-28A during the production run included the addition of two 26-gallon outboard wing-mounted fuel tanks, raising the total capacity to 177 gallons and increasing the still air range to 1175 nautical miles. The high-profile cockpit canopy was lowered 4 inches and the overturn pylon removed.

A wide variety of external stores could be carried for armament training. Either .30-caliber or .50-caliber gun pods were fitted, with 100 rounds of ammunition within the pod. A bomb rack of up to 500-pound capacity could be fitted under each wing, and could also accommodate various rockets. The front cockpit was fitted with a Mk 8 Model 5 gunsight and an armament control panel; a gun camera was mounted in the left wing leading edge to record armament training results.

T-28A production totaled 1194 aircraft, most of which were used by the USAF. A few were used by the U.S. Army for photographic work during parachute tests at Fort Bragg, North Carolina, during the 1960s. Approximately 150 T-28As were exported to Argentina, the Dominican Republic, Ecuador, Honduras, South Korea, Mexico, Nicaragua, the Philippines and Saudi Arabia. Civilian instructors under contract to train USAF pilots in the mid-1950s operated a few T-28As. Most of these civilian basic training facilities were in the southern part of the United States. T-28As were phased out of the Air Training Command in 1956, and were transferred to Air National Guard units to replace aging F-51 Mustangs. ANG units phased out the last T-28As in 1959 as they converted to jet fighter aircraft.

In the early 1950s the U.S. Navy once again gave serious thought to replacing the SNJ with a

T-28D

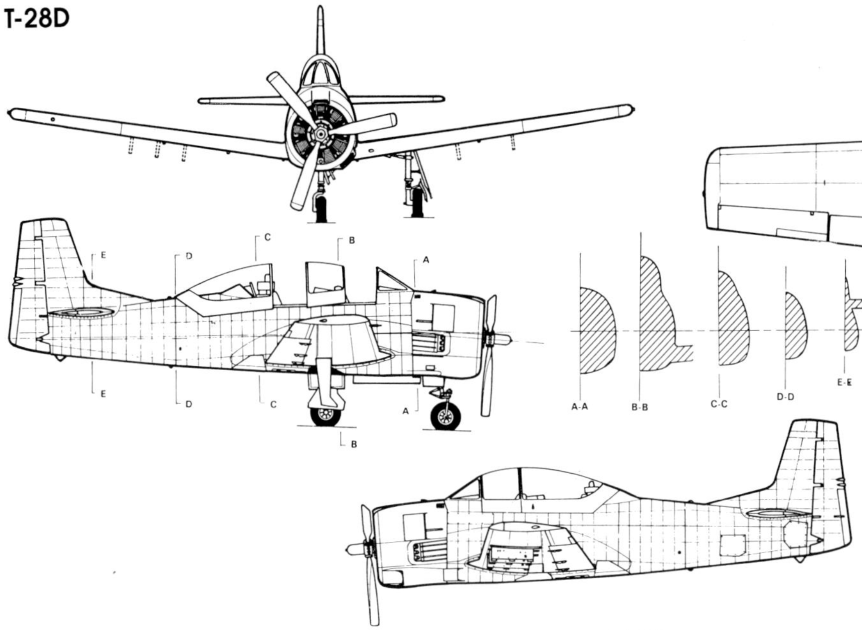

1/144 scale

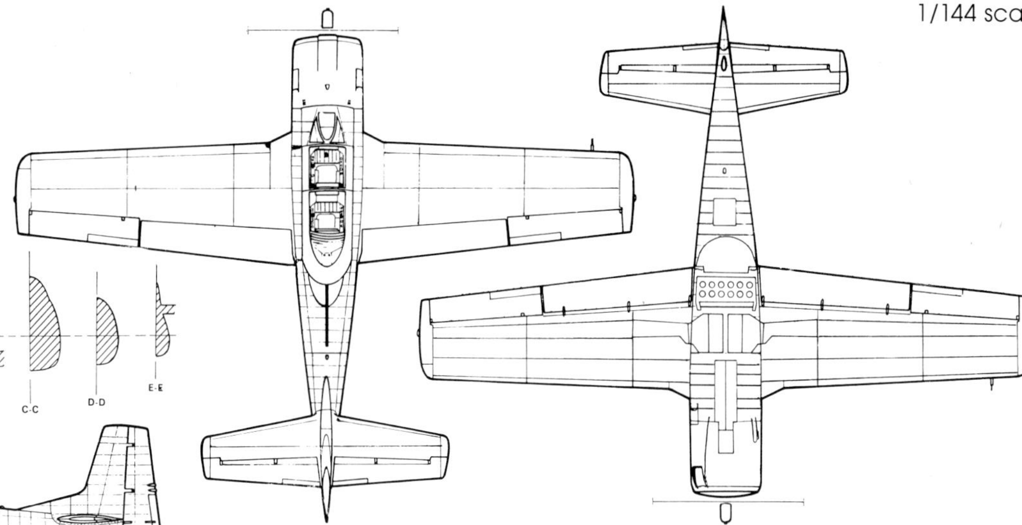

Bob Hunt drawings, originally published in Aviation News *in 1/72 scale*

T-28A

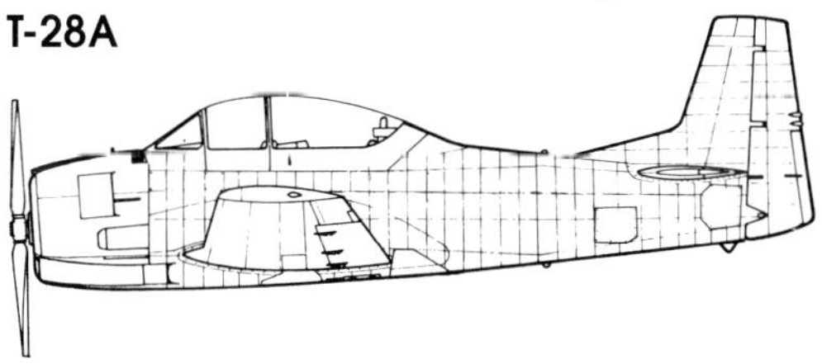

T-28B

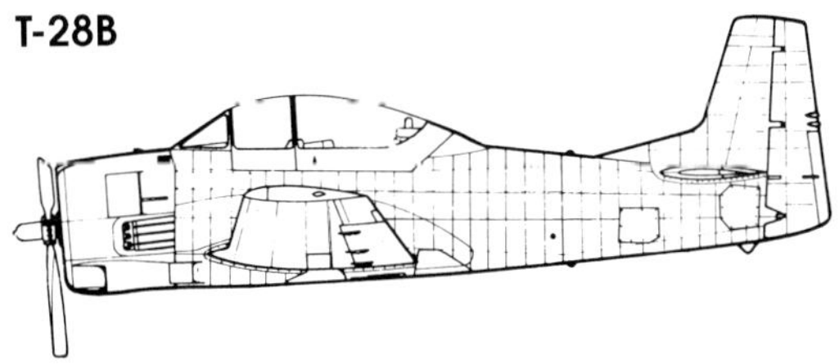

T-28C

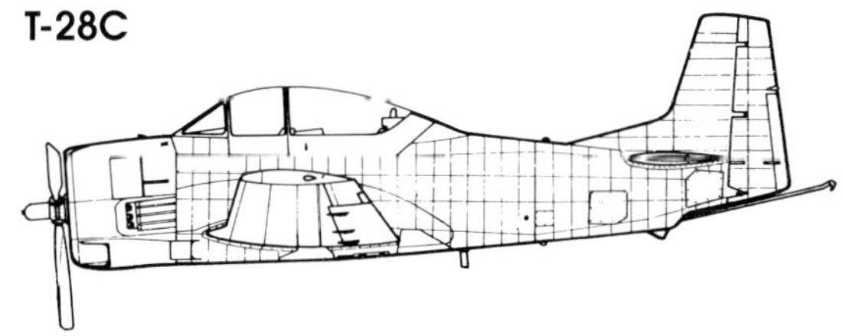

modern trainer. Several years had passed since the Navy issued the request for proposal that eventually led to the XSN2J-1 program. By then the SNJ was considerably outdated. Government policies concerning new aircraft funding and procurement had also changed, the focus now being on joint service procurement of a single type. In step with the times, a Navy test and training team evaluated the T-28A in early 1952. Following the tests, the Bureau of Aeronautics elected to power the Trojan with the 1425-horsepower Wright R-1820-86 engine; wartime experience with the R-1820 in the SBD and FM-2, and its selection as the powerplant for the recently procured S2F, led the evaluators to choose it for the Navy's version of the Trojan. The only other engine available in the 1200/1500-horsepower class was the twin-row Pratt & Whitney R-1830, too long and heavy for installation without major structural modifications to balance the aircraft.

NAA immediately went to work modifying a T-28A airframe to accept the larger engine. This aircraft, known in-house as the NA-199, was designated T-28B. The cowling underwent a major redesign, increasing the diameter to accommodate the R-1820 and the larger oil cooler duct. The split exhaust stack arrangement was revised for the nine-cylinder geometry. The T-28B was now fitted with a three-blade Hamilton Standard propeller of 10-foot 1-inch diameter. The speed brake originally used only on the XT-28 prototypes was standard on the T-28B. The nose gear was revised as per Navy requirements by using a smaller nose wheel with a high-pressure tire. Revised wingtips increased the span by six inches.

Engineering test pilot Bob Hoover flew the prototype T-28B from NAA's facility at Downey, California, on April 6, 1953. The annoying staccato sound of the T-28A's split-manifold 7-cylinder R-1300 was replaced by the R-1820's throaty truculence, and the 78-percent increase in power more than offset the 1700-pound gain in

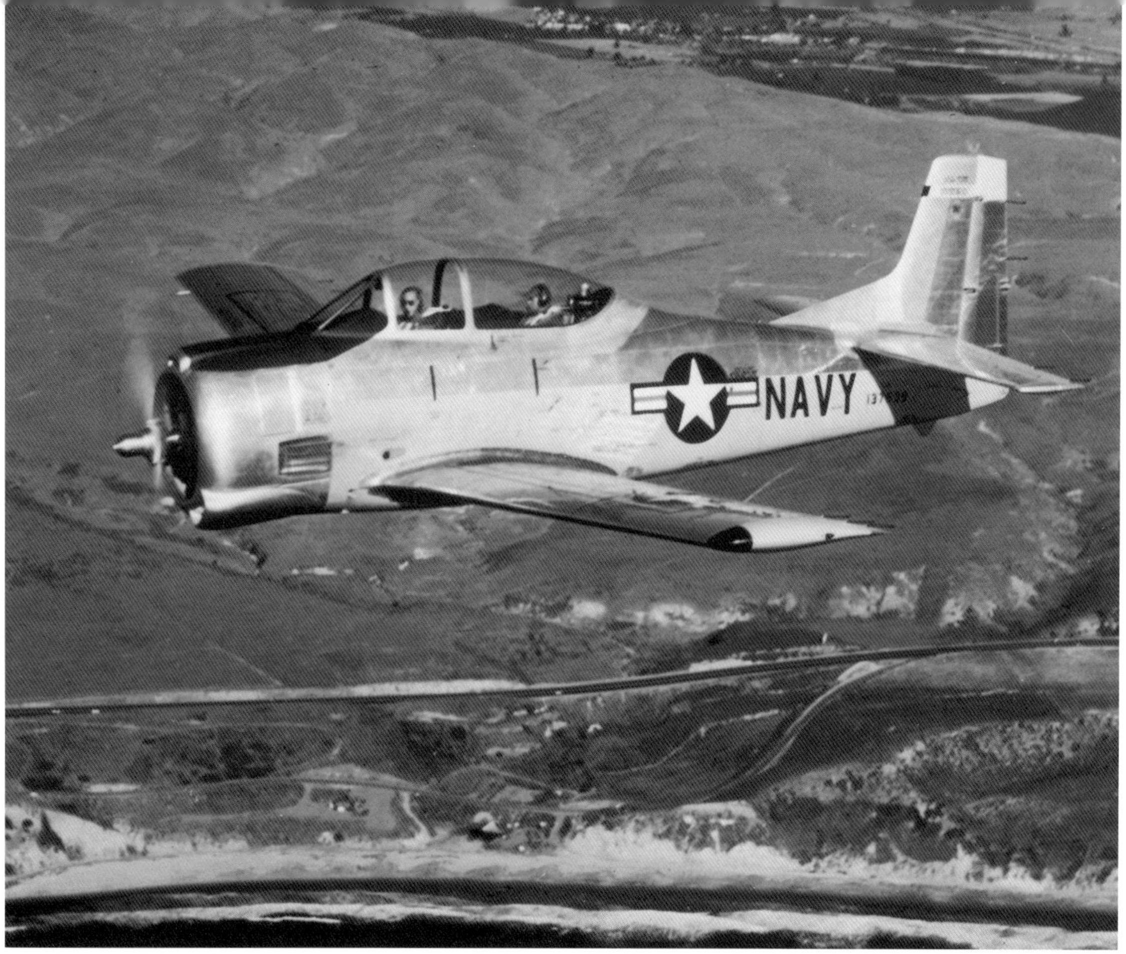

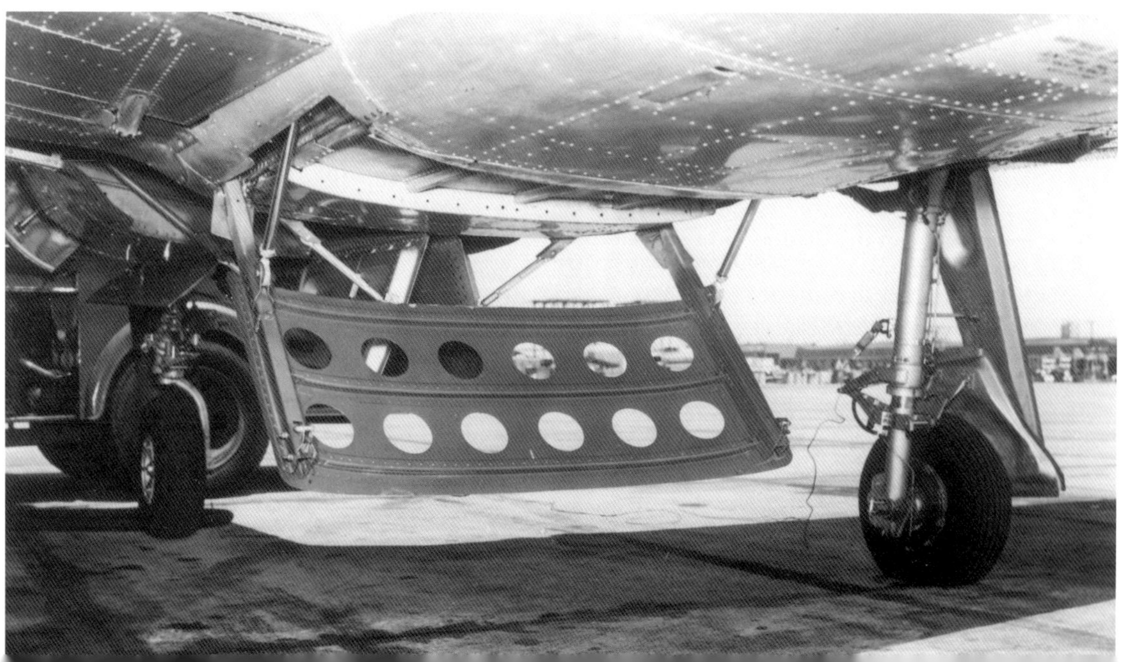

weight. The power loading was now 5.8, close to the P-51D Mustang's 5.6, and the T-28B was a real performer, especially in climb and turning performance. The speed was boosted to 346 miles per hour and the ceiling to over 37,000 feet.

Following manufacture of 440 T-28Bs at Downey, the production program, including the Los Angeles-based engineering, was moved to NAA's Columbus, Ohio, facility. There, an additional 51 B models were produced prior to the start-up of T-28Cs designed for carrier operations.

T-28B deliveries to Naval squadrons began in 1954, and the majority of these were used to replace the SNJs in primary and intermediate training roles. The two main training facilities were NAS Pensacola, Florida, and NAS Meridian, Mississippi. At the height of the Trojan's career a combined total of eleven training squadrons were based at these two Naval Air Stations.

T-28Bs were also used for various other duties, including as drone controllers at the Navy Missile Test Center at Point Mugu, California, and the Naval Weapons Test Center at China Lake, California. These aircraft, designated T-28BD (DT-28B after 1962), carried gun pods and were used to destroy runaway drones. The T-28B had the same external stores capability as the T-28A, but could also carry target-towing equipment and other loads. Several surplus Navy T-28Bs were eventually acquired by the USAF and U.S. Army during the late 1950s for various duties previously performed by T-28As.

A single T-28B was ordered from NAA by the Japanese Air Self Defense Force in June 1954 for use as a photo reconnaissance aircraft and was designated RT-28B. This aircraft had a camera bay installed in the underfuselage where the speed brake was normally located. The RT-28B served with the 501st Reconnaissance Squadron of the JASDF.

After its experience with the T-28B, the Navy determined that the Trojan would be even more

Above, the fourth production T-28B, BuAer 137639, shows the light gray paint along the fuselage behind the exhaust pipes. Left, a close-up of the air brake in extended position.

versatile if it were modified for carrier operations. It would also allow the training mission to operate with a single aircraft type, which is what the Navy had in mind ever since the XSN2J program was envisioned.

In September 1954, NAA began modifying two T-28B airframes to meet the naval carrier requirements. To absorb the stress of carrier landings, the rear fuselage structure required strengthening. A modified rear section was fitted with an arrestor hook and the rudder reshaped to allow for hook clearance. Designated NA-252, this configuration became the T-28C. The modification was completed in less than a year and the first T-28C flew on September 19, 1955.

The modifications for carrier operations increased the gross weight by approximately 450 pounds, with a negligible loss of performance. On production T-28Cs, the landing gear oleos were fitted with revised metering pins to better absorb the energy from high-rate-of-sink landings. The diameter of the Hamilton Standard propeller was decreased to 9 feet 4 inches to allow greater clearance for shipboard arrested landings. The Navy approved the modifications quickly.

With the T-28C in service, the Navy had a complete training program based around a single type of aircraft. T-28Bs continued to train student Naval Aviators in primary, intermediate and instrument instruction, while the T-28Cs were used for advanced, instrument and carrier qualification training.

In total, 299 T-28Cs were produced at the Columbus facility. They served with four squadrons in the Pensacola area, including VT-2, VT-3, VT-5 and VT-6. Instrument training was undertaken by VA-122 (Pacific Fleet) and VA-42 (Atlantic Fleet).

The T-28C had a long and successful career, lasting well into the late 1970s, when the T-34 Turbo Mentor and T-2 Buckeye programs began. The last T-28Cs were phased out in 1984 from

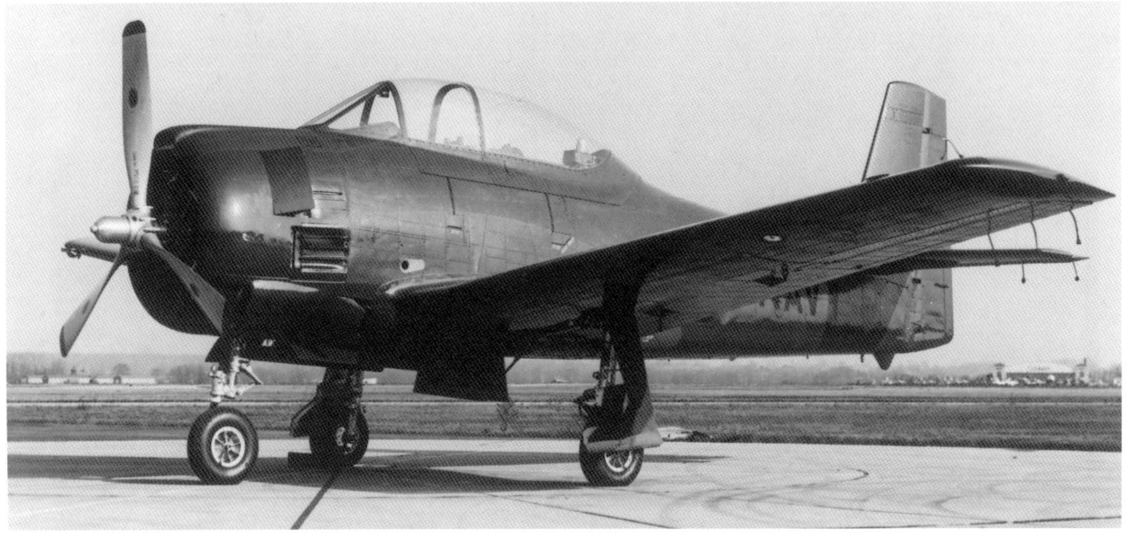

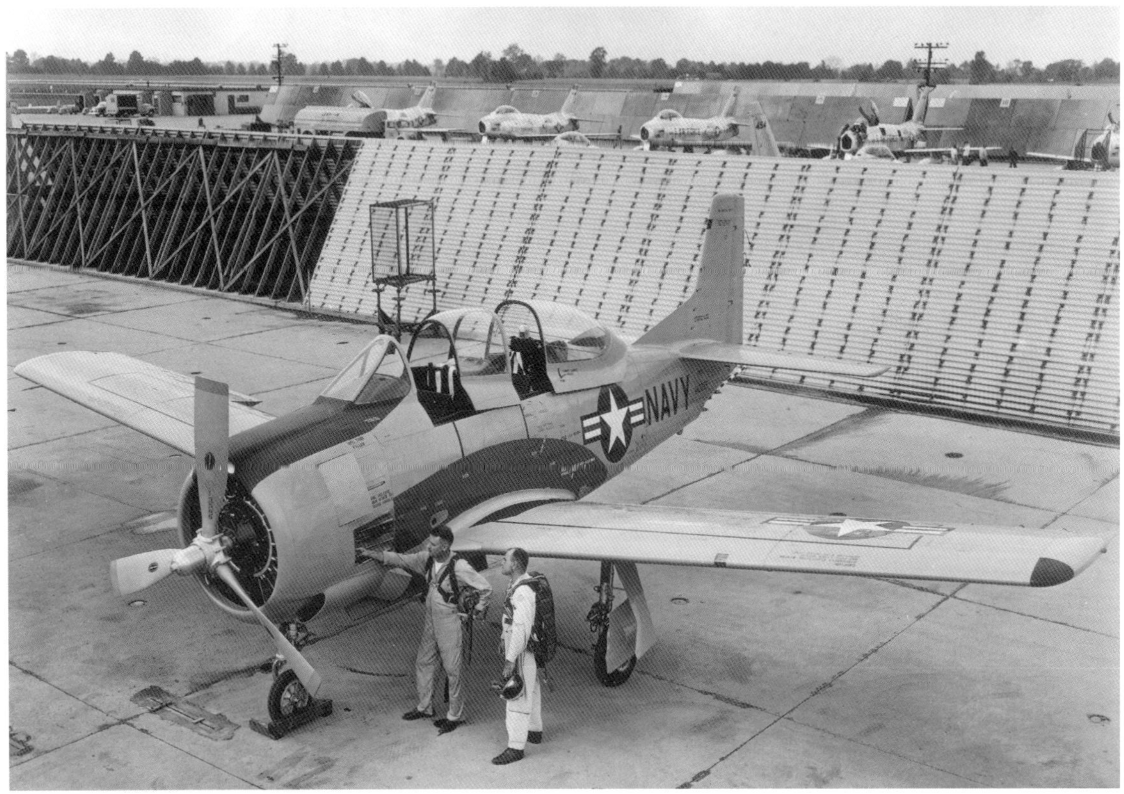

Top, the black markings are hardly visible on this T-28B, BuAer 138189, one of a very few painted dark blue overall. Below, the first production T-28C, BuAer 140053, in yellow.

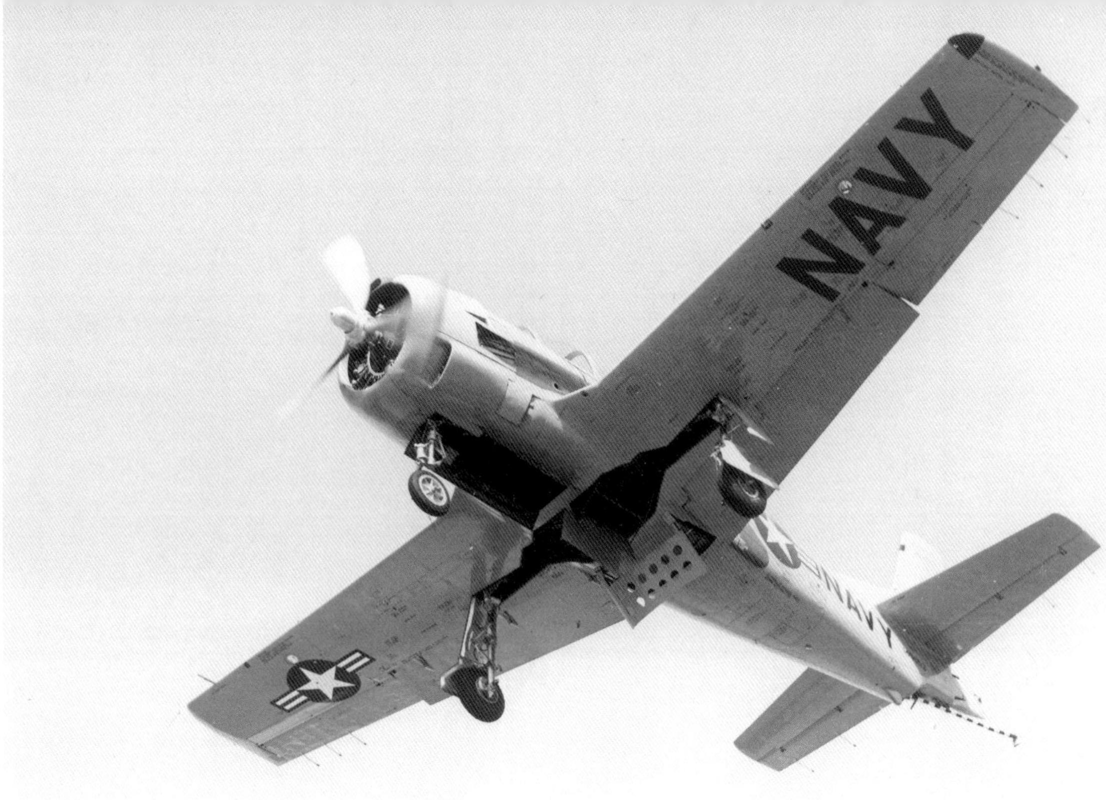

NAS Corpus Christi, Texas, ending nearly 30 years of active Naval service.

In October 1957 the Columbus group converted a surplus T-28A with a 1300-horsepower R-1820-56S engine as a flight demonstrator for commercial applications. Called the Nomad, it bore the civil registration N79Z. Contacting the Pacific Airmotive Corp. of Burbank, California, about a suitable engine/propeller combination, NAA learned that Pac Aero was also looking into the conversion of T-28As for foreign military counter-insurgency operations, in particular for Sud Aviation of France. NAA and Pac Aero made a working agreement for the former to provide conversion engineering and technical assistance and the latter to supply engines, propellers and accessories. Sud Aviation would purchase surplus T-28As from the USAF and set up a modification line at its St. Nazaire factory, rewiring the aircraft

Left, T-28Cs had tailhooks; here is one with its landing gear down and air brake extended. Below, T-28Cs, BuAer 146293 (the last) and 146282 in two-tone gray and white paint.

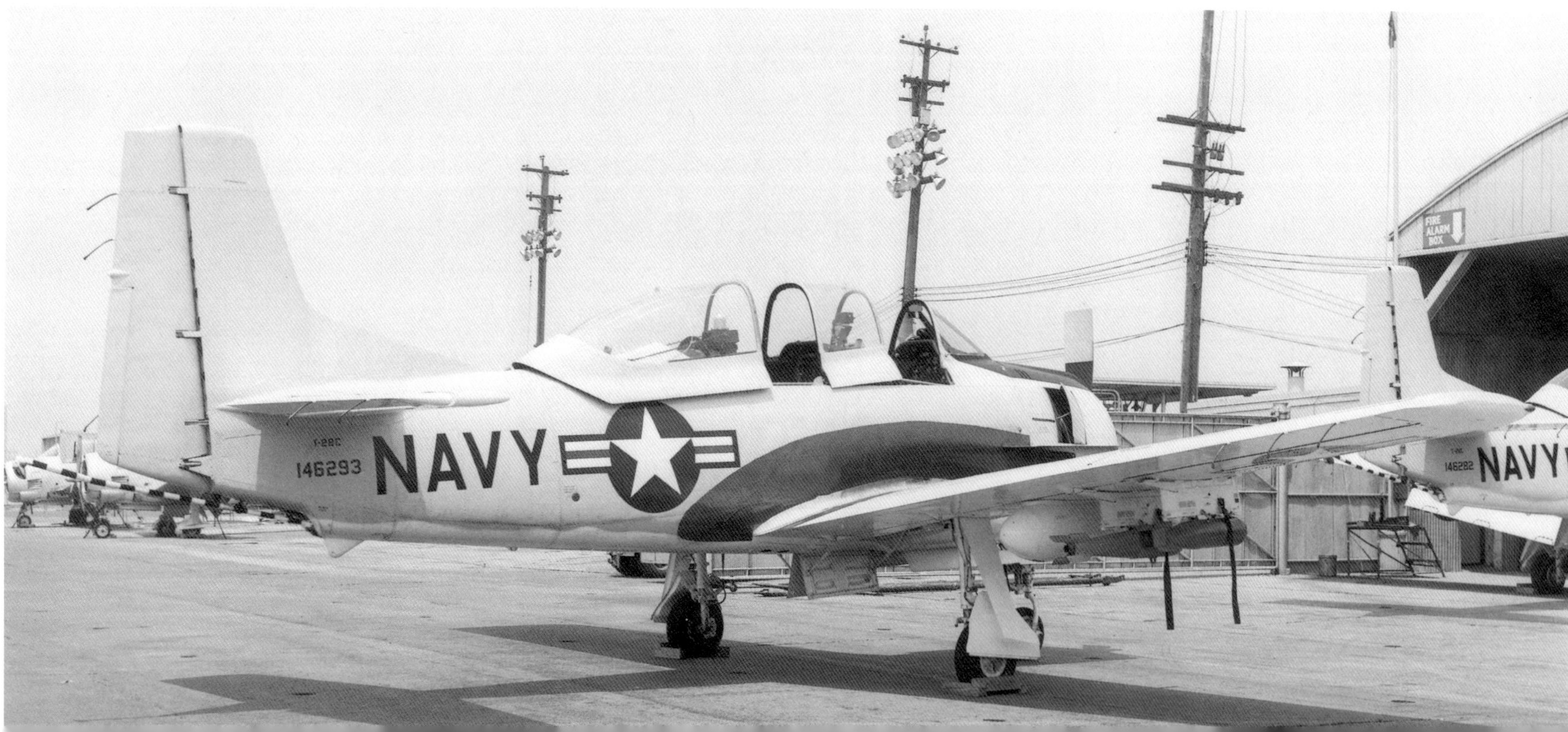

for French avionics and beefing up the wings for additional hardpoints carrying French ordnance.

On July 29, 1960, prior to the delivery of the NAA Nomad and three Pac Aero aircraft to France, FAA Type Certificate 1A18 was issued. Further marketing of the Nomad COIN concept was cut short when the USAF clamped down on T-28A sales. Sud eventually produced 245 reconnaissance/light attack aircraft, named Fennec (Desert Fox) by the *Armée de l'Air*. Fennecs based in Algeria were reassigned to ground-support squadrons in France until retirement in the late 1960s. The majority were sold to Nationalist China and to surplus aircraft dealers. Numerous Fennecs are currently in the U.S. civil registry.

In early 1961 U.S. Army special forces arrived in South East Asia to train elements of the South Vietnamese Army, while the USAF organized an Air Commando unit equipped with numerous aircraft, including T-28As. On October 13 detachment 2A of the Air Commando unit, tagged "Farm Gate," arrived in Vietnam and began modifying its T-28As for air-to-ground and reconnaissance training operations. Separately, the South Vietnamese Air Force operated "rice bowl Trojans" in air-to-ground combat.

Success with these operations prompted the USAF to procure a specialized version of the Trojan designated T-28D. This was essentially a remanufactured T-28A with the R-1820-86 power package, wing modifications for additional ordnance, installation of armor plate and self-sealing fuel tanks, and beefed-up center fuselage and horizontal tailplane. To fill the T-28D requirement a number of Navy T-28Bs and Cs, already equipped with R-1820 engines, were modified. North American remanufactured 313 T-28Ds and Fairchild Republic an additional 72.

The T-28D was deemed highly successful during its first two years of operations in Vietnam and Laos, but by the end of 1963 combat losses were exceeded by in-flight structural failures from metal fatigue. NAA was requested to propose improvements that would satisfy the survivability problem and meet the changing air-to-ground requirements. The outcome of this effort would be the YAT-28E.

The T-28Ds were withdrawn from Vietnam in 1964 and replaced by Douglas A-1 Skyraiders, although some USAF examples operated from Thailand on missions into Laos and Cambodia until 1972. T-28Ds were used by the Philippine, Congo, Thai and Bolivian Air Forces and by Honduras in its 1969 "soccer war" against El Salvador. The T-28D had proved itself as a ground-attack platform, was a joy to fly and was easy to service, with a history of low maintenance hours per flight hour operation.

In an effort to improve the overall performance of the T-28 as a COIN aircraft, the USAF awarded a contract to NAA to modify two T-28A airframes with the installation of the 2450-horsepower Lycoming YT-55L-9 turboprop engine in

With the canopies open for quick exit in case of emergency, a Naval Air Test Center T-28C leaves the deck. The photo plane is much farther to starboard of the carrier than the picture suggests.

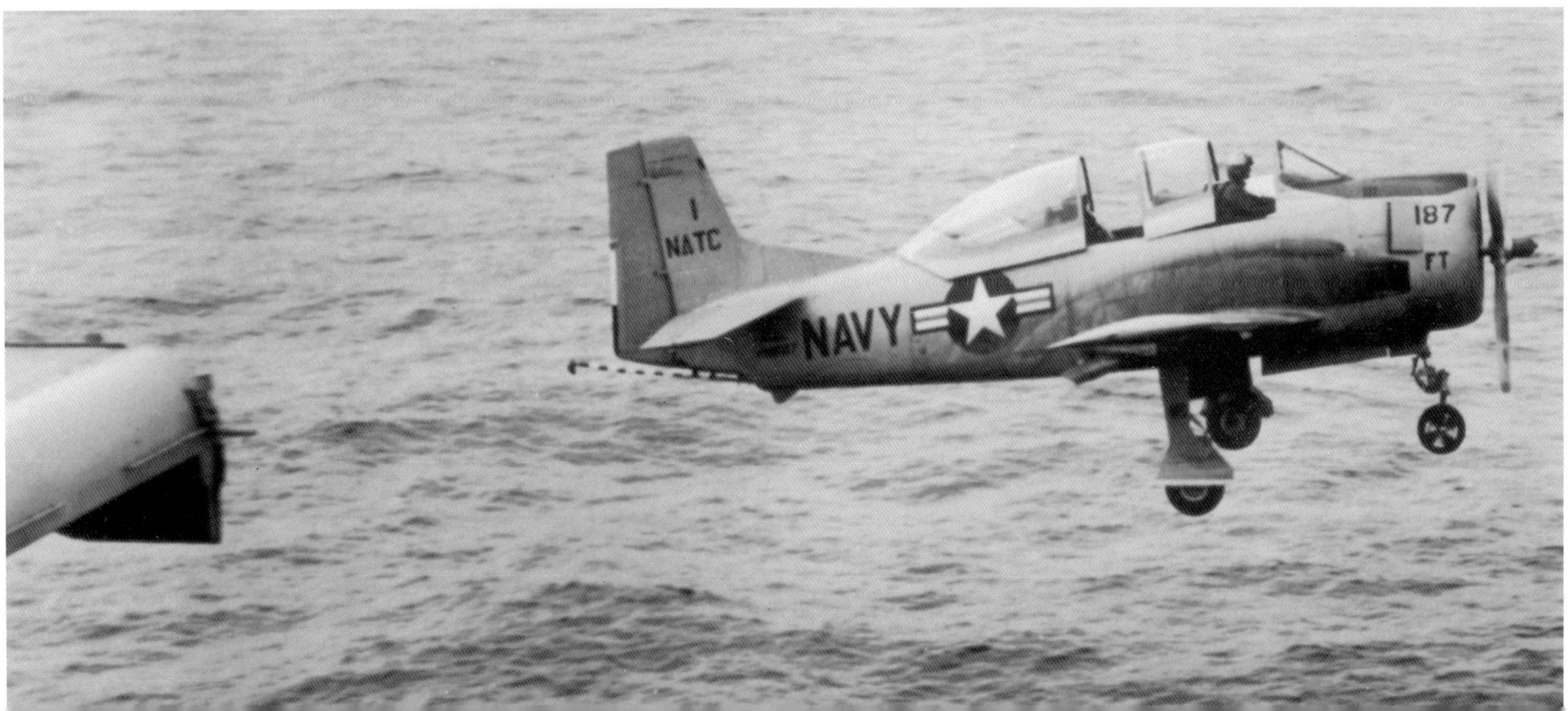

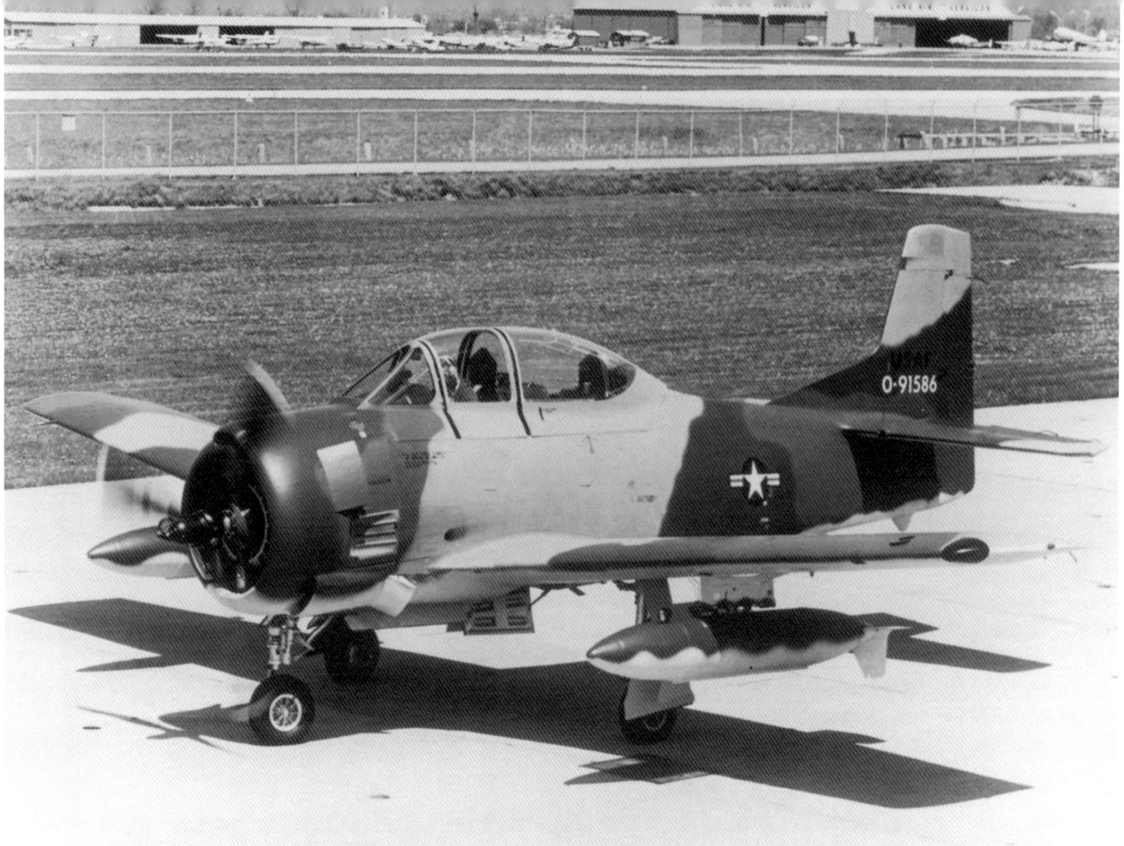

Above, a T-28A, AF49-1586, remanufactured as a T-28D with R-1820 engine, 3-blade prop, underwing stores and three-tone camouflage. Below is another, AF51-3480, assigned to 4407 CTS, 1 SOW, and photographed at Hurlburt Field in August 1970.

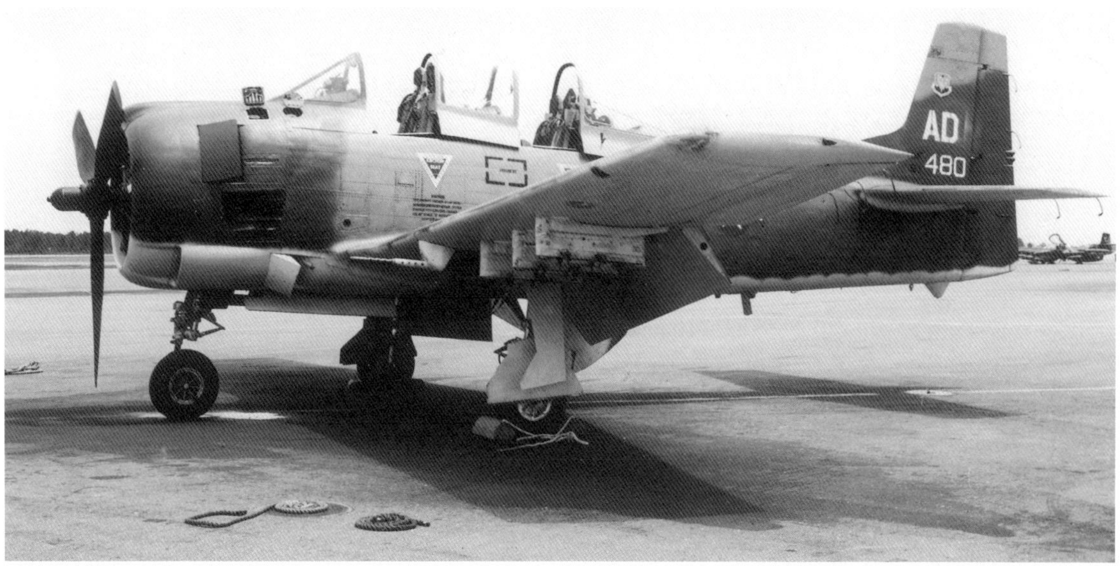

September 1962. The project, designated NA-284, was intended to improve the speed, over-the-target loiter time and hot-climate performance of the T-28 through lessons learned in Vietnam. Originally given the Air Force designation RA-28 (Reconnaissance Attack), the project was later redesignated YAT-28E and nicknamed Turbo Trojan.

The entire nose section forward of the firewall was modified to accept the T55 turboprop engine. The result was a longer, more streamlined nose. A single exhaust port on the left side was positioned just ahead of the wing root. The propeller was an 11-foot 6-inch, four-blade Aero Products/Allison unit. The wing was strengthened to accept twelve underwing pylons (up to 6000 pounds capacity), and larger flaps were fitted. The horizontal stabilizers were enlarged, and aileron and rudder travel was increased. The YAT-28E carried 250 gallons of fuel internally, with provisions for two 110-gallon tanks on the wing pylons. Maximum take-off weight was now up to 15,530 pounds.

The YAT-28E first flew on February 15, 1963, with NAA test pilot George Hoskins at the controls. Early flights showed outstanding performance, with a top speed of 360 miles per hour and a range of 2760 miles. On March 27, 1963, during the 13th test flight, the horizontal tail failed in high-g maneuvers and the aircraft entered a flat spin and crashed, killing Hoskins. Despite this tragedy, the Air Force was sufficiently impressed with its performance potential and ordered a third prototype. The second aircraft had a strengthened horizontal tail and an enlarged vertical stabilizer, and was fitted with North American LW-2C ejection seats, along with a reshaped canopy.

The third prototype was refined even further and closely resembled the proposed production configuration. Further Air Force evaluation revealed deficiencies, however, and with the advent of NAA's OV-10 Bronco, the USAF discontinued further YAT-28E testing in late 1964. The Navy evaluated the YAT-28E beginning in January 1966, but also found it inadequate for their needs and

Specifications: North American T-28A/B/C/D Trojan / YAT-28E Turbo Trojan

Description: U.S. Air Force and Navy primary and advanced trainer/attack aircraft. Single-engine low-wing monoplane of all-metal construction with retractable tricycle landing gear. Crew 2.

Engine: 800-hp Wright R-1300 Cyclone nine-cylinder radial air-cooled. (T-28B/C/D) 1425-hp Wright R-1820 Cyclone nine-cylinder radial air-cooled. (YAT-28E) 2445-shaft-hp Lycoming YT-55L-9 turboprop.

Dimensions: Span 40 ft 1 in, length 32 ft, height 12 ft 8 in, wing area 268 sq ft. (T-28B) Span 40 ft 7.25 in, length 32 ft 11 in, height 12 ft 8.5 in, wing area 271 sq ft. (T-28C) As T-28B, except: length 34 ft 4 in. (YAT-28E) As T-28B, except: length 36 ft 0 in.

Weights: 5111 lb empty, 6365 lb max. take-off. (T-28B) 6410 lb empty, 8038 lb max. take-off. (T-28C) 6424 lb empty, 8500 lb max. take-off. (T-28D) 6251 lb empty, 10,900 lb max. take-off. (YAT-28E) 7078 lb empty, 15,530 lb max. take-off.

Armament: T-28A/B/C: Two 0.50-in machine guns and two underwing pylons for bombs, rockets or flares. (T-28D) Two 0.50-in or four 0.30-in machine guns and various external stores mounted on four underwing pylons, including SUU-7B/A or SUU-14A bomblet dispensers, Mk 81/82 low-drag bombs, BLU-1C/B napalm bombs, BLU-10/B fire bombs, M1-A2 anti-personnel bomb clusters, Mk 24 flares and LAU-32A 2.75-in HVAR rocket pods. (T-28D-5) Six underwing pylons. (YAT-28E) Twelve underwing hardpoints for external stores.

Performance: Maximum speed 283 mph, service ceiling 24,000 ft, range 1350 miles. (T-28B) maximum 346 mph, ceiling 37,000 ft, range 1100 miles. (T-28C) maximum 343 mph, ceiling 35,500 ft, range 860 miles. (T-28D) maximum 352 mph, ceiling 37,000 ft, range 1184 miles. (YAT-28E) maximum 360 mph, range 2760 miles.

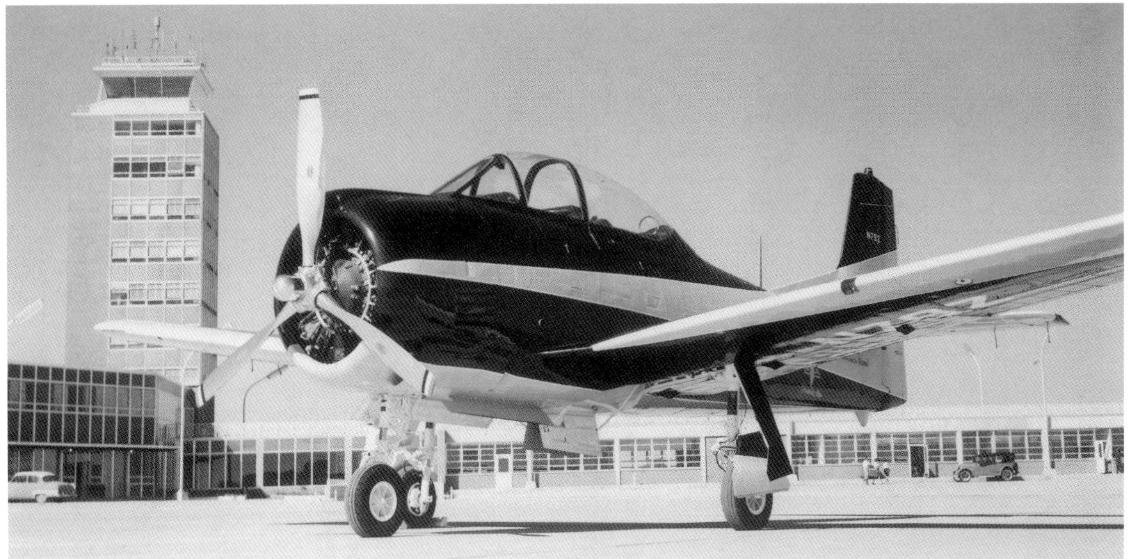

Two views of the NA-260 Nomad prototype, N79Z, built at Columbus, Ohio, to test the foreign market. NAA engineer Frank Compton is seen entering the front cockpit, below. Sud Aviation in France later rebuilt 245 T-28As as Fennecs.

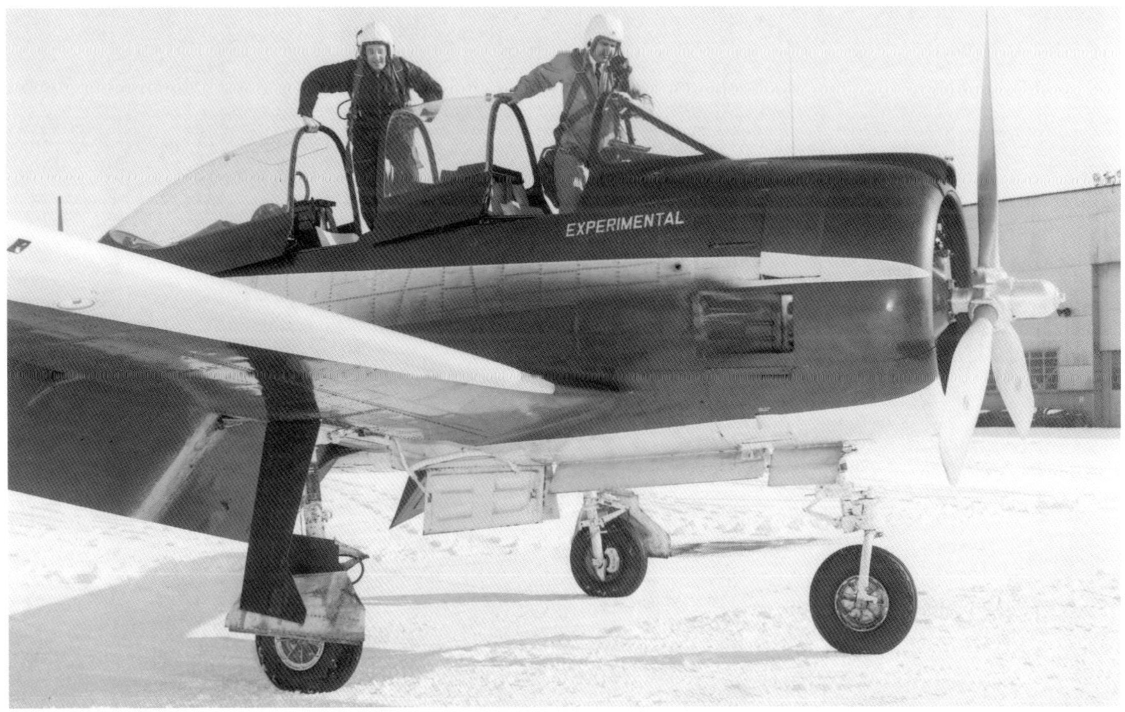

discontinued testing in May.

It should be noted that compared to the T-28A, the take-off gross weight of the YAT-28E had grown by a factor of 2.0, a king-sized achievement, to say the least.

Other T-28 derivatives included the Aero Industry Development Center T-CH-1. These were modified T-28As in service with the Chinese Nationalist Air Force and fitted with license-built Lycoming T53-L-701 turboprop engines of 1450 horsepower, similar in configuration to the YAT-28E. Fifty were built between 1976 and 1981.

The Hamilton Aircraft Company of Tucson, Arizona, offered T-28A conversions for civil and

Opposite page, the first turboprop YAT-28E, AF52-1242, above Ohio, carrying six bombs on its underwing pylons.

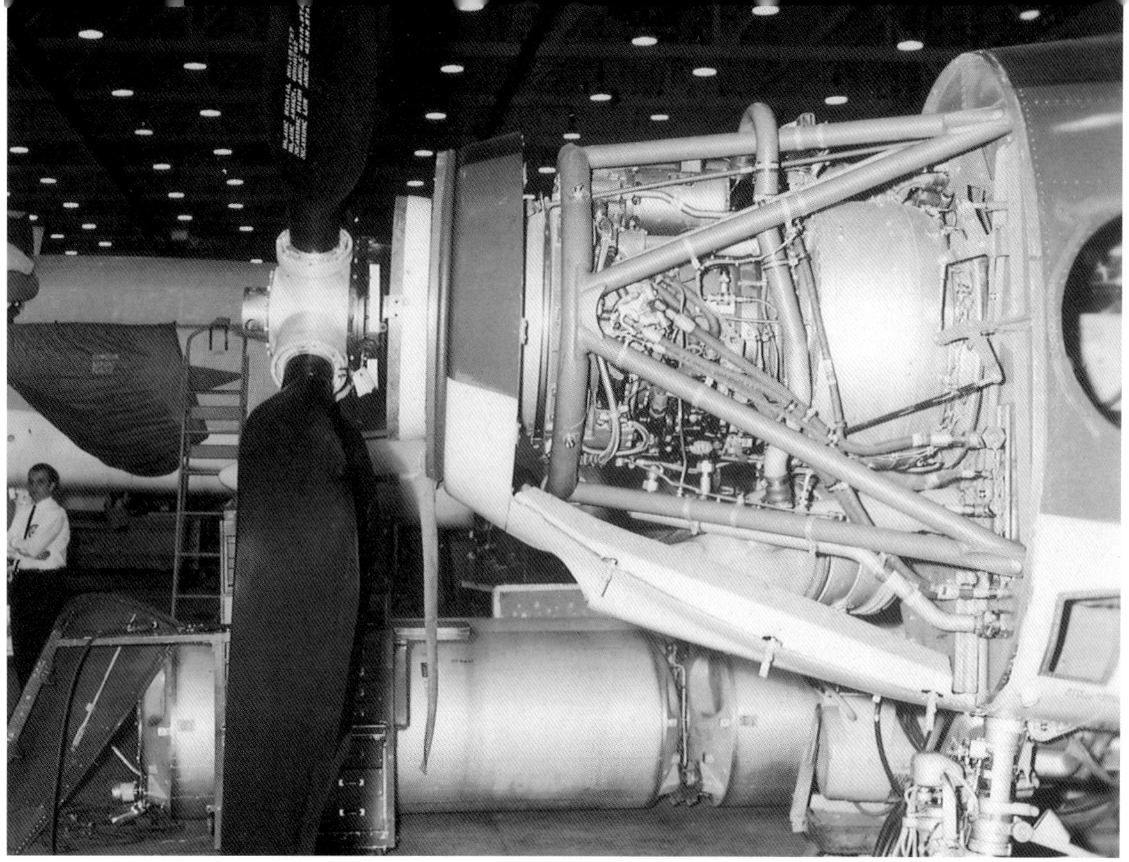

Left, the installation of the 2445-shaft horsepower Lycoming YT-55L-9 gas turbine in the YAT-28E. Compared to the T-28A from which it was modified, the length was increased by 4 feet, the empty weight by nearly 2000 pounds, the maximum weight by 7500 pounds, and the speed by 83 mph. But the OV-10A Bronco proved to be a more efficient light attack plane.

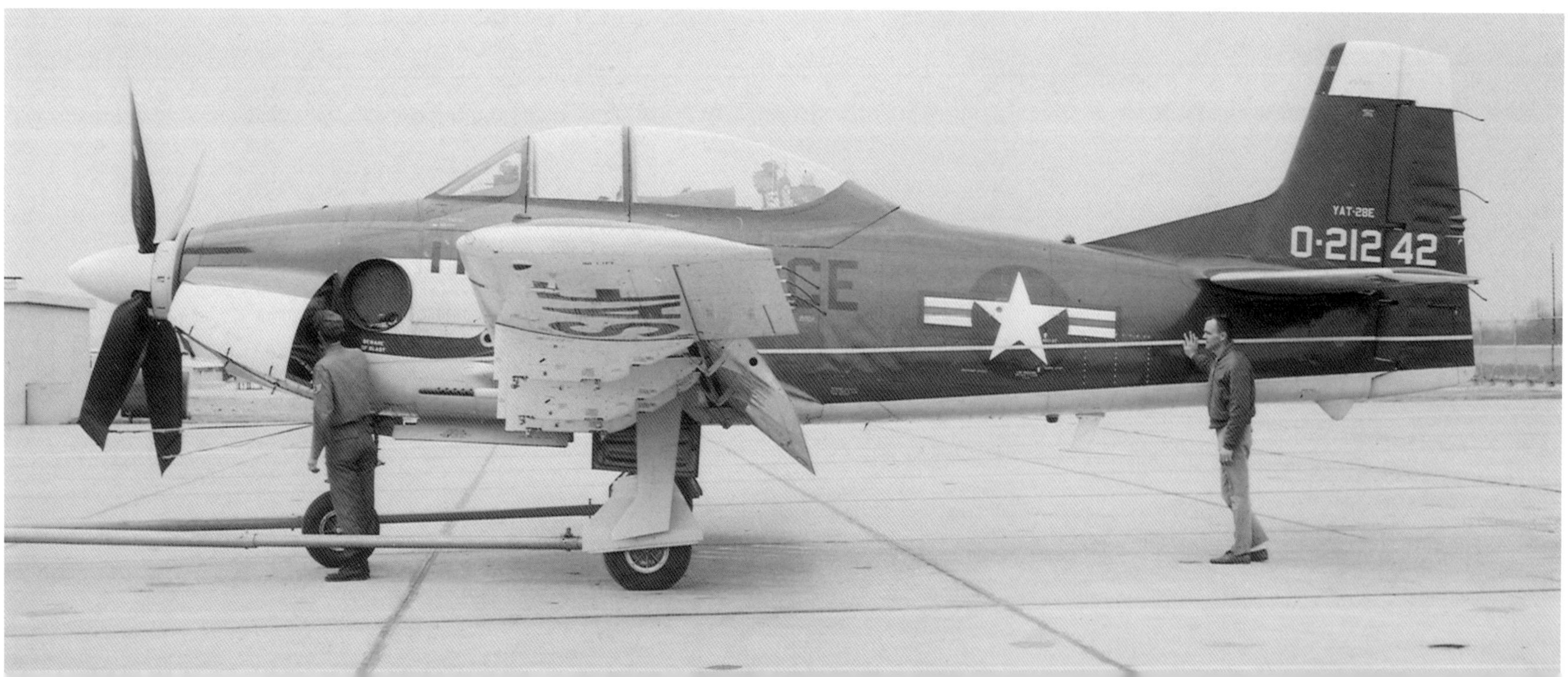

foreign military applications, using 1200-horsepower Wright 704-C9-GC1 engines. The Normair R-1 conversion for aerial photography was certified by the FAA in September 1958 and the five-seat Normair R-2, with fixed canopy and side door, was also offered. Six R-1s with tail hooks were purchased by the Brazilian Navy in 1962.

The Trojan served the USAF, U.S. Navy and foreign air forces in various capacities over three decades and proved to be among the most versatile aircraft of all time. There are currently 354 T-28s in the U.S. civil registry, plus another 50 in foreign ownership. They continue to be relatively cost-effective and simple to maintain.

Production: North American XT-28/T-28A/D Trojan / YAT-28E Turbo Trojan

NA-159	XT-28	AF48-1371/1372	2
	T-28A	AF49-1491 (static test)	1
	T-28A	AF49-1492/1756	265
NA-171	T-28A	AF50-195/319	125
NA-174	T-28A	AF51-3463/3796, -7482/7891	744
NA-189	T-28A	AF52-1186/1242, -3497/3498	59
NA-199	T-28B	BuAer 137636/137637	2
NA-200	T-28B	BuAer 137638/137810	173
	T-28B	BuAer 138103/138367	265
NA-218	RT-28B	(T-28B built for Japan)	1
NA-219	T-28B	BuAer 140002-140052	51
NA-225	T-28C	(T-28B BuAer 138185 and 138187 modified for carrier use)	(2)
NA-226	T-28C	BuAer 140053/140077	25
	T-28C	BuAer 140449/140666	218
NA-252	T-28C	BuAer 146238/146293	56
—	T-28D	(T-28A/B/C remanufactured for the Air Force)	(385)
NA-260	Nomad	(T-28A modified for commercial market with registration N79Z)	(1)
NA-284	YAT-28E	(T-28A AF51-3786, 51-3788 and 52-1242 modified)	(3)
NA-307	T-28C	(Canceled)	--
Total, XT-28/YAT-28E trainers and attack aircraft, 1949-1964			1987

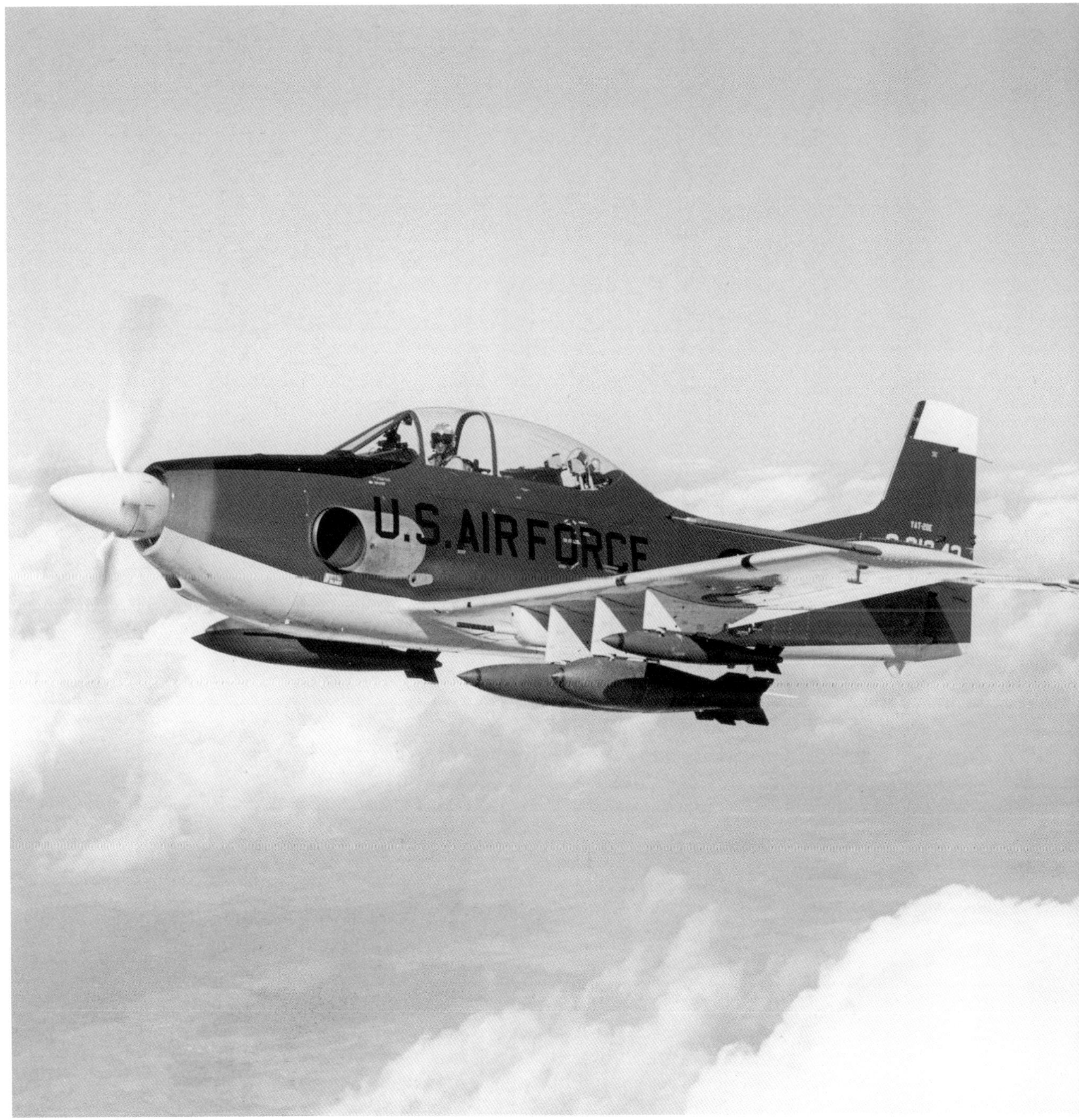

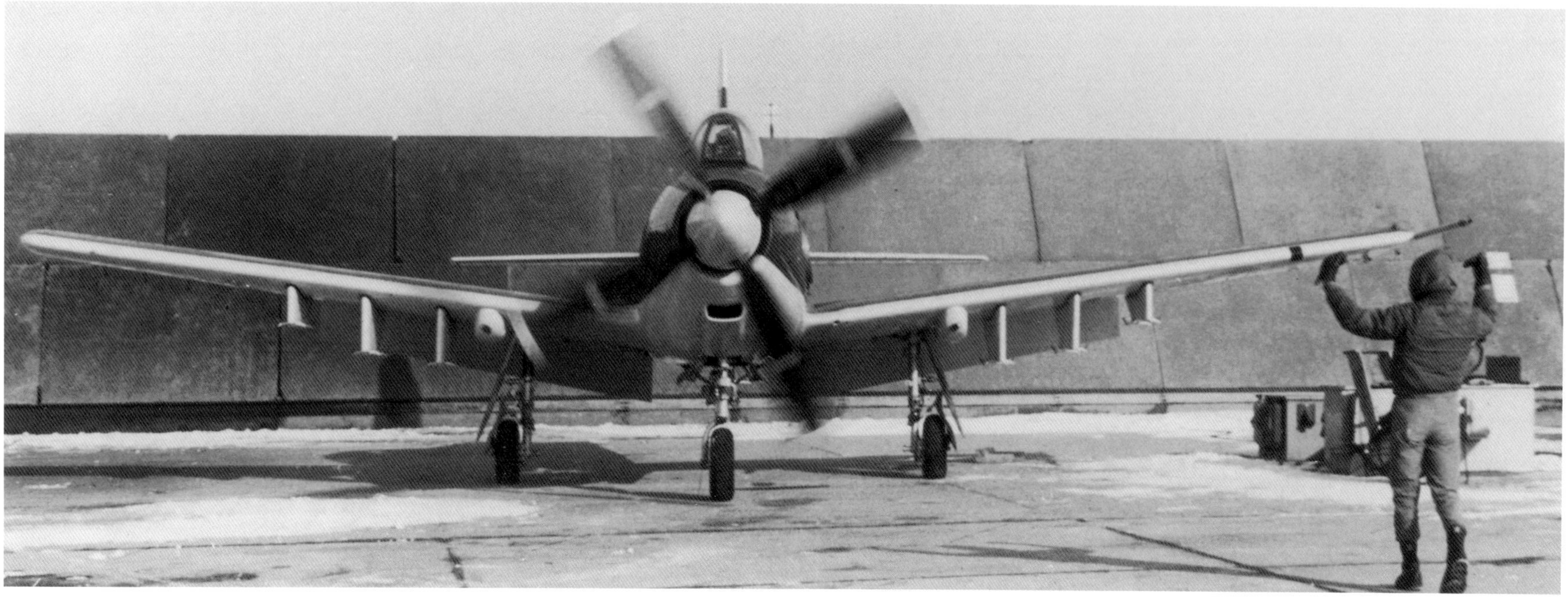

Shown above with six pylons and below with twelve (plus dummy wingtip Sidewinders), the YAT-28E could carry a payload of 7000 pounds. Opposite, 51-3786 in flight with a Cessna YAT-37D.

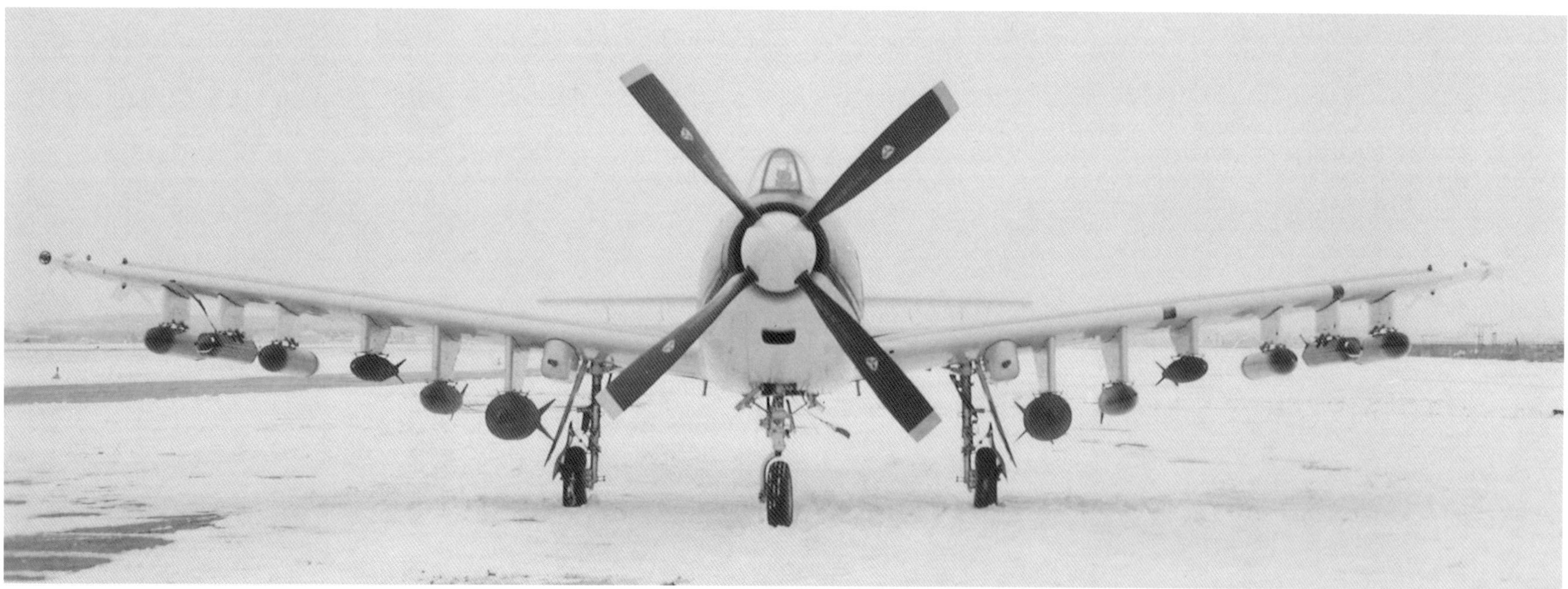

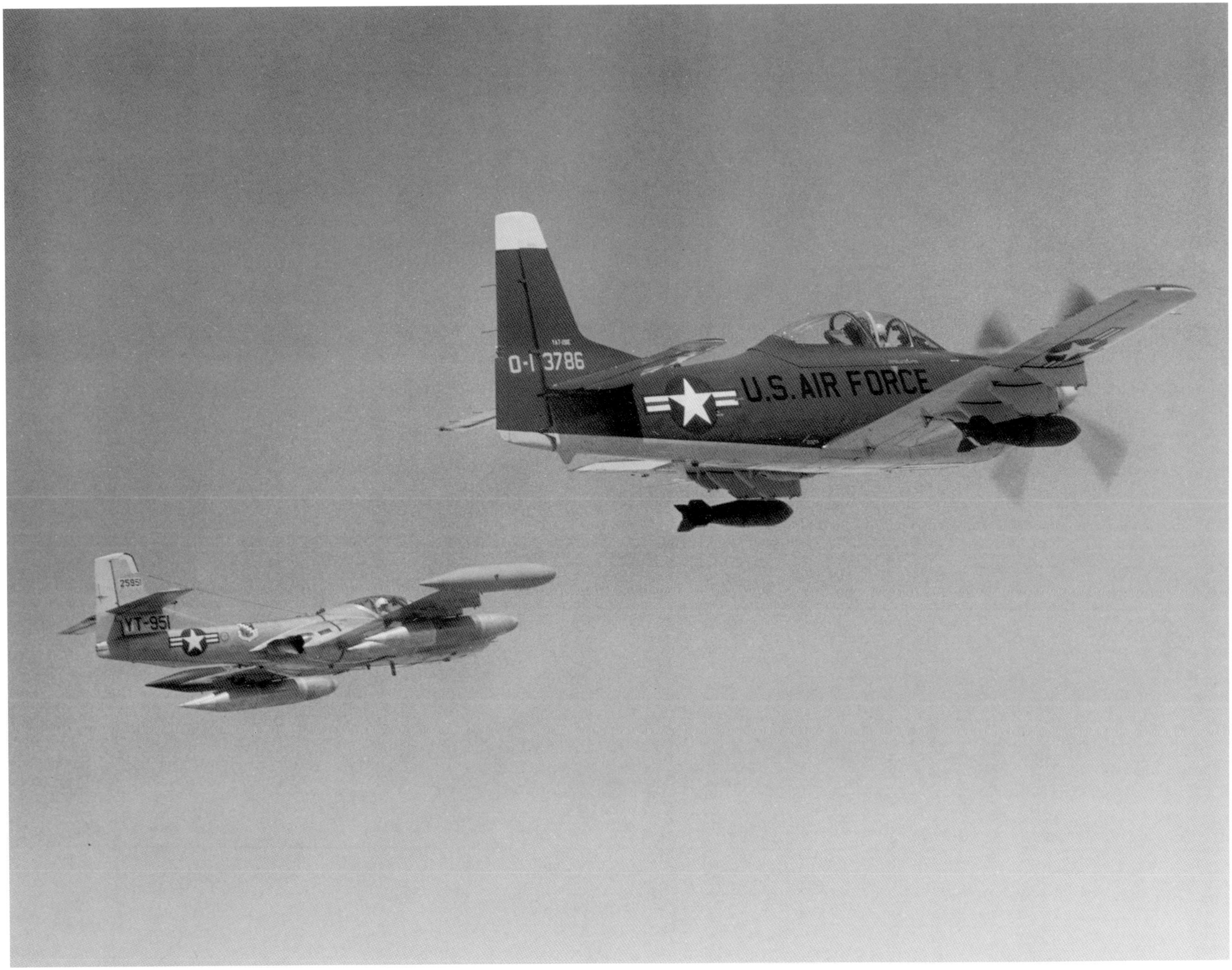

23

North American NA-143/-145 Navion / NA-154 L-17/U-18 / Ryan Navion / Tusco Navion Rangemaster / Twin Navion

WELL BEFORE the end of World War II, the U.S. government began to cancel many contracts that were currently held by the U.S. aircraft industry. The military felt that large numbers of aircraft were no longer needed, and this was the beginning of a huge cutback that affected every manufacturer in terms of productivity, profit and future projects. On VJ day in 1945, North American had orders for 8000 aircraft, reduced to just 24 by the end of the year.

All military contractors were forced to reassess their positions for a postwar aircraft market. What few military contracts lay ahead, particularly at the beginning of the jet age, would be fiercely contested among the bigger manufacturers, and the quantities of aircraft would not be as large as in wartime. Contractors were forced to reduce their manpower drastically, putting thousands of engineering and manufacturing personnel out of jobs. Including those who had worked at the recently closed Dallas and Kansas City plants, North American's work force dropped from 91,000 to only 5000 by VJ day.

To make use of its existing tooling and manufacturing machinery, NAA had to consider production of pots and pans and other metal houseware and hardware items. Hoping to avoid this, president and general manager Dutch Kindelberger, vice president Lee Atwood and engineers Ray Rice and John Young decided to build lightplanes for the general aviation market. The decision was necessary to retain the company's highly experienced engineering staff, which otherwise would be forced out of work.

Almost exclusively a military contractor up to that time, NAA had little lightplane experience to fall back on, the only project even remotely similar being the tandem two-seat NA-35 of 1939. Nevertheless, the engineers began the

The very first Navion, NX 18928, was expected to open up a large civil market for NAA. Although 1110 were built, high costs made the venture unprofitable. Opposite page, the second prototype.

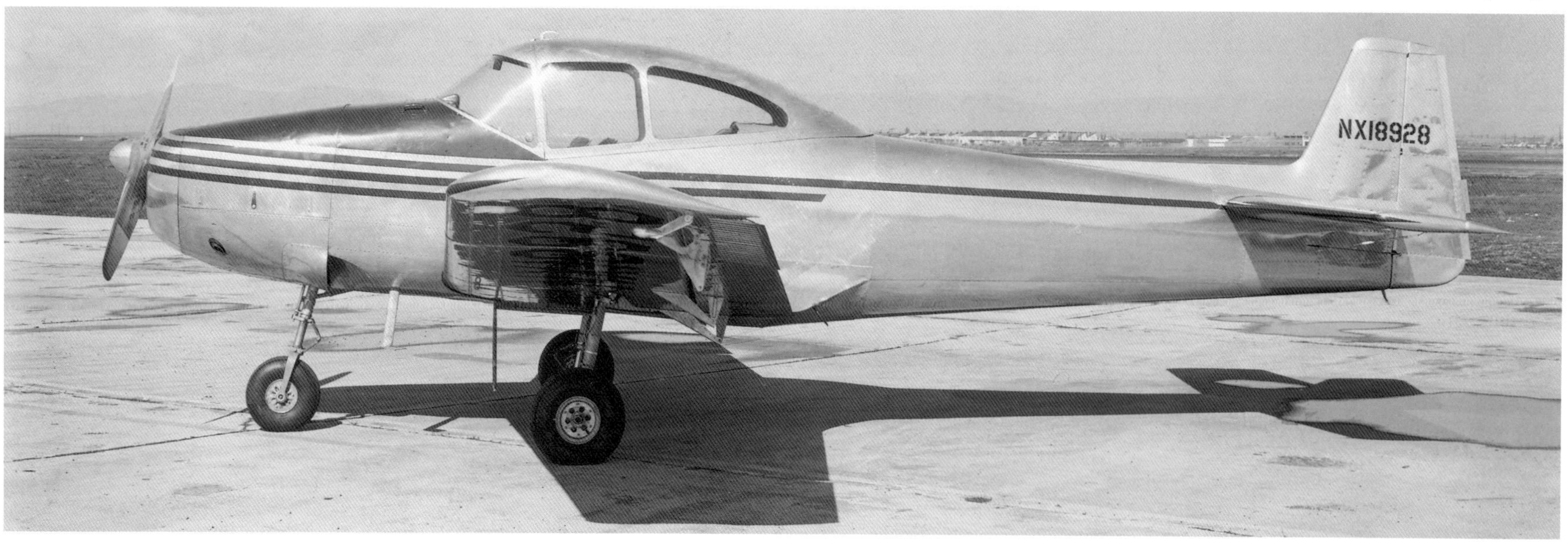

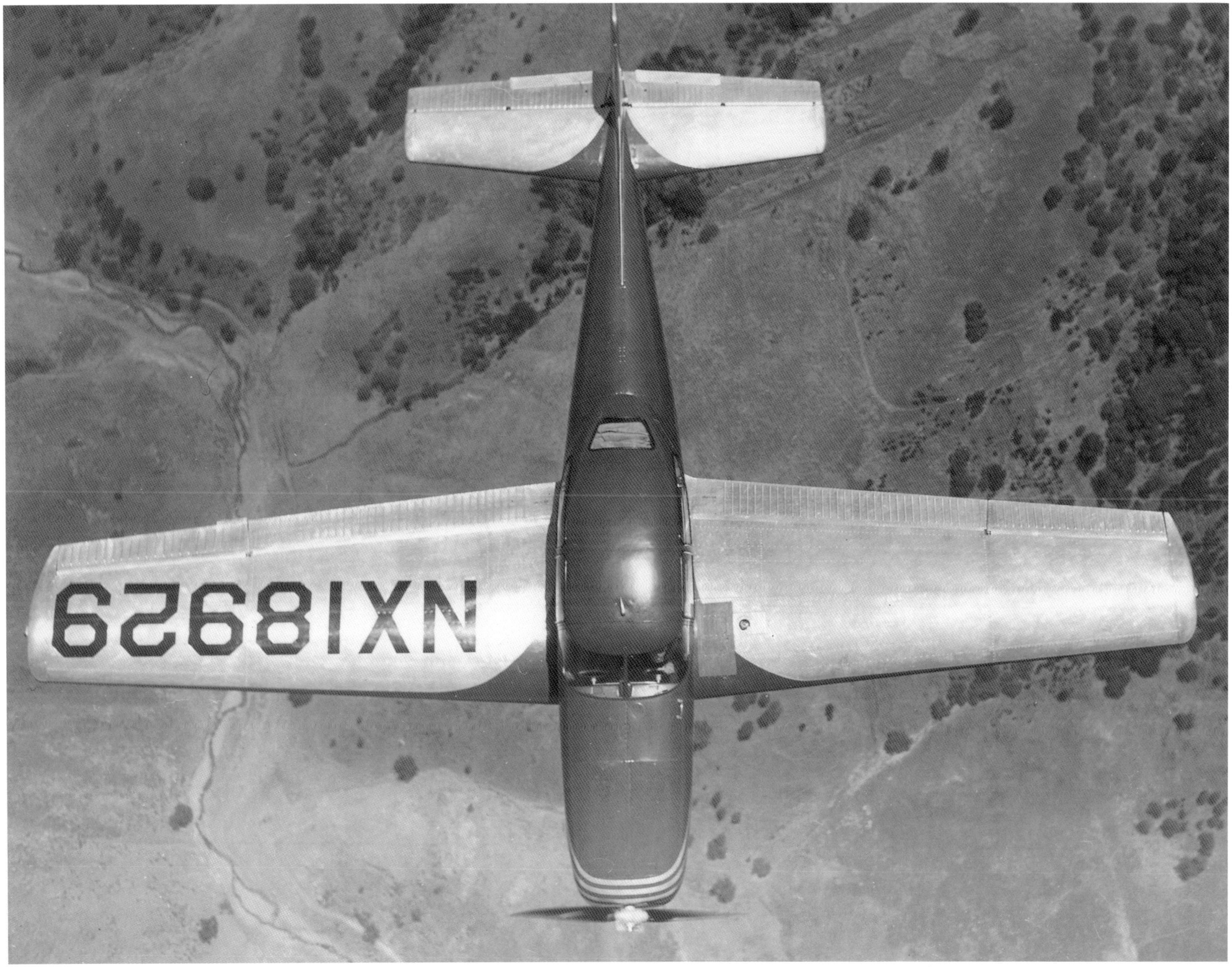
NX18929

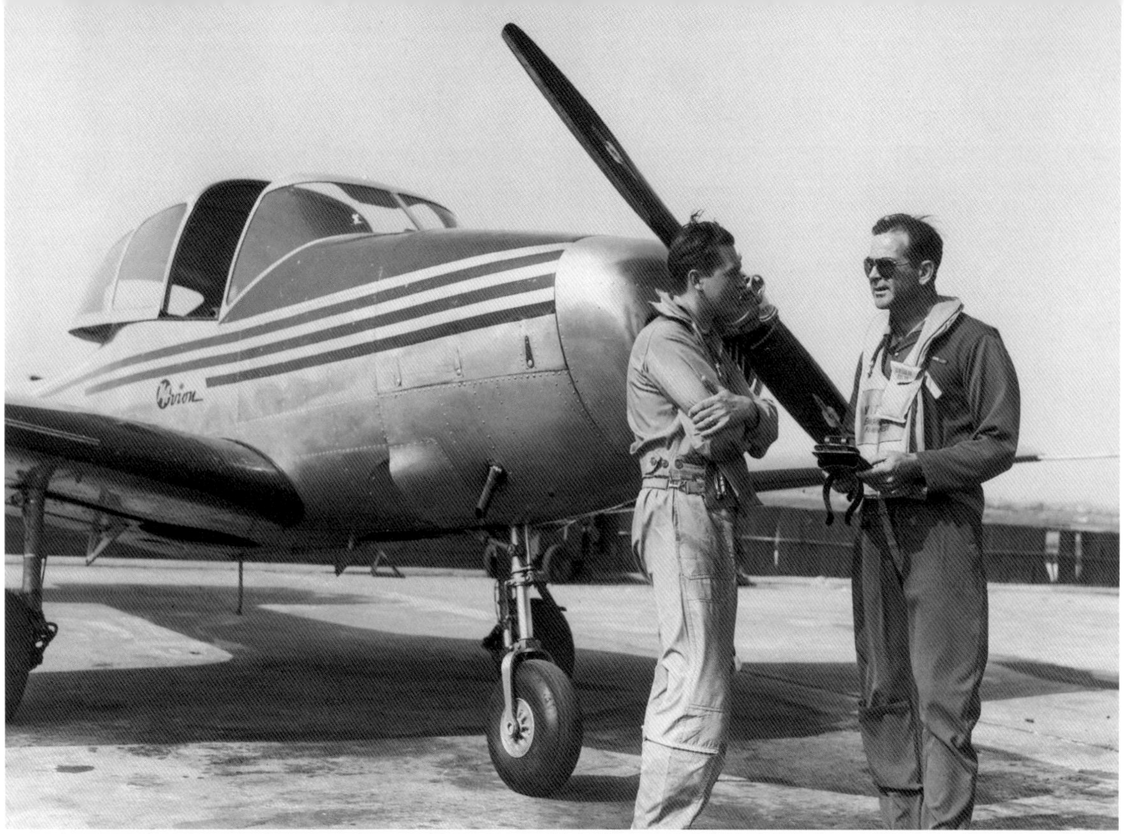

design of a four-place, all-metal low-wing lightplane with retractable landing gear in October 1945 under the charge number NA-143. It was powered by a Continental E-185-3 six-cylinder, horizontally-opposed air-cooled engine rated at 185 horsepower, and engineer Chandler King was given the responsibility for the powerplant installation and cooling ducting.

The basic layout of the aircraft was straightforward. Designed to achieve safe, predictable flying qualities and ease of maintenance, it had a very rugged structure, with higher-gauge materials than necessary employed throughout its structure. Not surprisingly, the designers came up with an airplane with an obvious family resemblance to the P-51D Mustang. Although the layout was conventional in many respects, it differed from most lightplane designs in featuring a large sliding canopy that allowed relatively easy

Left, NAA pilots Skip Ziegler and Ed Virgin after an early Navion flight. The 4-seat lightplane had a cruising speed of 150 mph. Below and opposite, two early production models.

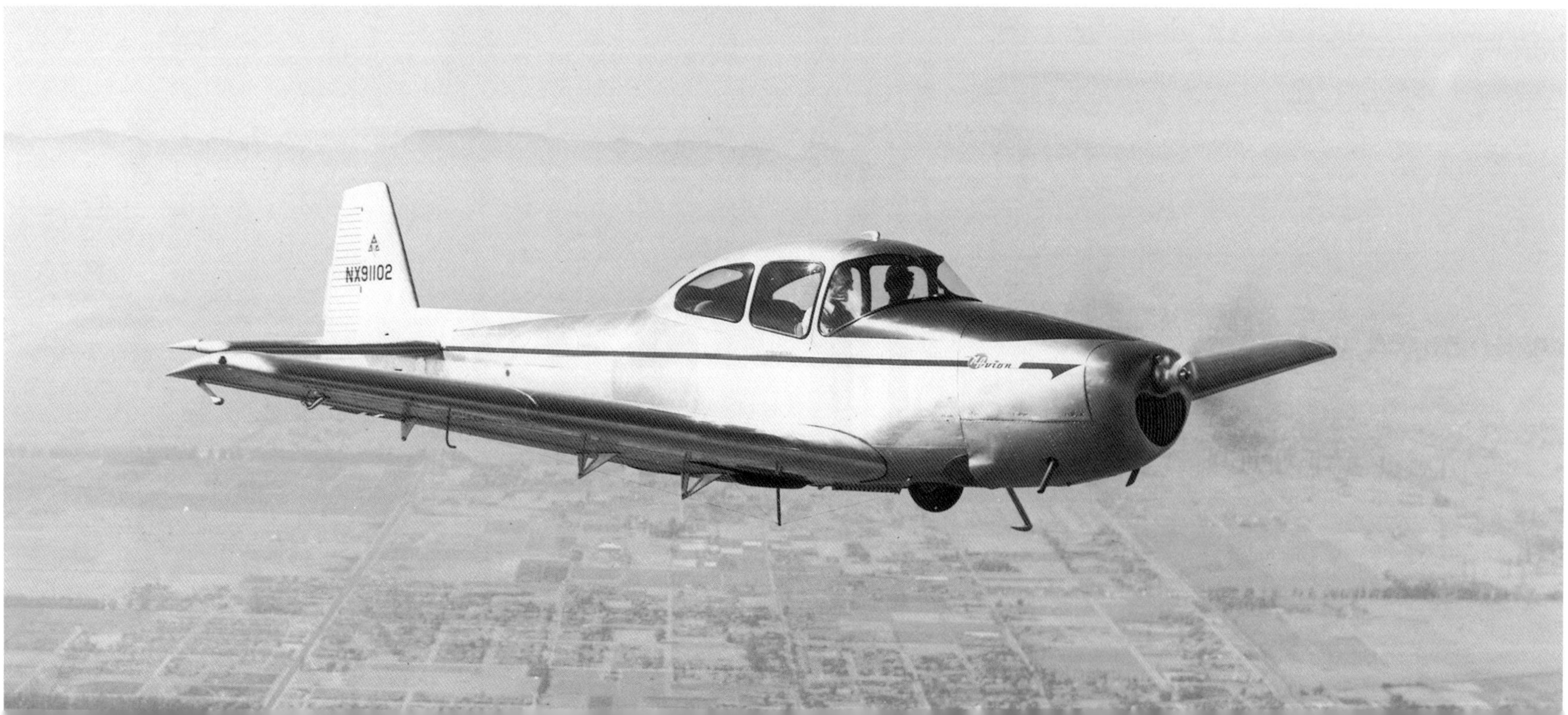

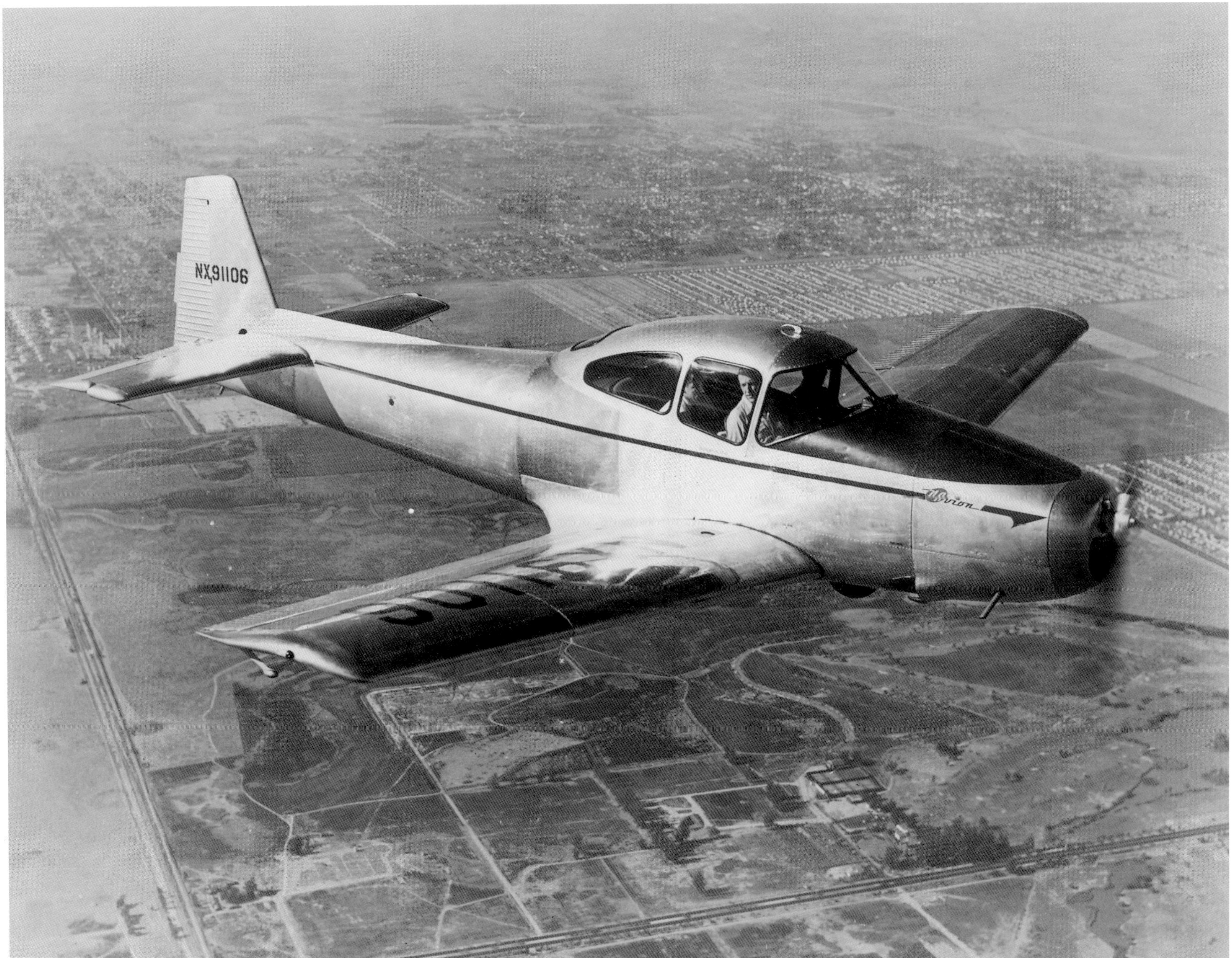
NX91106

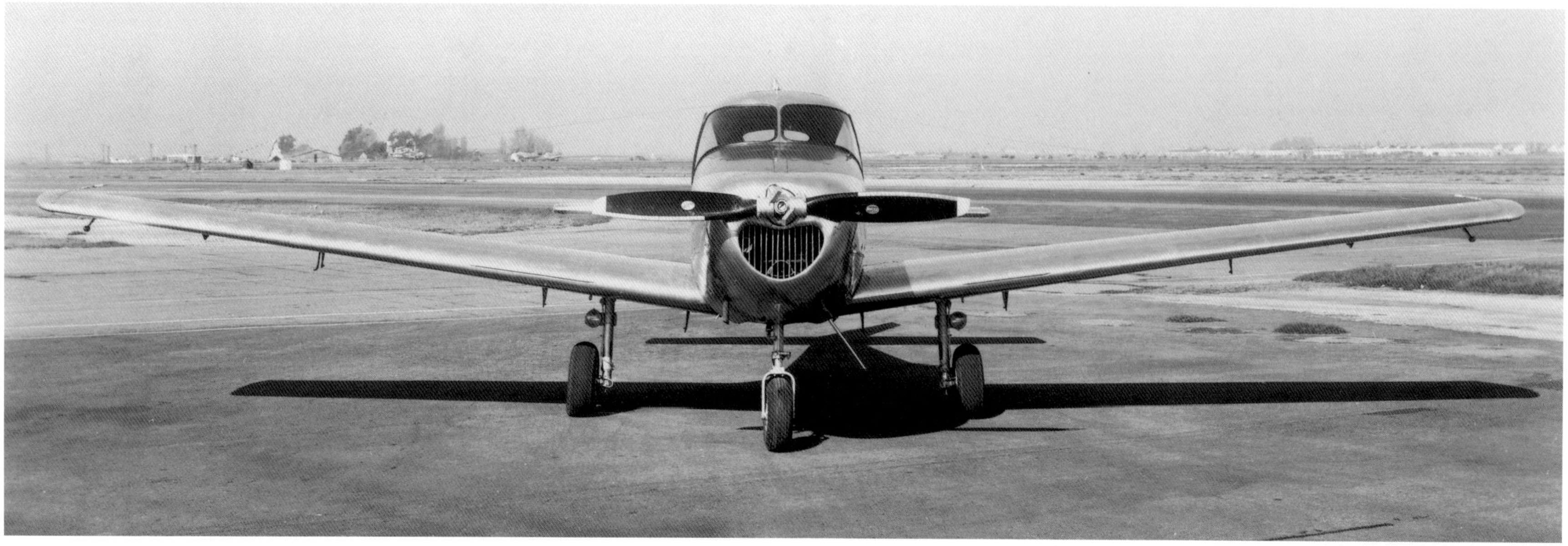

Front and rear views of NC 91139 show the Navion's basic simplicity. Much was made of its resemblance to the P-51; except for the tricycle gear, the overall proportions were certainly similar.

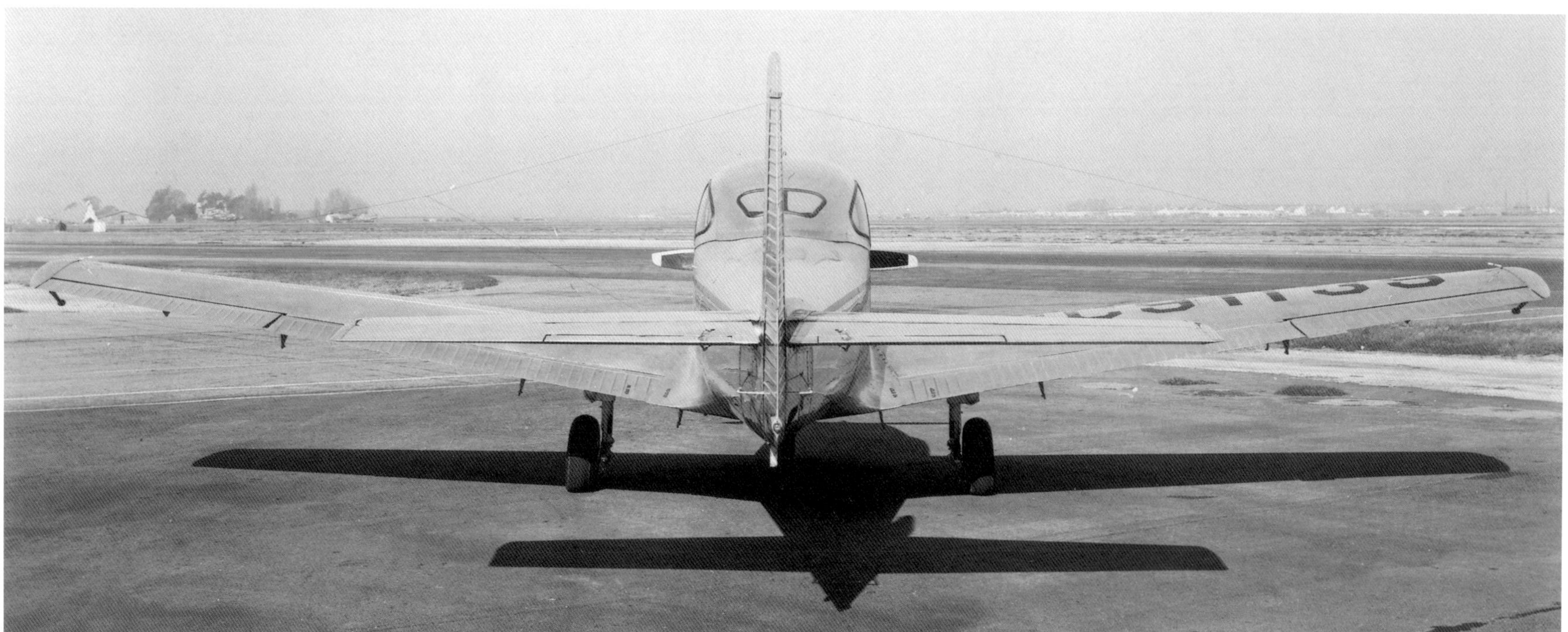

entry and exit for all four occupants. The cockpit featured dual control wheels and a relatively simple instrument panel. The new plane was named Navion, a combination of NA and *avion* (French for airplane). At the end of 1945 two NA-143 "simulated production" examples were built, with civil registrations NX18928 and NX18929, for testing and use as company demonstrators. NA-145 production models NX91100 and NX91101 also took part in the flight testing.

The first flight took place from Mines Field on January 15, 1946, with Ed Virgin and Bob Chilton at the controls of NX18928. It was a relatively trouble-free flight, with the exception of some engine cooling problems during low-speed climbs. After the engine ducting was revised to remedy the heat dissipation problems, the Navion went into production. It was advertised as having a cruising speed of 150 miles per hour, a range of over 500 miles and a useful load of just over 1000 pounds. At a fly-away price of $6100 plux tax, the plane featured a variable-pitch propeller, a steerable nosewheel and hydraulically actuated flaps, positioned either up or 40 degrees down.

Although the Navion was promising in terms of design and performance, it faced a major competitor in the Beechcraft Bonanza, another modern all-metal design that had better speed with a 165-horsepower Continental. But the major factor working against the Navion was its production cost per unit. Although the sales price was somewhat high for a private aircraft in 1946, it was reasonable for what the Navion offered in size and performance. The problem was that it initially cost NAA nearly $10,000 to build each example! Here was a situation where an established aircraft manufacturer was making an effort to compete in an entirely new segment and was taking a loss on each airplane.

On the market, the Navion was competitively priced and well received. Many were purchased as business aircraft to transport corporate executives efficiently on short to intermediate runs, particularly in the oil business, where survey trips were frequent. Between January 1946

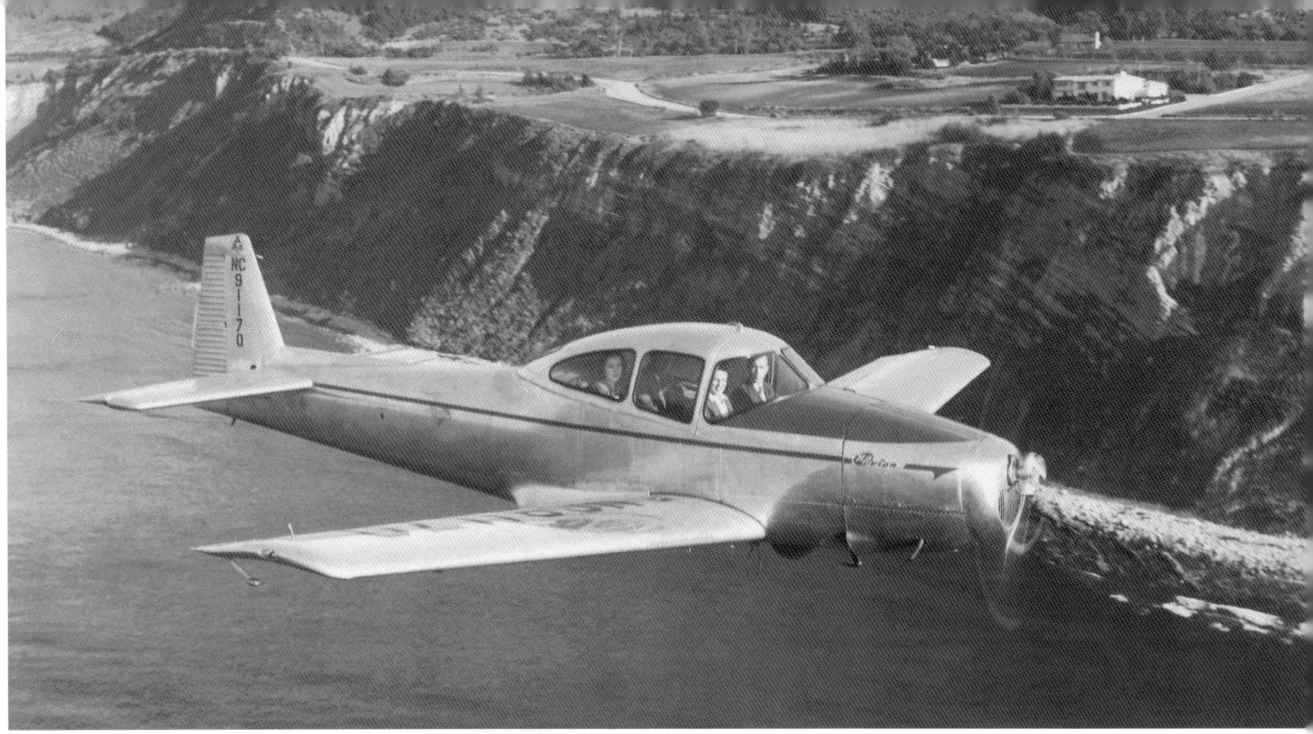

NC 91170 cruises the California coastline. Below, production Navions share the NAA flightline with XB-45 and P-82B.

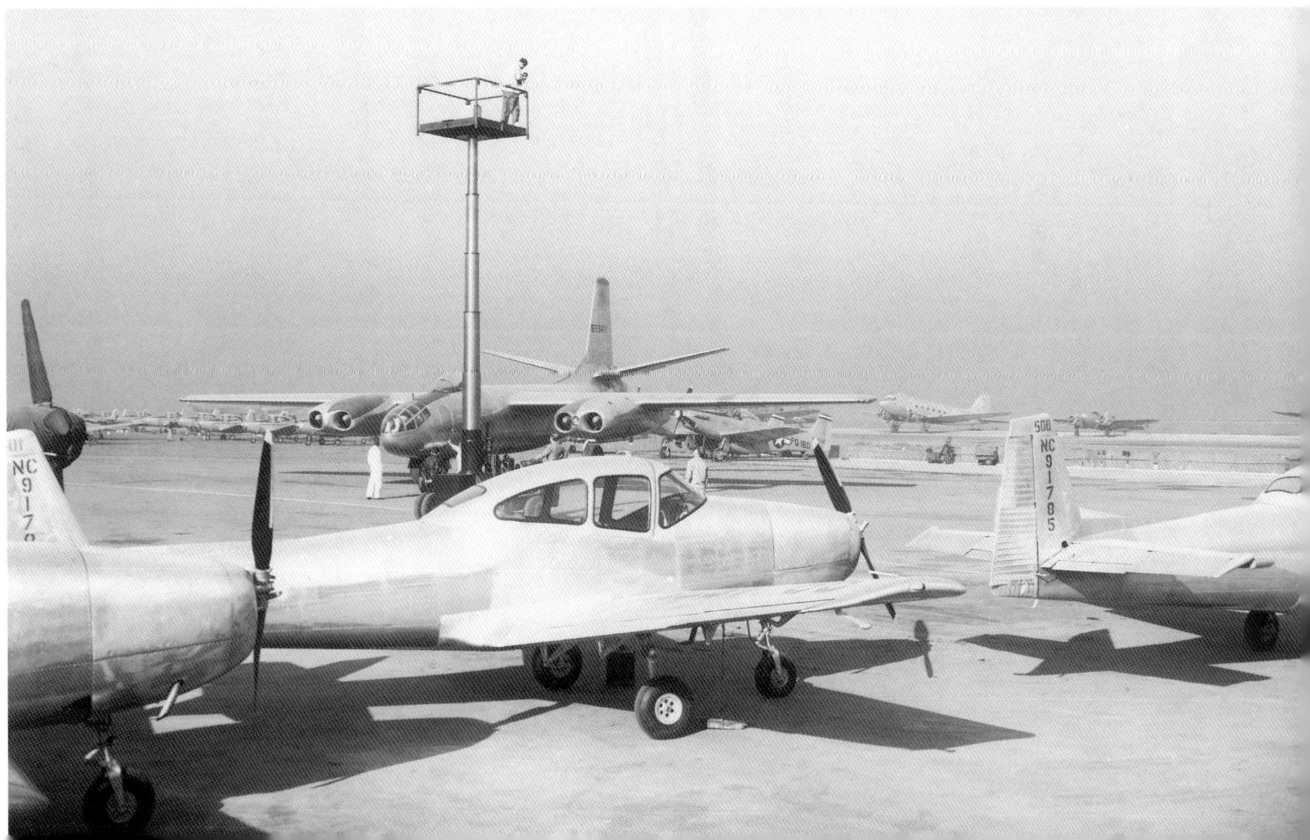

and mid-1947, 1027 production examples were built on NA-145. Some of these, designated Navion A, were fitted with an improved version of the engine, the E-185-9 with a crankshaft damper.

In June 1947 the Air Force ordered 83 Navions, with the charge number NA-153 and USAF designation L-17A, to be used as liaison aircraft. These examples, AF 47-1297 to 47-1379, were sold to the USAF at a price of approximately $10,000 each. Forty-seven of the L-17As were later transferred to Air National Guard units and a few were converted into QL-17 unpiloted drones. Many of these were transferred to the U.S. Army and were later designated U-18 (for Utility) under the tri-service designation system.

Between mid-1946 and 1947, NAA secured several new military contracts and it was apparent that the company was about to become busy again, producing F-82, B-45 and, especially, F-86 types. At that time NAA decided to sell the entire Navion program to the Ryan Aeronautical Corp. in San Diego, California. This included engineering documentation, jigs, fixtures, parts inventory and the Type Certificate.

Ryan began production of the Navion in 1948, and built a number of Navion A-models in either Utility or DeLuxe configuration. These were designated Navion 205, for the 205 horsepower the engine was producing for take-off. The base price for the Navion A in 1950 was increased to $9485, a much more profitable figure. That same year an upgraded Navion B, known as the Super 260, was introduced. This was fitted with the Lycoming GO-435-C2 engine rated at 260 horsepower. The top speed was now 174 miles per hour and the service ceiling was 18,000 feet. The cruising speed was also marginally better, but the range fell to 415 miles. Ryan production of civil Navions was 1077.

Ryan also manufactured an additional 163 L-17B and C liaison examples for the Air Force. These were nearly identical to the NAA-manufactured L-17As with the exception of revised avionics. A number of L-17Bs were used in the Korean War, one serving as General Douglas MacArthur's personal liaison aircraft.

NAA built 83 Navions as L-17As for the U.S. Army and National Guard. They performed liaison and light cargo duties.

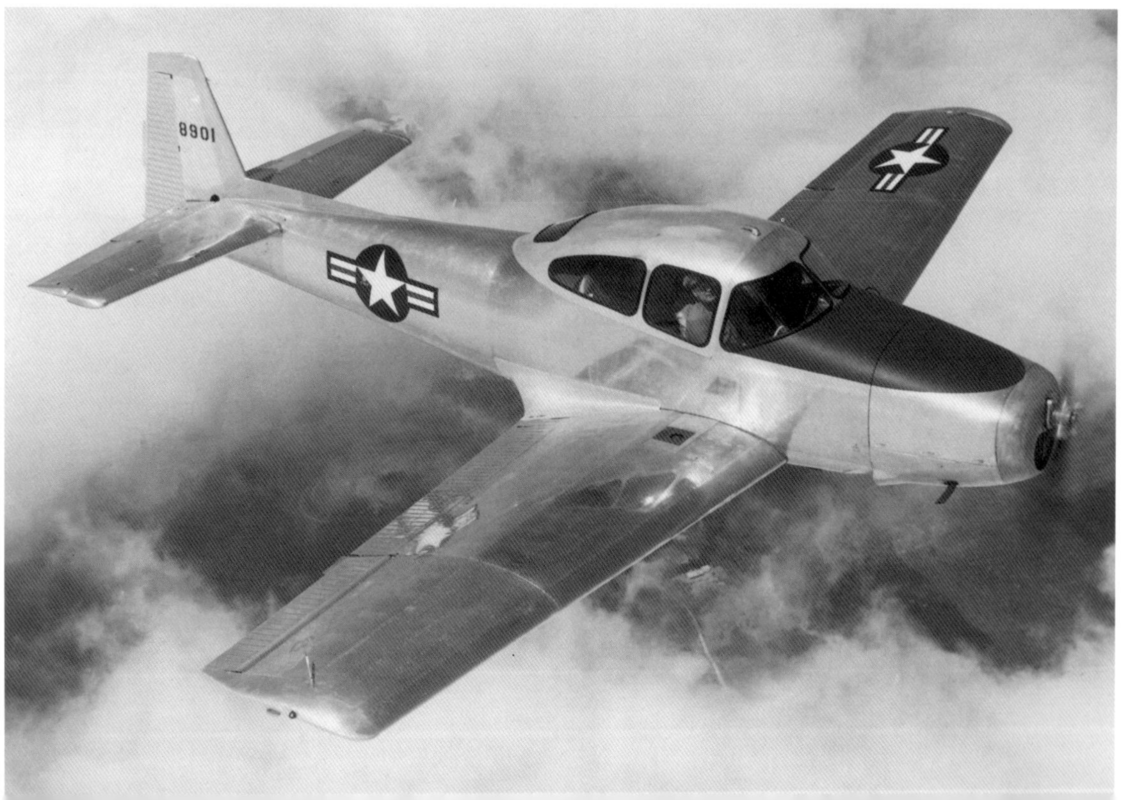

In 1953, one Ryan-built example was extensively modified to become a Navy trainer. This aircraft, known as the Model 72, N4860K, had a revised cockpit layout, with control sticks installed in place of the wheels, the two rear seats removed and a bubble canopy fitted. Also featuring modified wingtips, it was set up for aerobatics with a 6g load factor. Evaluated at NAS Pensacola, Florida, during 1953 and 1954, the Model 72 failed to secure a production contract, due in large part to the Navy's preference for tandem seating rather than side-by-side.

Ryan had built a total of 1240 Navions of various models by 1951, when the Korean War forced a shift in its production to KC-97 fuselage components and pilotless drones. The entire Navion program was again sold, this time to the Tusco Corporation in Galveston, Texas.

Tusco set up manufacturing as the Navion Aircraft Company and introduced several extensive modifications, including a redesigned fuselage with a side door and accommodation for five passengers, wingtip-mounted fuel tanks for increased range, and full IFR instrumentation. This model was renamed the Rangemaster.

The most extensive Navion modifications were the Riley twin-engine conversions performed by Temco Aircraft Corporation in Dallas, Texas. These were originally fitted with two 170-horsepower Lycoming engines and wingtip tanks and had a gross weight of 3600 pounds. Later twin-engine variants included the Camair 480, powered by two 240-horsepower Continentals and offering a cruising speed of 197 miles per hour. The gross weight increased to 4323 pounds and the aircraft was priced at $39,000 in the late 1950s.

Although the Navion was not a commercial success compared to contemporary civil aircraft, it was a sound design that developed a loyal following. In the late-1990s there were 1301 Navions with current FAA registrations, about 400 of them being early North American-built examples.

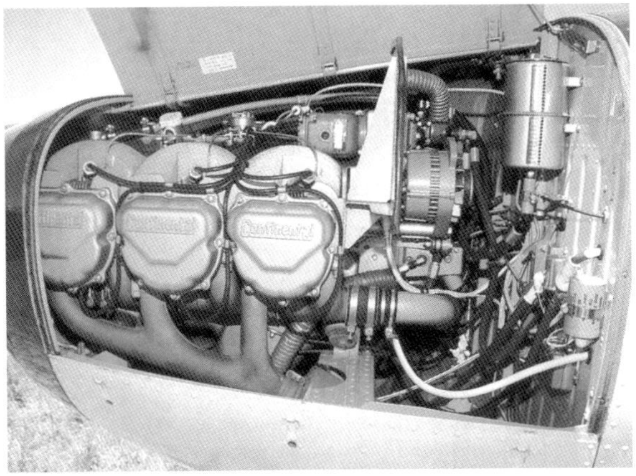

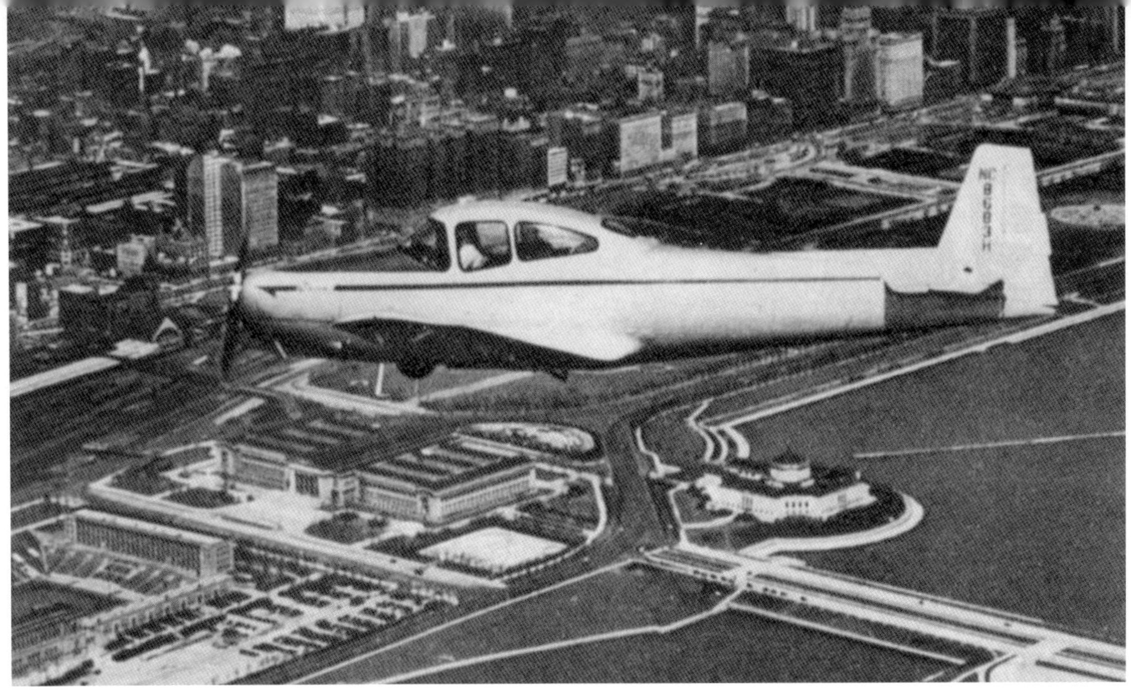

Above, the six-cylinder horizontally-opposed Continental engine installed in a restored Ryan Navion. Right, NAA Navion NC 8683H flying over Chicago in the late 1940s. The plane was used very effectively by business executives.

Navion 1/144 scale

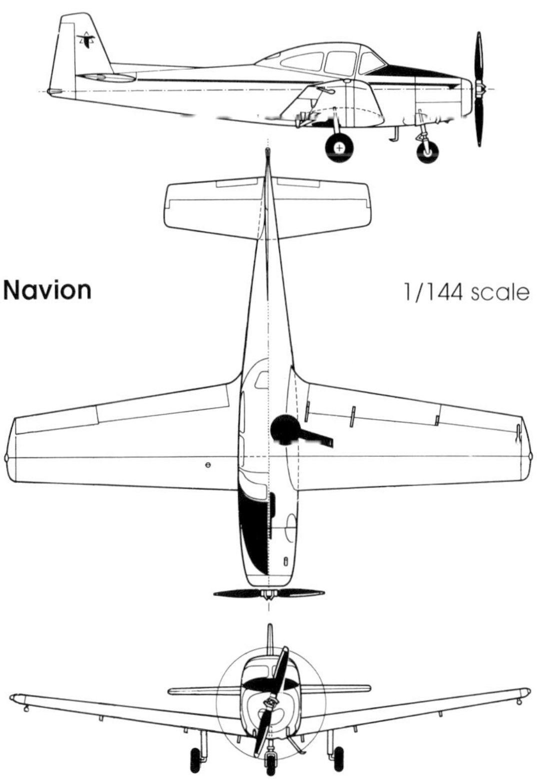

Specifications: North American NA-143/NA-145 Navion / NA-154/L-17A / Ryan Navion 205/ L-17B/C / Navion 260

Description: Private lightplane and military liaison aircraft. Single-engine low-wing monoplane of all-metal construction with retractable landing gear. Accommodation 4 (pilot and 3 passengers).

Engine: 185-hp Continental E-185-3/-9 six-cylinder horizontally-opposed air-cooled. (Navion 205/ L-17B/C) 205 hp for take-off. (Navion 260) 260-hp Lycoming GO-435-C2 six-cylinder horizontally-opposed air-cooled.

Dimensions: Span 33 feet 5 in, length 27 feet 6 in, height 8 feet 8 in, wing area 184 sq ft.

Weights: 1782 lb empty, 2750 lb gross. (Navion 205/L-17B/C) 1812 lb empty, 2905 lb gross. (Navion 260) 1860 lb empty, 3100 lb gross.

Performance: Cruising speed 145 mph, maximum 153 mph, service ceiling 13,000 ft, range 560-740 miles. (Navion 205/L-17B/C) Cruising speed 155 mph, maximum 163 mph, ceiling 15,600 ft. (Navion 260) Cruising 165 mph, maximum 174 mph, ceiling 18,000 ft.

Production: North American NA-143/NA-145 Navion / NA-154 L-17A / Ryan Navion / L-17B/C

NA-143	Navion	Production prototypes, factory serials 143-1 and -2, civil registrations NX18928 and NX18929 (later with production serials NAV-4-1 and -2)	2
NA-145	Navion	Production models, serials NAV-4-3/1027, civil registrations beginning with NX(NC)91100	1025
NA-154	L-17A	NAV-4-1028/1110 (154-1/83), USAF 47-1297/1379	83
Total, NAA-built Navion lightplanes, 1946-1947			1110
Ryan	Navion A	civil production	1077
	L-17B/C	USAF 48-921/1078, 49-1961/1965	163
Total, Ryan-built Navion lightplanes, 1948-1951			1240
Total, Navion lightplanes, 1946-1951			2350

North American NA-146>NA-184 XAJ-1/AJ-1/-2/-2P (A-2A/A-2B) Savage / NA-158/-163 XA2J-1

AT THE END OF WORLD WAR II, the United States Army Air Corps was the only service to possess atomic bombs and the means to deliver them: the Boeing B-29 Superfortress heavy bomber. When the U.S. Air Force was established in September 1947, it sought to maintain total control of the American nuclear arsenal and nuclear strike capability.

In a report dated December 17, 1947, U.S. Navy Rear Admiral Daniel Gallery described the deficiencies of the USAF nuclear strike capability, stating that the Navy could better perform the task. Known as the "Gallery Memorandum," the report described how inefficient the USAF B-29 and B-36 were in performance and suggested that a naval mission involving the use of aircraft carriers, each with twelve expendable aircraft, and submarines would be a more effective attack team.

Gallery theorized that the ideal attack aircraft would have a shorter range but much greater performance over the target than the USAF heavy bombers. To achieve this, the aircraft would be much smaller than a heavy bomber and have no landing gear. The idea was to transport the aircraft across the ocean aboard carriers and launch them within 1000 miles of the target. Taking off from large carriages on the flight deck, the attackers would penetrate the target area at high speed, deliver their payloads and then ditch in the ocean, where rescue submarines would retrieve the aircrews. To avoid detection, the carriers would make evasion maneuvers after launching the aircraft and the ditched planes would not lead enemy air defenses to the carriers.

The first of three XAJ-1 prototypes, BuAer 121460, was rolled out in June 1948. Note the jet engine intake above the wing.

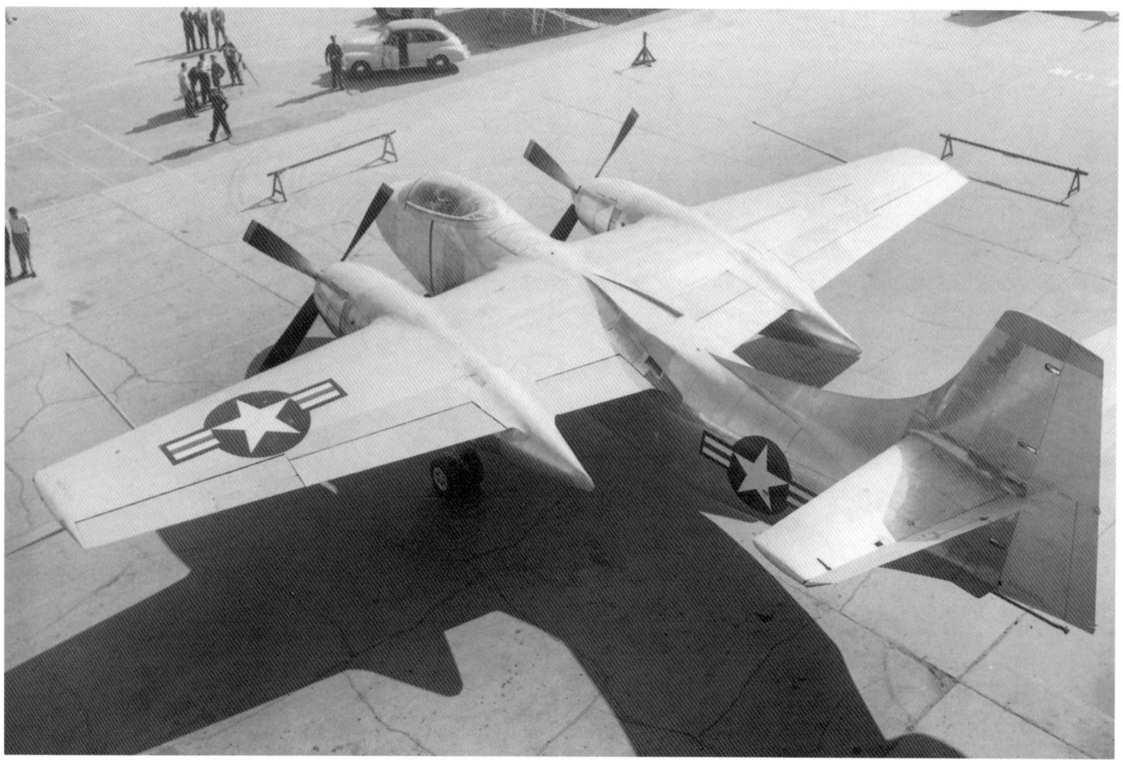

The plan was ambitious, and an expensive proposition in terms of expendable aircraft. In the spring of 1948, Secretary of Defense James Forrestal and the Joint Chiefs of Staff established the Key West Agreement, maintaining the USAF as the primary service to deliver nuclear arms, but allowing the Navy to develop its own nuclear strike force to support maritime operations.

The U.S. Navy had already laid plans for a nuclear attack aircraft three years before the Key West Agreement. On August 13, 1945, one week after the first A-bomb was dropped on Hiroshima, the Navy initiated a design competition for a carrier-based attack plane with a 10,000-pound payload. Later that year, Deputy Chief of Naval Air Operations Vice Admiral Marc Mitscher, Navy weapons specialist Captain William Parsons (who had been involved with the Hiroshima mission) and Commander John Hayward discussed the development of a carrier-based nuclear strike program. The "Fat Man" bomb used in the Nagasaki mission weighed about 10,000 pounds and the plan was for the new attack aircraft to carry one. Commander Fredrick Ashworth (weapons engineer on the Nagasaki mission) and Captain J. N. Murphy of the Navy Bureau of Aeronautics concurred that with modifications NAA's AJ-1 design could carry the new Mark Four atomic bomb. Although the Douglas company also submitted a proposal, the AJ-1 was eventually selected.

The design was unusual in that it featured two Pratt & Whitney R-2800-44W radial engines mounted in wing nacelles and an Allison-built General Electric J33 turbojet mounted in the fuselage. The 4600-pound-thrust jet engine was to be used during take-off, high-speed runs at low altitude and in the event of failure of one of the piston engines. The air intake for the jet engine was located just ahead of the vertical fin with a door that was opened during operation, and the exhaust was located beneath the vertical stabilizer at the rear of the fuselage. Both the R-2800s and the J33 burned the same high-octane aviation fuel and could operate from the same tanks.

The AJ-1 was a relatively large airplane, with a wingspan of more than 71 feet and a length of 63 feet. Height at the top of the vertical stabilizer was over 20 feet. On production models the wingtips and vertical stabilizer folded for carrier operations. The large diameter of the four-bladed propellers, combined with the landing gear configuration, resulted in little more than one foot of deck clearance for the blades. A crew of two sat side-by-side under a frameless sliding bubble canopy.

On June 24, 1946, NAA received a contract for three XAJ-1 (NA-146) prototypes and one static test airframe. An order for twelve production aircraft, subsequently named Savage, was placed on October 6, 1947, and in May 1948 an additional 28 were ordered. The first prototype, BuAer 121460, was rolled out in June and NAA test pilot Bob Chilton flew for the first time on July 3, 1948.

The flight test program was not without incidents. Although the first prototype enjoyed a long and successful career, the number two and three aircraft were lost in accidents due to struc-

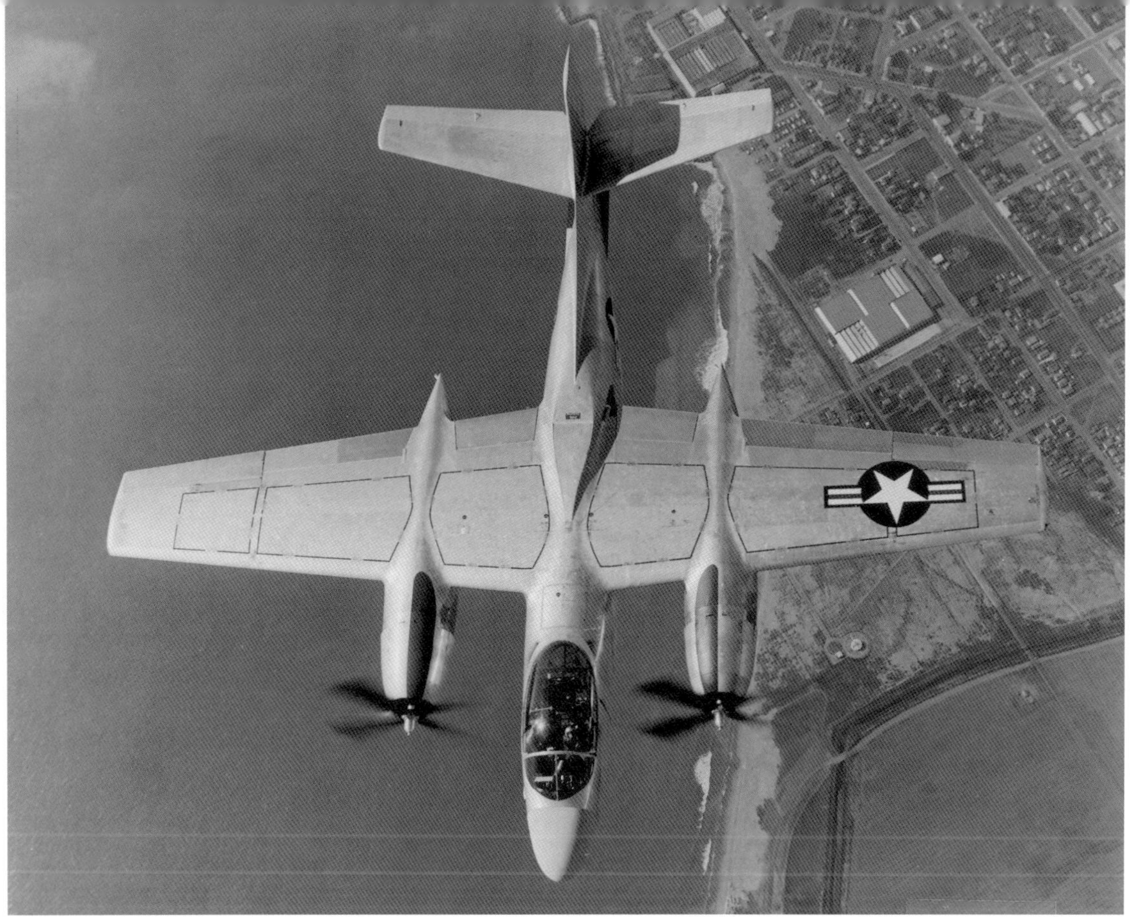

Although bulky, the AJ was an efficient design. The original canopy was unframed, propeller spinners omitted temporarily.

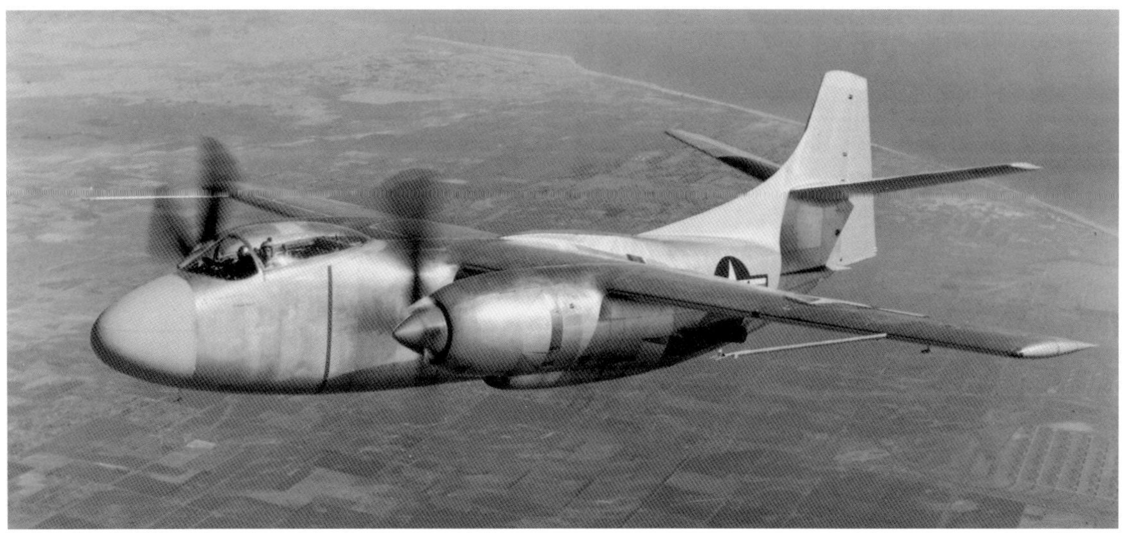

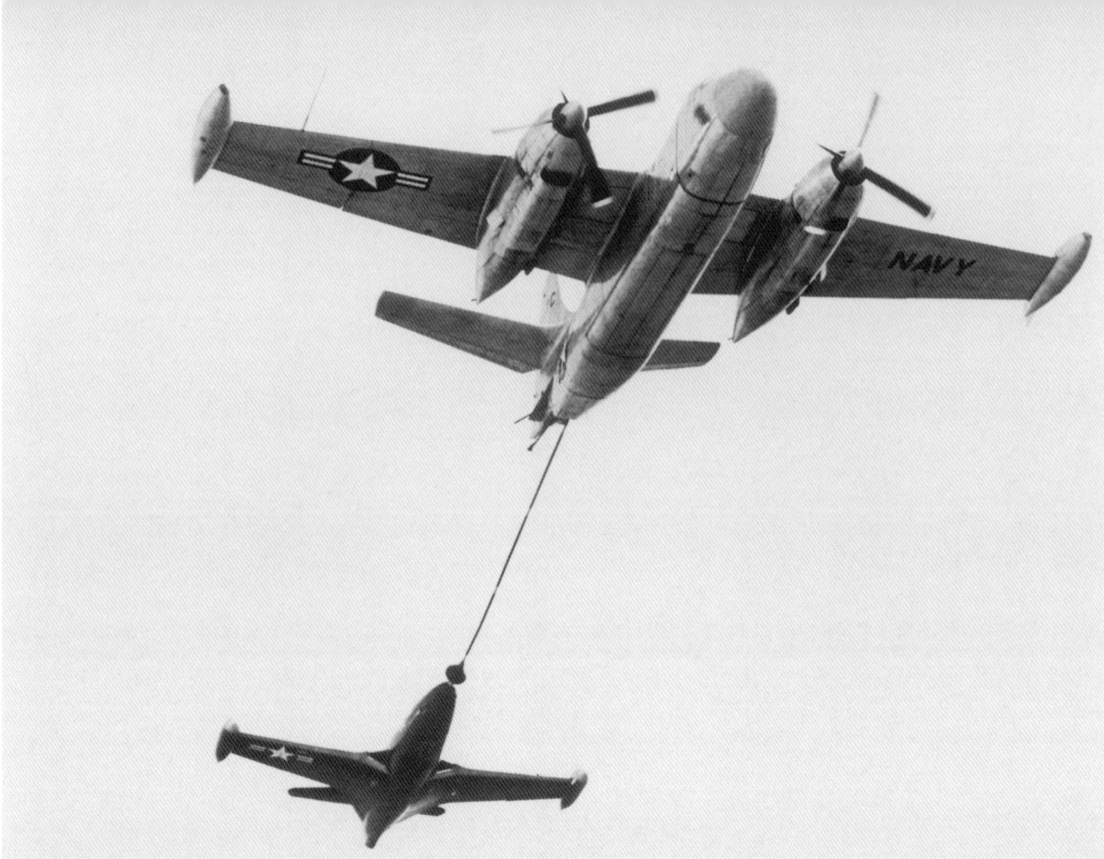

tural failure. The second prototype, BuAer 121461, lost its entire tail section during maximum rudder sideslip tests in February 1949 and crashed at sea near Los Angeles, killing test pilots Al Conover and Chuck Brown. The third prototype, 121462, crashed at Edwards AFB, California, after a wing failure, although the crew survived.

AJ-1 production began at Inglewood in early 1949, incorporating a number of significant changes. Structural improvements were made, and the early framed, sliding canopy was later changed to a fixed canopy. Although the canopy featured an emergency escape window, entry and exit were made through a door located on the right side of the fuselage. Wingtip-mounted fuel tanks were added to increase range. Production aircraft had a crew of three, including a bombardier, and the wingtip-mounted tanks were also used.

Above left, XAJ-1 was used by NATC for in-flight refuelling tests, shown here with a Grumman F9F Panther. This proved to be one of the Savage's major duties. Below, the second production AJ-1, BuAer 122591, in overall dark blue paint.

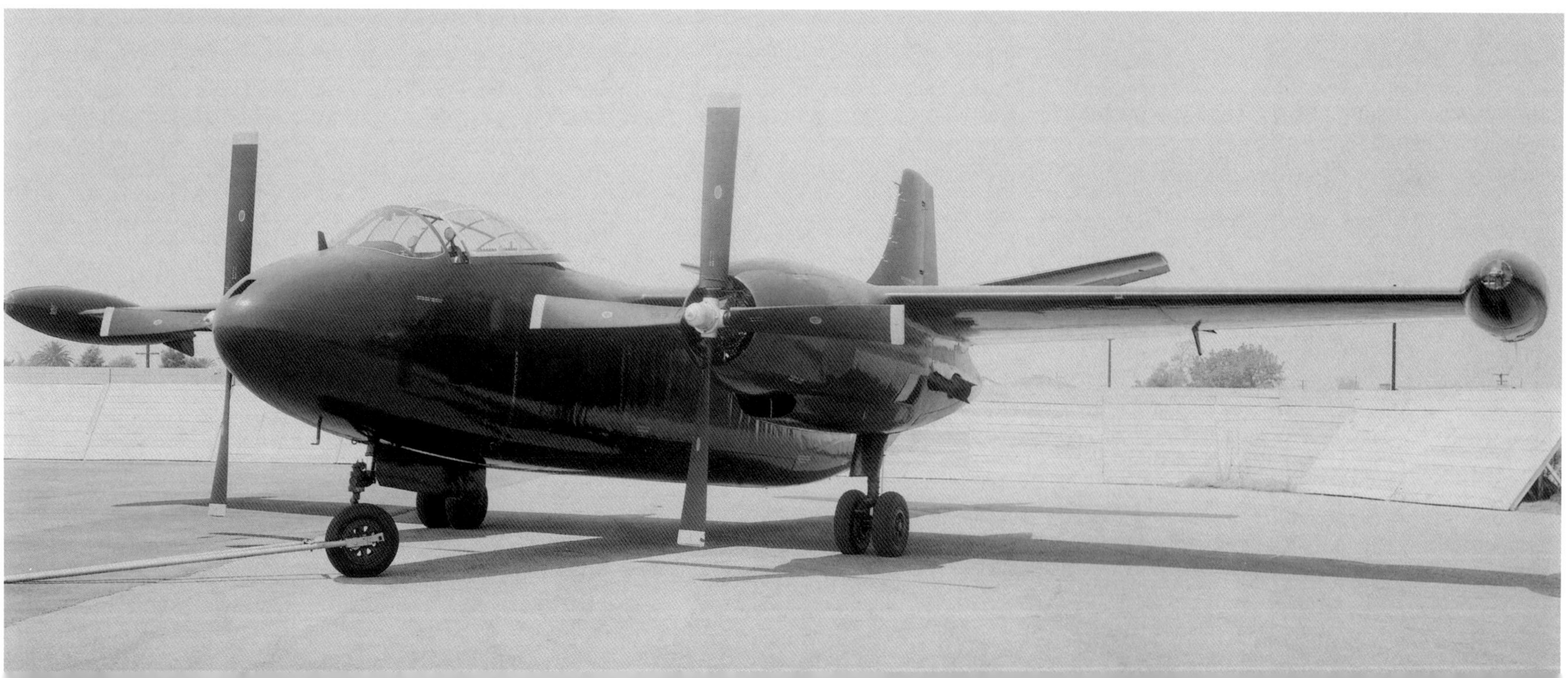

The first flight of a production AJ-1 took place in May 1949, and shortly thereafter several Savages were flown to the Naval Air Test Center, NAS Patuxent River, Maryland, and to NAS Moffett Field, California, where they underwent Board of Inspection and Survey (BIS) and Fleet Indoctrination Programs (FIP). VC-5, the first squadron to receive Savages, participated in the FIP carrier trials program. The first AJ-1 carrier take-off took place on April 21, 1950, from the USS *Coral Sea* (CVB-43) with Captain Hayward at the controls, and the first carrier landing was made on August 31. A total of 55 AJ-1s were built, serving with Composite Squadrons VC-5, VC-6, and VC-7 from a variety of carriers, including the *Coral Sea, Kearsarge, Oriskany, Yorktown* and *Wasp.*

Early in 1950 the Navy decided to produce an improved Savage for the photo reconnaissance role, given the initial designation AJ-1P. A

Right, a splendid view of a VC-5 AJ-1, flying above San Francisco in 1950; although dwarfed in this view, the JRM Mars had nearly three times the wingspan of the Savage. Below, a look into the AJ-1 jet tailpipe, not usually visible.

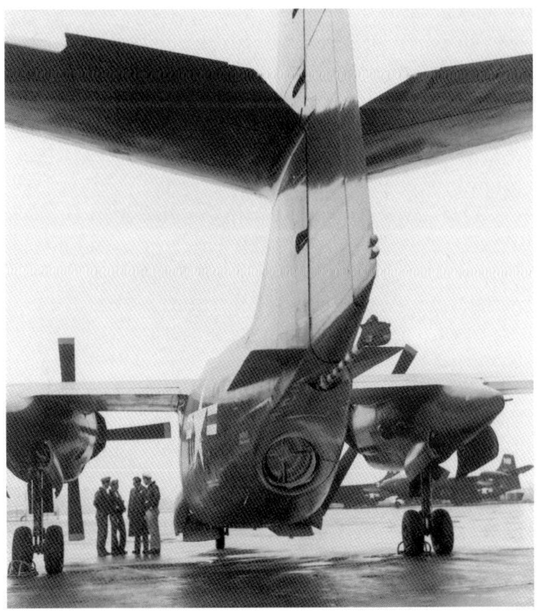

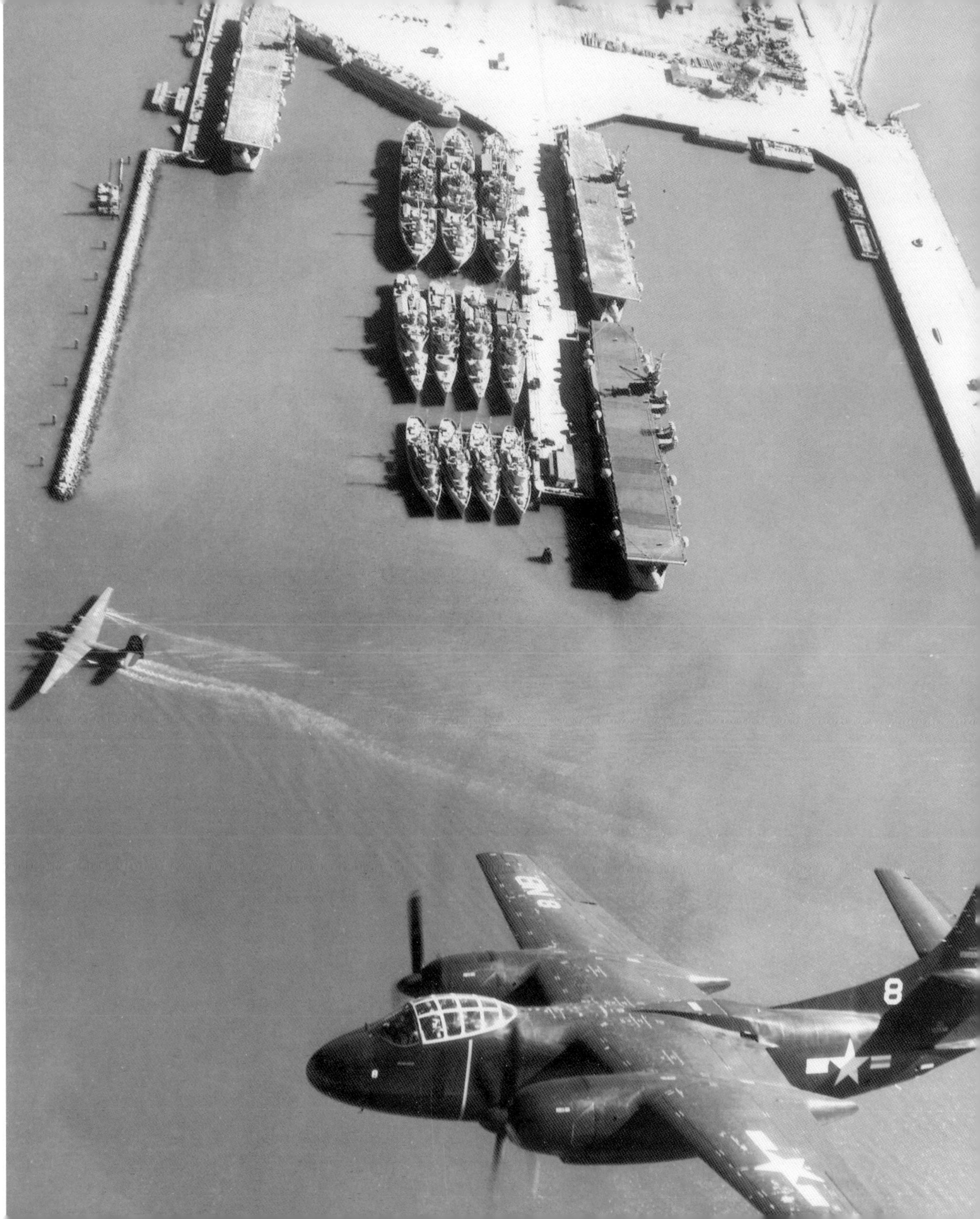

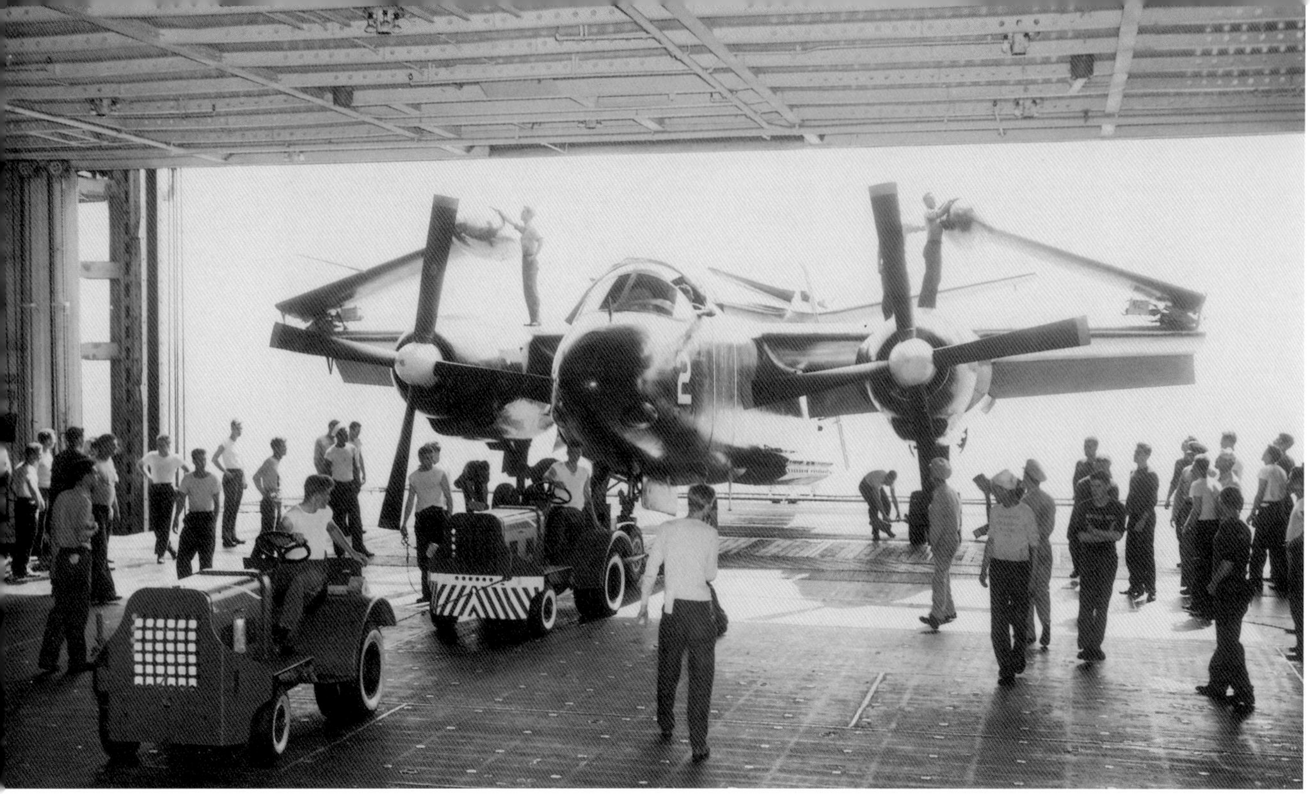

production AJ-1, BuAer 124850, was modified at NAA's Downey facility and an order for 23 was placed on August 18, 1950. These aircraft, designated AJ-2P, were built at NAA Columbus.

For the photo reconnaissance role, the nose was redesigned with a fold-down access bay for the K-38 camera and the bomb bay area was fitted with both down-looking and side-looking high-resolution camera equipment. Improved R-2800-48 and J33-A-10 engines were installed, and the tail section was redesigned. The vertical fin was increased in height by approximately one foot and lengthened slightly, while the dihedral tailplanes were replaced by a conventional horizontal stabilizer. An improved hydraulic system was used and the internal fuel capacity was increased.

The cockpit and crew configuration also differed from the AJ-1. A wheel now replaced the control stick. On earlier AJ-2Ps, the third crewman sat in a compartment forward of the bomb bay; on later versions, a crew of four was carried, with an extra seat in the upper deck, facing aft. The crew now consisted of a plane captain, photo plane commander, photo navigator and photo technician. Most of the AJ-2P crews were graduates of the Aerial Photo School at NAS Pensacola, Florida, having completed a 21-week course. AJ-2Ps served with VJ-61 at Miramar, California, and in the Philippines, with VAP-61 on Guam, and with VJ-62 and VAP-62 at Jacksonville, Florida.

The improvements implemented in the AJ-2P were also used to produce an improved attack version, the AJ-2. This differed in cockpit layout from previous versions, with the throttles moved to a center console where the other crew members could aid the pilot if necessary. The AJ-2 had a crew of three, all on the upper deck, with the third crewman facing aft.

On August 14, 1950, the Navy signed an or-

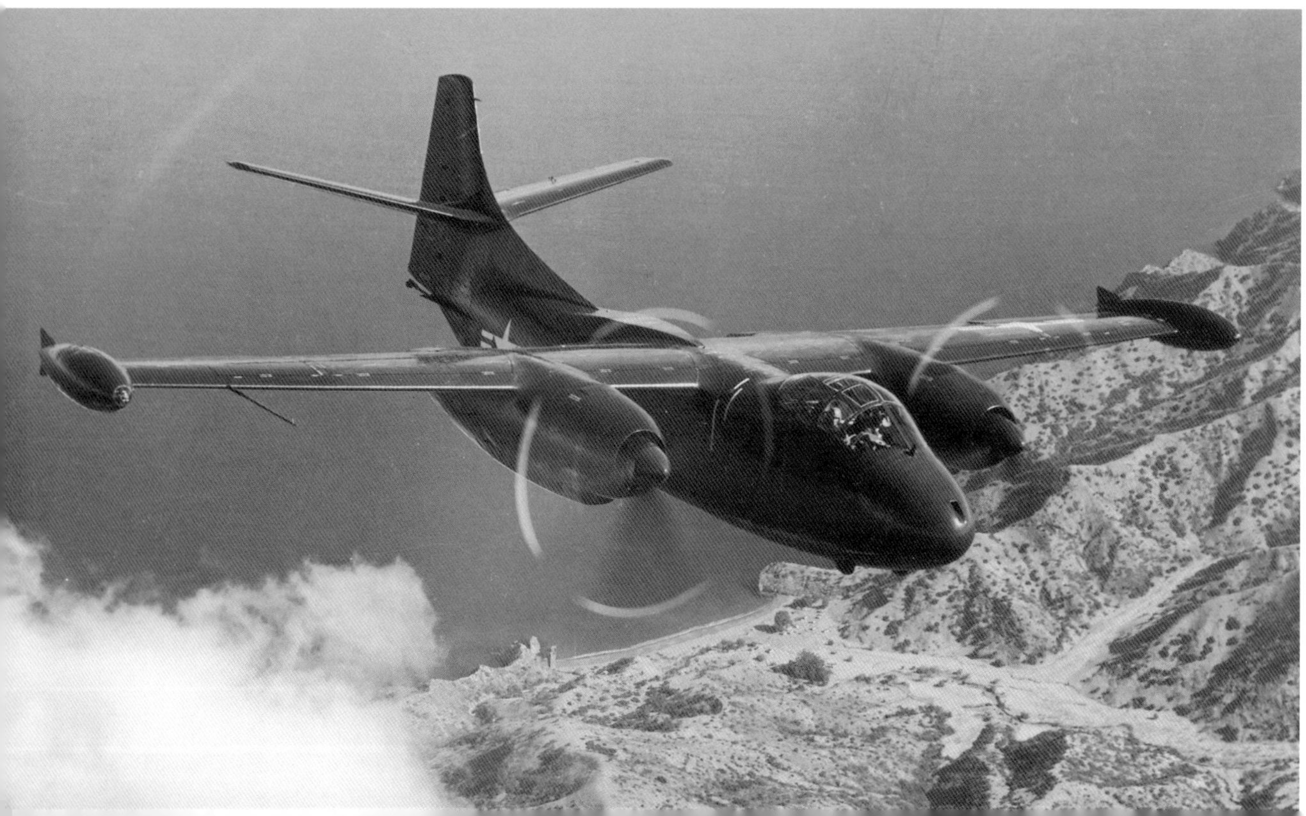

Top, an AJ-1 below the deck of the carrier USS Lake Champlain, *CVA-39. Left, an AJ-1 over one of the Channel Islands off Southern California. 300-gal. wing tanks were usually fitted.*

1/144 scale

Bob Hunt drawings, originally published in Aviation News *in 1/72 scale*

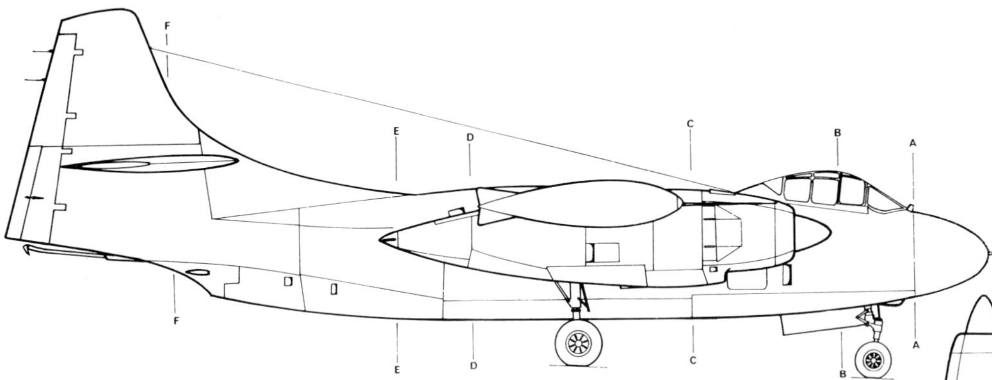

AJ-2

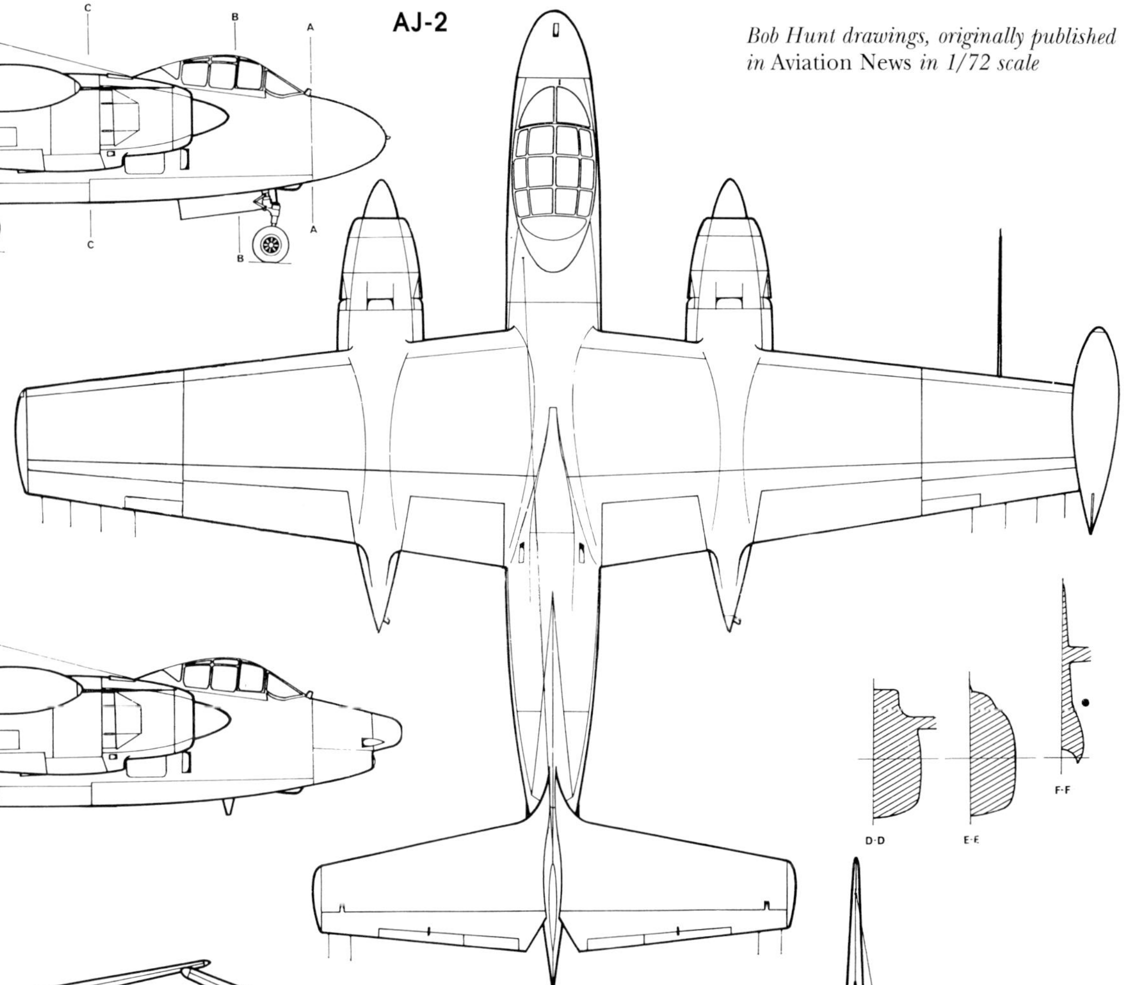

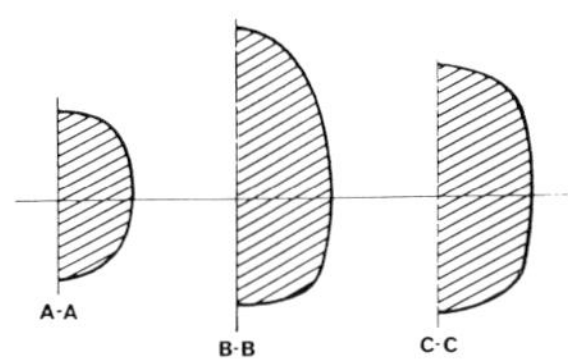

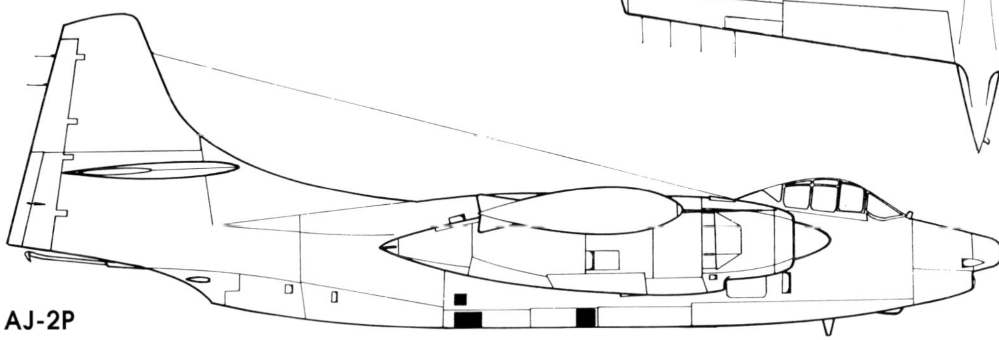

AJ-2P

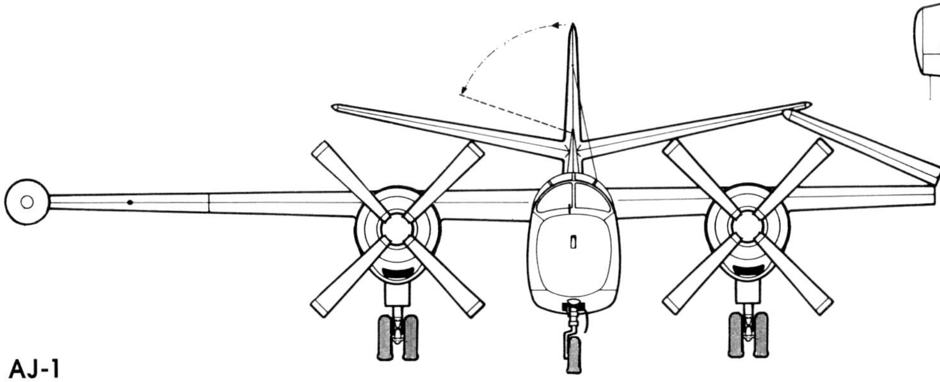

AJ-1

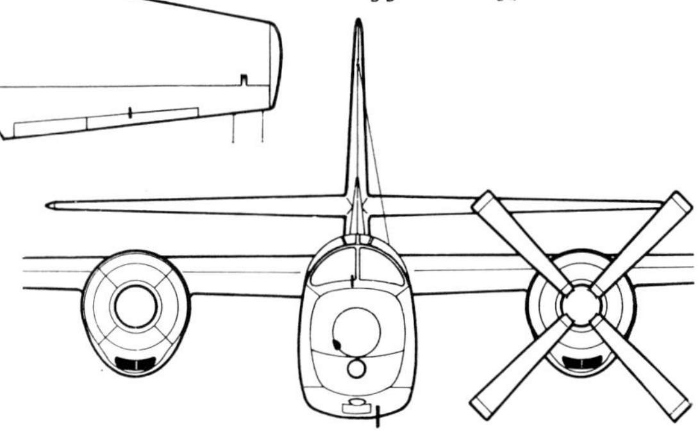

AJ-2P

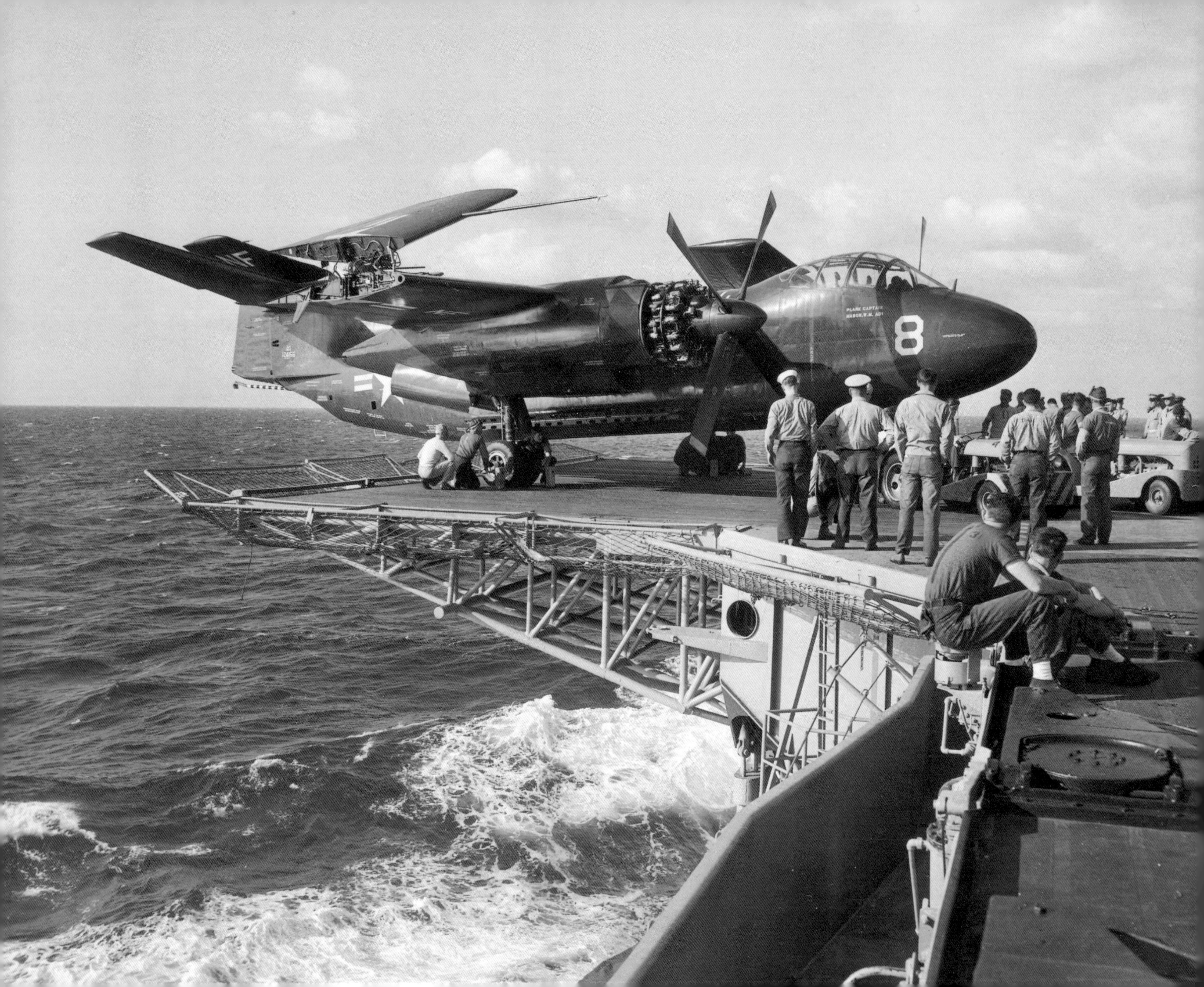
PLANE CAPTAIN
MASON, W.M. AD1
8

der to modify 42 of the original AJ-1s to AJ-2 specifications and ordered 55 new AJ-2s on February 14, 1951. AJ-2s served with Composite Squadrons VC-7, VC-8 and VC-9, and Heavy Attack Squadrons VAH-5, VAH-6, VAH-7, VAH-9, VAH-11, VAH-15 and VAH-16.

Although originally designed as an attack plane, the Savage spent much of its career as an in-flight-refueling tanker. Tanker tests were conducted with F2H Banshee and F9F Panther fighters as well as a variety of other types. Savages served as tankers with squadron VC-5 aboard the USS *Franklin D. Roosevelt,* VC-6, and with most of the Heavy Attack Squadrons.

Because of the Savage's size, some carrier handling difficulties were encountered. These led to a proposal for a smaller version of the aircraft, designated AJ-3. The changes included a reduction in wingspan and length, each by about five feet, but the AJ-3 proposal never got beyond the initial design stage.

Late in 1948, only four months after the first flight of the XAJ-1, the Navy had ordered two prototypes and one static test article of a significantly different attack plane, the XA2J-1 (NA-163). The XA2J-1 was powered by two 5035-shaft-horsepower Allison T40-A-6 turboprop engines driving four three-blade, contra-rotating propellers. To handle the power of these engines required an extensive redesign, which included a very large, foldable, swept-back vertical stabilizer. Twin radar-controlled 20-mm cannons were to be carried in a tail turret. The forward fuselage and cockpit area also differed considerably from the AJ-1 designs. Named the Super Savage, the XA2J-1, BuAer 124439, first flew from Edwards AFB on January 4, 1952. The Super Savage was very overweight and the performance with the turboprops was less than satisfactory. Troubles with the engines led to a very short test program, and the project was terminated without the second prototype ever flying. A smaller turboprop version known as the XA2J-2 was planned, but never built. It was intended to perform mine-laying and close air-support duties.

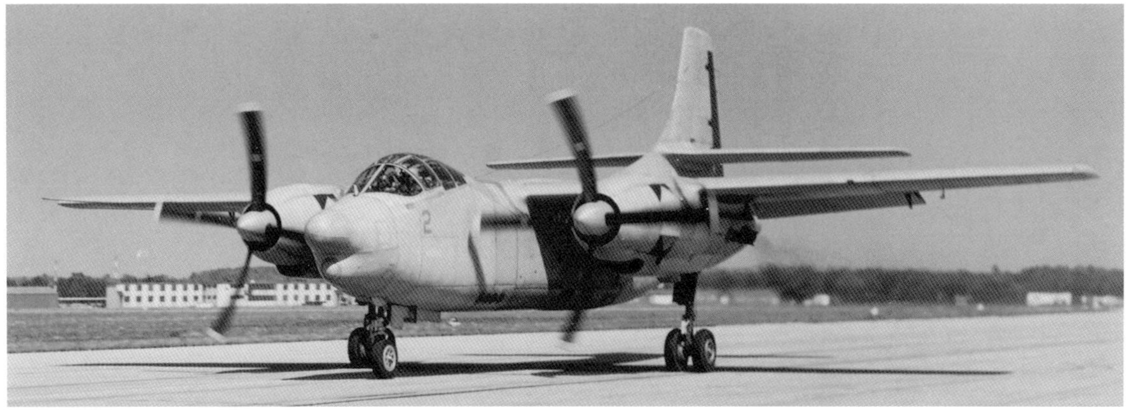

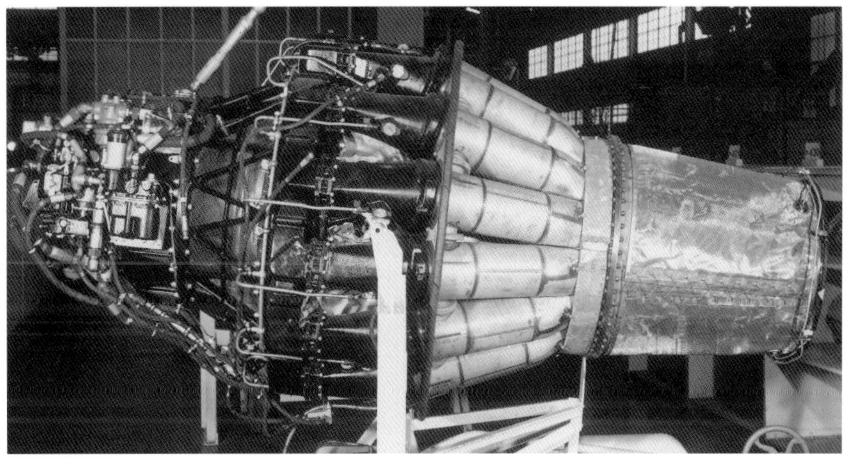

Above, still unpainted, the second photo-reconnaissance AJ-2P, BuAer 128044, taxying on NAA's Columbus runway. Left and below, close-ups of the AJ-2P's J33-A-10 jet engine, nose camera access and cockpit.

Opposite page, VC-6 Plane Captain W. M. Mason's AJ-1, BuAer 124166, with its wingtips and vertical tailplane folded, hanging out over the wake of the carrier USS Yorktown, *CVA-10.*

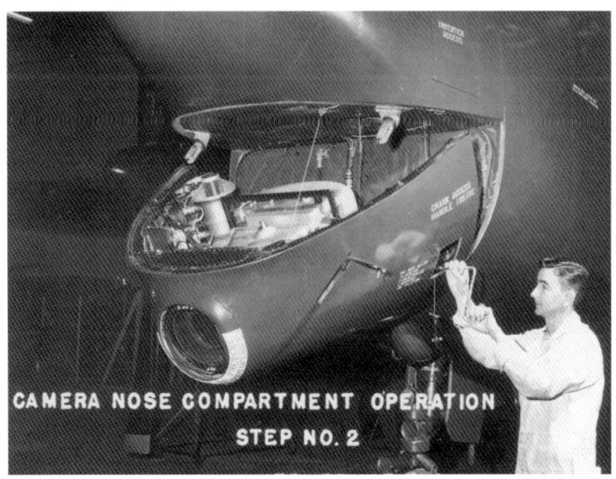

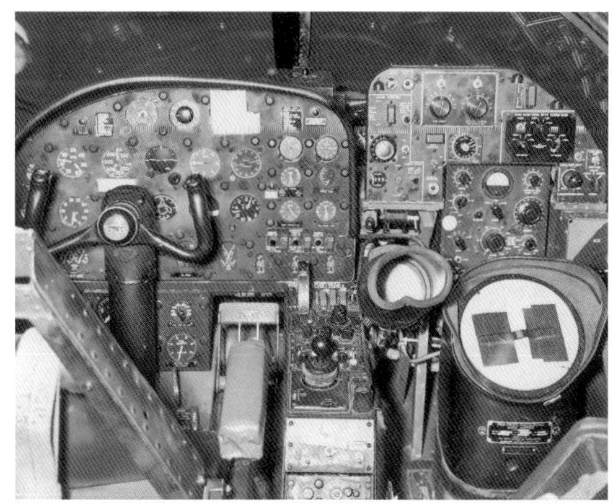

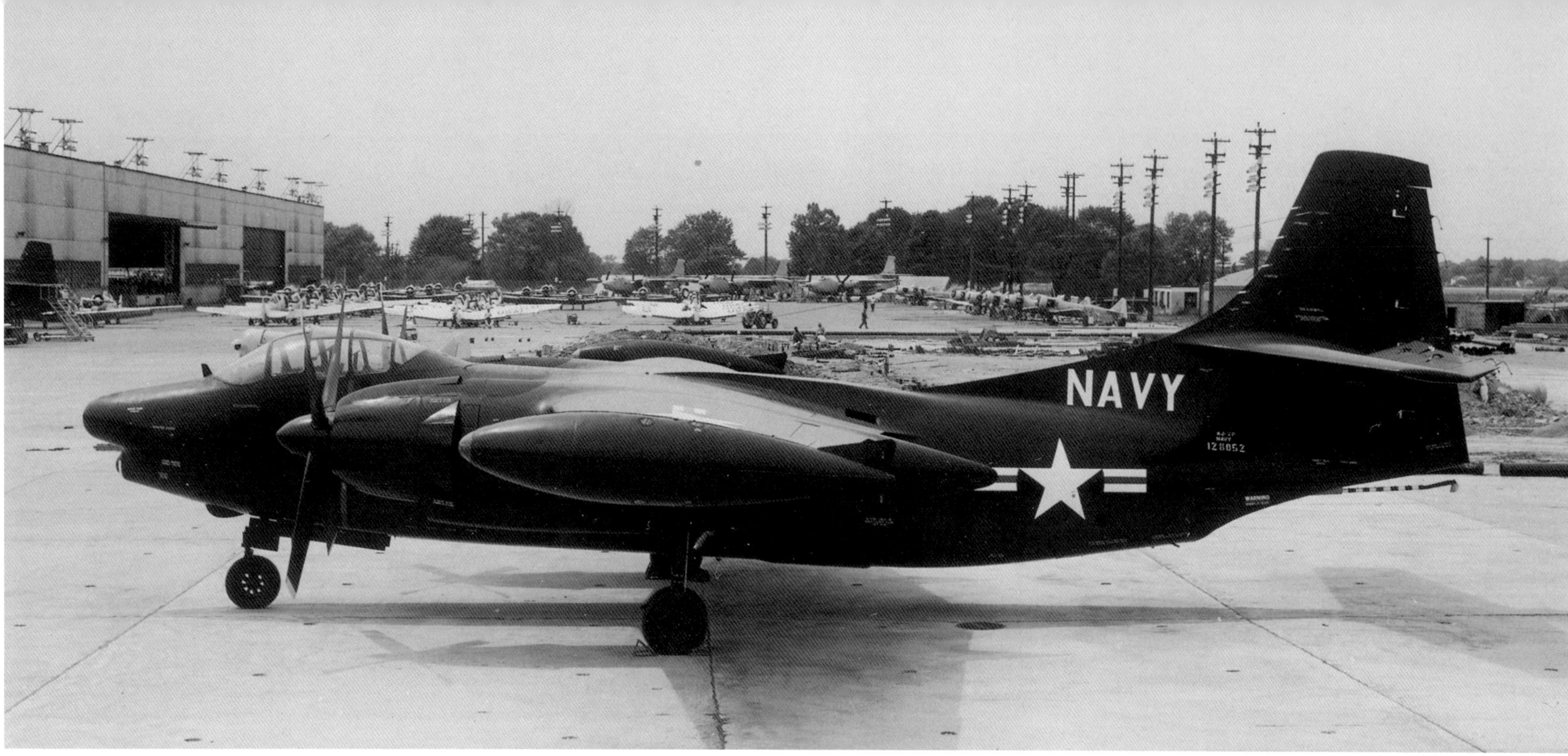

A recently completed AJ-2P, BuAer 128052, at North American's Columbus plant. Four unfinished AJ-2Ps can be seen in the background, as well as some 20 Texans being remanufactured as T-6Gs.

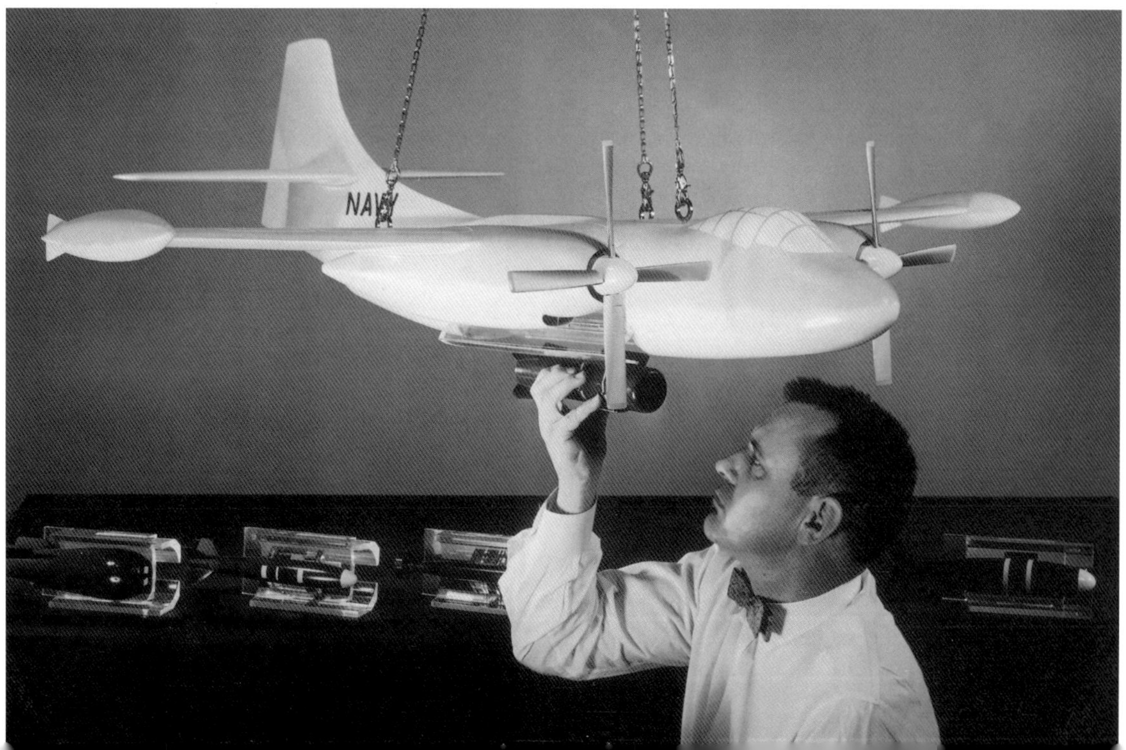

By 1960 most of the Savages were phased out of naval service and replaced by Douglas A3D Skywarriors. NASA acquired three AJ-2s and used them for weightlessness training between 1960 and 1964. One AJ-2 and an upgraded AJ-1 were used as forest fire bombers in California between 1961 and 1968; two other AJ-2s were sent to Washington state for the same duty but never saw action. The last flight of a Savage occurred on May 9, 1984, when an AJ-2 used by Avco-Lycoming as a testbed was delivered to NAS Pensacola. Repainted in Navy colors, it has served as an exhibition piece. The Savage's career was necessarily short: with the Navy's transition to jet aircraft, piston-engine aircraft quickly became obsolete.

Columbus engineer Frank Compton demonstrates nuclear bomb loads envisioned for the Savage. The horizontal tailplane, lacking dihedral, indicates that the model is the AJ-2 version.

Specifications: North American NA-146>184 XAJ-1/AJ-1/-2/-2P (A-2A and A-2B) Savage

Description: U.S. Navy carrier-based attack bomber and refuelling tanker. Three-engine (two piston, one jet) high-wing monoplane of all-metal construction with retractable landing gear and folding wings. Crew 3. (AJ-2) Modified as twin-engine refuelling tanker. (AJ-2P) Photo reconnaissance with 5 cameras in nose.

Engines: Two 2300-hp Pratt & Whitney R-2800-44W eighteen-cylinder radial air-cooled and one 4600-lb-thrust Allison J33-A-10 turbojet. (AJ-2) R-2800-48s and J33-A-10. (AJ-2 tanker) Jet engine removed.

Dimensions: Span 75 ft 2 in, length 63 ft 1 in, height 21 ft 5 in, wing area 836 sq ft.

Weights: 30,776 lb empty, 50,963 lb gross, 54,000 maximum. Fuel 1217-2457 gal., including two 300-gal. wingtip tanks. (AJ-2) 2517 gal.

Armament: One 8500-lb Mk-6 or 7600-lb Mk-15 nuclear bomb or six 1600-lb conventional bombs. (AJ-2 tanker and AJ-2P) None.

Performance: Cruising speed 270 mph, maximum 449 mph, landing 120 mph, climb 2900 ft/min, service ceiling 40,800 ft, range 1732 miles, 2993 miles ferry. (AJ-2) Cruising speed 260 mph, landing 119 mph, climb 3050 ft/min, ceiling 41,500 ft,range 2475 miles, 3056 miles ferry.

Production: North American NA-146>184 XAJ-1/AJ-1/-2/-2P (A-2A and A-2B) Savage

NA-146	XAJ-1	BuAer 121460/121462	3
NA-155	AJ-1	Study and mock-up	
NA-156	AJ-1	BuAer 122590/122601	12
NA-160	AJ-1	BuAer 124157/124184	28
NA-169	AJ-1	BuAer 124850/124864	15
NA-175	AJ-2P	BuAer 128043/128054, 129185/129195	23
NA-183	AJ-2P	BuAer 130422/130425, 134073/134075	7
NA-184	AJ-2	BuAer 130405/130421 134035/134072	55
Total, AJ-1/-2 attack planes, 1946-1954			143

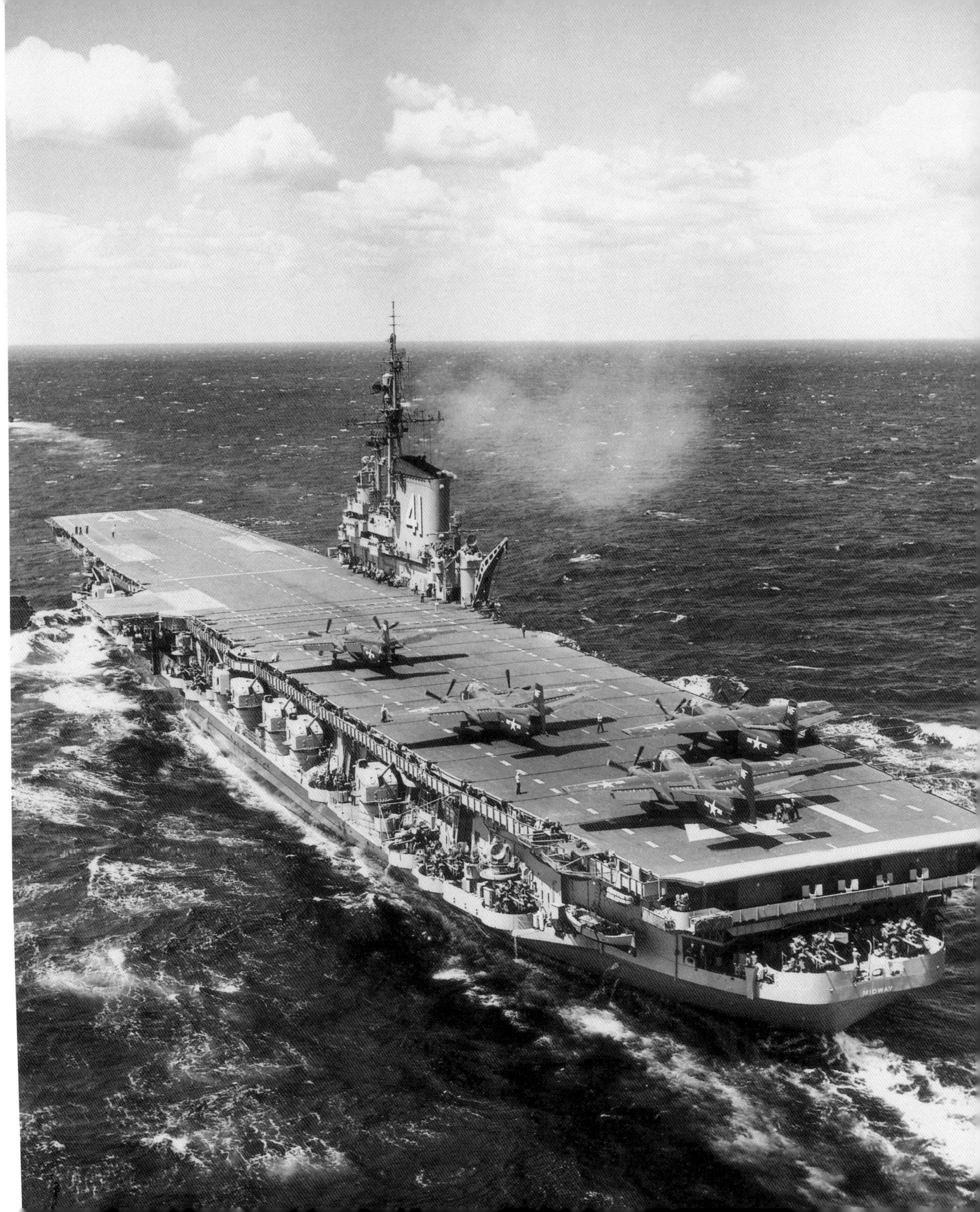

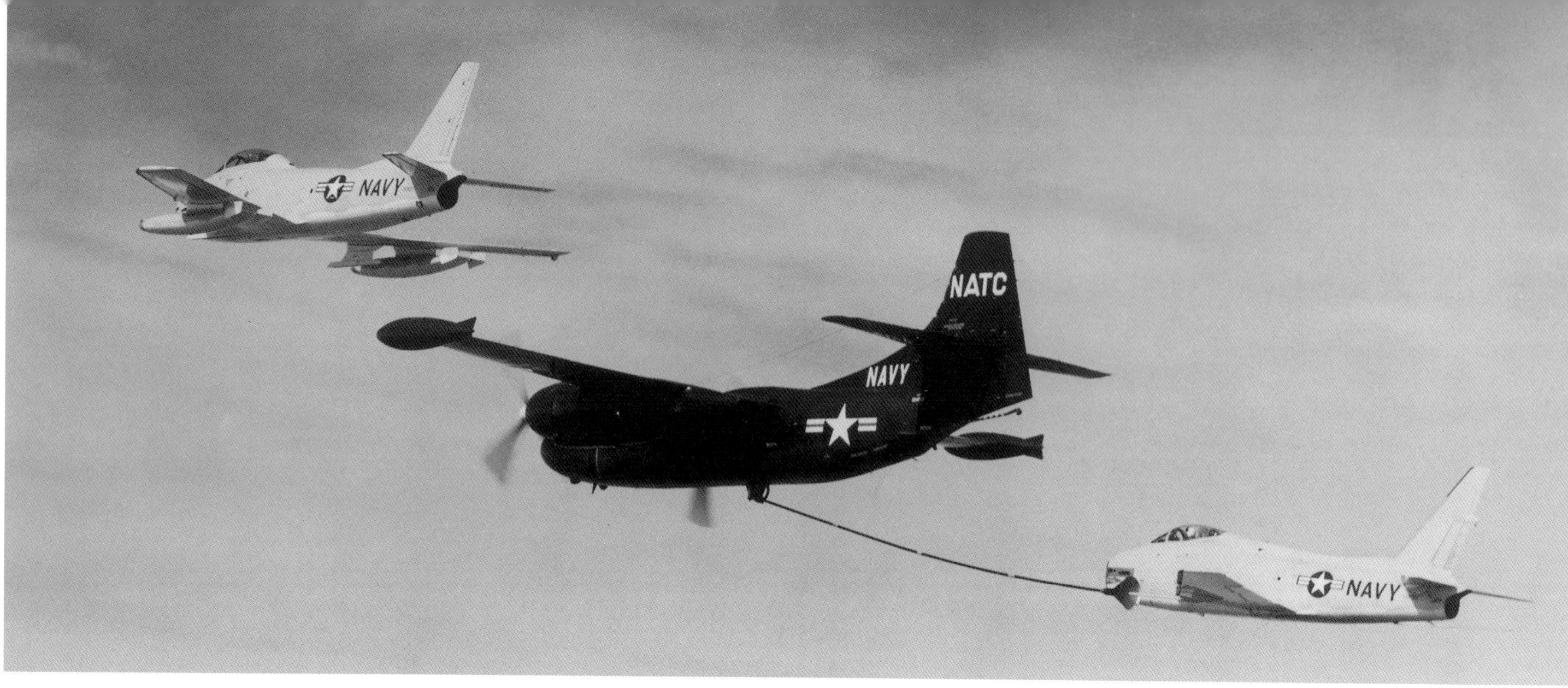

42

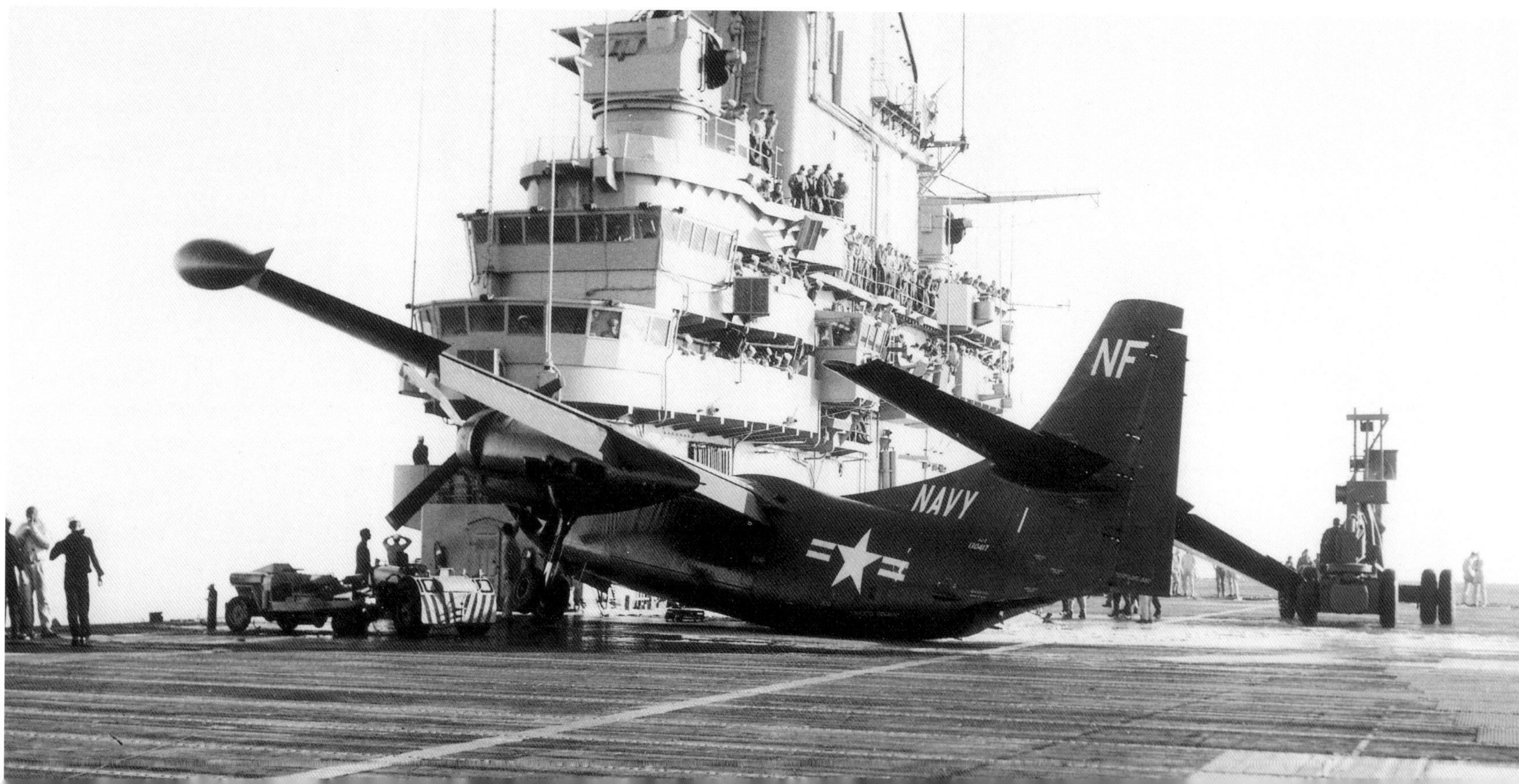

Opposite page top, an NATC AJ-2 tanker is about to refuel an FJ-4 Fury while an FJ-3 holds station to port. Below, a VC-6 AJ-2 has come to grief on the deck of the USS Lake Champlain*; note the foam below the collapsed starboard landing gear.*

Specifications: North American NA-158/163 XA2J-1 Super Savage

Description: U.S. Navy carrier-based attack-bomber. Twin-engine high-wing monoplane of all-metal construction with retractable landing gear and folding wings and tail. Crew 3.

Engines: Two 5035-hp plus 1225-lb-thrust Allison T40-A-6 turboprops with 6-bladed contra-rotating propellers.

Dimensions: Span 71 ft 6 in (46 ft folded), length 70 ft 2.5 in, height 24 ft 2 in (16 ft with tail folded), wing area 836 sq ft.

Weights: 35,354 lb empty, 46,882 lb gross, 61,170 lb maximum. Fuel 1802-2620 gal.

Armament: (Projected) 10,500 lbs of bombs and two radar-aimed 20-mm cannon in tail.

Performance: Cruising speed 400 mph, maximum 451 mph, stall 100 mph, climb 6820 ft/min, service ceiling 37,500 ft, range 2180 miles with 8000-lb bomb load.

Production: North American NA-158/163 XA2J-1 Super Savage

NA-158	XA2J-1	Engineering	—
NA-163	XA2J-1	BuAer 14439/124440	2
Total, XA2J-1 prototypes, 1948-1952			2

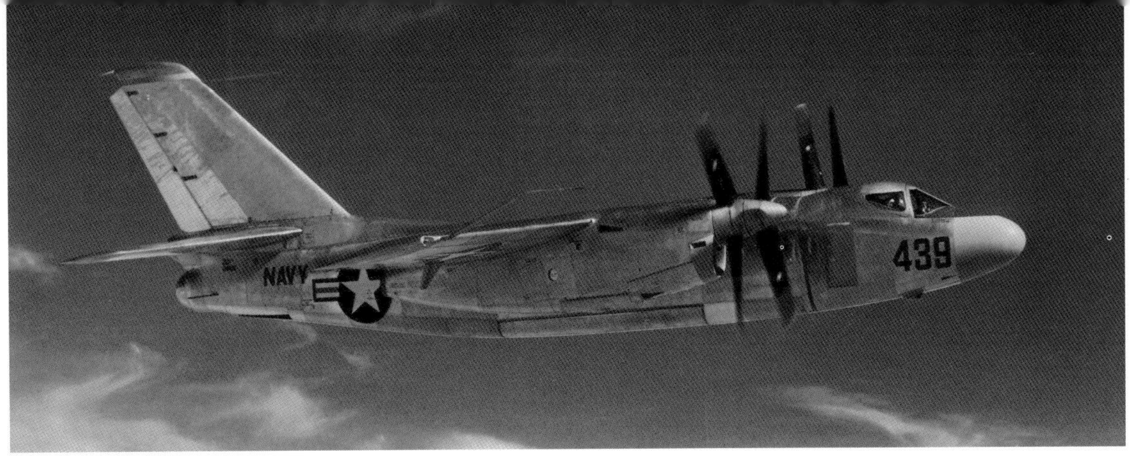

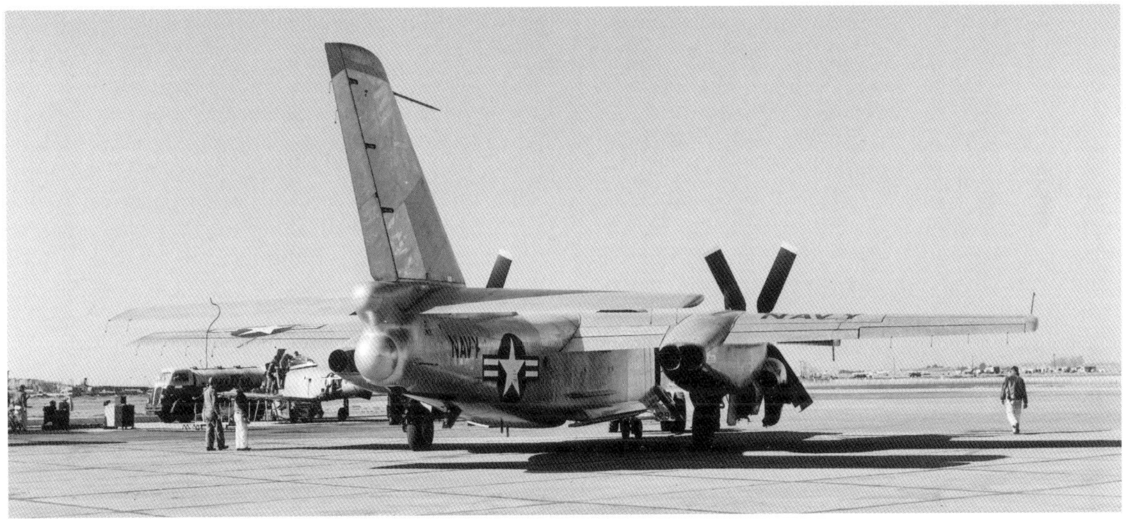

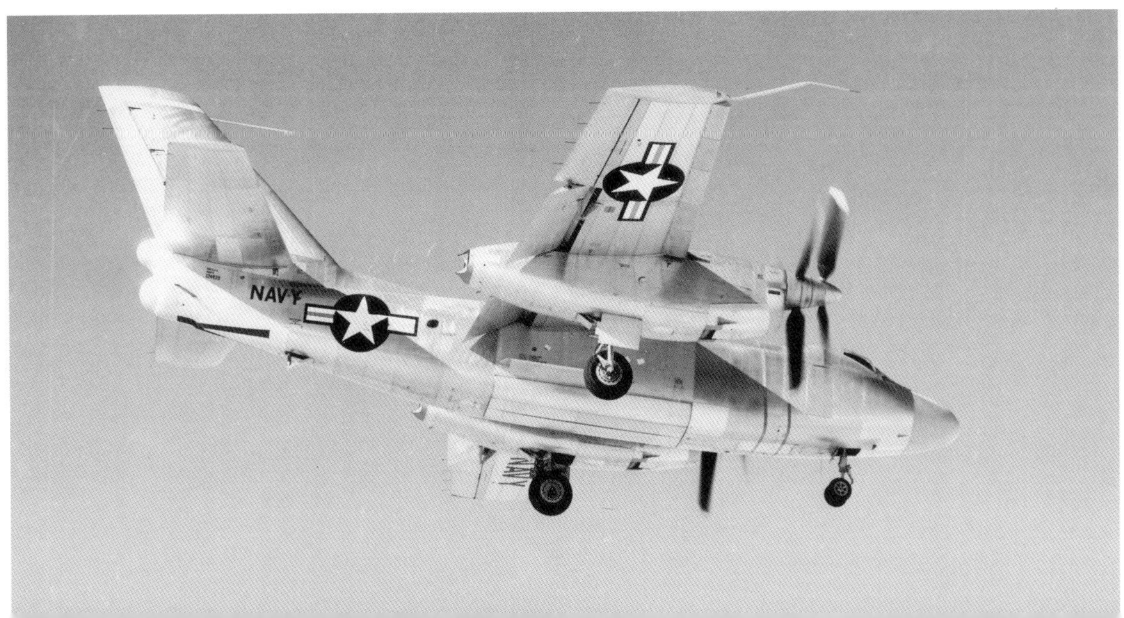

Right, three views of the first prototype XA2J-1 attack bomber. With two 5035-horsepower Allison T40-A-6 gas turbines and contra-rotating propellers, the Super Savage promised much improved performance but was overweight and suffered from numerous engine problems. The second XA2J-1 never flew.

NORTH AMERICAN NA-180>NA-262 YF-100A / F-100A/C/D/F Super Sabre

WHEN THE F-86 ENTERED SERVICE with the USAF in 1949, North American was already developing a follow-on aircraft to the Sabre. The F-86 had achieved Mach 0.87 in September 1948 while setting a world speed record, and the aim of the new project was to exceed Mach 1 in an operational air-superiority fighter. The task of developing and building supersonic aircraft was a major undertaking at the time, because the only planes to have flown at these speeds were experimental types such as the Bell X-1 and Douglas D-558-1 Skystreak, and data gathering was still in its infancy. NAA president Lee Atwood knew that achieving Mach 1 would require the use of new systematic instrumentation and wind tunnel technology, as well as an entirely new design. Modifications to the existing F-86 airframe simply would not achieve the projected performance.

The Allison Division of General Motors had made recent advances in jet propulsion and was the first to offer potential powerplants for the new design. The 9000-pound-thrust rating of the Allison J35 with afterburner was about 50 percent greater than that of the General Electric J47 then used in the F-86, and this encouraged NAA engineers to begin new design studies in April 1949.

In August, General Electric proposed an afterburning version of the J47, delivering up to 13,000 pounds of thrust, and on September 14 NAA began studies and drawings with this engine as a starting point. After a series of major wing and fuselage design changes to the F-86D and E models, a revised study known as Sabre 45 (named after its 45-degree wing sweep), became the true basis for the F-100. Designed around the new Pratt & Whitney J57 engine that was expected to deliver 15,000 pounds of thrust with afterburner, NAA's initial proposal, dated May 14, 1951, guaranteed a top speed of Mach 1.3 in level flight at 35,000 feet. The planned armament was four 20-mm cannon.

The Air Force Council took six months to decide how many Sabre 45s to order, and in the meantime NAA was funding its own engineering

The YF-100A, AF52-5754, made its first flight on May 25, 1953, a foggy morning at Inglewood. The Super Sabre never appears to be centered in photographs when the extremely long nose probe is included. Opposite page, the same plane, with YF-100 painted large on the nose and looking much shinier, in flight above the Air Force Flight Test Center at Edwards Air Force Base, California.

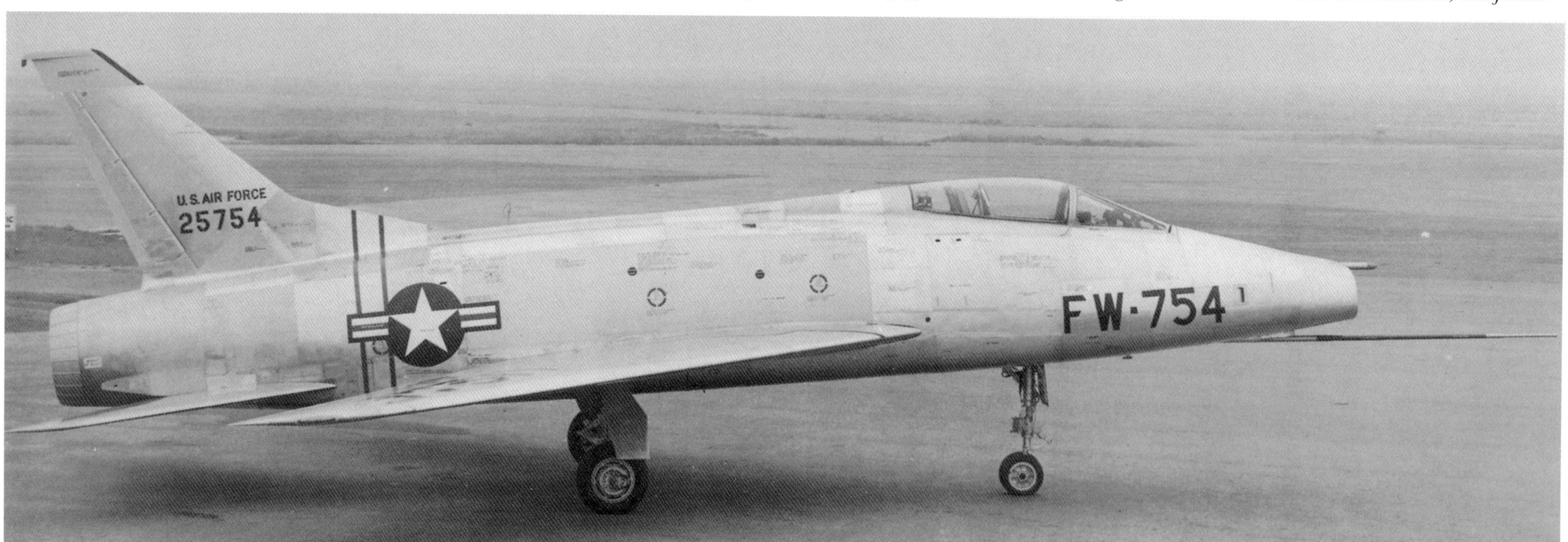

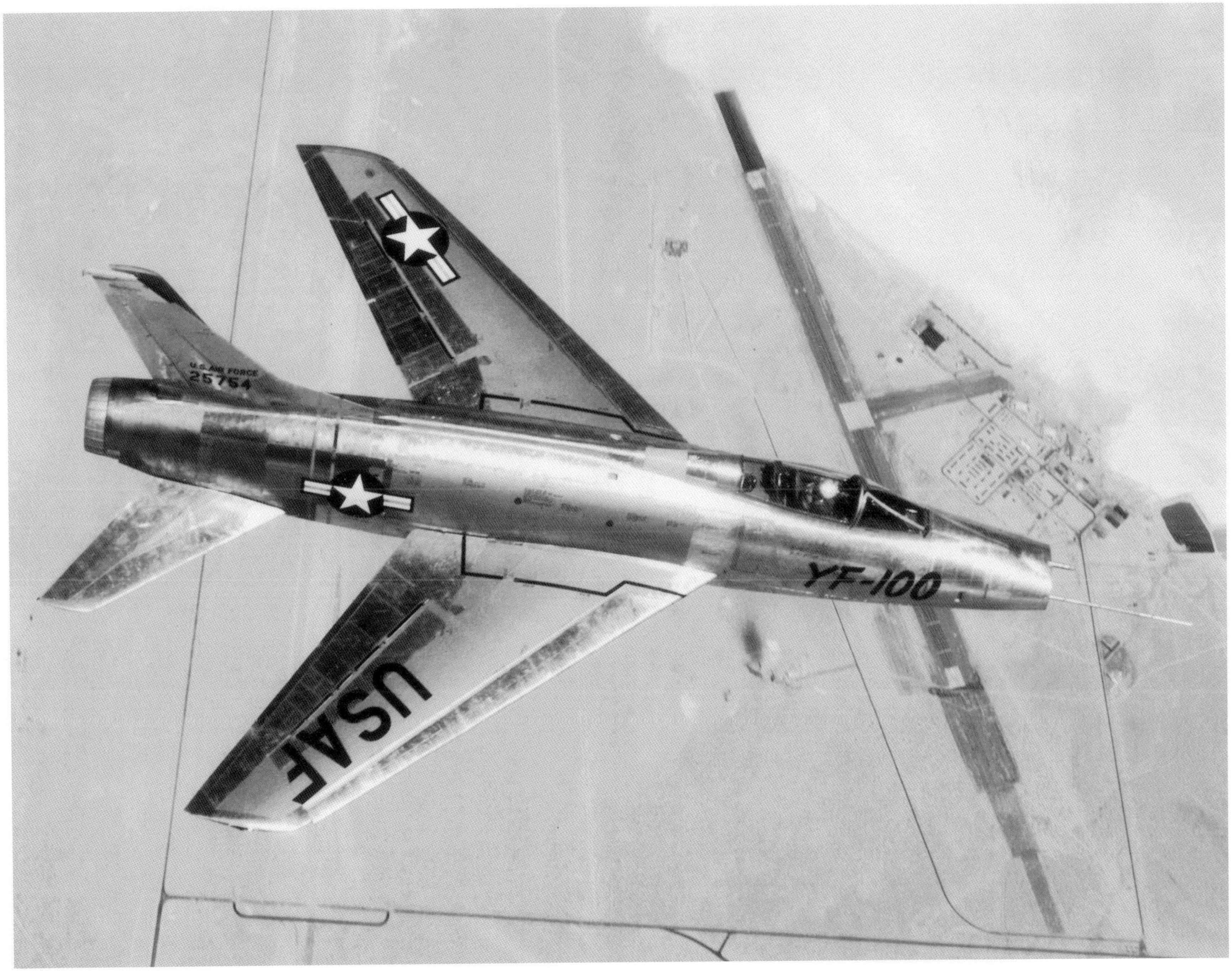
U.S. AIR FORCE
25764
YF-100
USAF

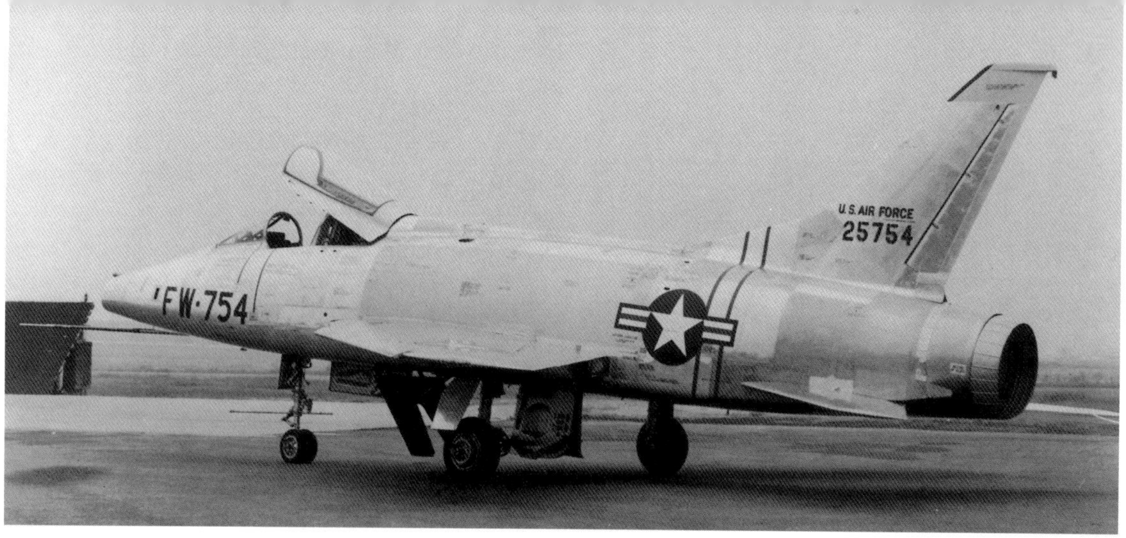

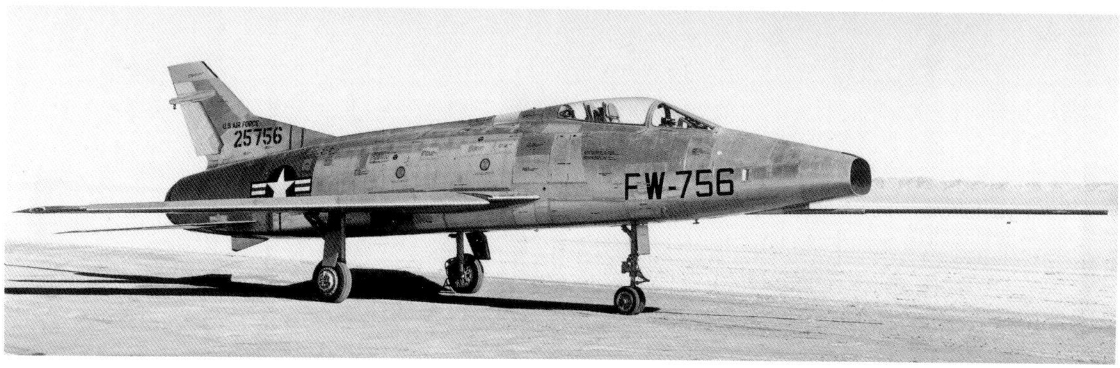

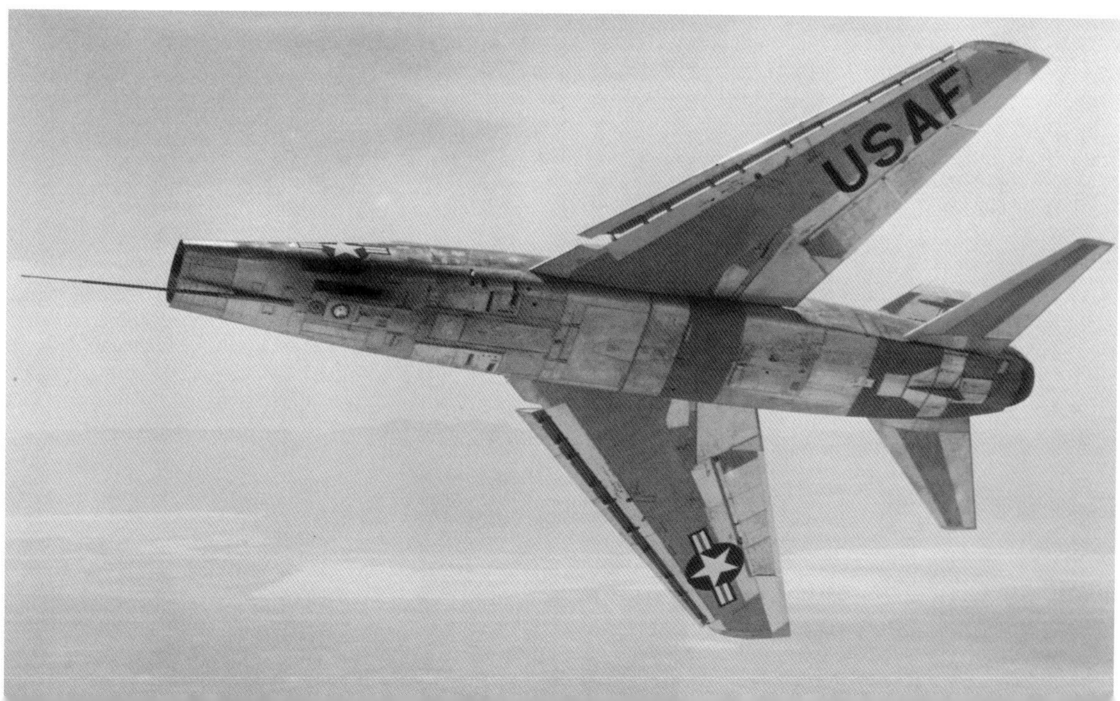

Left, from top, the first YF-100A, with the cockpit canopy open and air brake extended; the first F-100A-1 with its short vertical fin; and the sixth F-100A-1, showing the wing slats fully open.

costs. Finally, on November 1, 1951, the Air Force issued contract AF 33 (600)-6545, authorizing NAA to build two prototypes, one for aerodynamic testing and one for armament development, and to initiate engineering design and tooling for ninety-four production aircraft. When Air Force officers formally inspected the mock-up on November 9, 1951, they recommended three significant changes: single-point refueling, modifications to the control system and an increase in the number of rounds for each 20-mm cannon from the original 200 to 275.

NAA's Mock-Up Board decided against the change to the fuel system, made continued studies regarding the artificial feel of the control system and mandated the increase in ammunition. Separation of hydraulic systems, to minimize the chance of simultaneous systems failure resulting from battle damage, was also mandated on the mock-up. On December 7, 1951, the Air Force officially designated the Sabre 45 the F-100A, the first of the "Century Series" of fighter aircraft. As a successor to the F-86 it was named Super Sabre.

A series of further modifications to the design, including a reshaped fuselage and canopy and a lowered horizontal stabilizer, were approved by the Air Force in June 1952. NAA had also been told to study the possible installation of under-wing stores and to delete self-sealing fuel tanks to save weight. Valuable data gathered at the NACA wind tunnel led to further refinements to the fuselage fineness ratio and allowed the thickness/chord ratio of the wing and the stabilizers to be optimized. The wing incorporated inboard ailerons to reduce wing twist and conventional flaps were not used. Instead, the leading edge was fitted with a five-segment slat arrangement that automatically adjusted to air loads. The slats increased lift and decreased wing buffet near stall speed, allowing tighter turns.

The fuselage featured a belly-positioned speed brake, which when used not only slowed the aircraft, but further enhanced the turning radius. Other notable fuselage characteristics included a very thin-lipped oval nose inlet that optimized high-energy airflow to the engine, improved the pilot's forward view and reduced drag, increasing the top speed by nearly 50 miles per hour. A drag chute slowed landing roll-out speed.

The Air Force increased its order to 273 aircraft in August 1952, and as the first flight drew closer, both NAA and the Air Force agreed that two prototypes would not be enough for the projected test program. Therefore, the first ten production F-100A aircraft were allocated along with the two YF-100A prototypes for the estimated 540 hours of flight testing that lay ahead. The primary assignment of the two YF-100As was to obtain initial engine performance, to check overall stability, control and handling characteristics, and to evaluate and refine the armament installation. Production aircraft 1, 2 and 3 (the third, fourth and fifth test aircraft) would test the production powerplant installation, stability and control, as well as overall performance and cruise control.

In April 1953 the first NA-180 YF-100A prototype, USAF 52-5754, underwent final preparations for its first flight at North American's facility at the North Base of the Air Force Flight Test Center at Edwards AFB. NAA test pilot George Welch flew it for 57 minutes on May 25 and exceeded Mach 1 in level flight. In doing so, the YF-100A became the first aircraft to break the sound barrier on its maiden flight. Welch made a second flight later that day and again went past Mach 1. In its enthusiasm, NAA claimed the F-100 to be the world's first operational supersonic aircraft, although the definition "operational" was somewhat premature.

On May 29, during the Super Sabre's eighth flight, a slight rudder flutter was sensed; hydraulic rudder dampers were used after the ninth flight and the aircraft was later dive-tested at Mach 1.1 without subsequent flutter. As flight

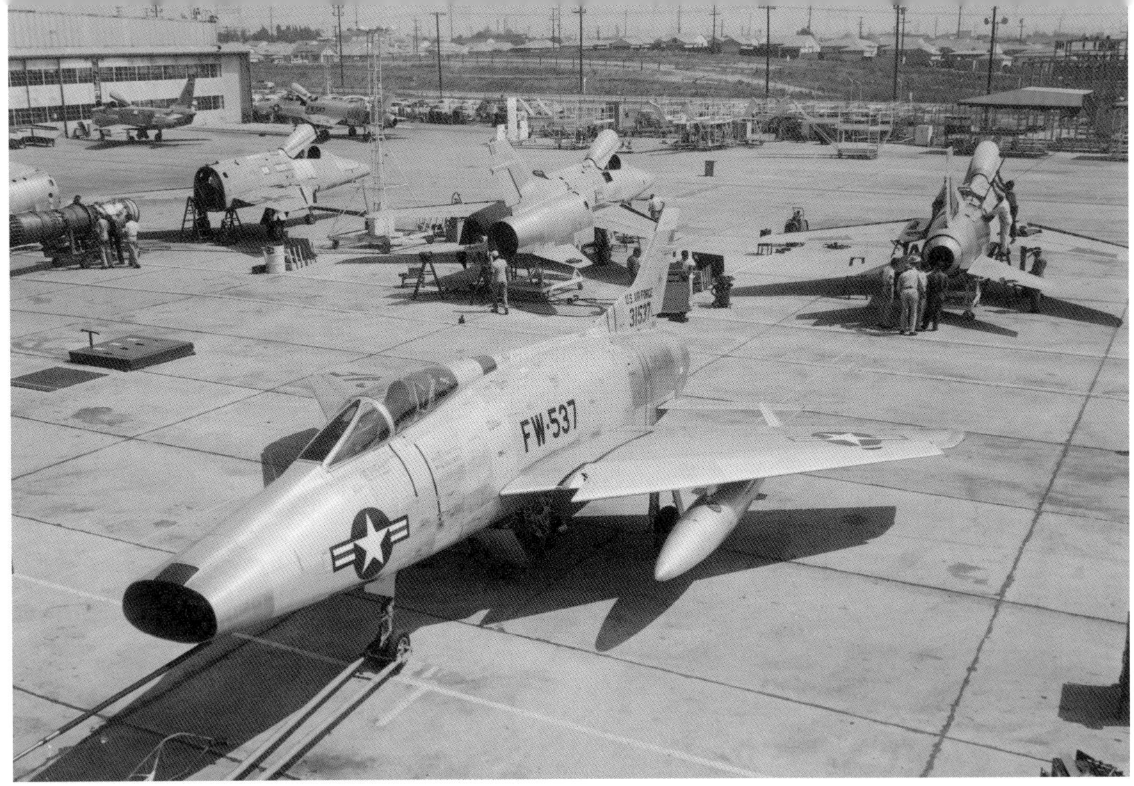

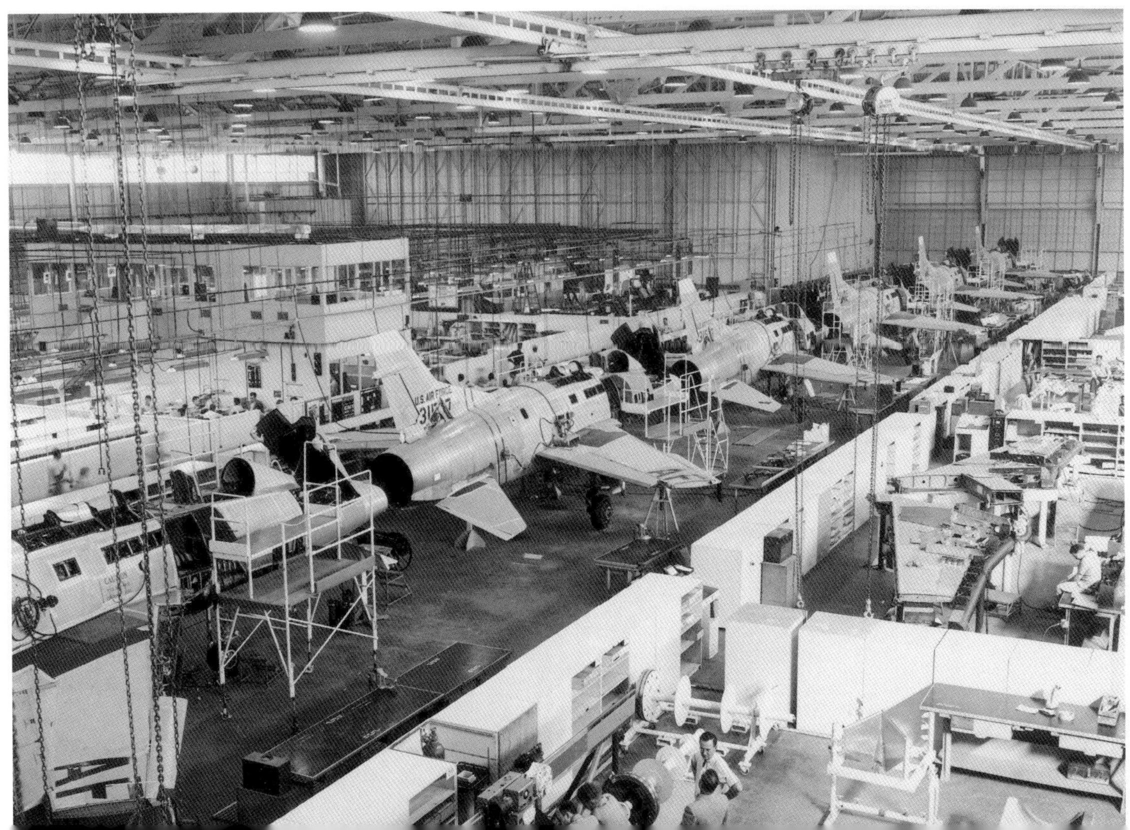

Above right, outdoor completion of F-100A-10s at Inglewood on August 19, 1954. Below, the last six F-100A-10s on the regular assembly line, photographed in November of that year.

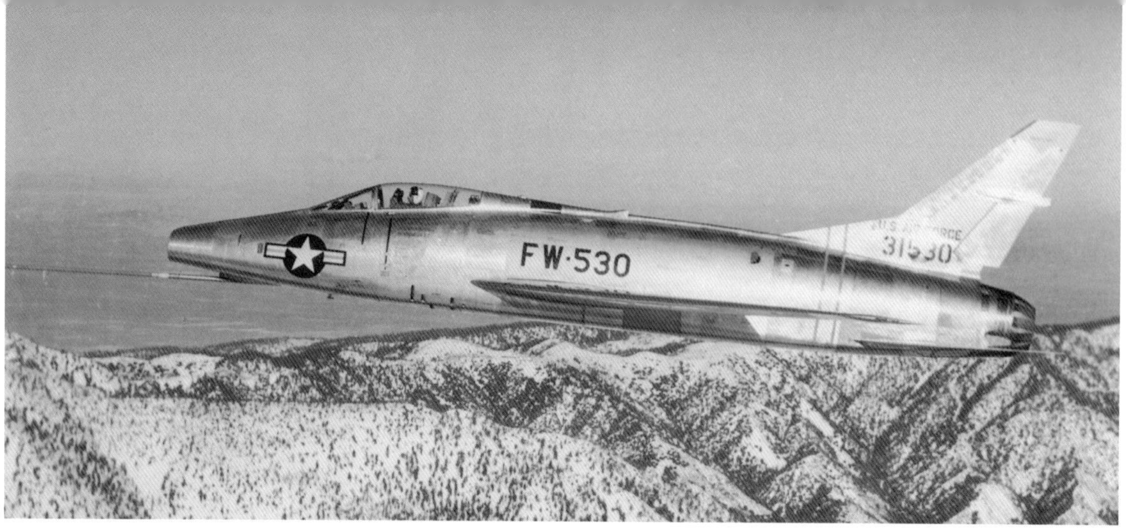

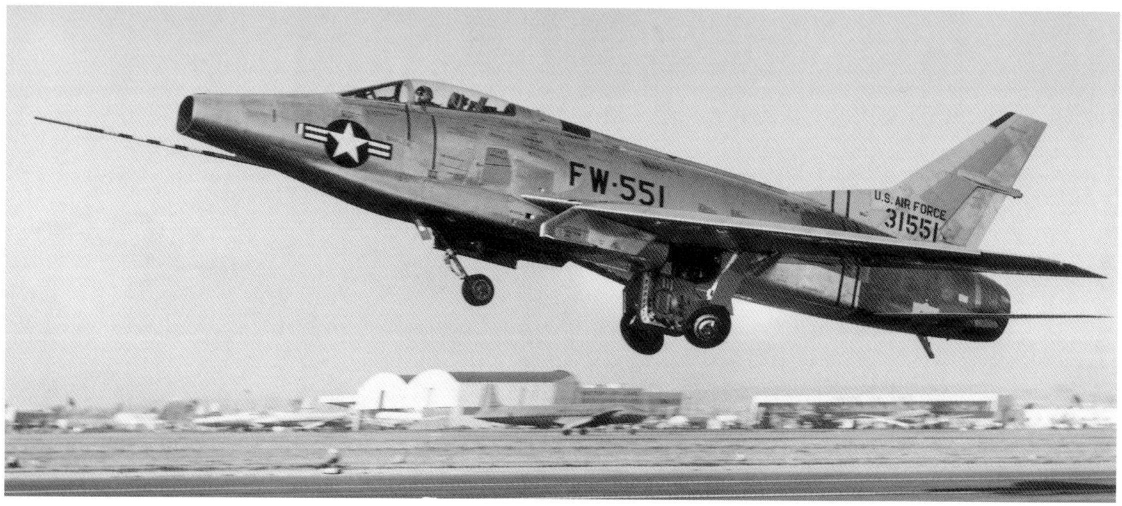

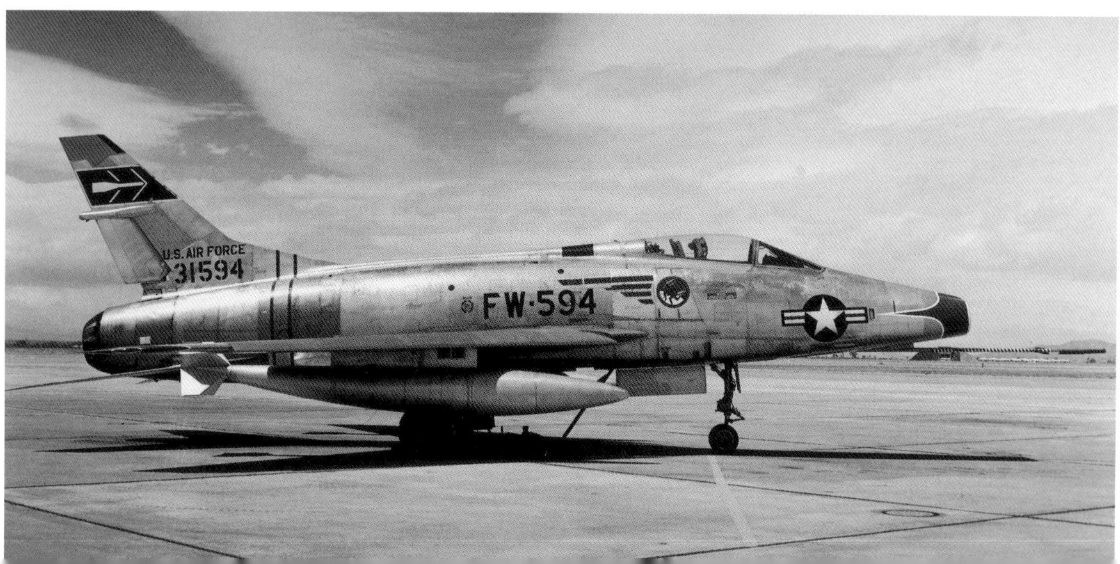

testing continued through June 1953, a preliminary evaluation of dynamic stability was made. Although the program had been relatively trouble-free, top Air Force officials had reservations concerning the validity of the data NAA provided. The Phase 1 testing period had been shorter than average, and although excellent performance was reported, including Mach 1.4 in dive tests, the Air Force wanted to be sure that any problems were uncovered and remedied prior to full-scale production. During Phase 1 the YF-100A was flown by other NAA pilots, including Daniel Darnell and Joseph Lynch.

In August, Major General Albert Boyd became the first Air Force F-100 pliot. Subsequent USAF flights were made by Lieutenant Colonels William F. Barnes, Richard Johnson and Frank K. "Pete" Everest, who would be the project pilot for the upcoming Phase 2 tests. Second-phase testing would also include Colonel Horace A. Hanes and Majors Arthur Murray, Robert L. Stephens and Charles E. "Chuck" Yeager.

Conducted during September 3-17, 1953, Phase 2 consisted of thirty-nine flights and more than 19 hours of flight time. The Air Force assessment was not overly positive. Its pilots stated that the F-100 outperformed any contemporary production fighter and had potential, but they found it unacceptable for service. Limited forward vision during takeoff and landing, slow climb rate, poor longitudinal stability in high-speed level flight, questionable handling at low speeds, slow control response and difficulty in landing were just a few of Everest's findings.

North American, not at all pleased by the criticism, suggested that the future users of the aircraft, Tactical Air Command pilots, fly the YF-100A and give their impressions. The TAC evaluation was much more positive. The Air Force decided that the initial evaluation by Everest's team had been overly critical and gave NAA the go-ahead for production. The company im-

From the top, the second F-100A-10, 53-1530, the twenty-third, 53-1551, and F-100A-15, 53-1594, all showing the taller vertical fins added to early short-tailed A-series planes.

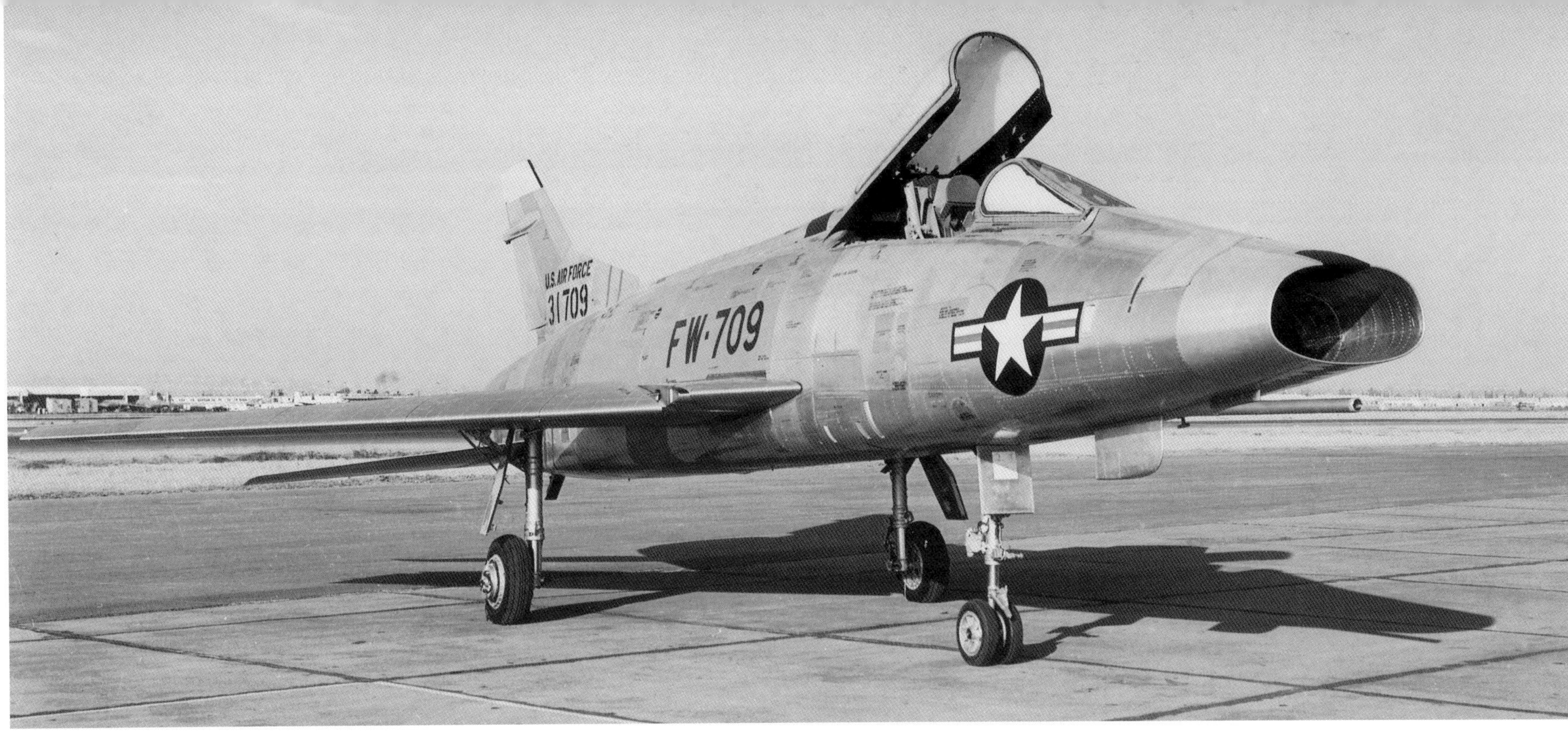

Above, the first F-100C-1, 53-1709, which flew in March 1954, and below, 53-1733 using its drag chute. With more-powerful J57 engines, F-100Cs were 70 mph faster than the A models. 49

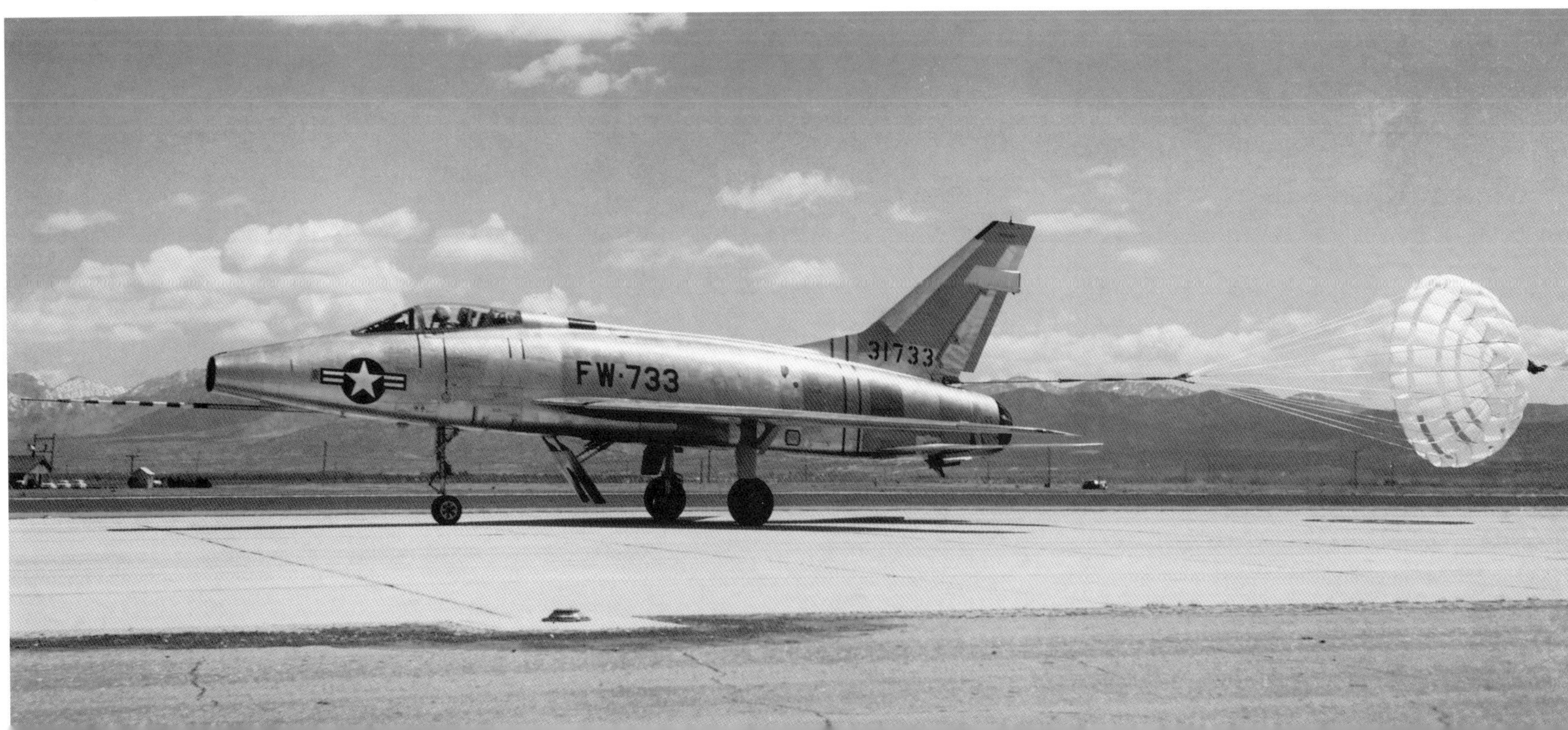

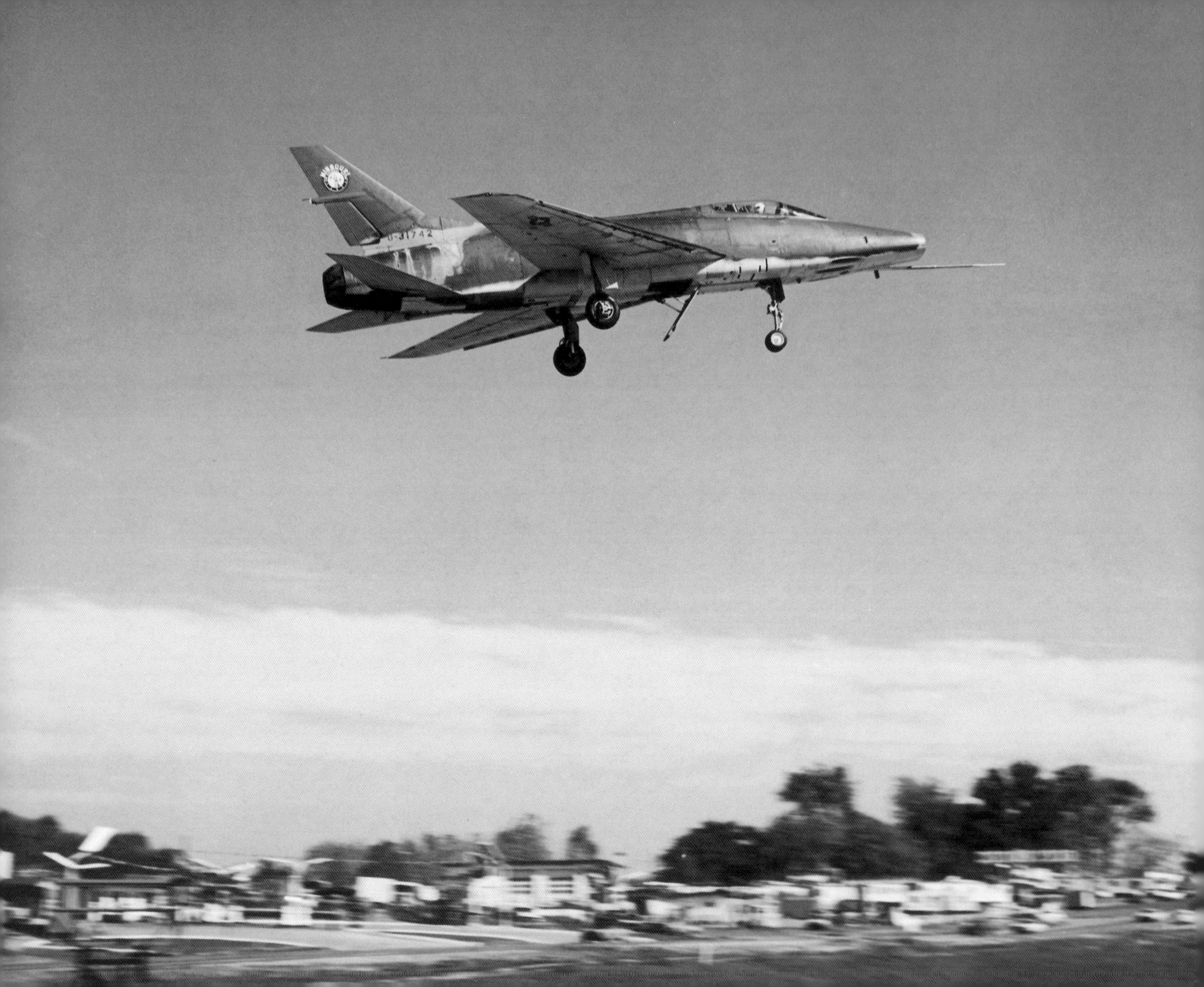
MISSOURI
0-31742

mediately began a development program for production components and modifications. The first production F-100A was delivered to the Air Force on October 26, 1953, and flew three days later.

On that day, October 29, Everest flew the first YF-100A to a new world speed record of 755.149 miles per hour at Salton Sea, California, beating the previous record held by Navy Lieutenant Commander James B. Verdin with a Douglas XF4D-1 Skyray. On December 8, 1953, the first production F-100A achieved Mach 1.34 at 35,000 feet, meeting its original performance guarantee.

After the Korean War ended, high-ranking Air Force officers had made an assessment of the conflict and developed new ideas for the uses of air power. In future warfare, they decided, air superiority, interdiction and close support of ground forces were the tasks that air power would perform. Further analysis dictated the need for nuclear weapon-delivery capability by TAC in addition to the Strategic Air Command (SAC). At the time, TAC had no fighter-bomber capable of nuclear weapon delivery (the Republic F-84F Thunderstreak was not yet available due to engine-related delays), so on December 31, 1953, the Air Council decided to modify the last 70 production F-100As to fighter-bomber specification. These aircraft were designated F-100C. The F-100B was never put into production; the designation was used initially for the project that later became the F-107. The first F-100C would use a dry wing (no fuel carried within it), and the fourth production F-100A would serve as prototype for this configuration.

As F-100A production commenced, a redesign of the horizontal stabilizer using a 3.5-percent thickness/chord ratio was phased in. There was some concern over a possible tendency of the new tail to flutter, but several tests of this configuration on a rocket sled at Edwards confirmed the soundness of the design. This was also

Opposite, a camouflaged F-100C-1, 53-1742, of 110 TFS, Missouri ANG, landing at St. Louis on October 23, 1969. The wing slats are open and the air brake is fully extended.

F-100A

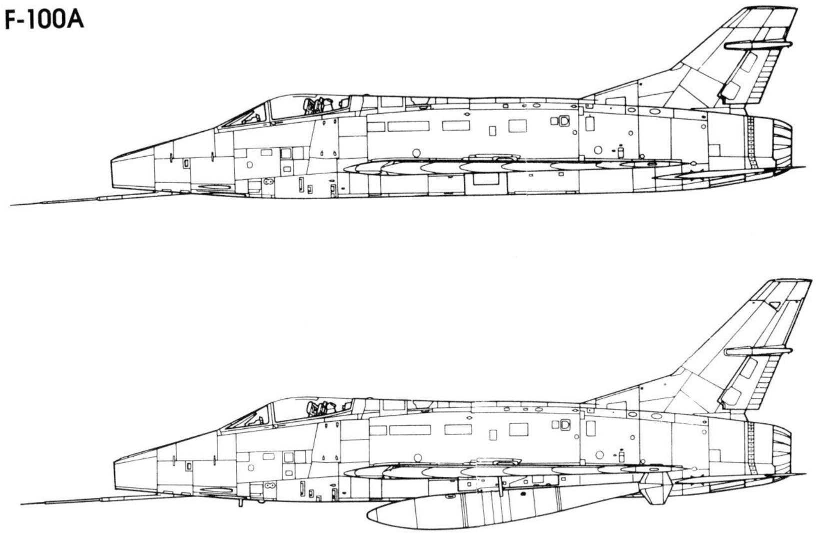

F-100C

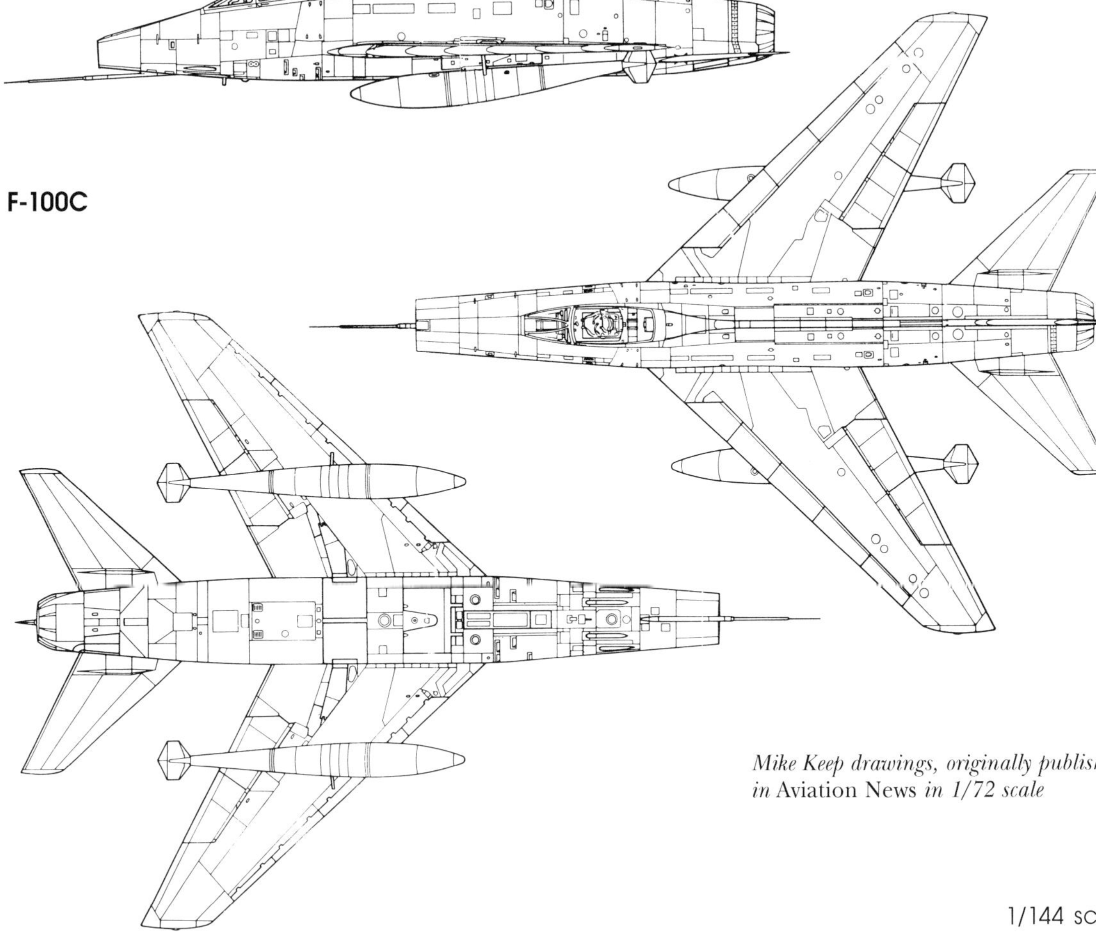

Mike Keep drawings, originally published in Aviation News *in 1/72 scale*

1/144 scale

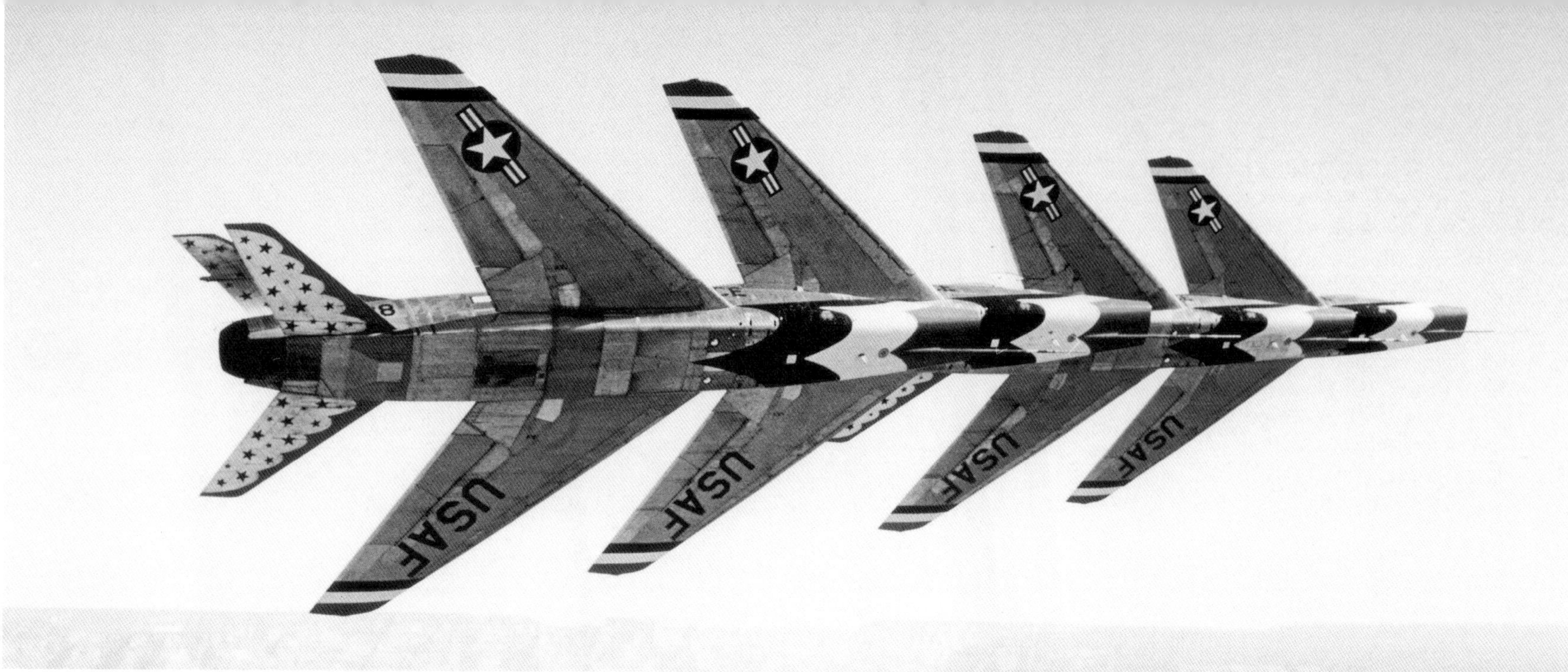

Above, four F-100C-10 Thunderbirds in perfect formation, 1957. Below, a close-up of F-100C-10, 55-2728 (the fourth plane above), showing the upper markings. Colors appear on page 65.

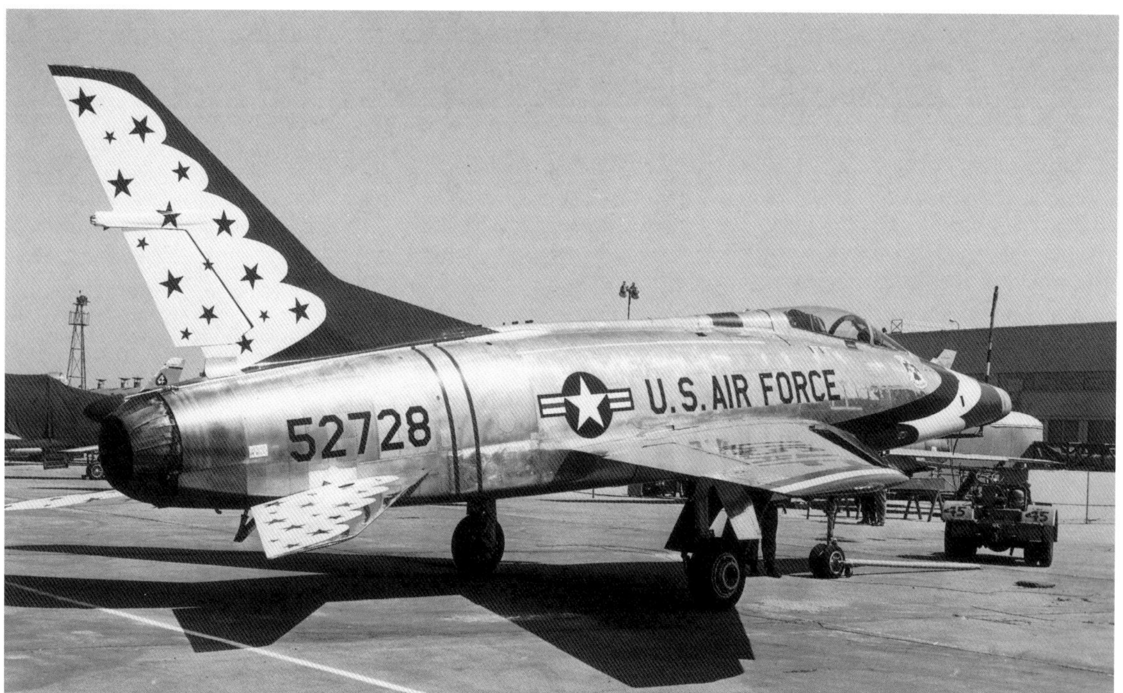

proved in dive tests to Mach 1.55 on the first production F-100A after the tail was retrofitted to it.

Six F-100A production models were taken from the assembly line and modified during March and April 1955 to a tactical reconnaissance configuration by removing the 20-mm cannon and ammunition-feed devices and installing five cameras facing forward and sideways in the nose. Designated RF-100A and nicknamed "Slick Chicks," these aircraft saw limited use in Europe and later the Far East on various photo-reconnaissance duties.

After TAC took delivery of its first F-100As with the 436th Fighter-Day Squadron (479th Fighter-Day Group) at George AFB, California, several reports of directional instability began to surface. The vertical tail on production A-models was shorter than that of the YF-100A, and it soon became obvious that more fin/rudder area was needed. After the catastrophic disintegration of the ninth production aircraft claimed the life of George Welch on October 12, 1954, and two more accidents occurred within a month, the Air

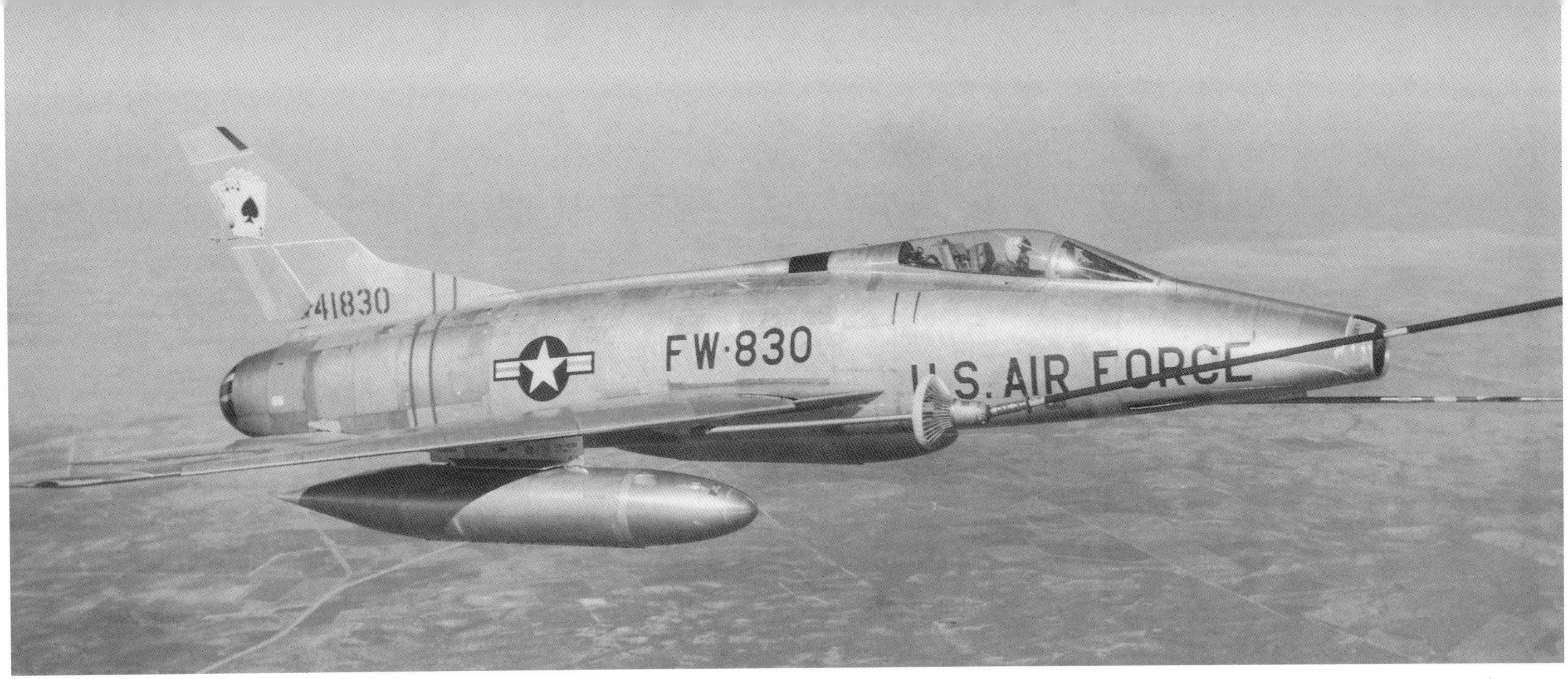

Above, an F-100C–15, 54-1830, of 322 FDW, based at Foster AFB, Texas, refuelling in flight. Below, 54-144, a camouflaged F-100C–15 of 120 TFS, based at Andrews AFB in 1970.

Force decided to ground all F-100 operations on November 10, 1954.

Following a series of inquiries into these accidents by both NAA and the Air Force, and after all research and data were amassed, an entirely new vertical stabilizer of 27-percent greater area was put into production and retrofitted to all existing short-tailed F-100As. The wingtips were extended to increase span and area and to compensate for a change in the aircraft center of gravity. Bob Baker, now chief test pilot at NAA, worked with USAF Captain Milburn Apt and NACA's Joe Walker in an exhaustive test program on the upgraded aircraft, analyzing a large amount of data concerning supersonic flight characteristics. After all the reports were completed, NAA distributed this knowledge to all U.S. aircraft manufacturers in the interests of progress and safety. Super Sabres were flying again by February 1955.

The F-100As serving with the 479th Fighter-Day Wing achieved initial operational capability in September 1955 and continued their duties

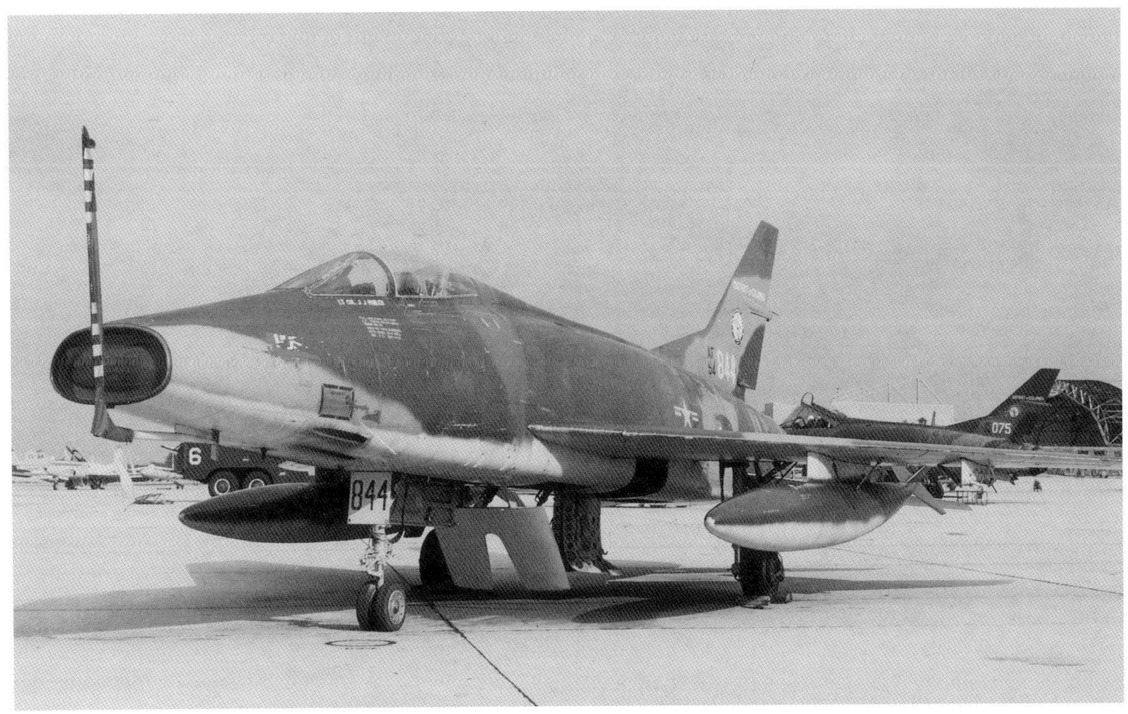

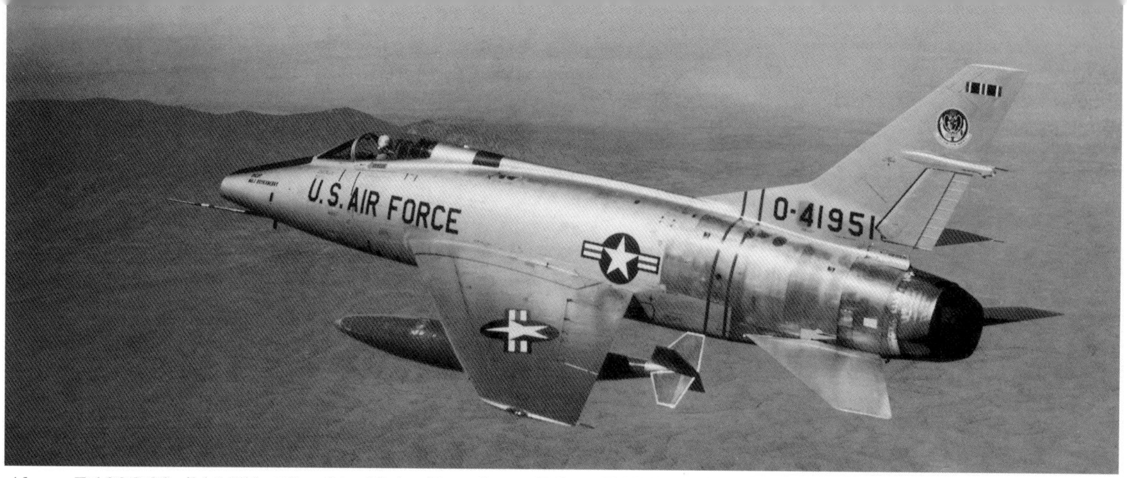

Above, F-100C-20, 54-1951, piloted by Major Rosenberry. Below, F-100C-20 54-1966, converted into a two-seat trainer.

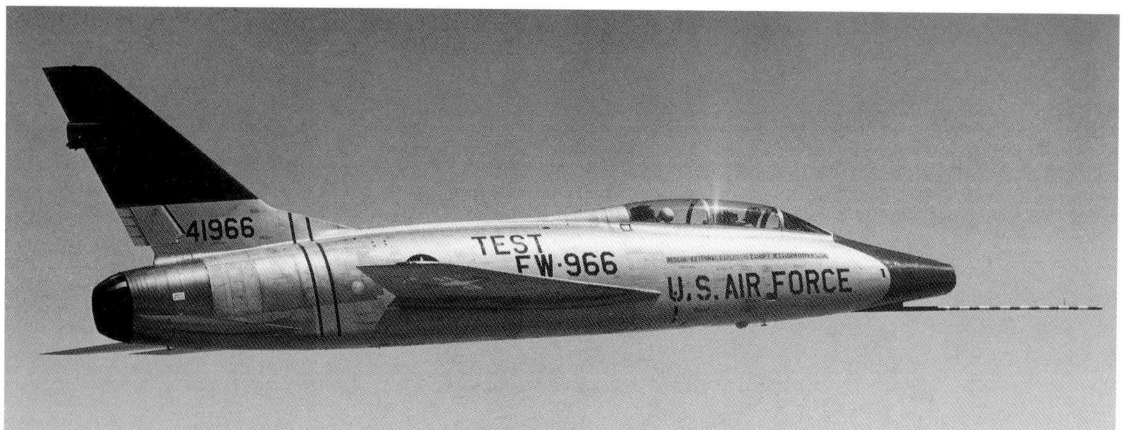

Below, in Thunderbird-like markings, a Skyblazers F-100C-25, 54-2010, of 50 TFW, based at Hahn, Germany, in 1961.

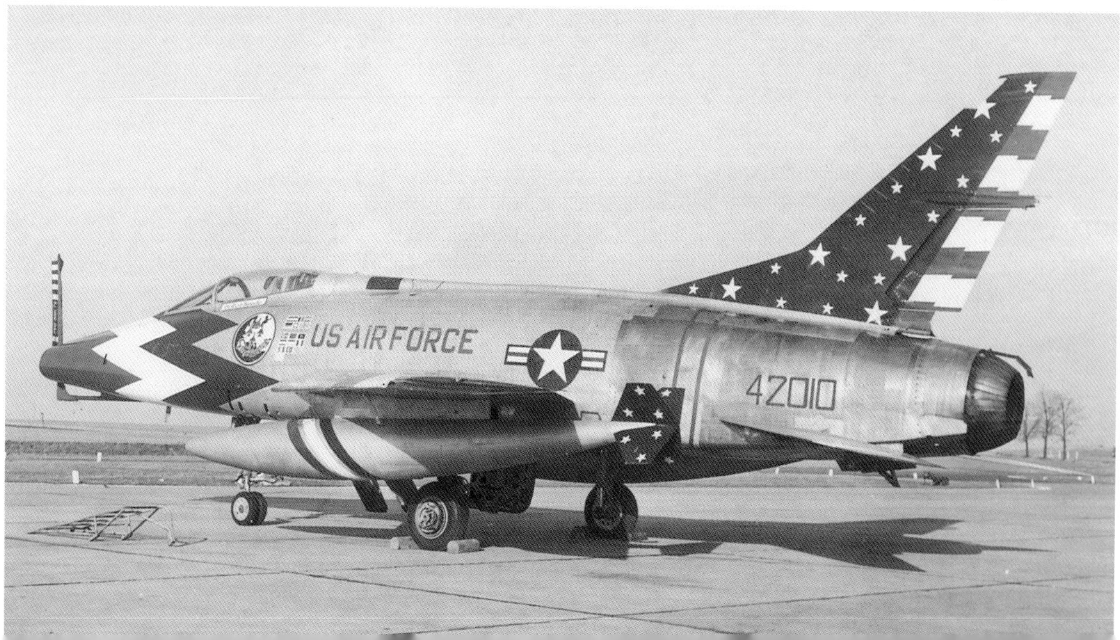

until early 1958 when they were phased out of active USAF service and put into Air National Guard (ANG) units or into storage. A total of 203 F-100As had been accepted by the USAF; eventually 118 were transferred to Taiwan. Approximately 50 aircraft were destroyed in accidents and about 35 were used for crew training until early 1970 when the last F-100As left the USAF active inventory.

The F-100C was an answer to Air Force requests for ground-attack capability and increased fuel capacity. Although the C-model specification called for a "wet" wing (fuel to be stored there as well as in the fuselage), the prototype was actually a dry-wing aircraft, like the A-models. The F-100C production aircraft had a wet wing with an internal capacity of 451 U.S. gallons and was strengthened internally to handle the shock of releasing the stores that it would carry on six under-wing hardpoints. These stores included external fuel tanks (of which several different types were tried, eventually standardizing on two 335-gallon units), bombs ranging from 500-2000 pounds, up to twelve five-inch HVAR (high-velocity air rocket) weapons and provision for special stores, including nuclear weapons.

The first F-100C flew in March 1954, and on May 27th a total of 564 were authorized for production. Later this was amended to build the last 224 to F-100D specification. The first production F-100C flight came on January 17, 1955, with NAA's Alvin S. White at the controls. The C-model used an upgraded J57-P-21 engine, increasing the speed by about 35 knots and reducing time to climb to 35,000 feet by ten percent. On August 20, 1955, Col. Horace Hanes set a new world speed record in an F-100C at Palmdale, California, recording 822.135 miles per hour. Deliveries to TAC began in April 1955, and F-100Cs became operational with the 450th Fighter-Day Squadron at Foster AFB, Texas. North American built a total of 476 F-100Cs, including 25 at NAA's Columbus, Ohio, facility. Eventually, about one-third of that total ended up in USAFE (U.S. Air Forces in Europe) squadrons. The Skyblazers, an aerobatic flight demonstration team

1/144 scale

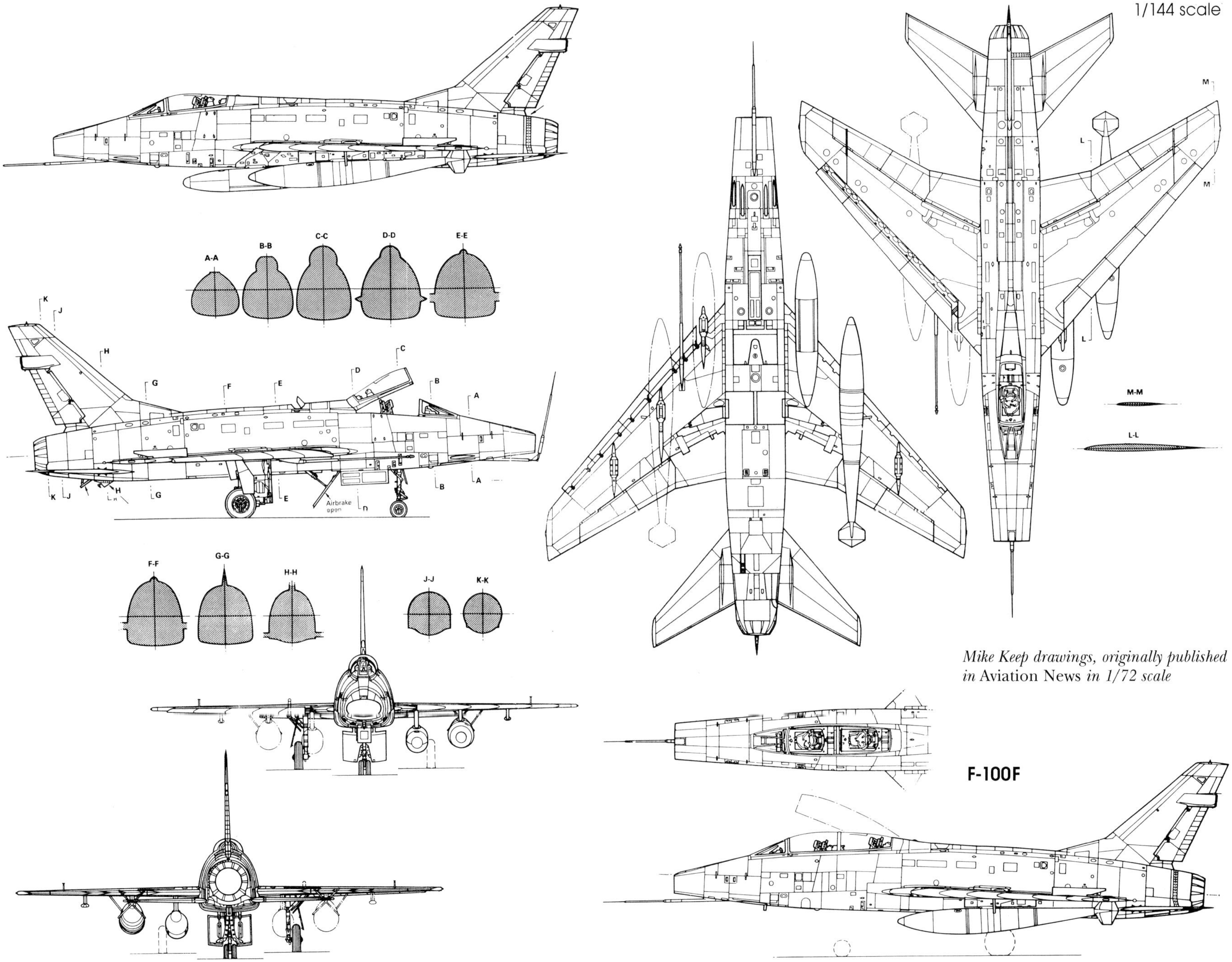

Mike Keep drawings, originally published in Aviation News *in 1/72 scale*

F-100F

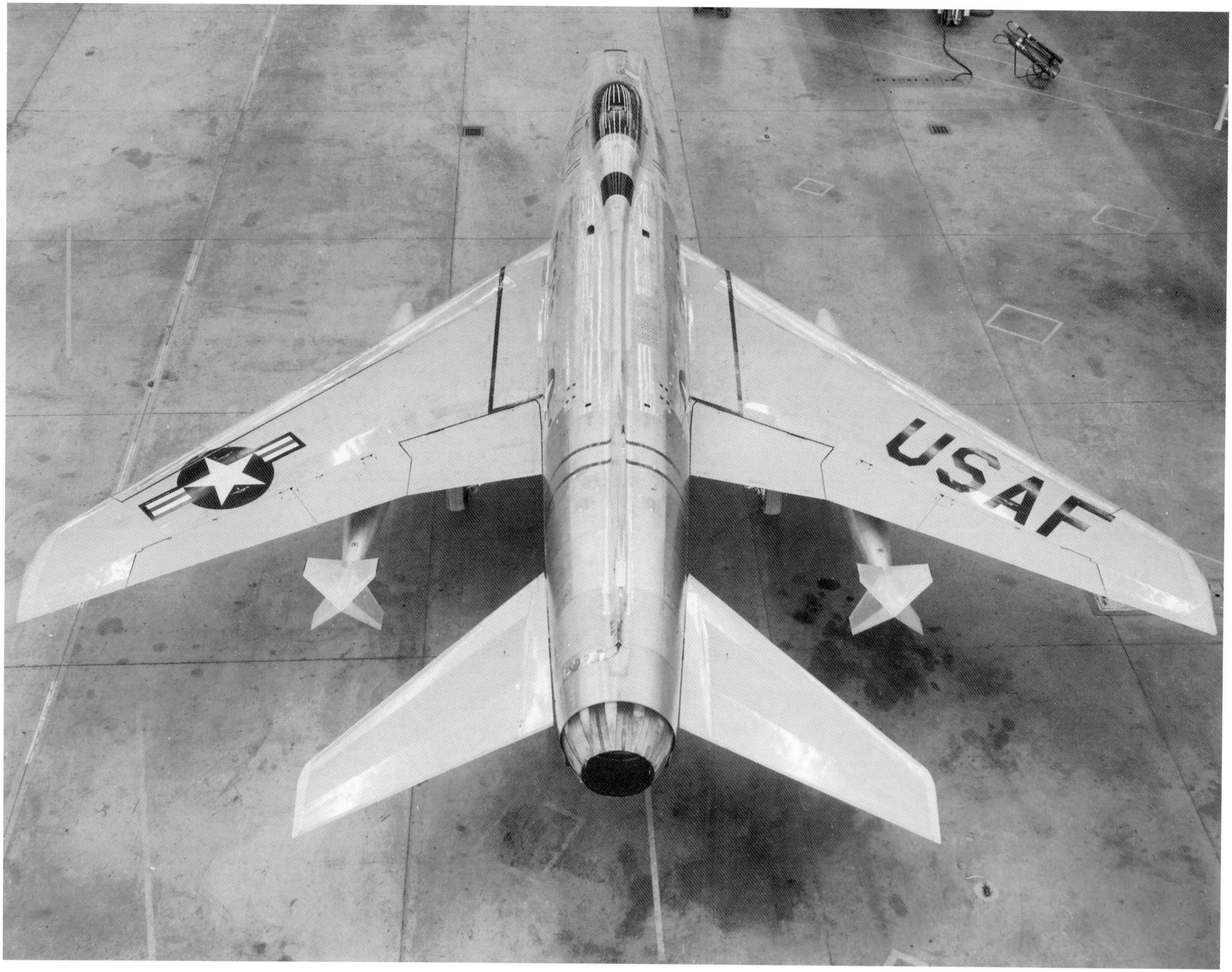
USAF

formed by the 36th Fighter-Day Group in Europe, used F-100C Super Sabres in their public air show routines until late 1961, while the USAF Thunderbirds used the F-100C from 1956 until 1964. The TAC and USAFE aircraft were transferred to ANG units beginning in 1959, reaching full strength by 1966. Like the F-100A, the F-100C had a relatively short service life, flying its last USAF mission in 1970.

The definitive version of the Super Sabre was realized with the F-100D, which was finally the true fighter-bomber that the Air Force needed. With a vastly improved wing that now had conventional flaps and with yet another slight increase in vertical stabilizer area, the D-model was a dedicated fighter-bomber that featured an inflight refueling system and a force-release system for under-wing store drops. The F-100D first flew on January 24, 1956, and deliveries to the 405th Fighter-Bomber Wing, TAC, based at Langley AFB, Virginia, followed in September. Project High Wire, a program of standardizing and upgrading F-100Ds (and also some F-100Fs) to extend the variety of non-nuclear weapons that could be carried (as well as upgrading wiring and standardizing the cockpit arrangement) was begun in April 1962 and completed in mid-1965 at a cost of $150 million.

In 1964, the Thunderbirds converted to F-100Ds, and would use them until 1969. During an October 1967 airshow, the solo Thunderbird F-100D shed both its wings after a high g-load maneuver (estimated at 6.5g) on pull-up, while the pilot ejected safely. F-100Ds in combat also had this problem, so ordnance delivery passes were restricted to 4gs until a modification could be made to reinforce the wings.

In Vietnam F-100Ds were called upon for a variety of ground-attack roles, all of which they performed effectively. One hundred of the lighter F-100Cs also served in South East Asia with called-up ANG units and did an outstanding job. By the time the Super Sabre was flying routine missions in Vietnam, it had become affectionately known as "The Hun," as in one hundred, and many Super Sabre pilots adopted the nickname.

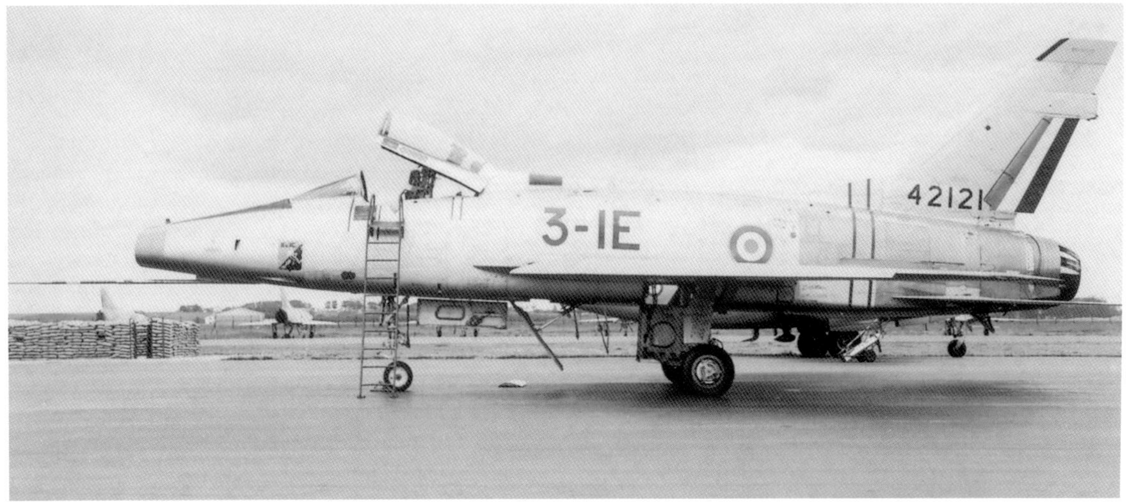

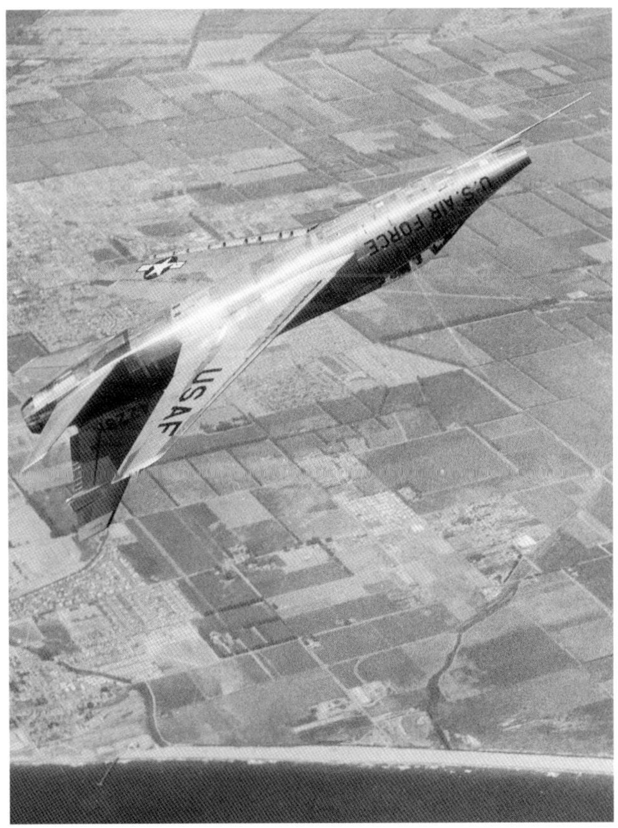

Opposite page, factory view of F-100D-1. Above, the first F-100D-1, 54-2121, in Armée de l'Air *service.*

Left and below, F-100D-15, 54-2281, loops for camera.

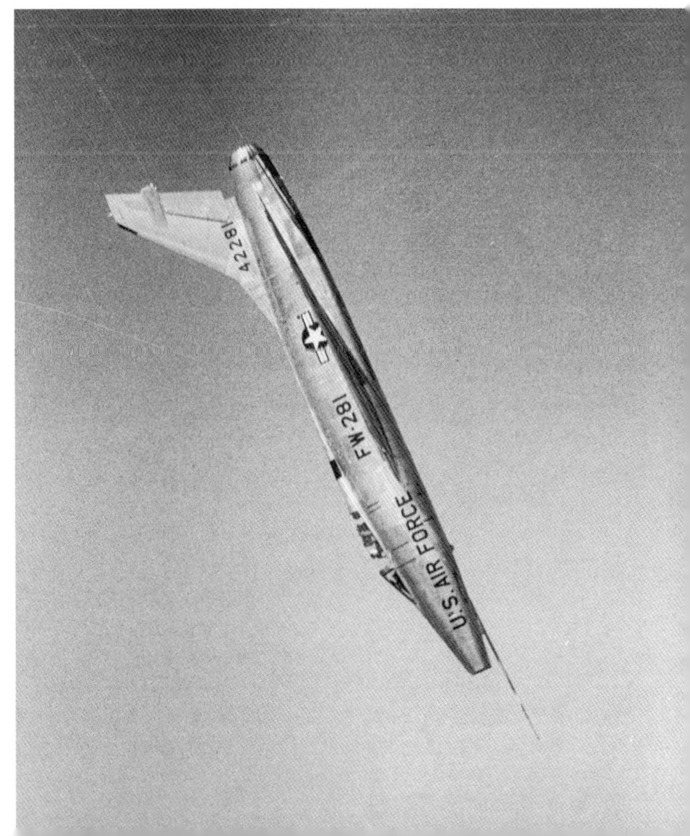

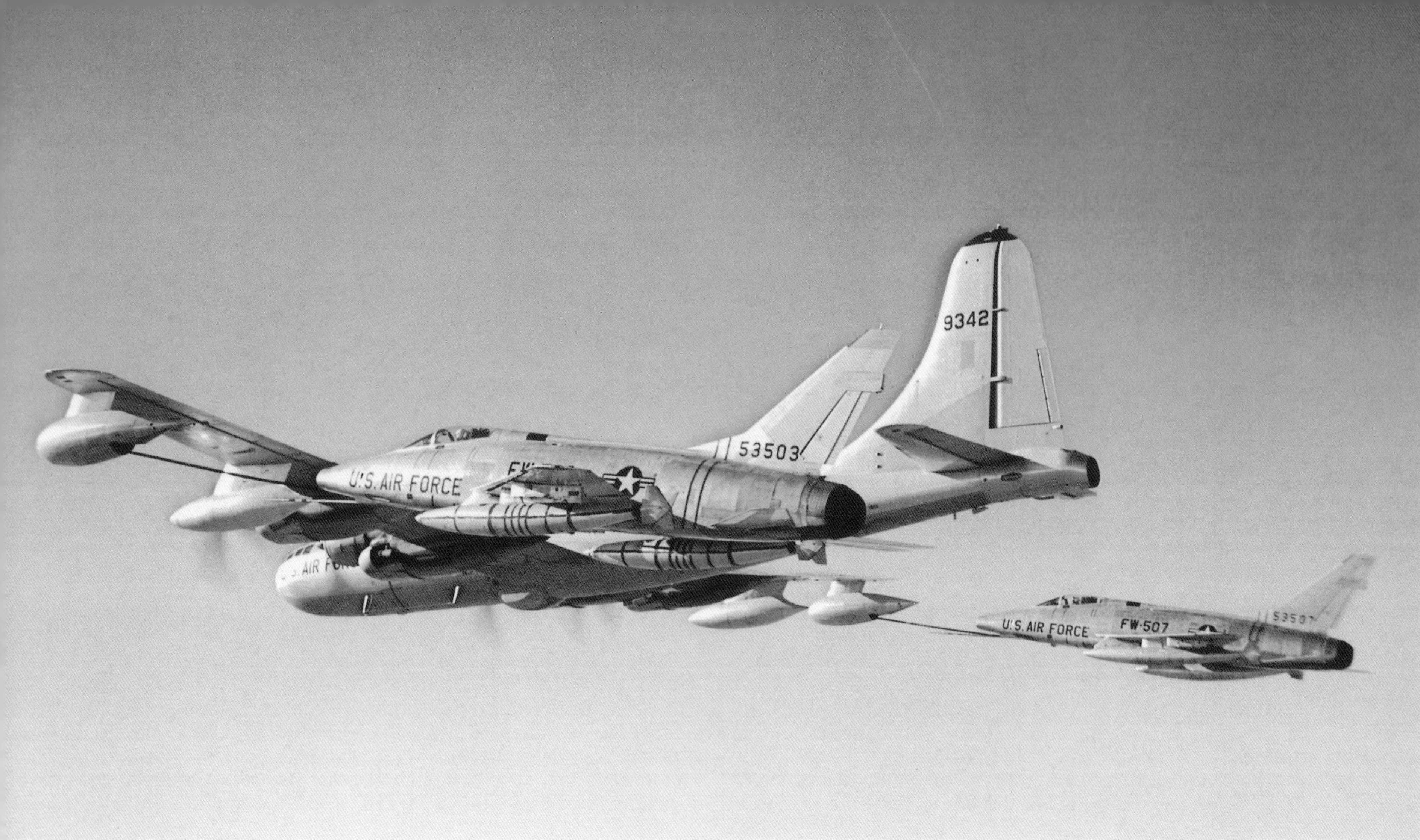

Boeing KB-50D-120 Superfortress, 49-342, refuels two F-100D-20s, 55-3503 and 55-3507, from tanks below wingtips. Later KB-50s added two J47 engines to boost maximum speed to 440 mph.

The F-100Ds used in Vietnam had a variety of ground attack weapons at their disposal. Most often used were the 750-pound iron bombs, although the 500-pound types were also used, depending on mission requirements. F-100's also delivered napalm as well as 2.75-inch folding-fin air rockets (FFAR). The Super Sabre proved to be an ideal fighter-bomber, optimized for close-support activities in Vietnam, and continued in that role until U.S. involvement in South East Asia began to wind down. By 1972 most of the combat aircraft served in ANG units stateside.

The F-100F was a two-seat trainer development, designed to familiarize pilots with the handling characteristics of the Super Sabre. Modified as a two-seater, the TF-100C first flew on August 3, 1956, and the first production F-100F flew on January 7, 1957. The F-100F also saw service in South East Asia as a "Wild Weasel" low-altitude ground attack aircraft with the specific job of destroying enemy surface-to-air missiles.

Other developments included the use of F-100D and F models as remote-controlled drones, designated QF-100D and QF-100F. These performed as targets for air-to-air or surface-to-air weapons tests. In 1985 a QF-100D was shot down by an unarmed AMRAAM missile launched from an F-16 over the White Sands Missile Range in New Mexico.

The Super Sabre was also used to test the

Zero-Launch (ZEL) system, originally intended as a means of sheltering aircraft from enemy attack and launching them without a runway should the airfield sustain damage. The test aircraft were fitted with Rocketdyne solid-propellant rocket engines delivering 130,000 pounds of thrust, enabling the platform-mounted F-100s to accelerate from zero to 275 knots in four seconds. NAA engineering test pilot Al Blackburn performed sixteen manned launches and Captain Robert F. Titus flew four. Although the ZEL tests were successful, the Air Force lost interest in the idea and the last ZEL flight was made on August 26, 1959.

Four foreign air forces used Super Sabres, including more than one hundred F-100As that ended up with the ROCAF in Taiwan. Many of the A models there had D vertical tails fitted and the ROCAF also received F-100Fs. The *Armée de l'Air* operated one hundred Super Sabres (85 F-100Ds and 15 F-100Fs), the first of which arrived in France in May 1958 and remained in service until being replaced by Sepecat Jaguar fighter-bombers in 1975.

Det Kongelige Danske Flyvevapen, the Danish Air Force, received 58 Super Sabres, the last batch arriving in early 1961. These replaced Republic F-84G Thunderjets and were assigned to the same close-support and maritime attack missions. The Super Sabres suffered a high attrition rate; accidents claimed thirty-five of the aircraft and the Danish government asked the U.S. for more F-100Ds but was turned down due to the American need for aircraft in Vietnam. The Danish fleet was replaced by General Dynamics F-16s in 1985.

In Turkey, the *Turk Hava Kuvvetleri* (THK) received Super Sabres starting in October 1958, a mix of F-100C, D and F models, and their eventual inventory totalled 206, including aircraft from U.S. surplus and a few acquired from Denmark.

The remaining ANG F-100s were retired in 1979. The Super Sabre was the last production fighter built by North American. Of 2294 built, approximately 30 remain, most on display in AFB and private museums.

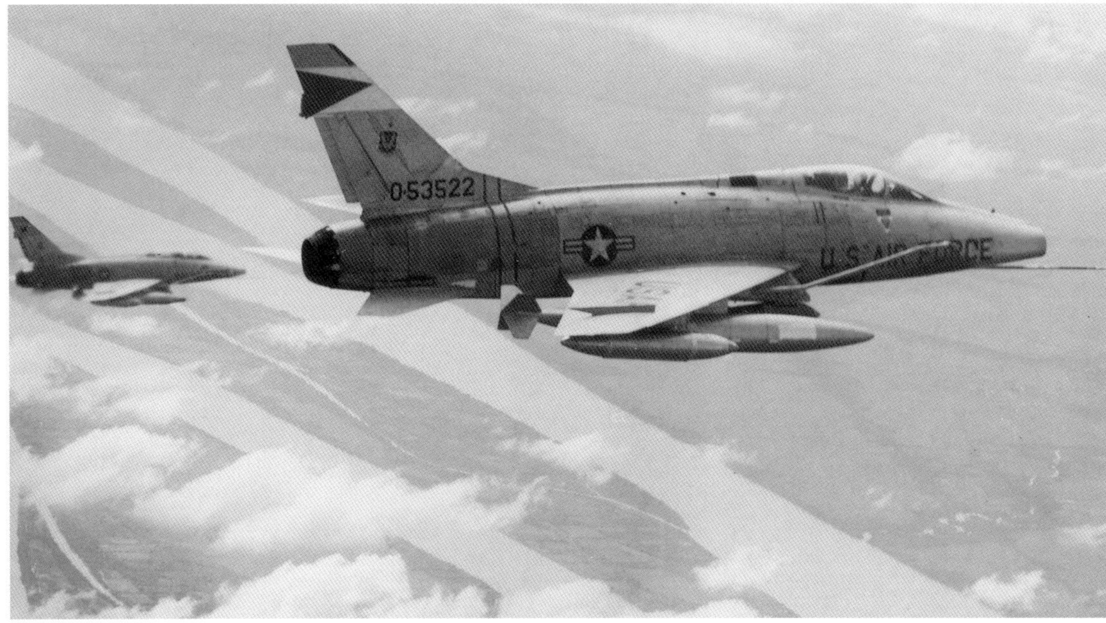

Bare metal and camouflaged Super Sabres operated in South East Asia. Above, F-100D-20, 55-3522, of 523 TFS over Vietnam; below, F-100D-25, 55-3658, drops napalm bombs on Vietnamese targets. The refuelling probes on starboard wings were standard equipment.

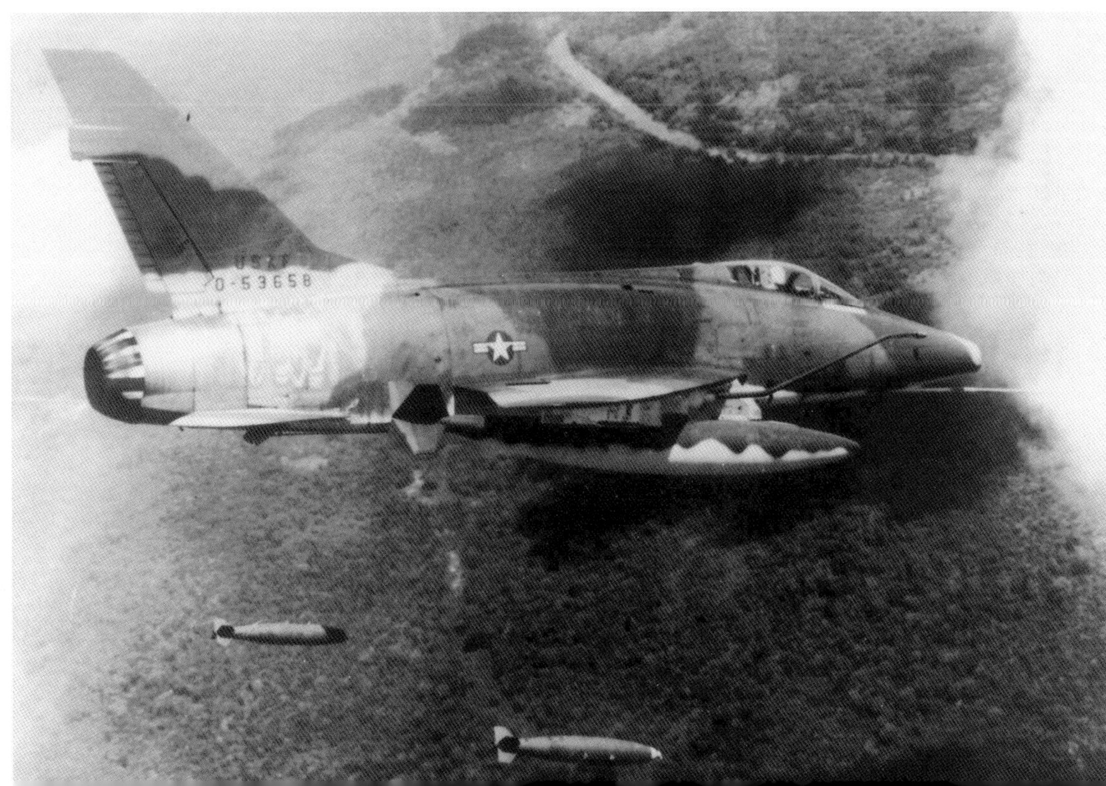

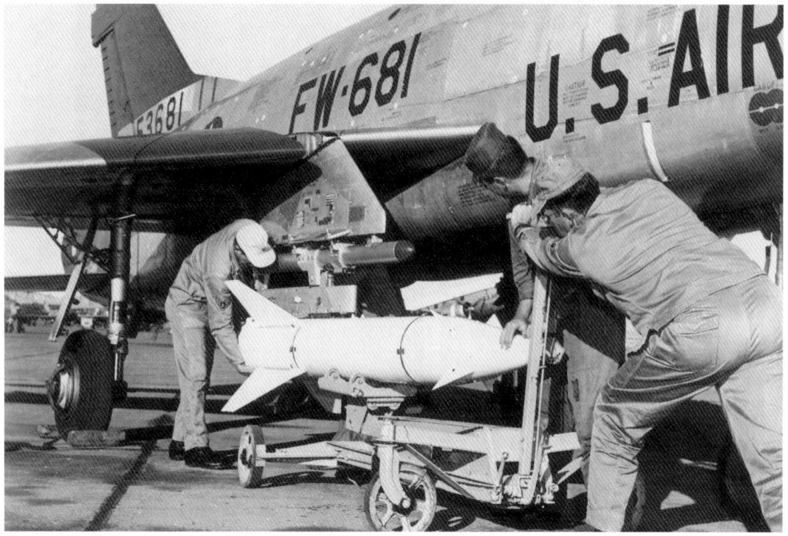

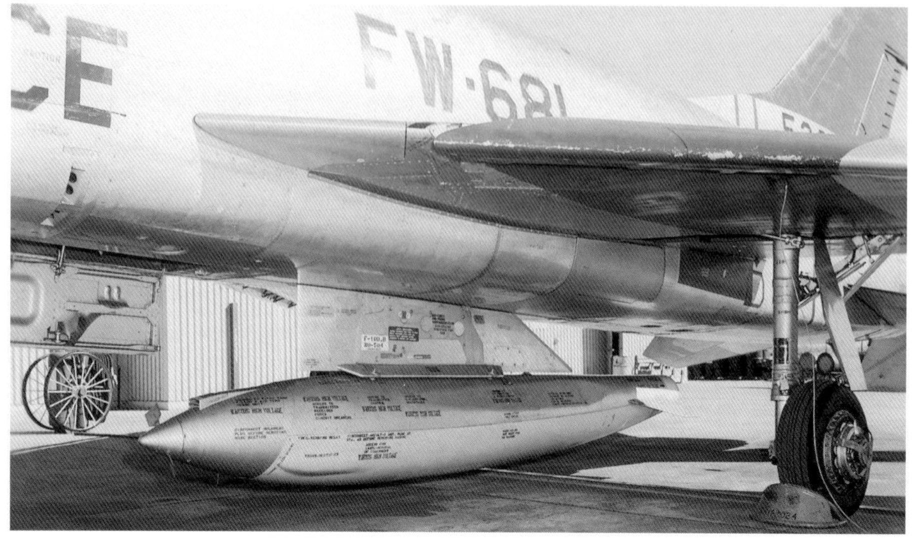

Above, USAF crew loads missile for testing on F-100D-25, 55-36811, and right, an ALQ-31 pod installed under the fuselage of the same plane in February 1958. Below, an F-100D-30 Super Sabre, 55-3765, of 21 TFW in company with a Convair F-102A Delta Dagger, 56-0960, of 4 FIS in flight over northern Japan in 1962. Both units operated from Misawa Air Base.

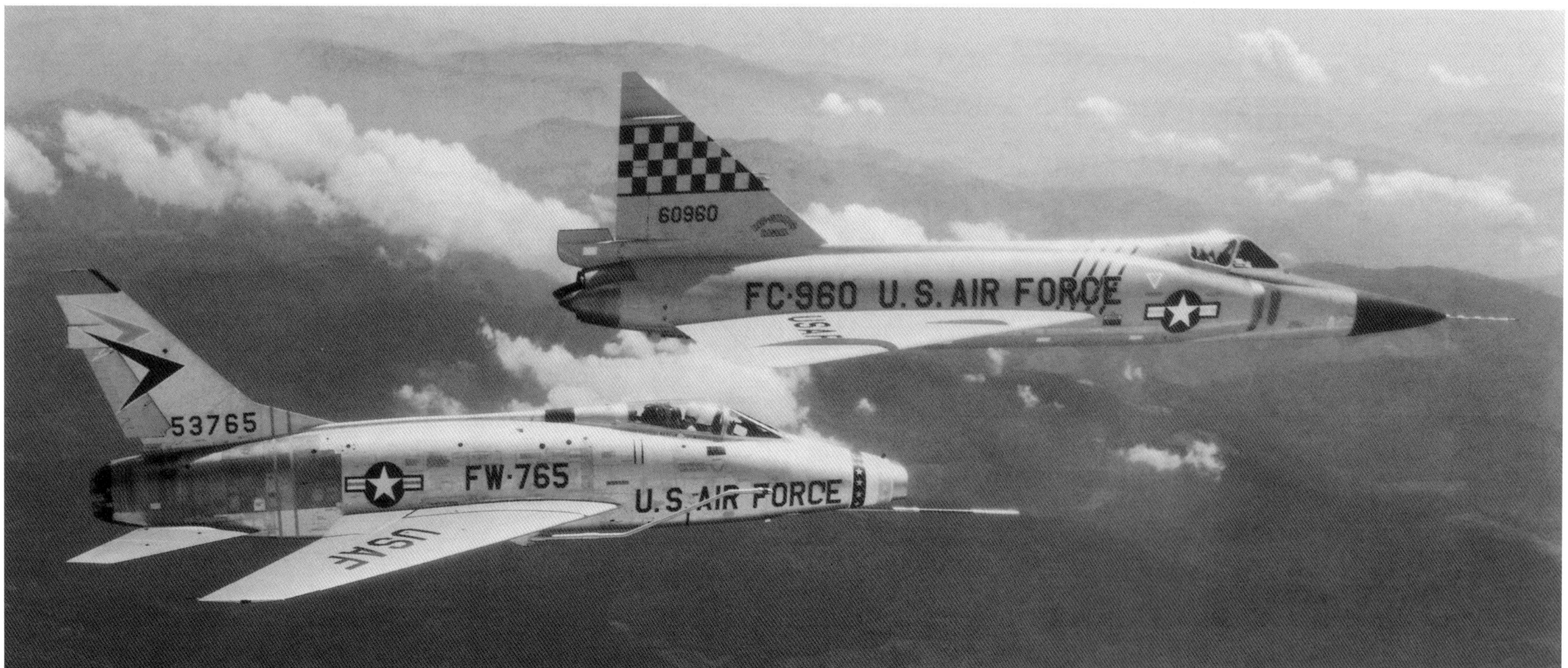

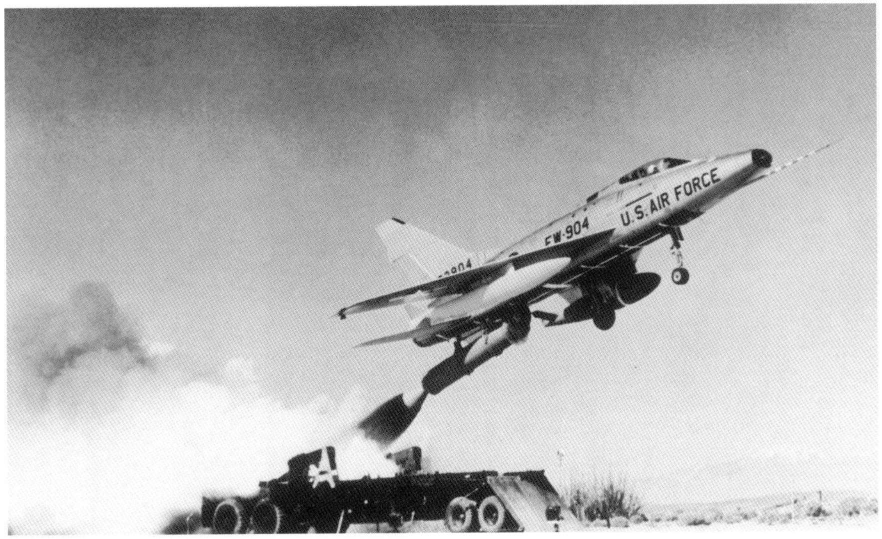

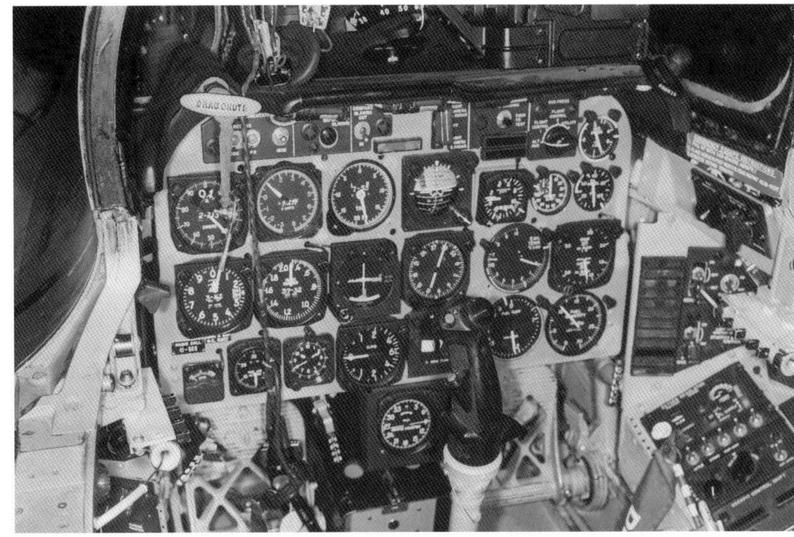

Above, powered by an NAA Rocketdyne solid-propellant rocket of 130,000 pounds/thrust, engineering test pilot Al Blackburn zero-launches in the first F-100D-90, 56-2904, with gear down, standard for the ZEL operation, and fuel tank and dummy nuclear bomb under the wings. Below, a two-seat F-100F-1, 56-3735, photographed in June 1957. Above, the F-100F front cockpit.

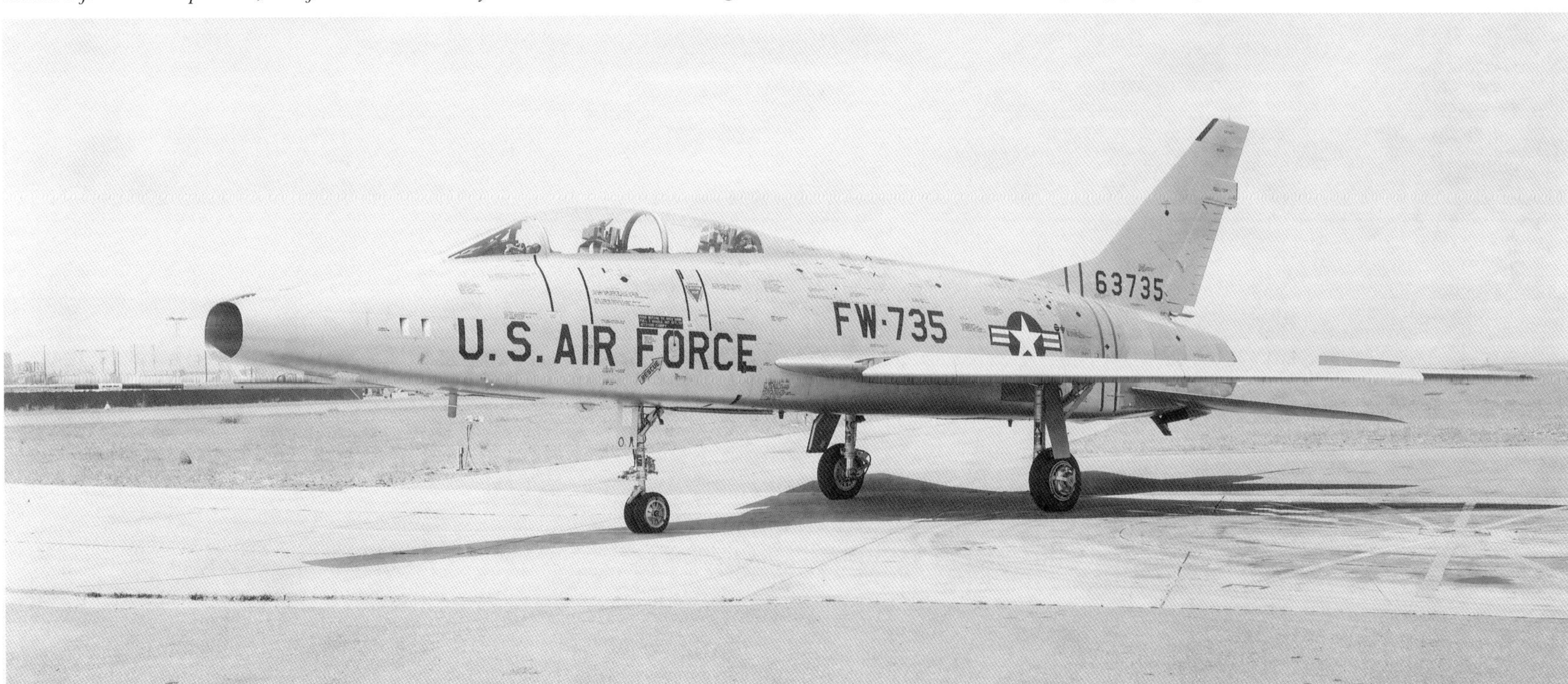

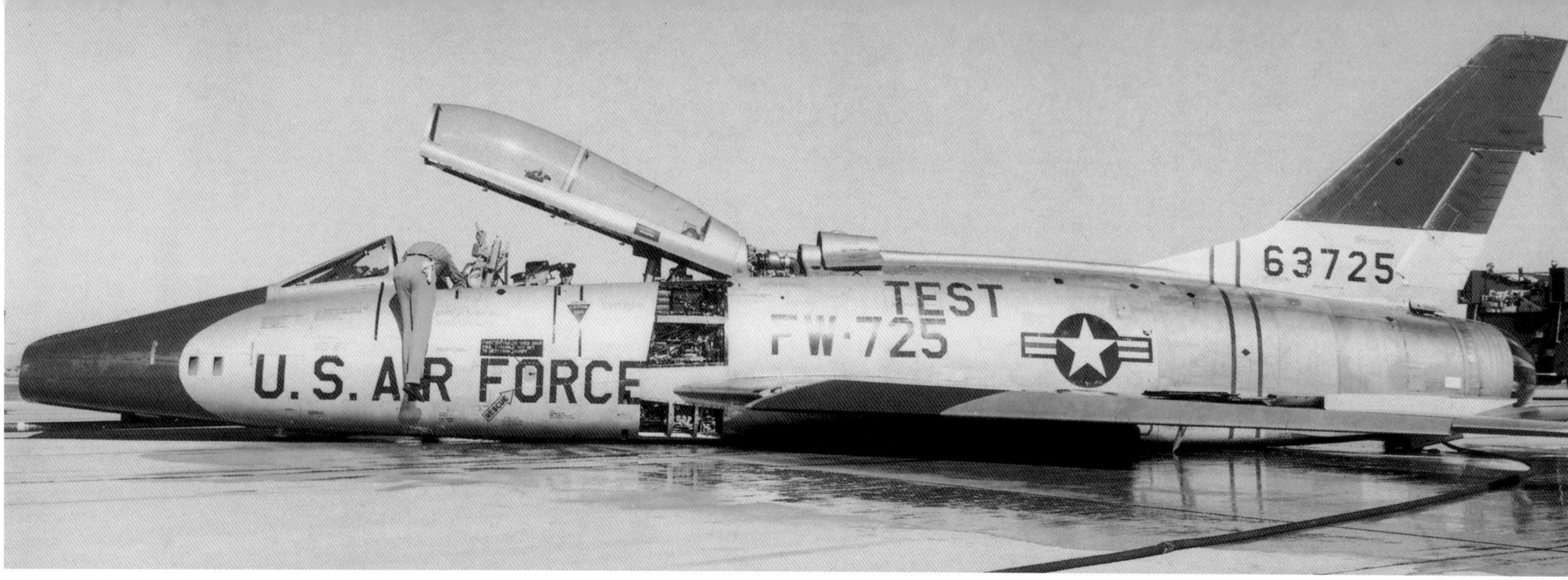

Above, the first F-100F-1, 56-3725, has bellied in without major damage after one of a continuing series of test flights. Below, the same plane with a thrust-reverser mounted at the tail.

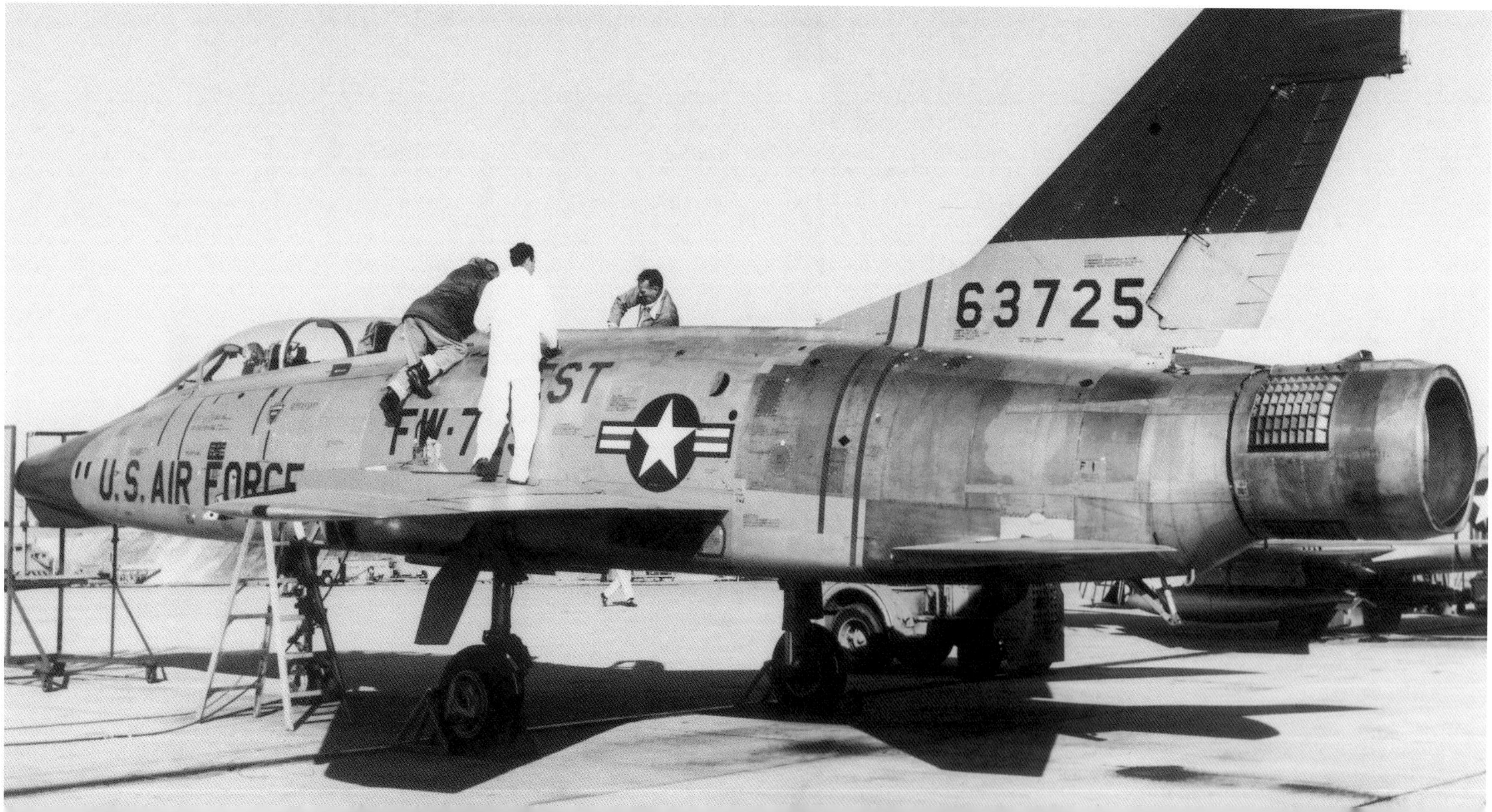

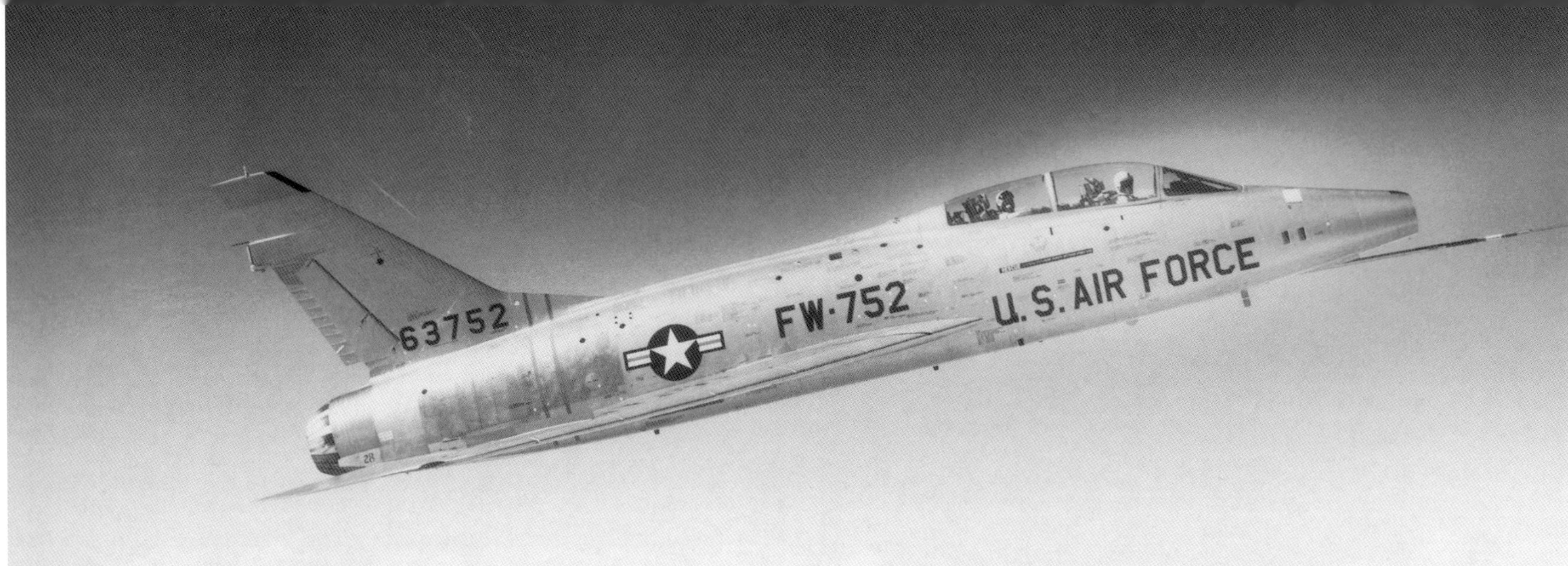

Above and below, two distinct aspects of the thirteenth F-100F-5, 56-3752. Altogether, 339 two-seaters were built; used mainly as trainers, they also performed "Wild Weasel" missions in Vietnam.

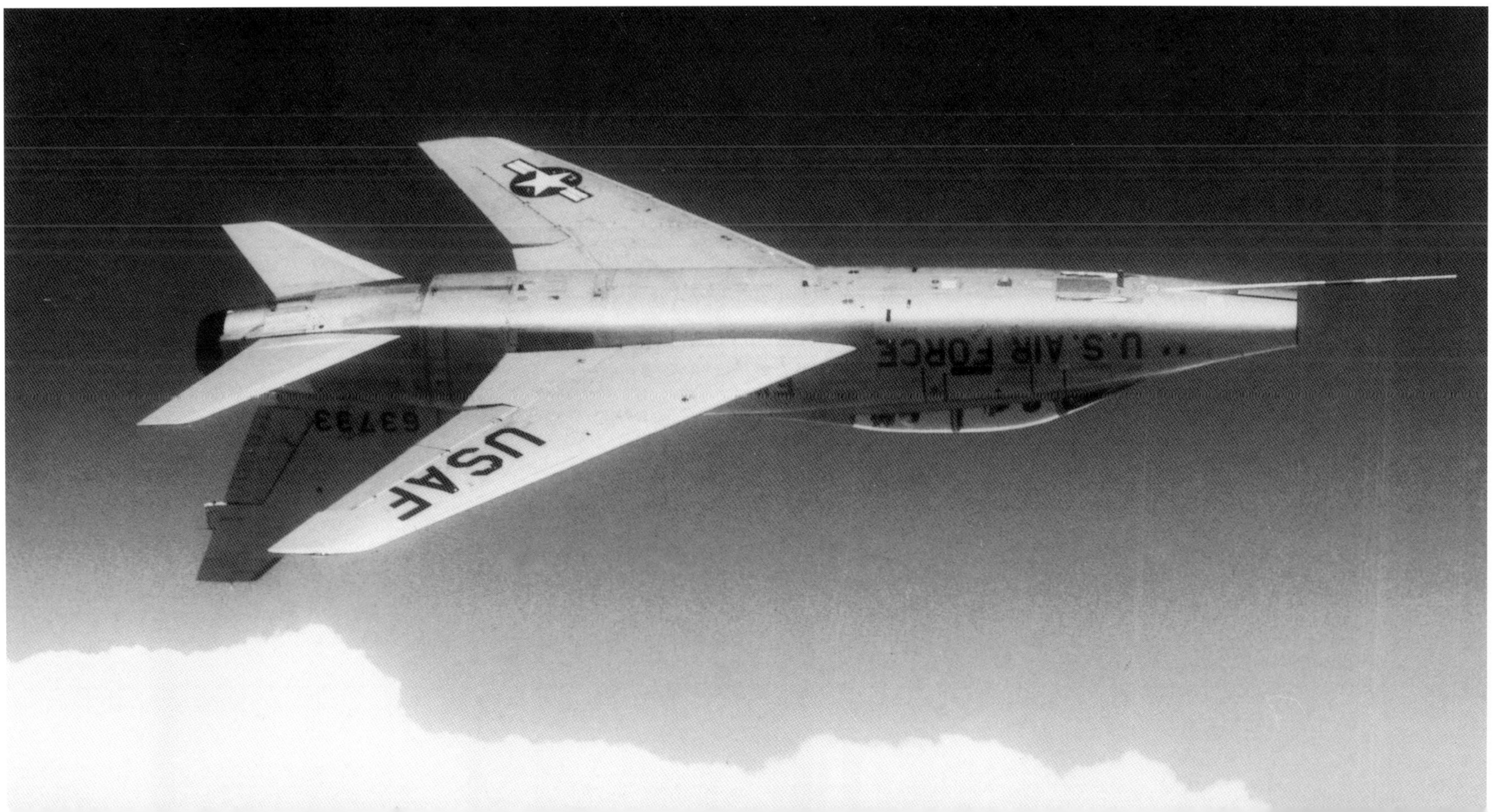

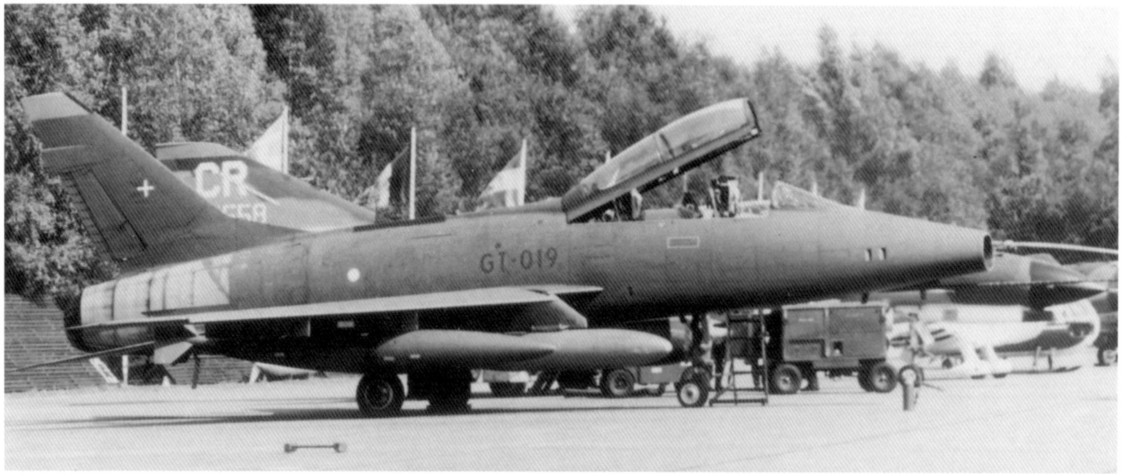

F-100F, GT-019, one of 58 Super Sabres acquired by the Danish Air Force by 1961. More than half were destroyed in accidents.

Production: North American NA-180>NA-262 YFG-100A / F-100A/C/D/F Super Sabre

NA-180	YF-100A	AF52-5754/5755	2
NA-192	F-100A-1	52-5756/5765	
	F-100A-5	52-5766/5778	
	F-100A-10	53-1529/1568	
	F-100A-15	53-1569/1608	
	F-100A-20	53-1609/1708	203
	RF-100A	Modified photo version	(6)
	F-100B	Redesignated YF-107A	—
NA-214	F-100C-1	53-1709/1778	70
NA-217	F-100C-1	54-1740/1769	
	F-100C-5	54-1770/1814	
	F-100C-15	54-1815/1859	
	F-100C-20	54-1860/1969	
	F-100C-25	54-1970/2120	381
NA-222	F-100C-10	55-2709/2733	25
NA-223	F-100D-1	54-2121/2132	
	F-100D-5	54-2133/2151	
	F-100D-10	54-2152/2221	
	F-100D-15	54-2222/2303	
	F-100D-20	55-3502/3601	
	F-100D-25	55-3602/3701	
	F-100D-30	55-3702/3814	496
NA-224	F-100D-35	55-2734/2743	
	F-100D-40	55-2744/2783	
	F-100D-45	55-2784/2863	
	F-100D-50	55-2864/2908	

	F-100D-55	55-2909/2954	221
NA-230	TF-100C	(54-1966)	(1)
NA-235	F-100D-60	56-2903	
	F-100D-90	56-2904	
	F-100D-60	56-2905/2946	
	F-100D-90	56-2947	
	F-100D-60	56-2948/2962	
	F-100D-65	56-2963/3022	
	F-100D-70	56-3023/3142	
	F-100D-75	56-3143/3198	
	F-100D-90	56-3199/3346	444
	F-100E	Designation not used	—
NA-243	F-100F-1	56-3725/3739	
	F-100F-5	56-3740/3769	
	F-100F-10	56-3770/3919	
	F-100F-15	56-3920/4019	295
NA-245	F-100D-80	56-3351/3378	
	F-100D-85	56-3379/3463	113
NA-254	(Thirteen F-100D canceled)		---
NA-255	F-100F-20	58-1205/1233	29
NA-261	F-100F-15	58-6975/6983	9
NA-262	F-100F-15	59-2558/2563	6
	DF-100F	(56-3725)	(1)
Total, F-100A/F fighters, fighter-bombers and trainers,1949-1957			2294

Specifications: North American NA-180>NA-262 YF-100A/F-100A/C/D/F Super Sabre

Description: U.S. Air Force air-superiority fighter. Single-jet mid-wing monoplane of all-metal construction with retractable landing gear. Crew 1. (F-100D) Fighter-bomber. (F-100F) Fighter trainer. Crew 2.

Engine: (YF-100A) 13,200-lb-thrust Pratt & Whitney XJ57-P-7 axial-flow turbojet with afterburner. (F-100A) 14,800-lb-thrust J57-P-7/P-39. (F-100C/D/F) 16,000-lb-thrust J57-P-21.

Dimensions: (YF-100A) Span 36 ft 9 in, length 45 ft 7 in, wing area 385 sq ft. (F-100A/C) Span 38 ft 9 in, length 47 ft 1 in, wing area 400.2 sq ft. (F-100D) As F-100C, except: Length 47 ft 5 in. (F-100F) As F-100D, except: Length 50 ft 4 in.

Weights: (YF-100A) 18,135 lb empty, 24,789 lb gross, 28,561 maximum take-off. (F-100A): 18,185 lb empty, 24,996 lb gross, 28,971 lb maximum. (F-100C) 19,270 lb empty, 28,700 lb gross, 36,549 lb maximum. (F-100D) 20,638 lb empty, 30,061 lb gross, 38,048 lb maximum. (F-100F) 21,712 lb empty, 31,413 lb gross, 39,122 lb maximum.

Armament: Four 20-mm cannon with 200 rounds each. (F-100C) Provision for six external stores. (F-100D) 500-750 lb bombs. Provision for AIM-9 Sidewinder missiles, 2.75-in or 5-in rockets, napalm or cluster bombs.

Performance: (YF-100A) Cruising speed 600 mph, maximum 660 mph, climb 12,500 ft/min, service ceiling 52,600 ft, combat radius 422 miles, ferry range 1410 miles. (F-100A) Cruising 589 mph, maximum 852 mph, climb 23,800 ft/min, ceiling 44,900 ft, radius 358 miles, ferry 1294 miles. (F-100C) Cruising 593 mph, maximum 924 mph, climb 21,600 ft/min, ceiling 38,700 ft, radius 572 miles, ferry 1954 miles. (F-100D) Cruising 590 mph, maximum 864 mph, climb 19,000 ft/min, ceiling 36,100 ft, radius 534 miles, ferry 1995 miles.

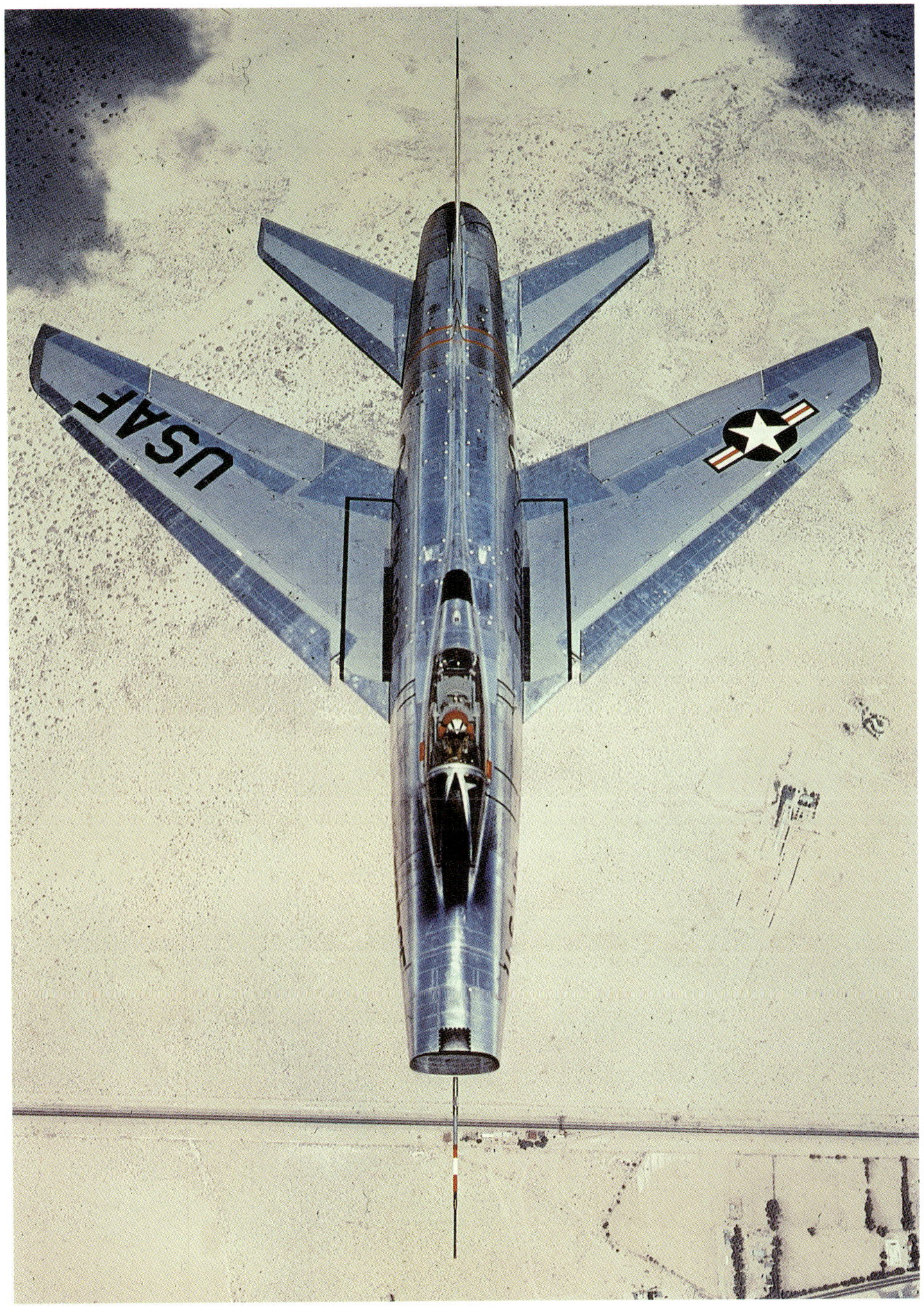

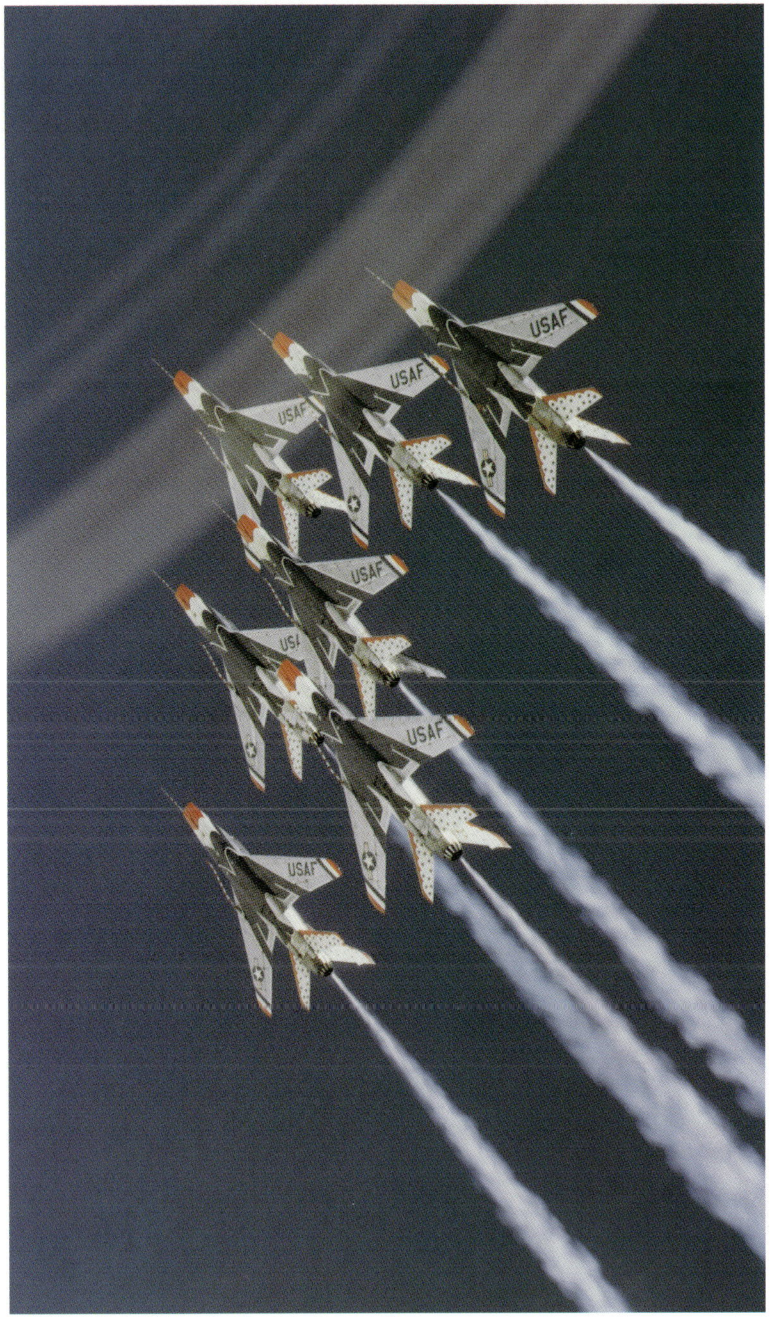

Above the Mojave desert near Edwards AFB, F-100C, USAF 54-2099, looks elegantly clean. Right, seven F-100D Thunderbirds in tight formation. Note that the lead plane does not smoke.

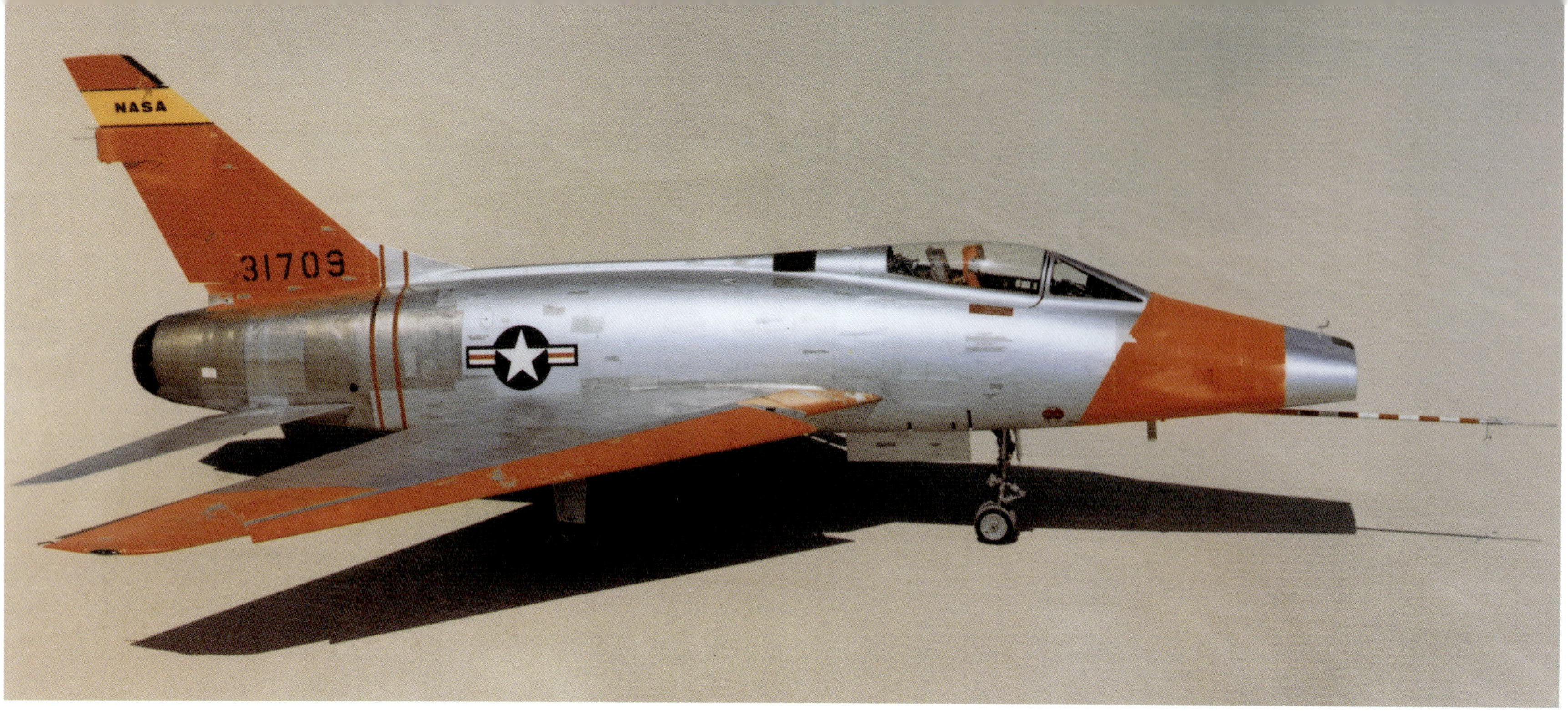

Above, JF-100C, 53-1709, used by NASA for X-15 stability programs. Below, taking off from NAA Palmdale plant, is F-100D-90, 56-3277. Camouflage—even with wavy line on probe!—is for Vietnam operations. Opposite page, USAF T-28A, 51-3763, on a typical NAA publicity photo flight, heading south along the California coast. Note gray paint above wing, behind exhaust.

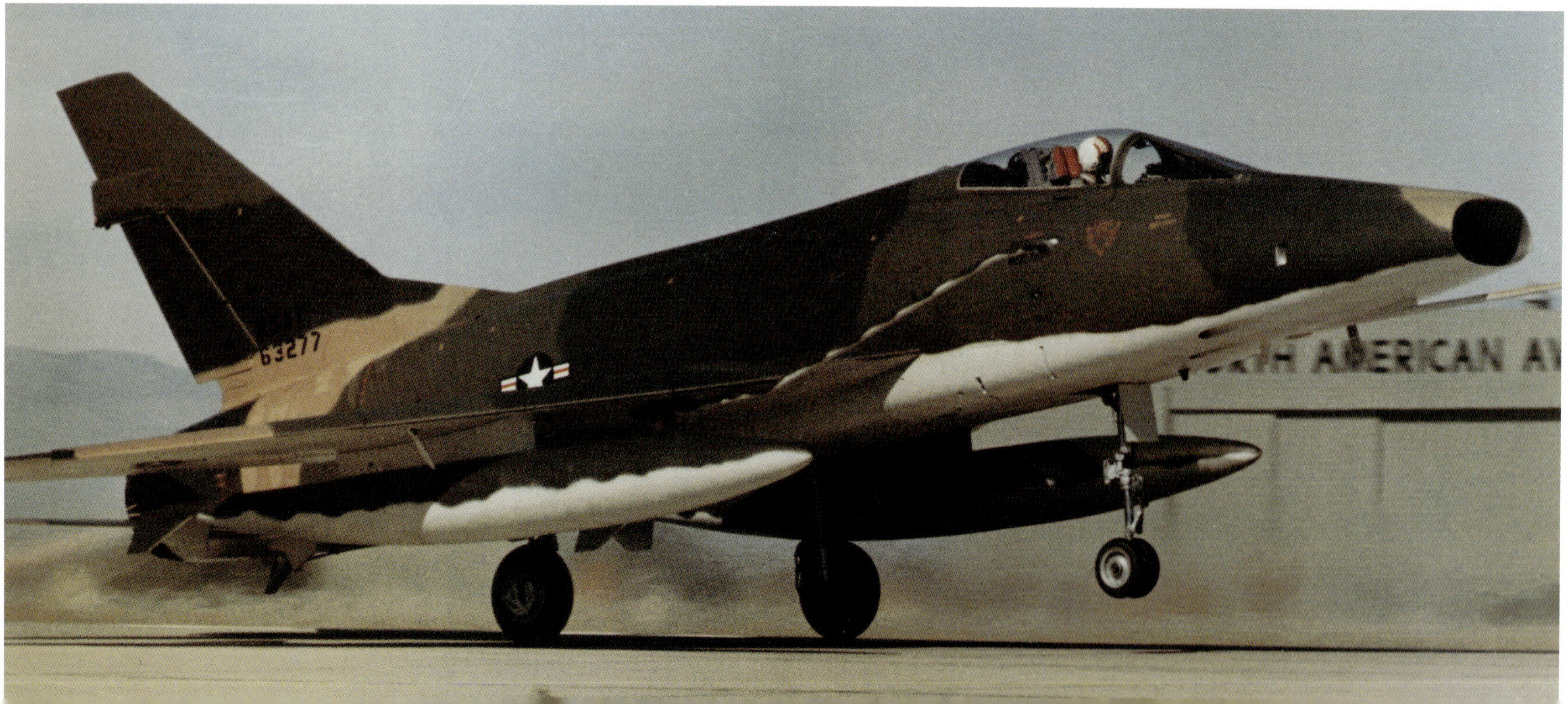

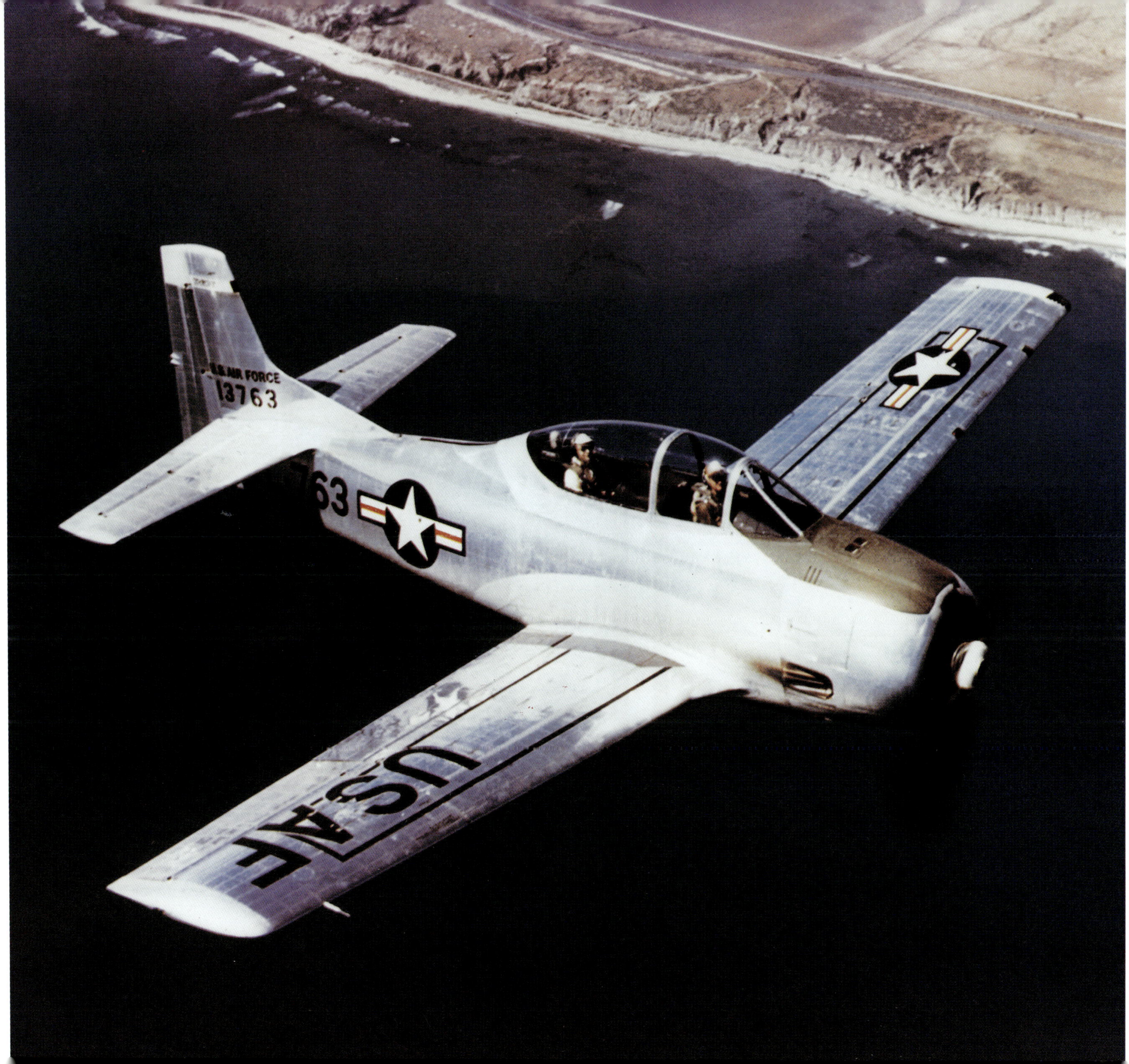
U.S. AIR FORCE
13763
763
USAF

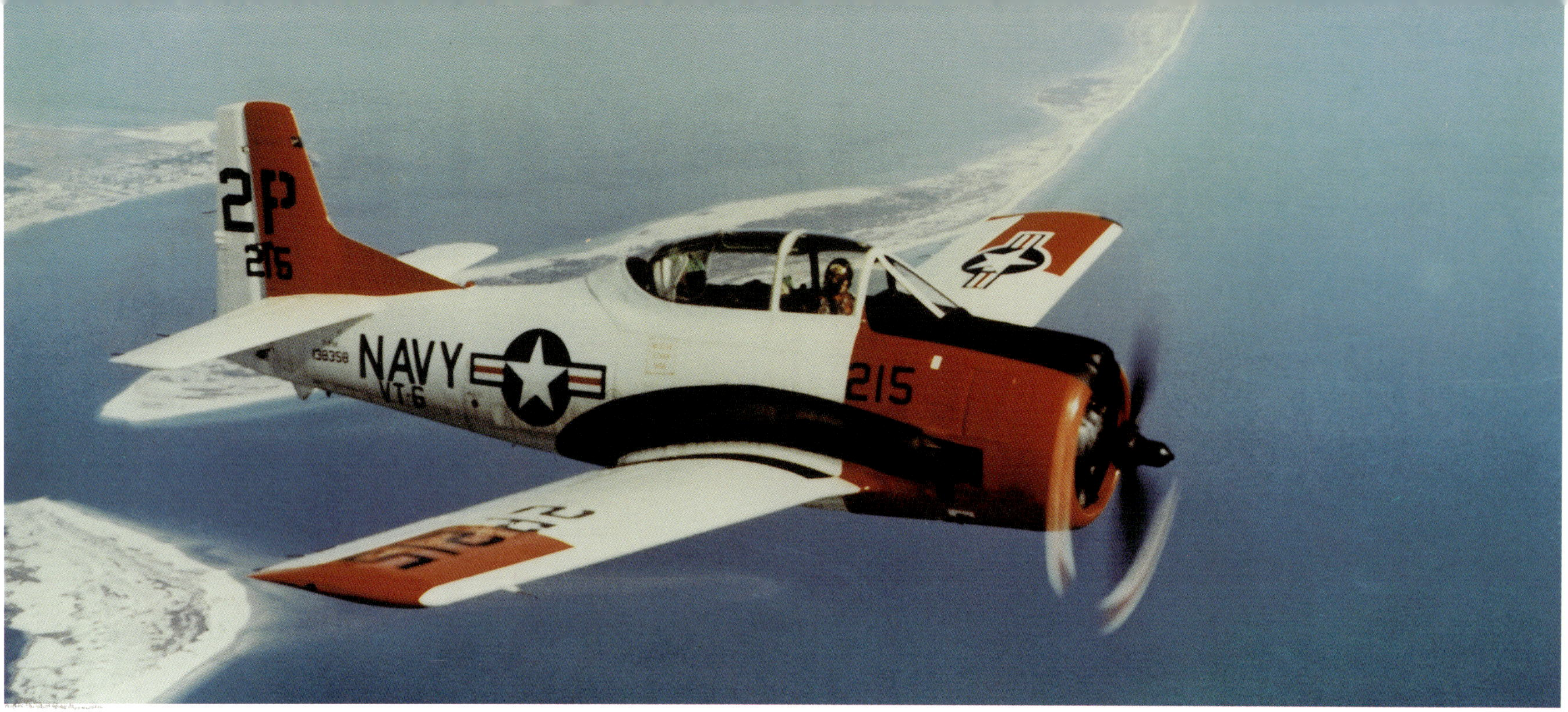

In flight above the Gulf coast, Navy T-28B, BuAer 138358, of VT-6, stationed at NAAS Whiting, near Pensacola, with new 1959 paint. Below, YAT-28E, 52-1242, at Columbus in 1963. Outboard of gun pods are twelve weapons pylons plus mock-up tip rockets. Opposite page, NX 18928, the first of two NA-143 Navion prototypes, in the very handsome original paint scheme.

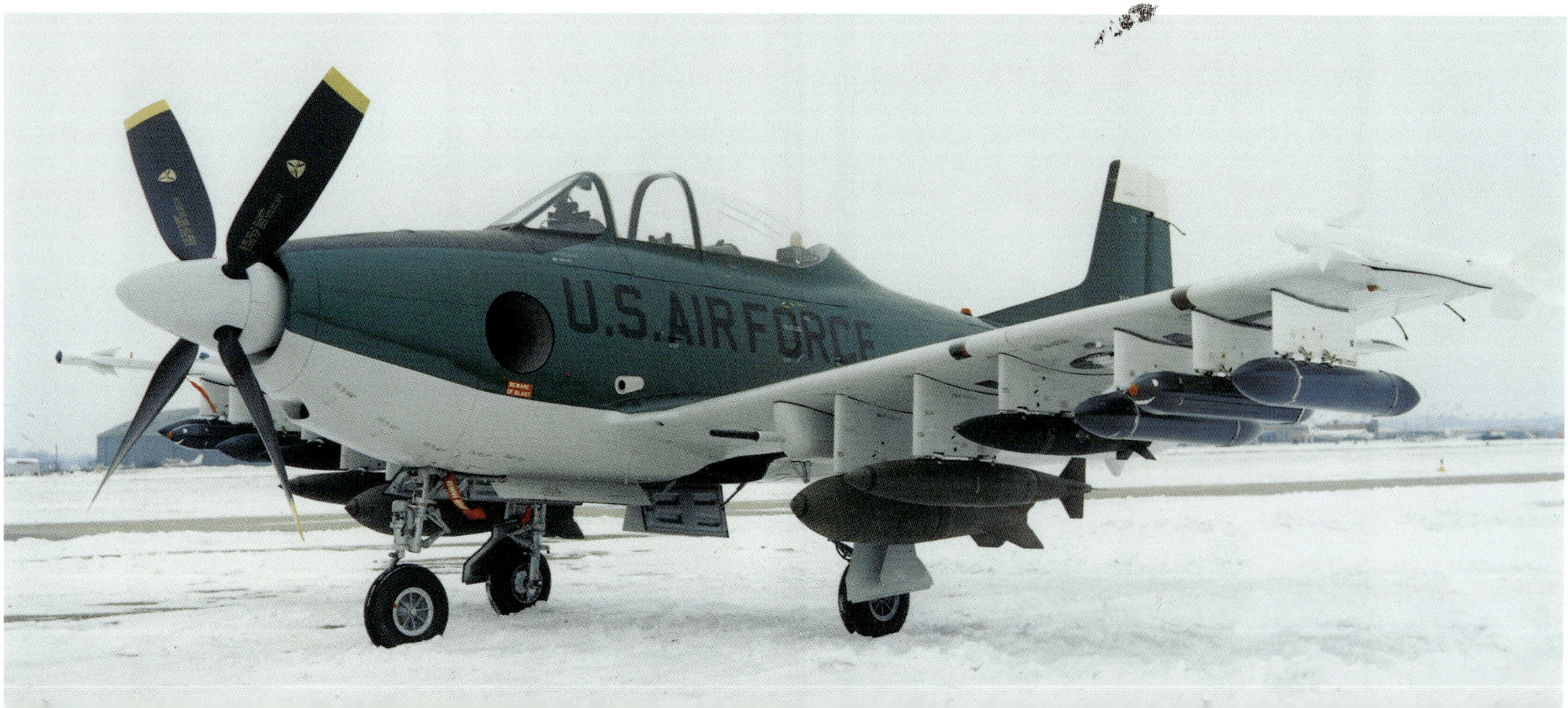

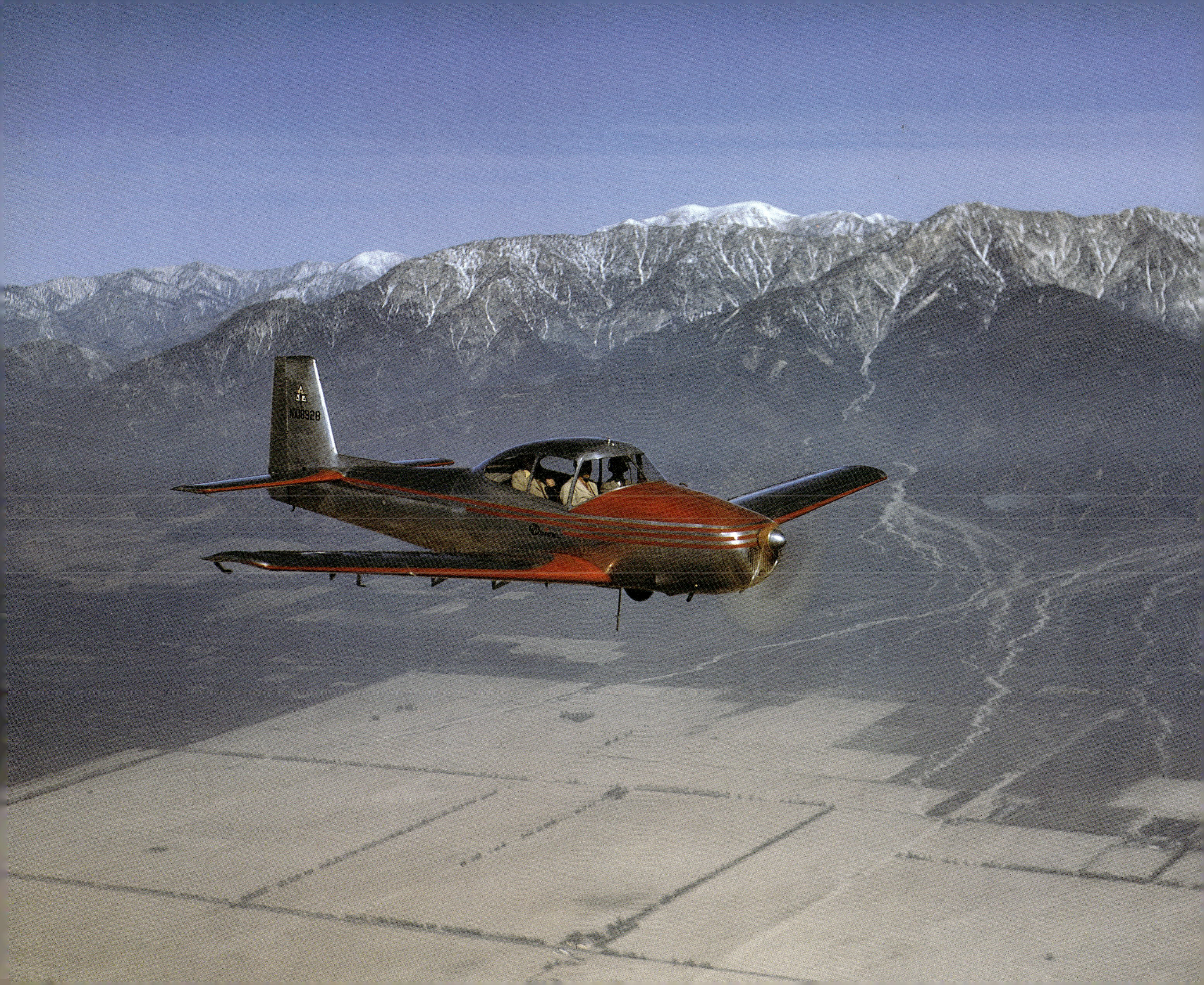
NX18928

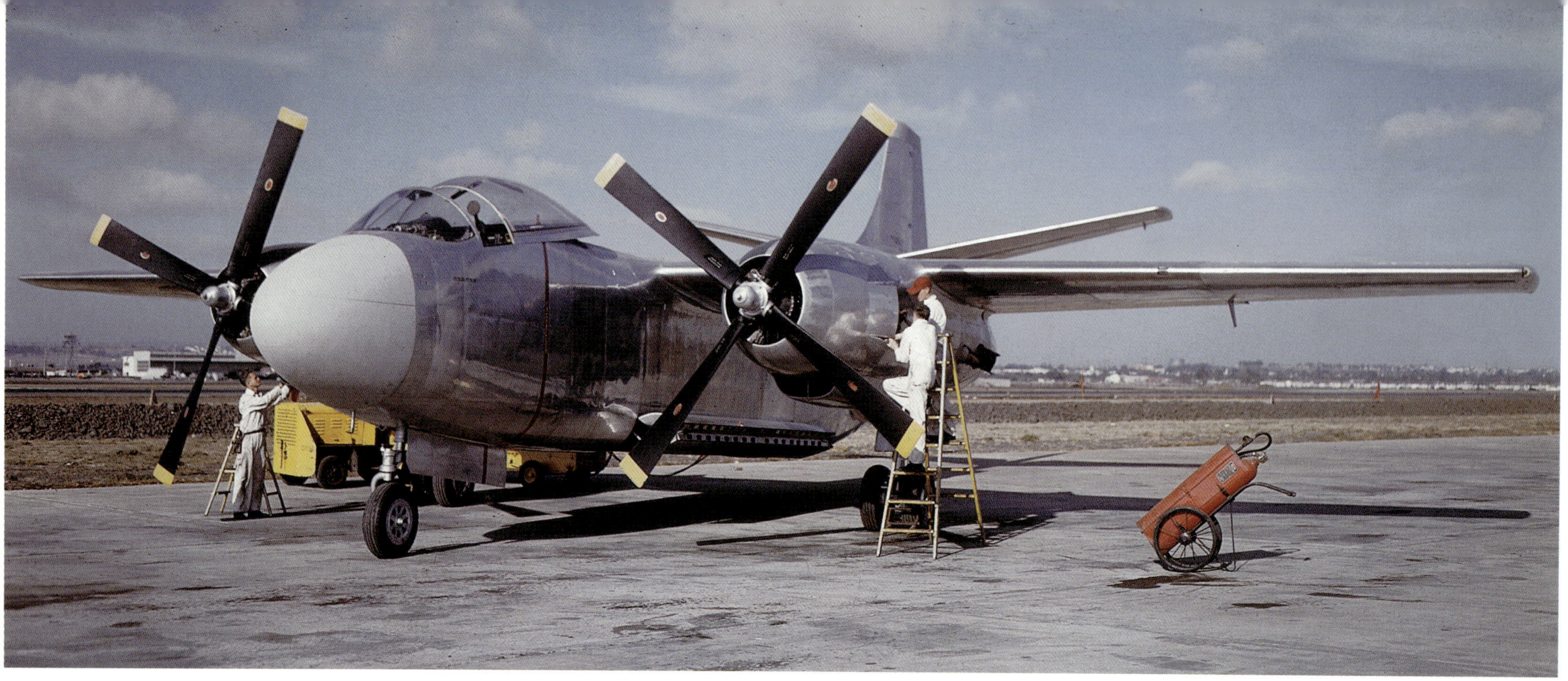

Above, the prototype XAJ-1, BuAer 121460, after the removal of the propeller spinners. Below, serving with photo squadron VJ-62, based at NAS Sanford, is AJ-2P, BuAer 128046. Opposite page, USS Forrestal *on its 1956 shakedown cruise. An F7U takes off from the angled deck as one gray and two blue Savages fly overhead. The AJ-2s were used to refuel the fighter aircraft.*

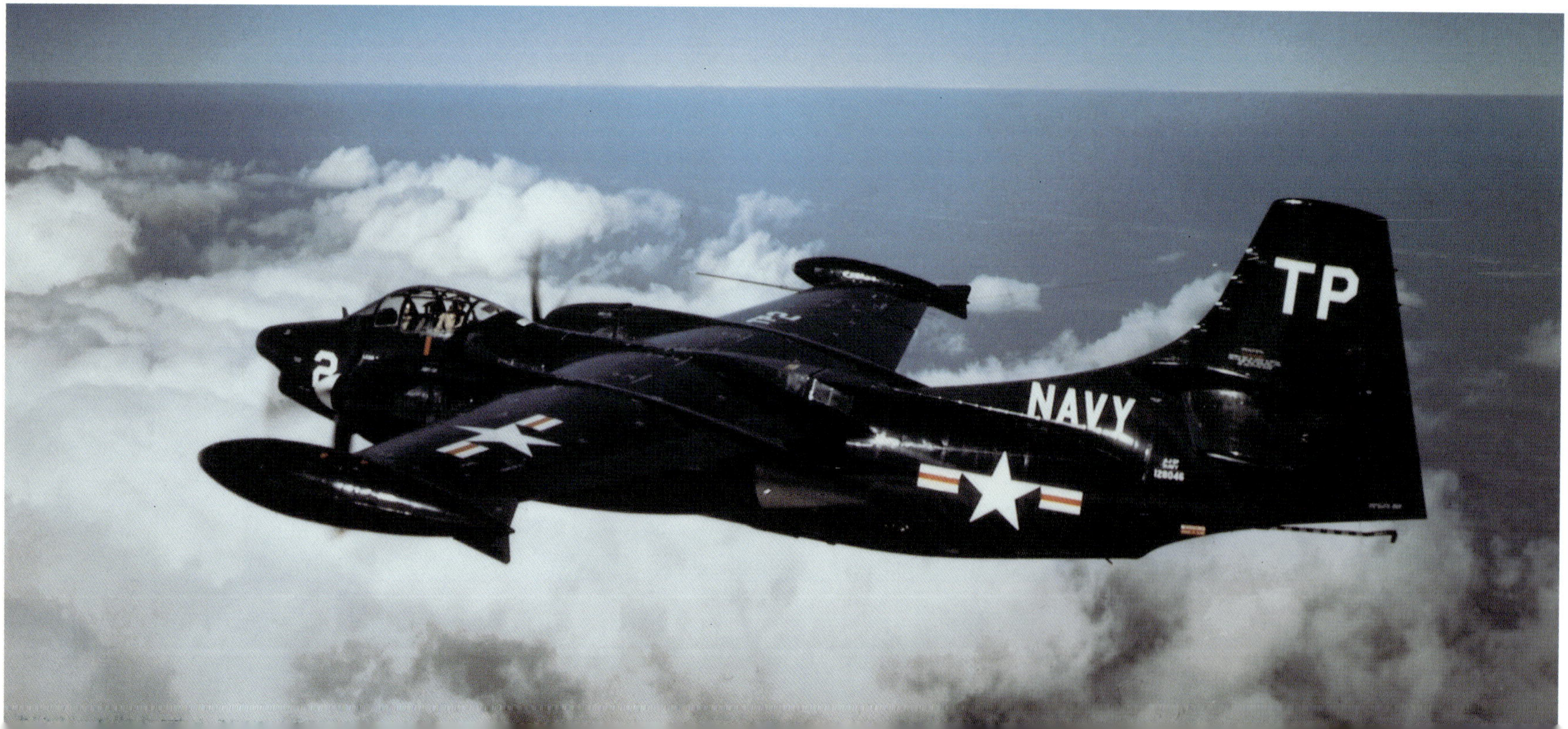

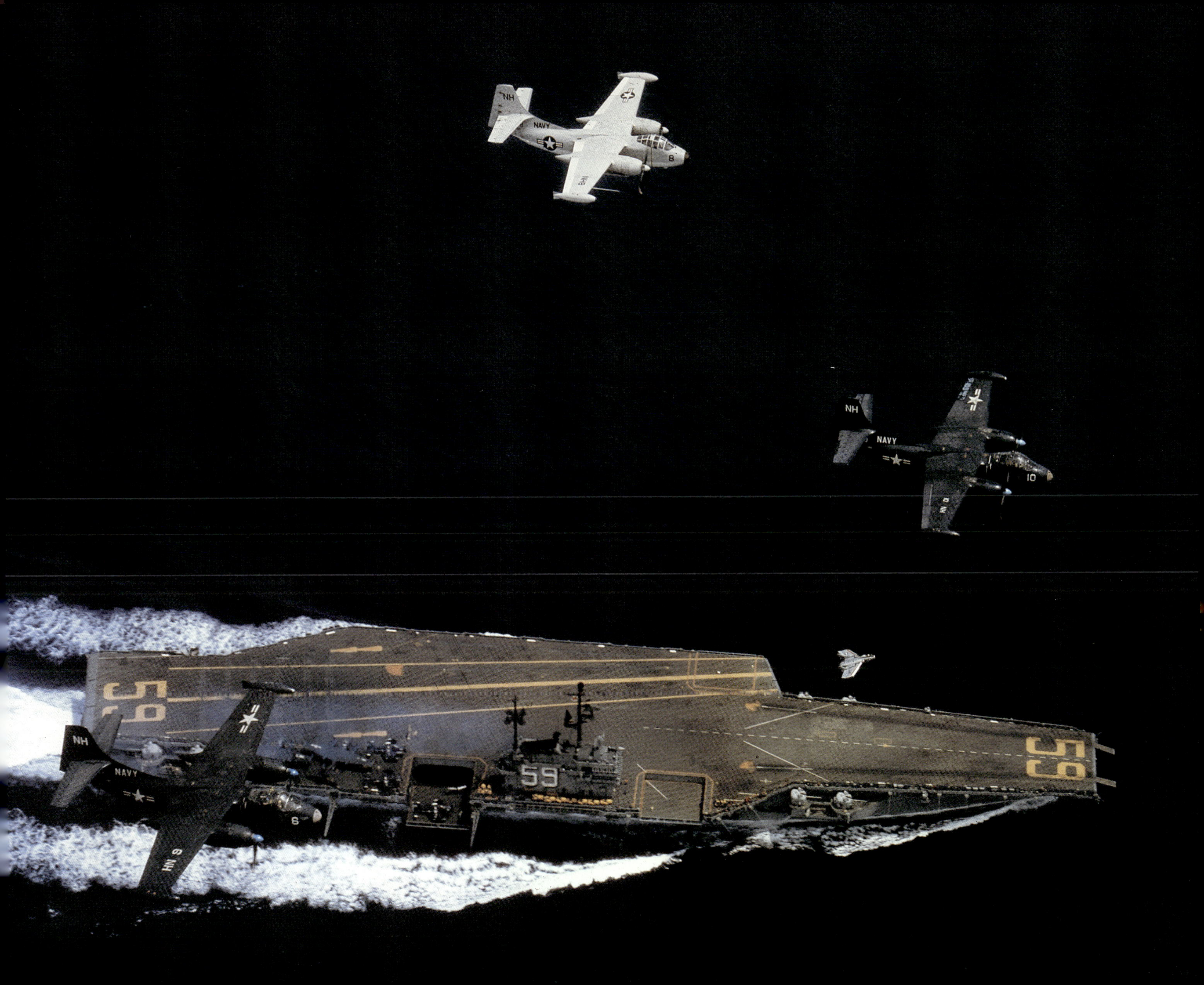

NH
NAVY
8
NH
NAVY
10
59
59
59
NH
NAVY
6

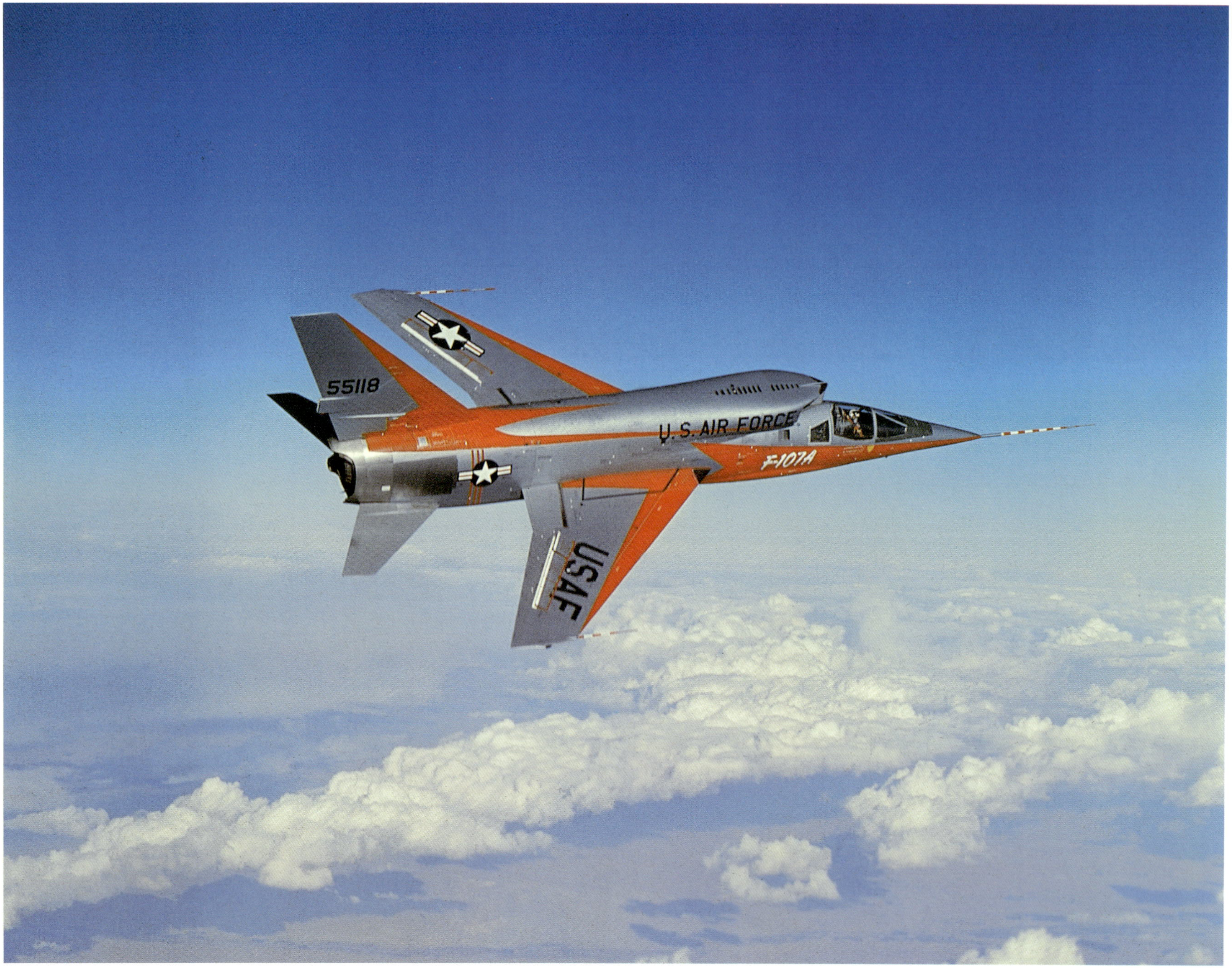
55118
U.S. AIR FORCE
F-107A
USAF

North American NA-212 YF-107A / NA-257 F-108 Rapier

NORTH AMERICAN began design studies for an all-weather interceptor version of the Super Sabre in June 1953. But with the advent of Convair's F-102 Delta Dagger, which entered full-scale production that same year, the Air Defense Command informed North American that another such aircraft would not be needed.

However, on August 4, 1954, the Tactical Air Command placed an order with the company for an all-weather fighter-bomber development of the F-100 that would incorporate an air-to-air track-and-find radar. The Air Materiel Command had diverted some $93 million to fund the project, which was designated Air Force Weapon System 305B, and the North American entry would eventually compete head-to-head with the Republic F-105 Thunderchief for the production contract.

Initially, the designation YF-100B was to be used for the prototypes, but as the aircraft's mission would dictate the need for an entirely new design, the charge number NA-212 and the designation F-107A were given to the project. Preliminary design studies incorporated a radar in the nose section with the air intake positioned directly below, similar to the F-86D/K/L series. When this initial configuration, with the intended semi-submerged weapon layout in the lower fuselage, was later found to be aerodynamically unsuitable, the final design study used a novel approach, positioning the air intake atop the fuselage. To optimize intake airflow to the new Pratt & Whitney J75 engine throughout the aircraft's performance envelope, the unique layout featured a two-position inlet duct on the first two YF-107A aircraft, and a variable air inlet duct (VAID) on the third. Because of the position of the inlet, the fuselage, which incorporated the "area rule" concept, had a sharp nose profile housing the radar and a very tidy installation of the weapon stores and four 20-mm cannon.

The vertical stabilizer and the horizontal stabilizers were one-piece, all-moving surfaces. The wing, although similar in sweep and planform to that of the F-100, was a completely new design that featured spoilers on the upper and lower surfaces instead of ailerons.

Originally, the Air Force contract called for a total of nine flight-test and one static-test aircraft to be built. By September 1954 the full-scale mock-up was inspected and approved by the Air Force, and production design began on May 1, 1955. In January 1957, however, the contract was revised to just three aircraft. The flight test program was comprised of Phase 1, company evaluation; Phase 2, USAF evaluation; and Phase 3, joint NAA/USAF fire control system evaluation. Each of the three YF-107As was assigned to one test phase.

On September 10, 1956, the first aircraft, USAF 55-5118, took off from the Air Force Flight Test Center, Edwards AFB, with NAA test pilot Bob Baker at the controls. Overall, the first flight was very successful, although a brake chute malfunction on landing resulted in nose gear failure. The damage was slight and after repairs to the fuselage and fuel system the aircraft was flying again in early October. The first Mach 2 flight was made on November 3, 1956, with NAA's J. O. Roberts the pilot. By November 17 thirty flights had been made and Phase 1 testing was complete after 17 hours and 10 minutes of flying time, including evaluation flights by Major General Albert Boyd and Colonel Horace Hanes, respectively the commander and director of the AFFTC.

The second aircraft, USAF 55-5119, flew on November 28, followed by the third, 55-5120, on February 18 of the following year. Obtaining performance and stability data, as well as pilot comments on overall handling, were the main objectives of the Phase 2 test program, which encompassed a total of thirty-two flights between December 3 and February 15. Both 55-5118 and -5119 were used in the second phase, which proved the handling to be predictable and the performance sufficient for Tactical Air Command requirements.

A series of zoom-climb tests were conducted immediately after Phase 2 was

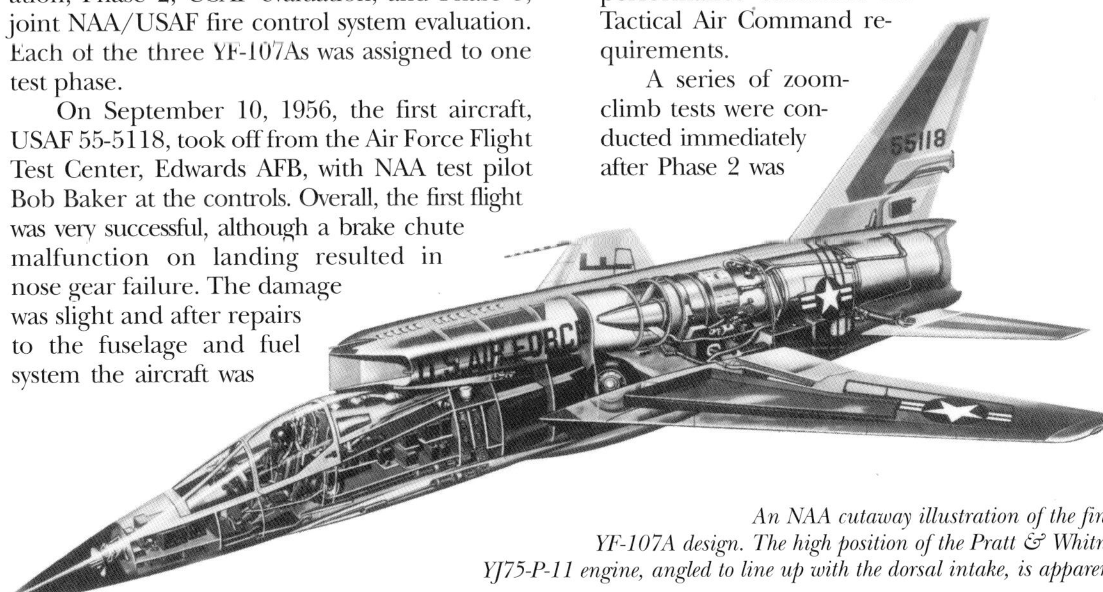

An NAA cutaway illustration of the final YF-107A design. The high position of the Pratt & Whitney YJ75-P-11 engine, angled to line up with the dorsal intake, is apparent.

In flight, with its highly polished aluminum exterior and a striking red paint scheme, the YF-107A was most impressive. This is the first of the three prototypes, USAF 55-5118.

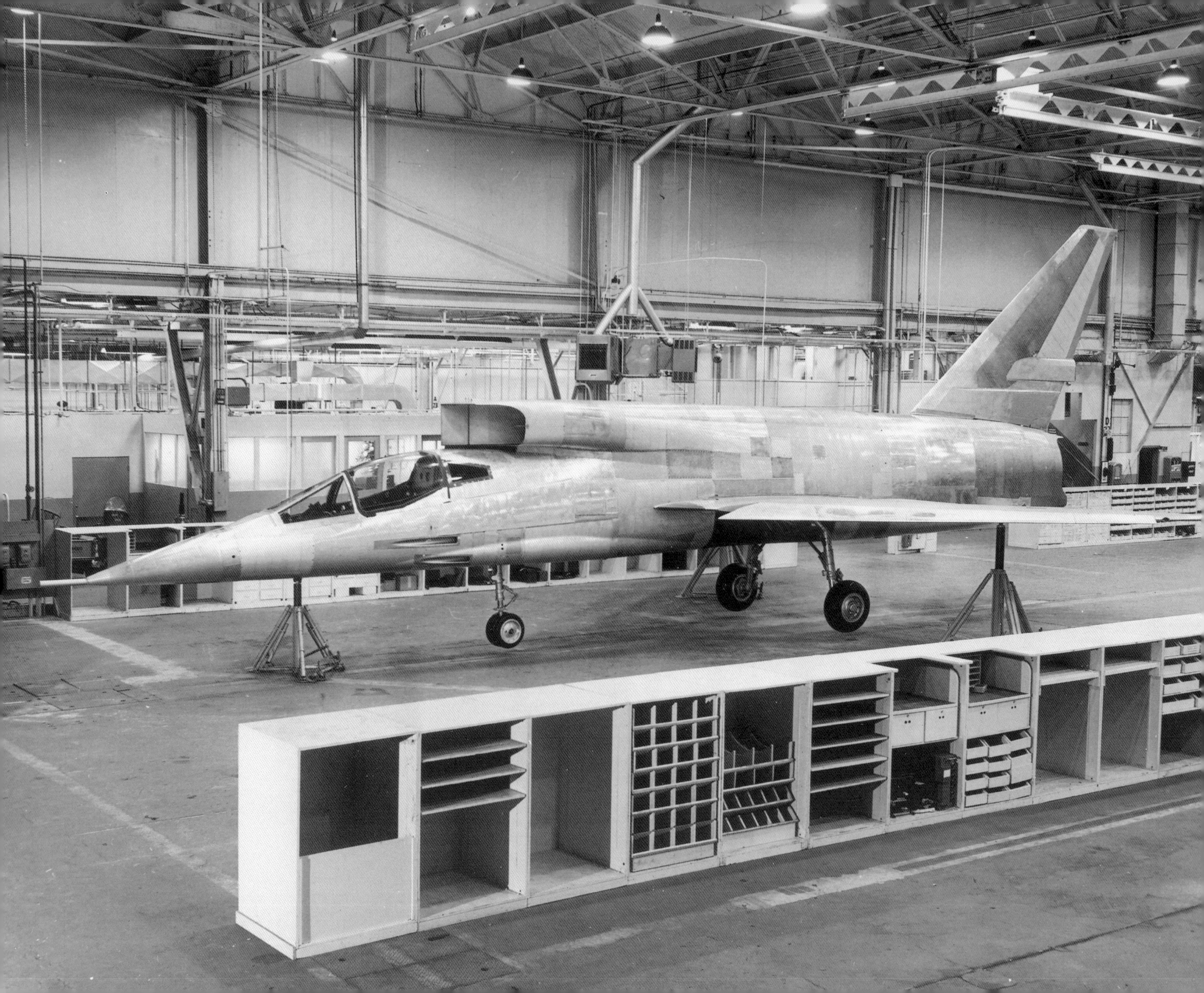

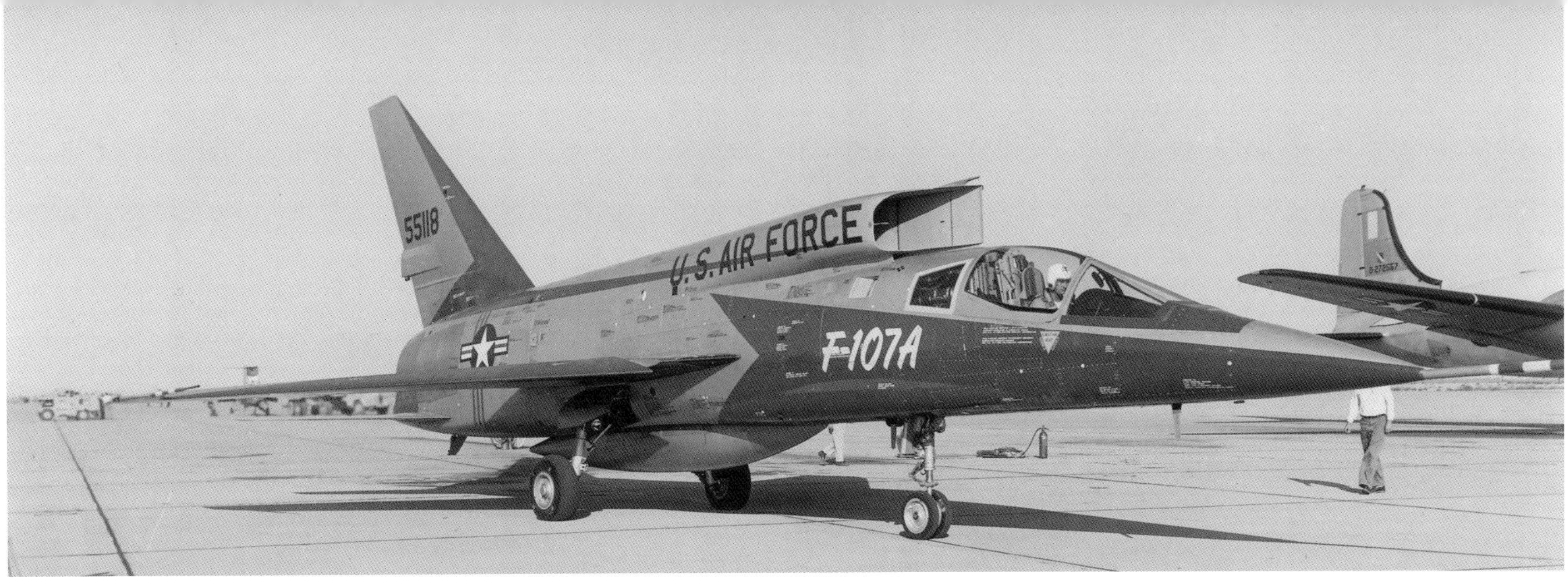

Opposite, the first YF-107A under construction. Above and below, the flashy paint scheme was part of a high-level publicity campaign for a fighter that was not destined for production.

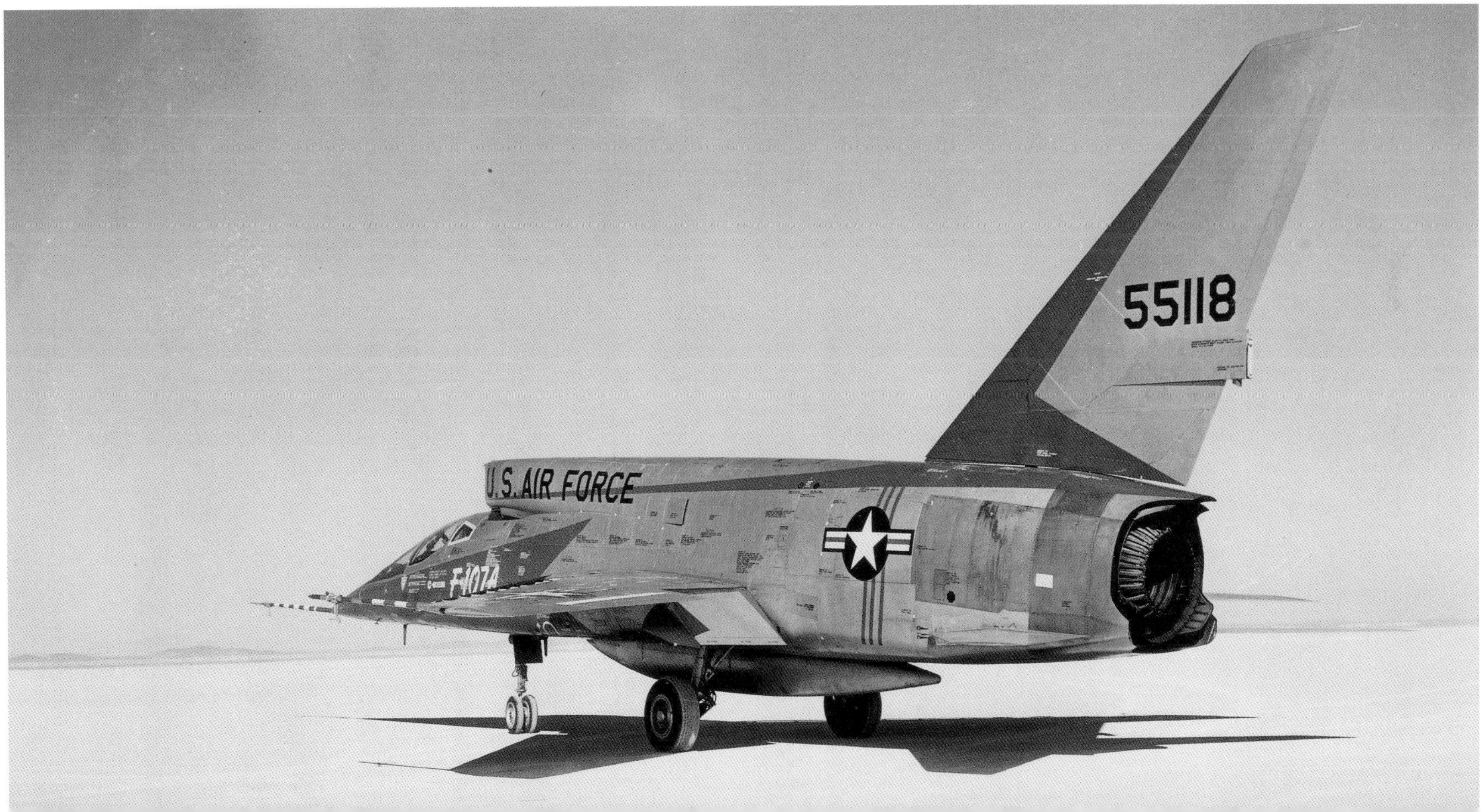

completed, with both Roberts and Al White piloting. These tests were terminated after the third attempt, due to compressor stalls that could delay the test program if engine damage ensued. Also, the Lockheed NF-104 was performing zoom-climbs in altitude-record attempts and the Air Force did not want publicity for the NAA design. The tests did show that the YF-107A could climb vertically at speeds in excess of Mach 1.

White was chosen as the chief pilot for the final weapons-system test phase, flown out of Air Force Plant 42 at Palmdale. The drop sites were NAS China Lake and the Salton Sea test range, and both the LABS (Low Altitude Bombing System) and level-flight drops from higher altitudes were explored. Various stores including the Sandia and TX-28 were tested and the drops ranged from speeds of Mach .9 to Mach 2 and altitudes of 8000 to 35,000 feet. The results proved the overall versatility of the F-107 program; although the competition with the F-105 was very close, Tactical Air Command ultimately rejected the F-107A in favor of the Republic design, the latter's concealed bomb bay being the deciding factor. Two of the YF-107As were subsequently used by NACA, contributing to the development of the side-stick control system for the X-15 program. The last F-107 flight was made by Scott Crossfield on August 1, 1959.

Currently, aircraft number 1 resides at the Pima County Air Museum, Tucson, Arizona, and number 2 at the USAF Museum at Wright-Patterson AFB in Dayton, Ohio. The third YF-107A was cut up at Edwards AFB in 1963 and shipped to Shepard AFB, Texas, to be used as a fire-fighting training aid.

This close-up shot makes the gun fairing and semi-submerged weapons bay apparent, as well as the position of the spoilers, flush with the lower surface of the wing, outboard of the flap.

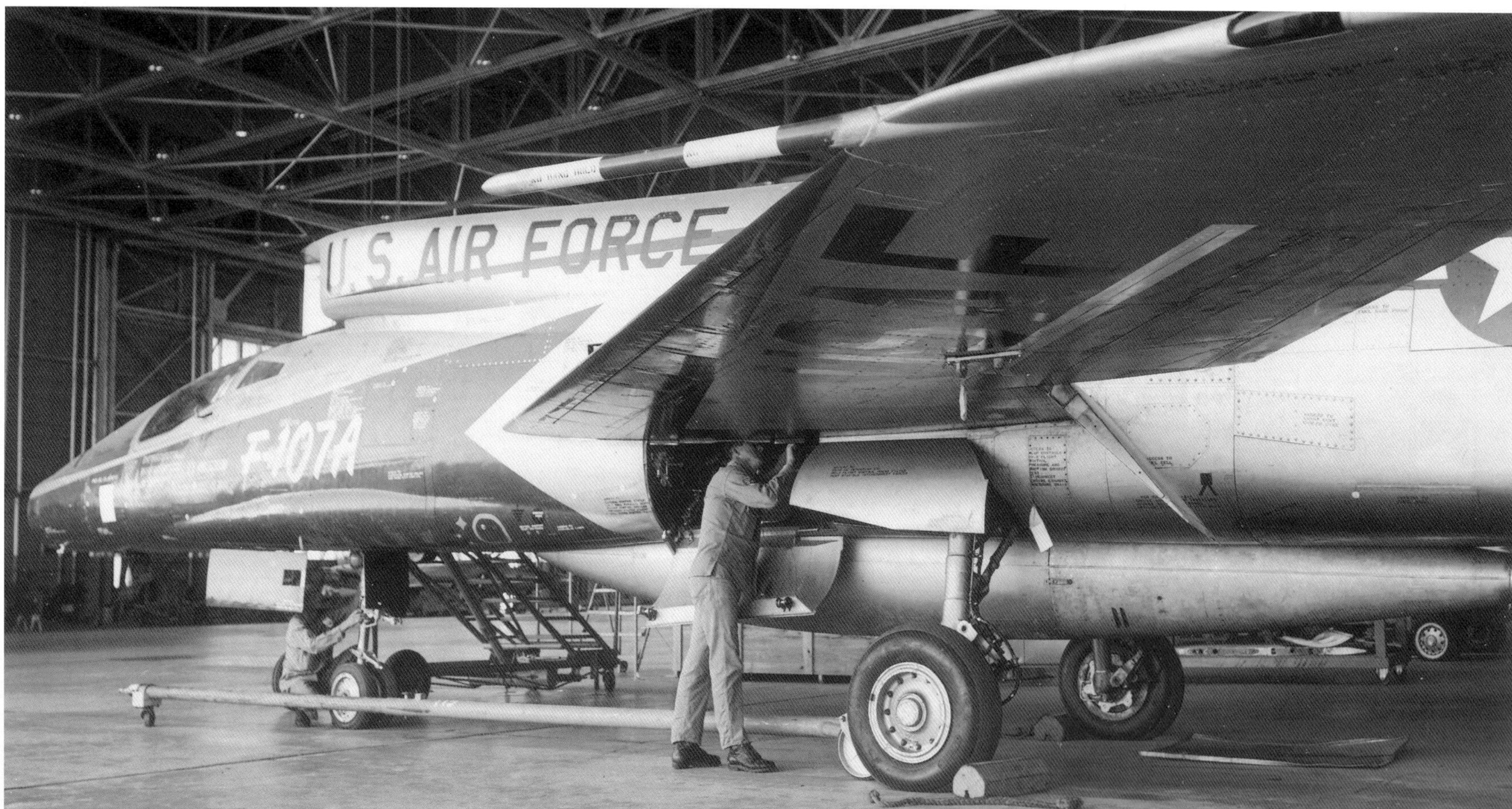

Specifications: North American NA-212 / YF-107A

Description: U.S. Air Force fighter-bomber. Single-jet mid-wing monoplane of all-metal construction with retractable landing gear. Crew 1.

Engine: 15,500-lb-thrust Pratt & Whitney YJ75-P-11 axial-flow turbojet, 23,500-lb-thrust with afterburer.

Dimensions: Span 36 feet 7 in, length 60 feet 10 in, height 19 feet 6 in, wing area 376.02 sq ft.

Weights: 25,144 lb empty, 37,153 lb combat, 42,600 lb with 500-gal. and 275-gal. drop tanks, 49,000 lb gross. Fuel 650/2000 gal.

Armament: Four 20-mm M-39 automatic guns with 200 rounds each, various centerline stores and two under-wing pylons for bombs or rockets.

Performance: Cruising speed 598 mph, maximum 1295 mph, climb 39,900 ft/min, service ceiling 53,200 ft, combat radius 788 miles, ferry range 2428 miles.

Production: North American NA-212 / YF-107A

NA-212	YF-107A	AF 55-5118/5120	3
Total, (YF-100B) YF-107A fighter prototypes, 1954-1959			3

The YF-107A cockpit, set far forward, was necessarily small.

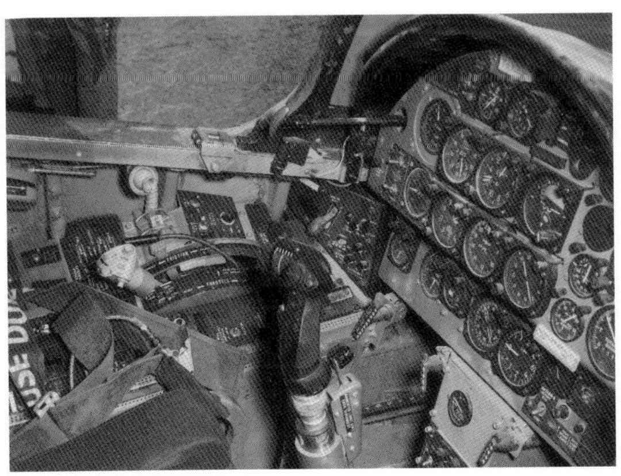

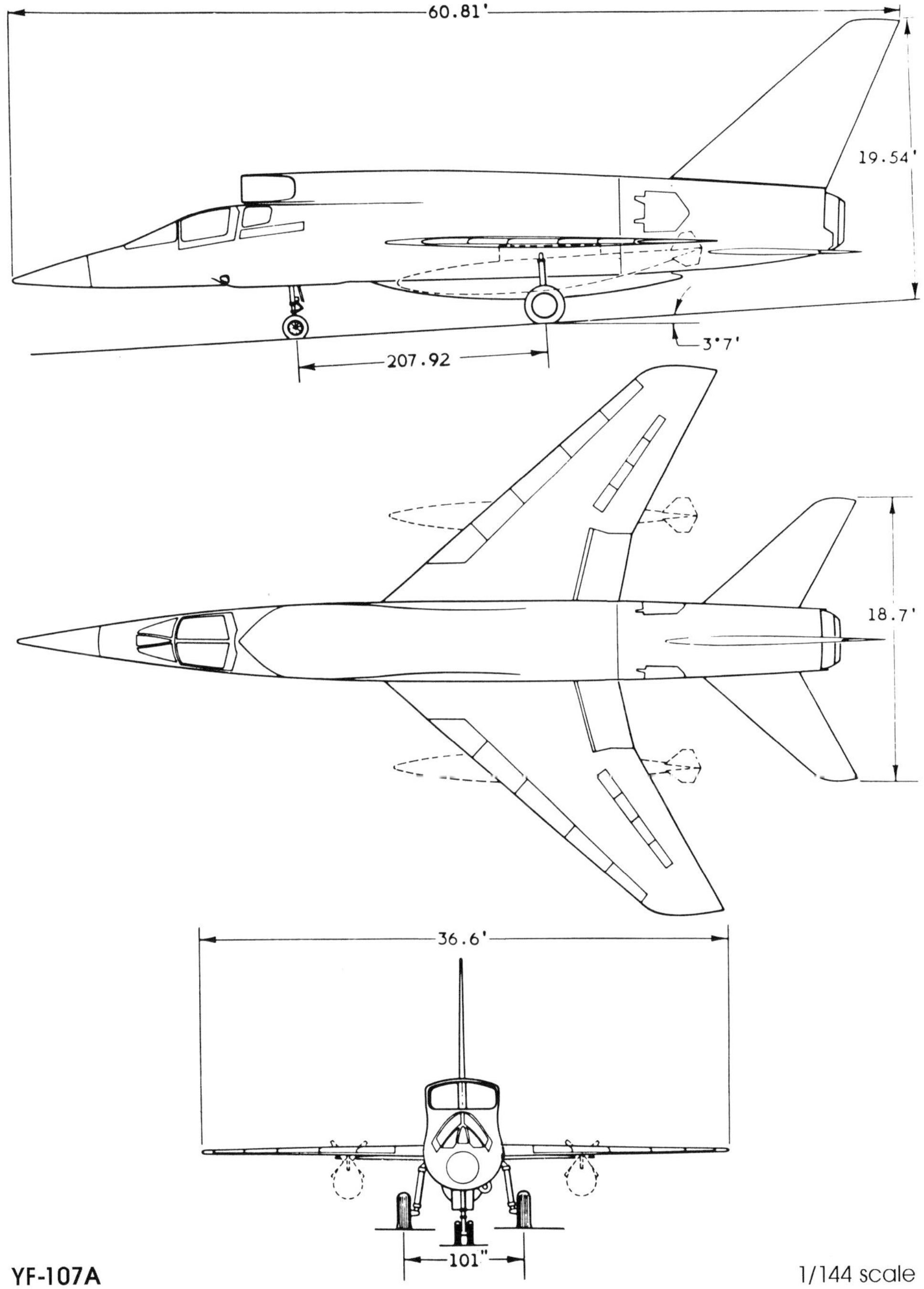

YF-107A

1/144 scale

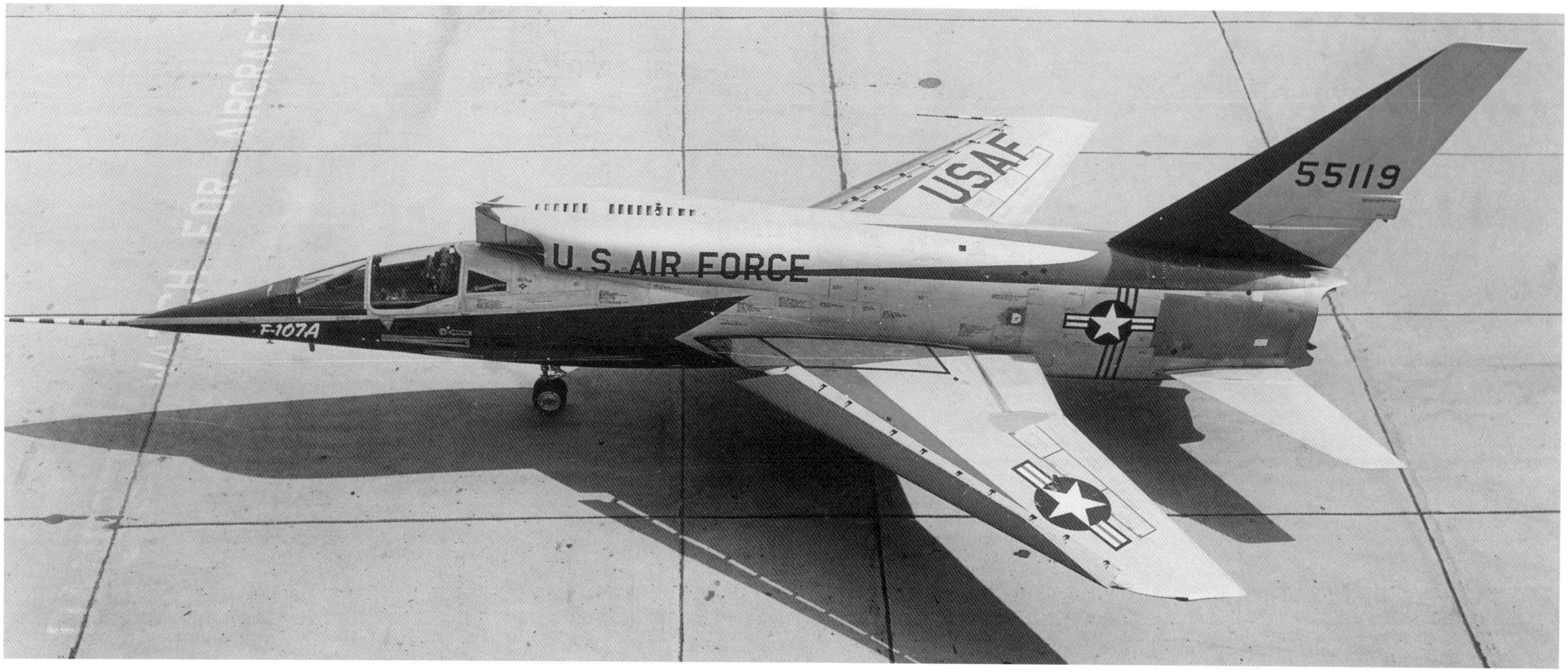

Above, the second YF-107A, 55-5119, was essentially similar to the first. Below left, the wing spoilers are shown in their open position, and right, the all-moving horizontal stabilizer in full up position. On the opposite page, the third prototype, 55-5120, proves itself as agile as an F-100. Two of the YF-107As were used by NASA until 1959 to test side-stick controls for the X-15.

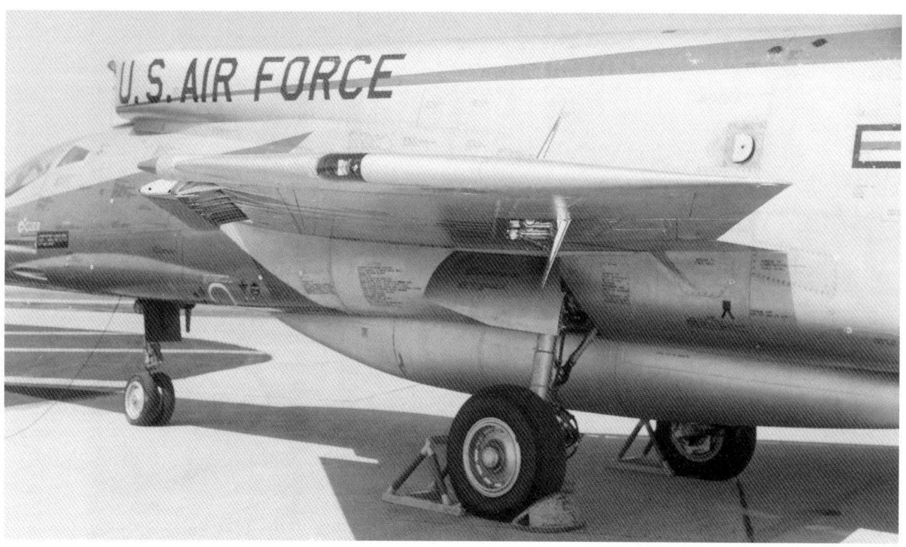

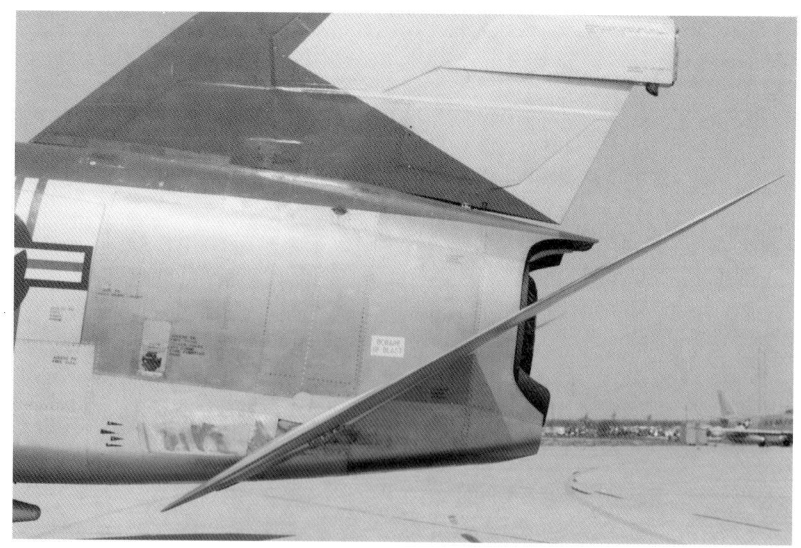

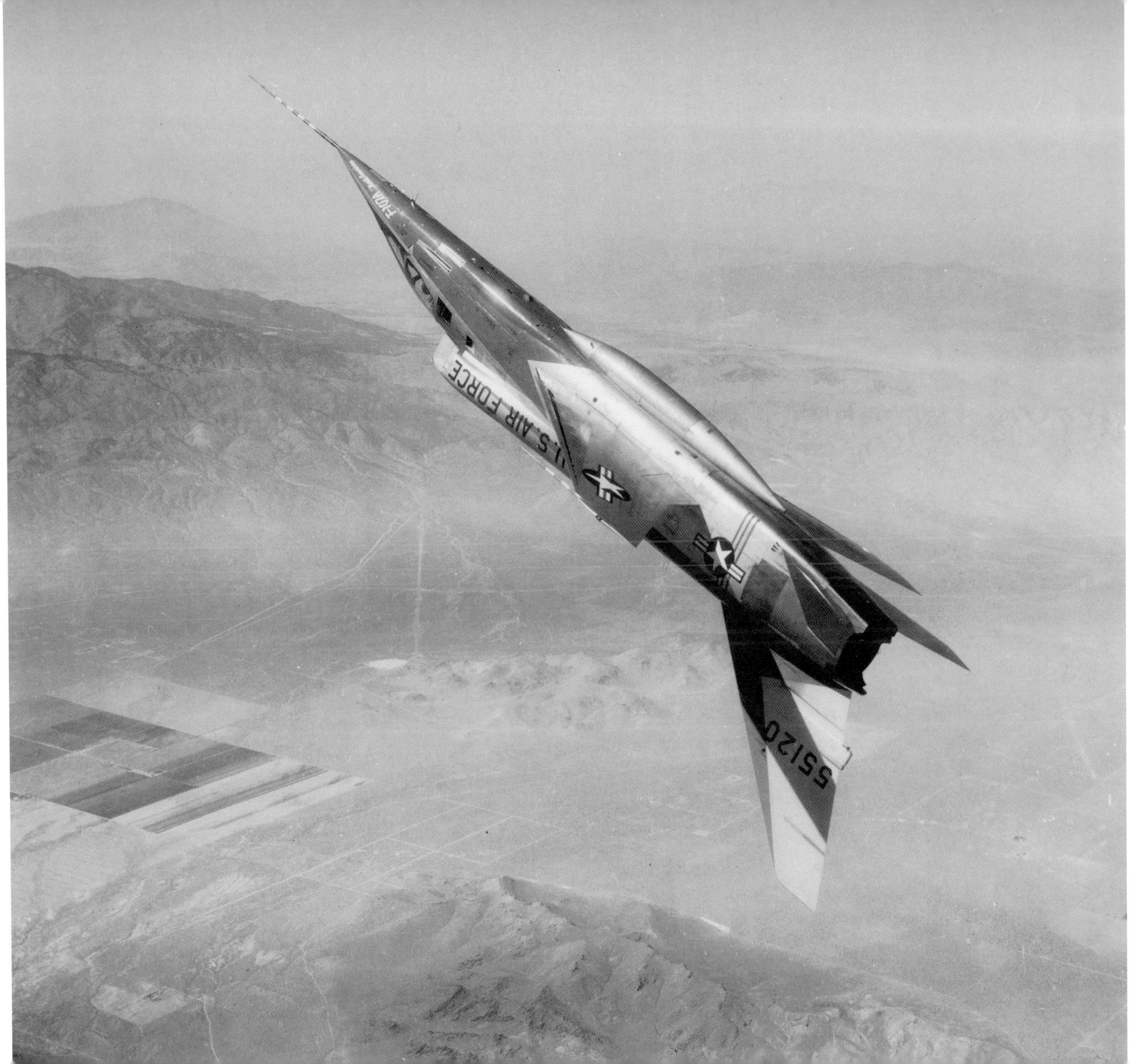
U.S. AIR FORCE
55120

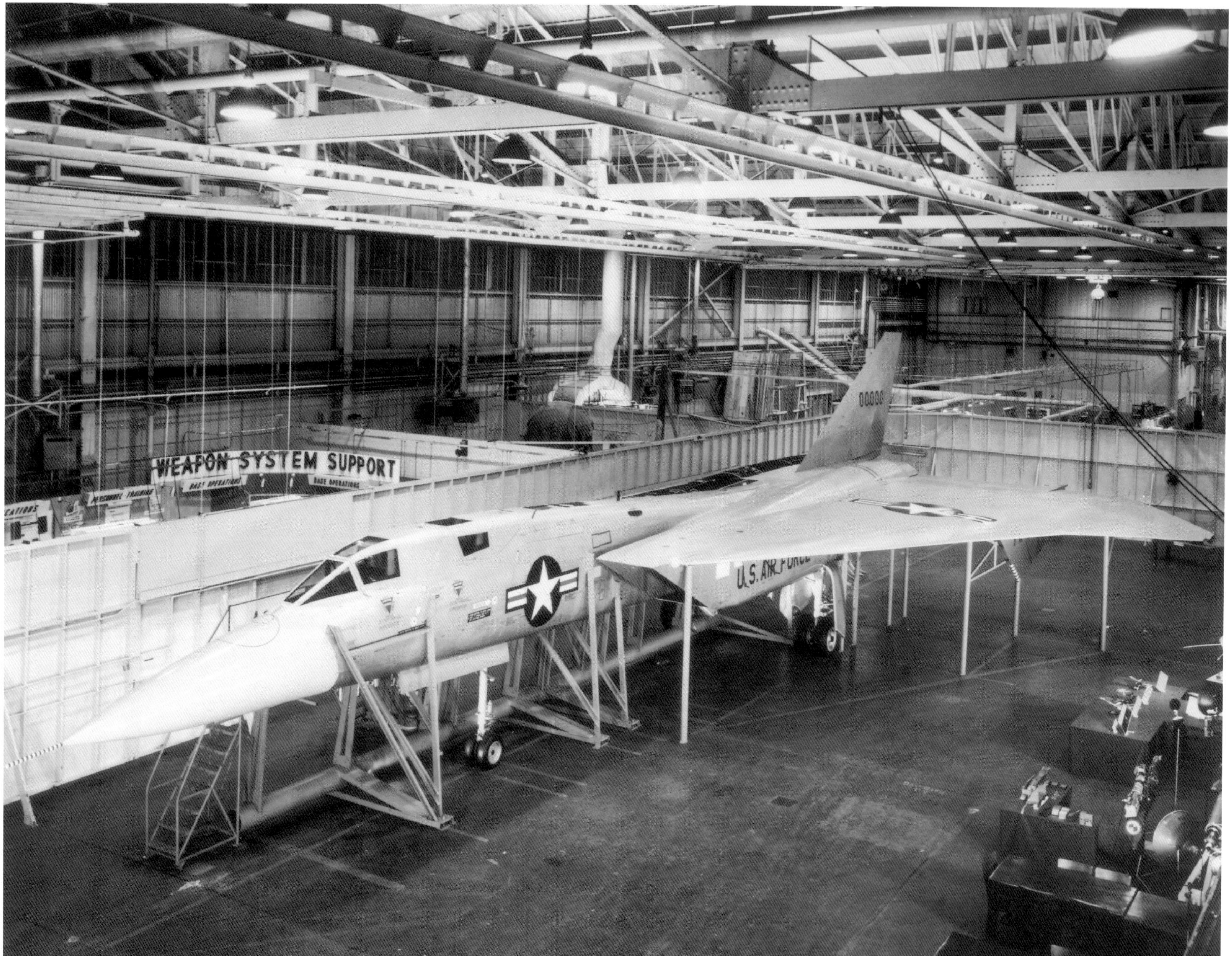
WEAPON SYSTEM SUPPORT
BASE OPERATIONS
BASE OPERATIONS
00000
U.S. AIR FORCE

NA-257 / F-108 Rapier

IN 1957 NAA received a contract to build two prototypes for a proposed Mach 3 long-range interceptor to be designated F-108. The proposed two-man fighter was intended to escort the Strategic Air Command's B-70 Valkyrie intercontinental bomber, of which NAA would also build two prototypes. The F-108 was in some respects a scaled-down B-70, having a large delta wing and long, protruding nose. Power was to be provided by two General Electric YJ93 engines (of which six were used in the B-70).

By 1960 the Air Force had decided that Intercontinental Ballistic Missiles (ICBMs) were a more cost-effective method of delivering nuclear weapons and certainly less risky to aircrews, as the Soviet Union had recently made tremendous advances in its air defense system. This left the B-70 without a mission; both the bomber and the F-108 were subsequently canceled. Although the two XB-70A prototypes would be built, the F-108 got only as far as the full-scale mock-up stage, leaving the three YF-107As as the last fighters actually built by North American.

Specifications: North American NA-257 / F-108 Rapier

Description: U.S. Air Force long-range interceptor. Twin-jet mid-wing monoplane with retractable landing gear. Crew 2 (pilot and fire-control officer).

Engines: Two 29,300-lb-thrust General Electric J93-GE-3AR turbojets with afterburner.

Dimensions: Span 57 feet 2.5 in, length 89 feet 2.5 in, height 22 feet 1.2 in, wing area 1865 sq.ft.

Weights: 50,907 lb empty, 76,118 lb combat, 102,533 lb maximum take-off. Fuel 7109 gal.

Armament: Three AIM-47 Falcon air-to-air missiles.

Performance: (estimated): Maximum speed 1980 mph at 76,500 ft, climb 39,000 ft/min, service ceiling 77,400 ft, range 1150 miles, ferry range 2488 miles.

Production: North American NA-257 / XF-108

Engineering and full-scale mock-up only. --

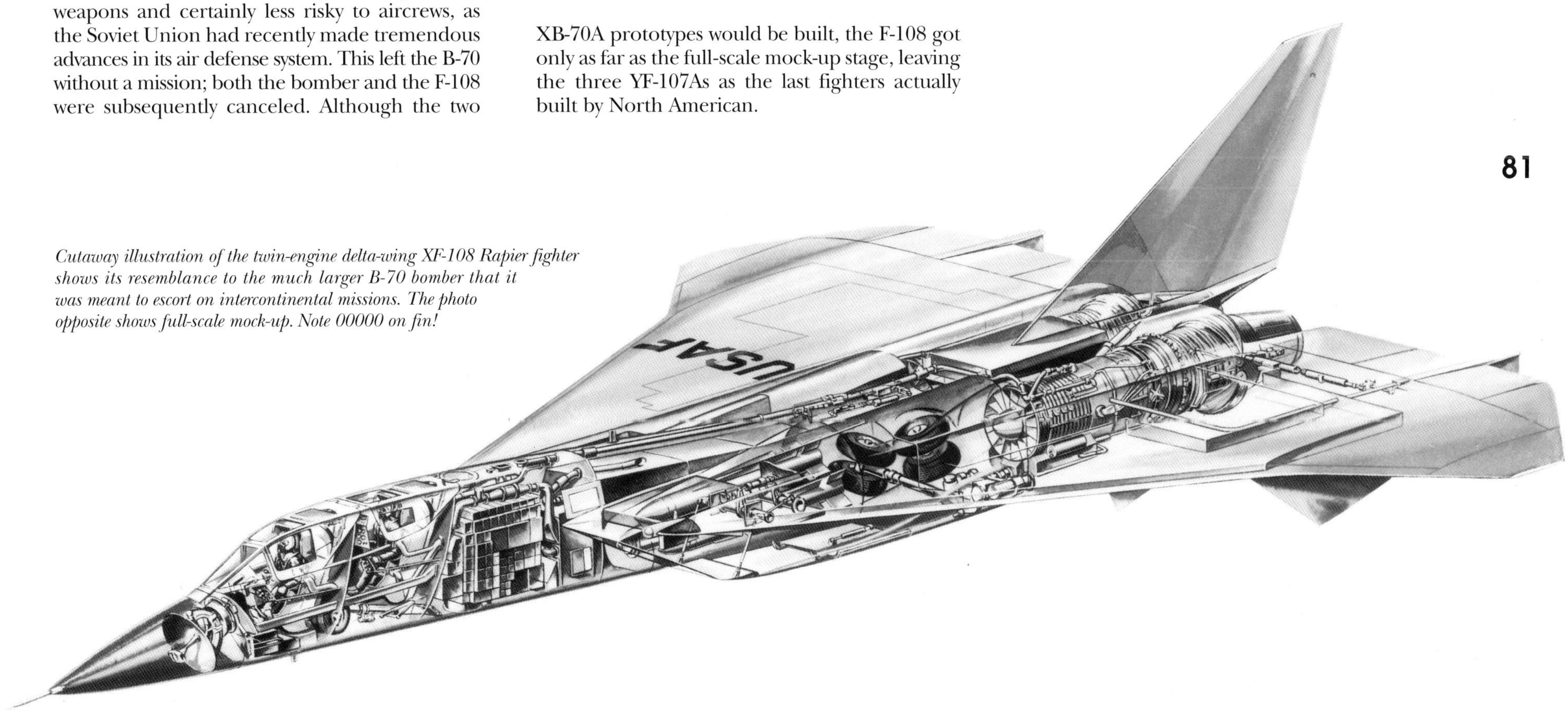

Cutaway illustration of the twin-engine delta-wing XF-108 Rapier fighter shows its resemblance to the much larger B-70 bomber that it was meant to escort on intercontinental missions. The photo opposite shows full-scale mock-up. Note 00000 on fin!

NORTH AMERICAN NA-233>NA-316 A3J-1/-2 A-5A/-5B/RA-5C Vigilante

NORTH AMERICAN AVIATION made a corporate decision late in 1952 to transfer the responsibility for acquisition of Navy-let programs from the Los Angeles Division to the Columbus Division. George Gehrkens, chief engineer at Columbus, requested Raymond Rice, the LAD chief engineer, to provide preliminary design and aerodynamics personnel to support the operational change. Early in 1953 Frank Compton assumed the title of weapons system advanced design manager and Mac Blair became the manager of advanced aerodynamics.

Technical teams were assembled from the ranks of the two divisions to take on forthcoming projects, including a request for proposal for a carrier-based advanced trainer, a study for the replacement of the F3H carrier-based fighter, and technical input to the Bureau of Aeronautics for an all-weather low-altitude attack aircraft system. Over the next four years the Gehrkens/Compton/Blair team pursued the three projects strenuously, their efforts culminating in contracts for the T2J/T-2 trainer, the FJ-4 fighter (described in Volume 1) and the NAGPAW (North American General Purpose Attack Weapon) A3J all-weather attack system.

Preliminary design was well underway on the weapons system by late 1953 and a design was presented to BuAer in mid-1954. NAA's proprietary concept led to a NAGPAW unsolicited proposal in 1955. Following a series of design iterations, including a major change to a two-place high-altitude penetration configuration, the Navy negotiated a contract with NAA for two

The first YA3J-1 Vigilante, BuAer 145157, was completed at Columbus in May 1958. Following its current practice, North American painted the name big on the nose for publicity photos.

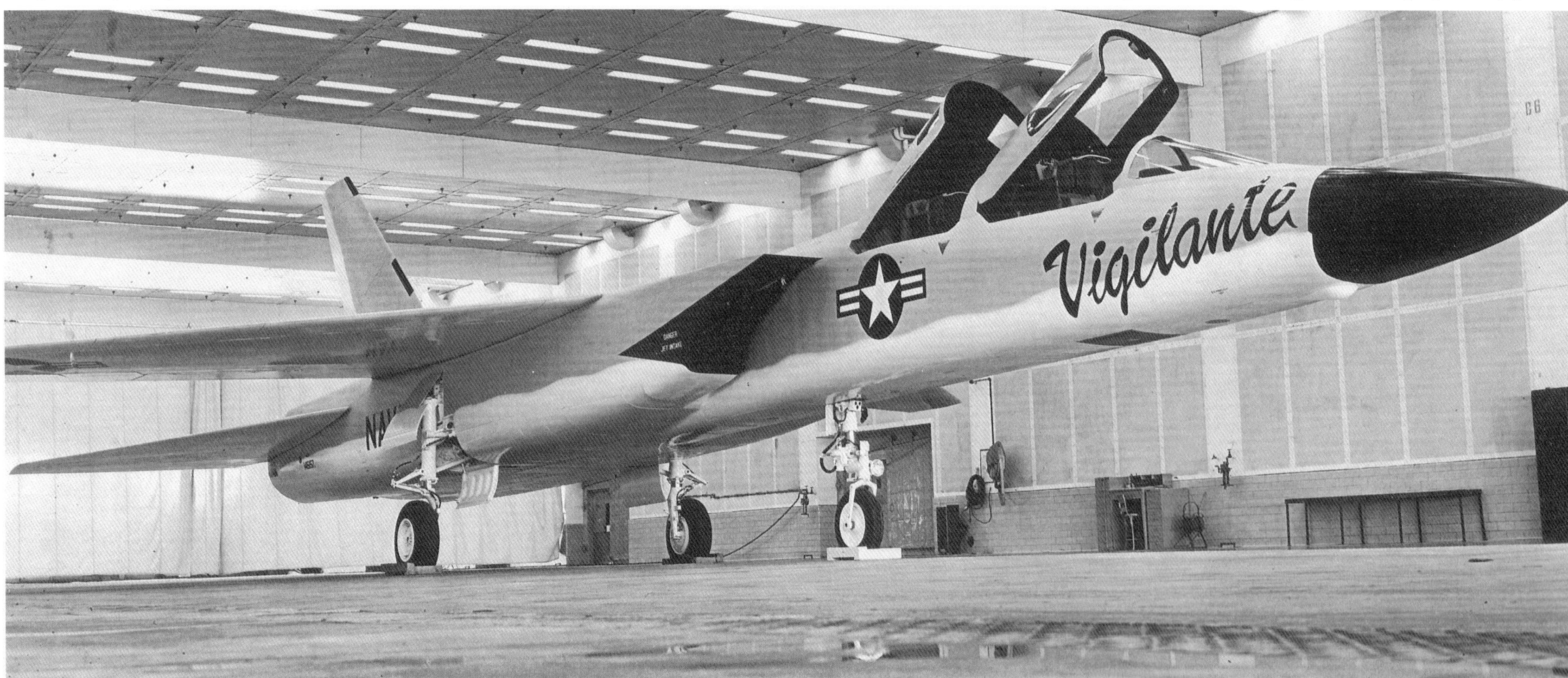

YA3J-1 prototypes. John Fosness was appointed A3J program manager and Reggi Clark, vice president of manufacturing, began preparations for construction of the prototypes.

The NAGPAW/A3J design was a major leap forward in technology, including the following firsts:

1. Ku-band multi-mode monopulse radar with terrain avoidance.

2. Airborne digital computer for navigation-bombing calculations.

3. Nav/Bomb system with inertial auto-navigator coupled to the radar, and television sight for check point and initial point acquisition.

4. Fully integrated autopilot and air data system for weapon release solution.

5. Production use of heads-up display.

6. Production use of fly-by-wire control system.

7. Production use of aluminium lithium alloy in aircraft construction.

8. Variable inlet using horizontal ramp geometry.

9. Slot-deflector spoilers for roll control (a joint development with LAD's F-107 program)

10. Application of H-11 hot work tool steel in fuselage machined frames.

11. STRUX high-speed steel for landing gear struts.

Other new features included a highly loaded swept wing with blown flaps, a system using engine air bleed to the flap upper surface to increase the lifting capability of the wing. The wing also featured a leading-edge flap arrangement, or droop section, for increased lift and ease of handling at low speeds. For carrier storage, the wingtips folded upward and internal fuel was carried in the wing main box all the way to the hinged portion. Roll control was effected by three sections of spoilers and deflectors on each wing, together with differential movement of the tailplanes (stabilators). The two inboard spoilers were conventional, deflecting upward, and the outboard was inverted, deflecting downward. The vertical stabilizer, originally designed as a twin-tail arrangement, was later changed to a single, all-moving surface with a folding tip for carrier hangar clearance.

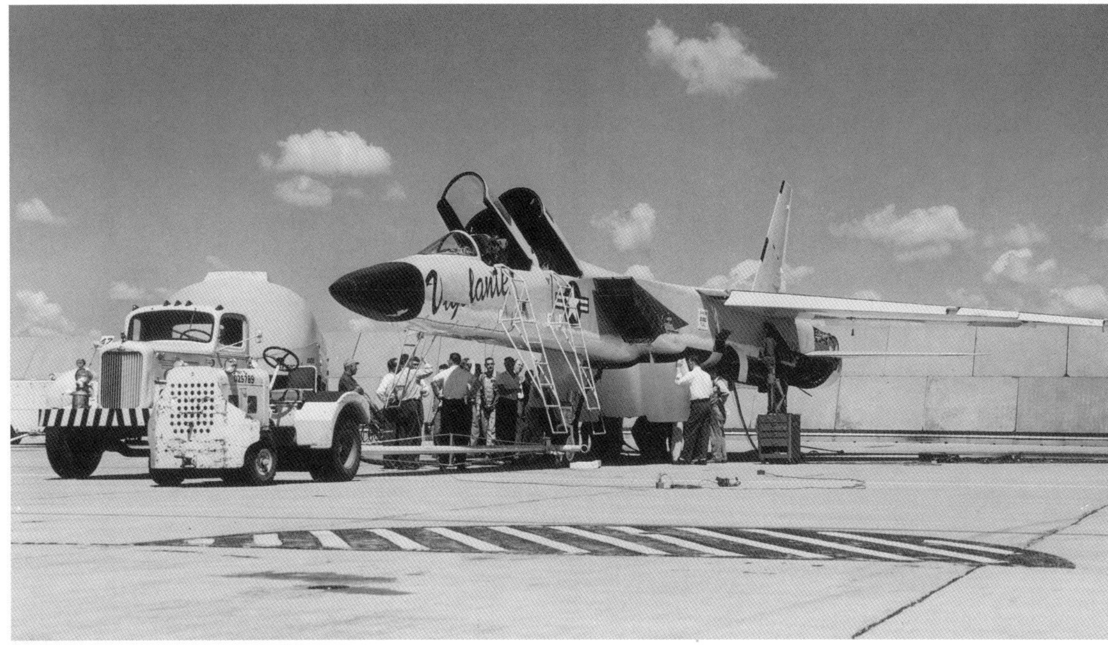

The Vigilante made its first flight on August 31, 1958. For a carrier-borne attack plane, it was huge–more than 76 feet long.

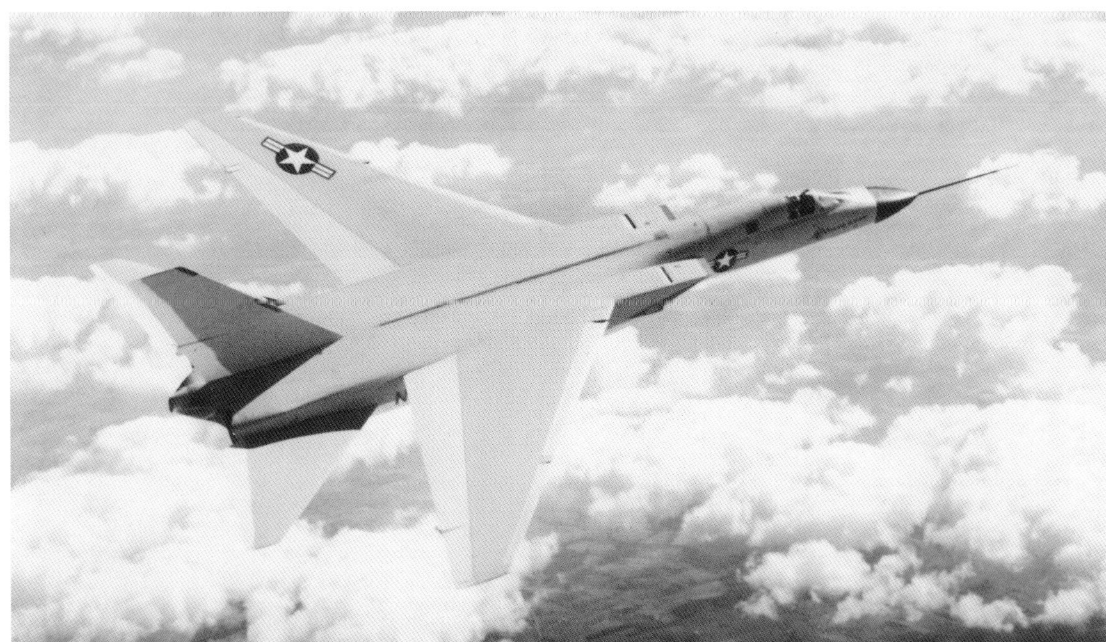

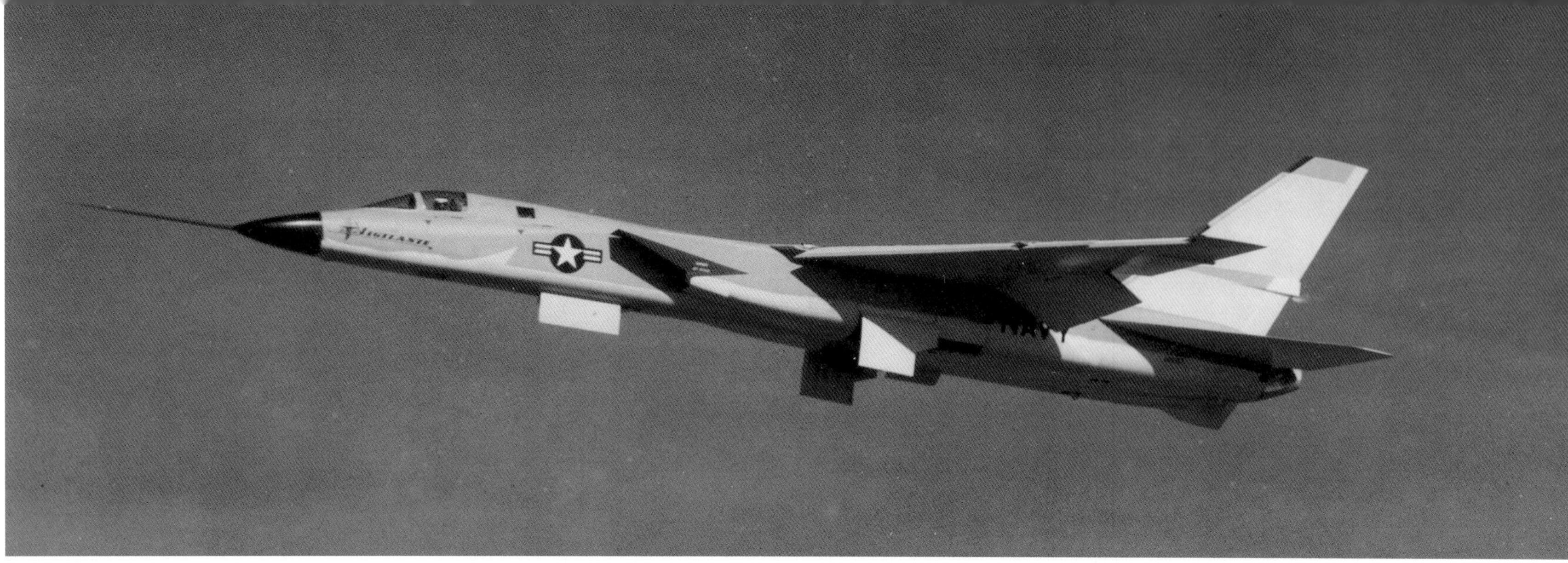

Above, YA3J-1 gear is up but doors are still open. Below, the same plane repainted after modifications; Navy weapons testing began at Palmdale and China Lake in California in October 1958.

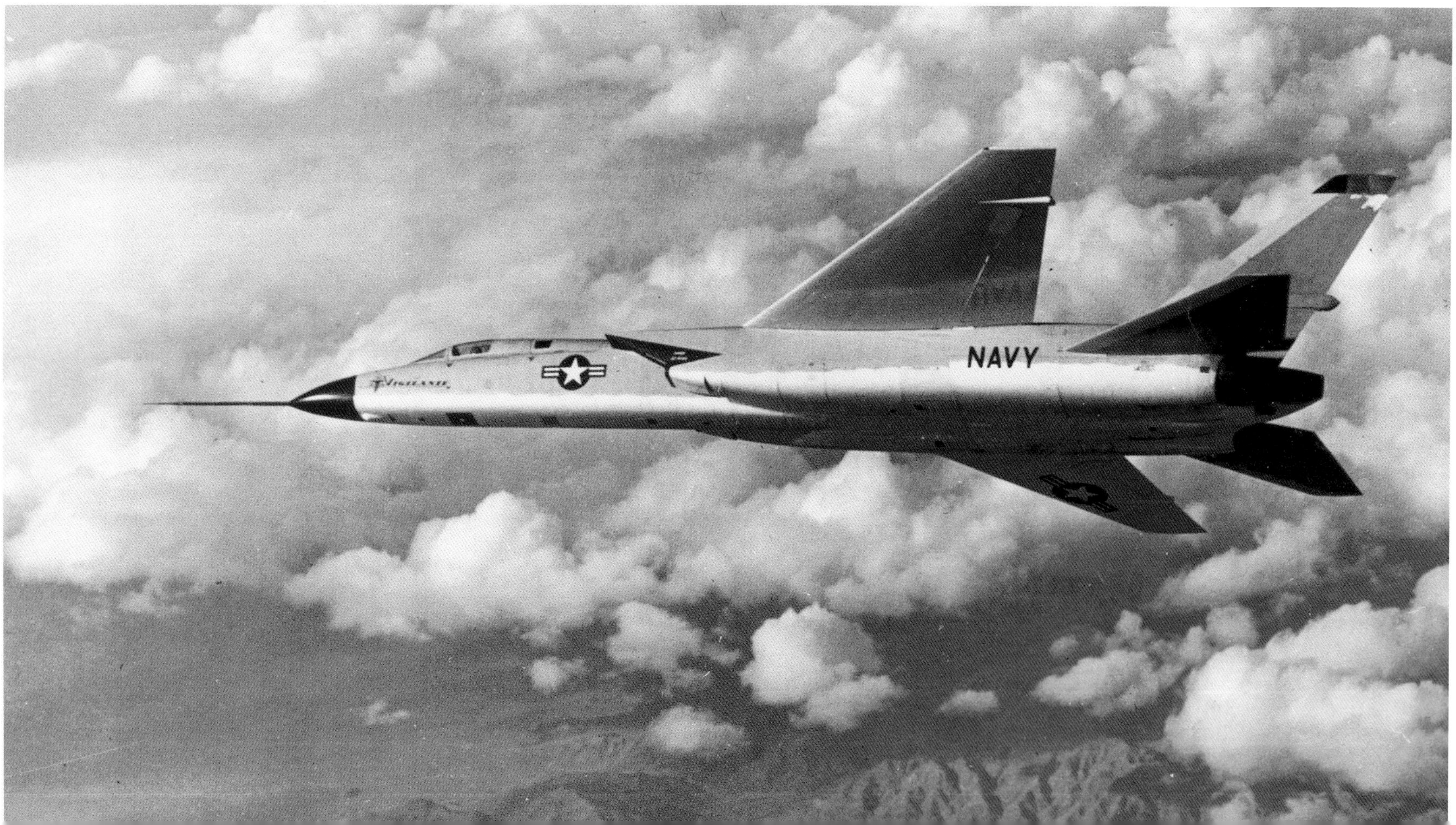

The variable inlet system adjusted both profile and area to suit the engine requirements at different flight Mach numbers. Power was provided by two General Electric YJ79-GE-2 turbojet engines of 15,150-pound-thrust with afterburner.

Although the Navy originally had specified the ability to attack below the level of enemy radar detection, the aircraft had a minimal radar cross-section, the slim fuselage having minimum drag at high subsonic and supersonic Mach numbers, particularly for a bomber. An advanced ASB-12 radar-equipped inertial navigation system (REINS), in conjunction with a versatile digital analyzer (VERDAN), was developed by NAA's Autonetics division and featured the very first use of a digital computer in an aircraft. The YA3J-1 also pioneered the electric fly-by-wire control system with a mechanical backup.

Within the lower fuselage and between the two engines was an internal bay designed to house a Mk 28 nuclear weapon. Activated by a solid-fuel cartridge catapult system, the bomb, integrated with two jettisonable fuel storage tanks, was to be released by rearward ejection through an opening between the two engines. This novel weapons-delivery system was designed to overcome the difficulties of weapons release at twice the speed of sound.

Due to the heat generated by Mach 2 flight, much of the internal structure was made of titanium. The engine bays were sprayed with a gold coating, and a power system using pure nitrogen instead of hydraulics was planned but later dropped in favor of a conventional hydraulic system. A liquid oxygen system was developed for the two-man crew, two converter bottles providing a supply time of 8 hours. The in-flight refueling probe retracted into the forward fuselage.

The roll-out of the first YA3J-1, BuAer 145157, took place on May 16, 1958, and during that ceremony the aircraft was named Vigilante. Dick Wenzell, chief test pilot of the Columbus Division, made the first flight on August 31, 1958. Other than a few minor systems problems,

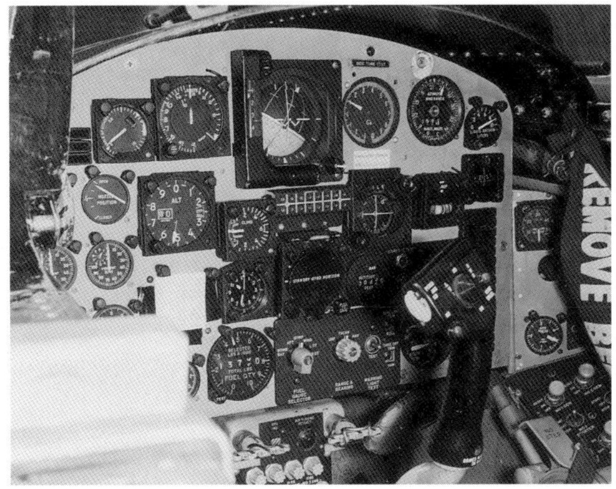

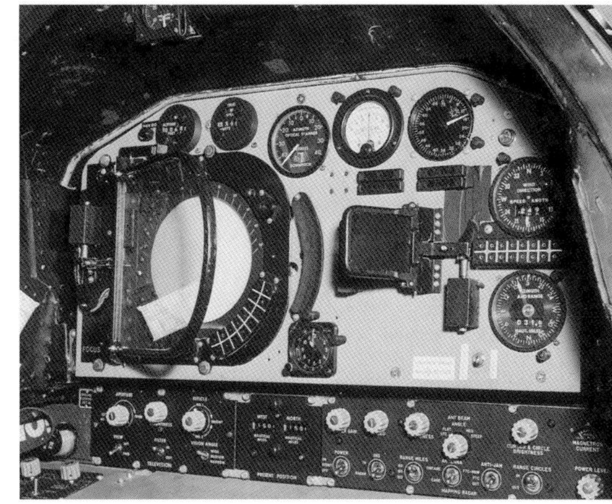

Above, the instrument panels of the fourth production A3J-1, 146697: the pilot's at left and the navigator's at right. Below, the third A3J-1, 146696, looking almost motionless in low-speed flight, with all wing surfaces open and landing gear extended.

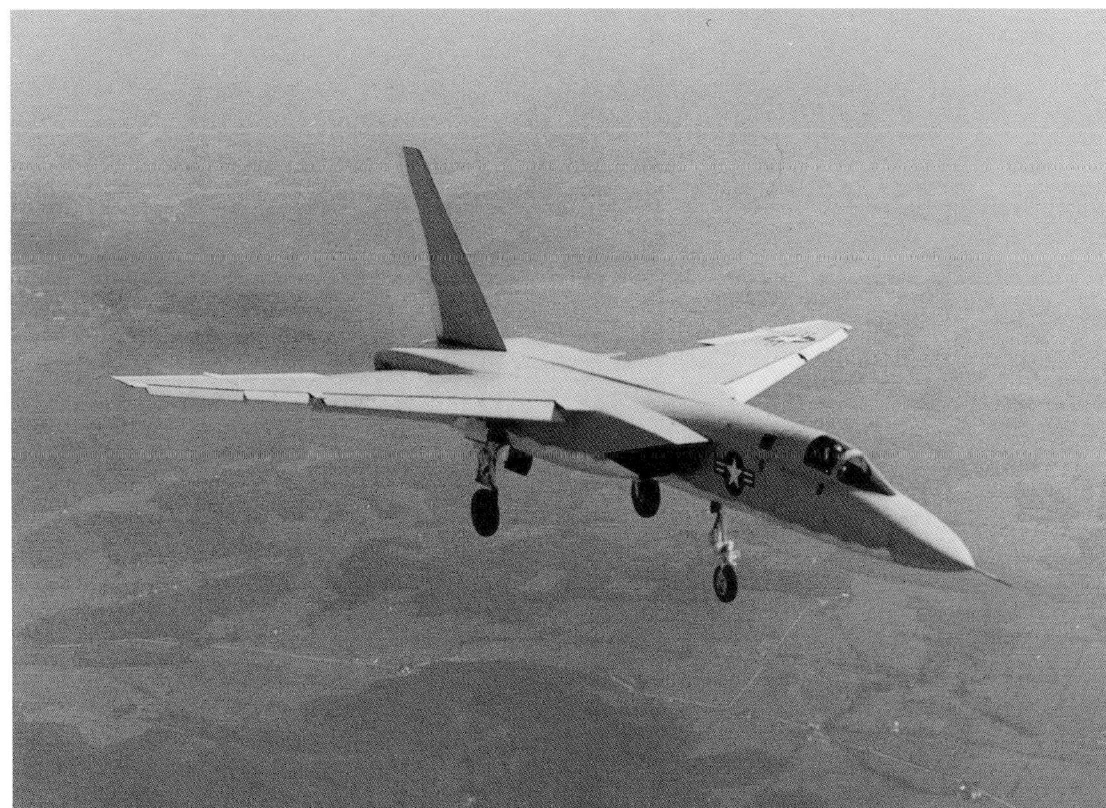

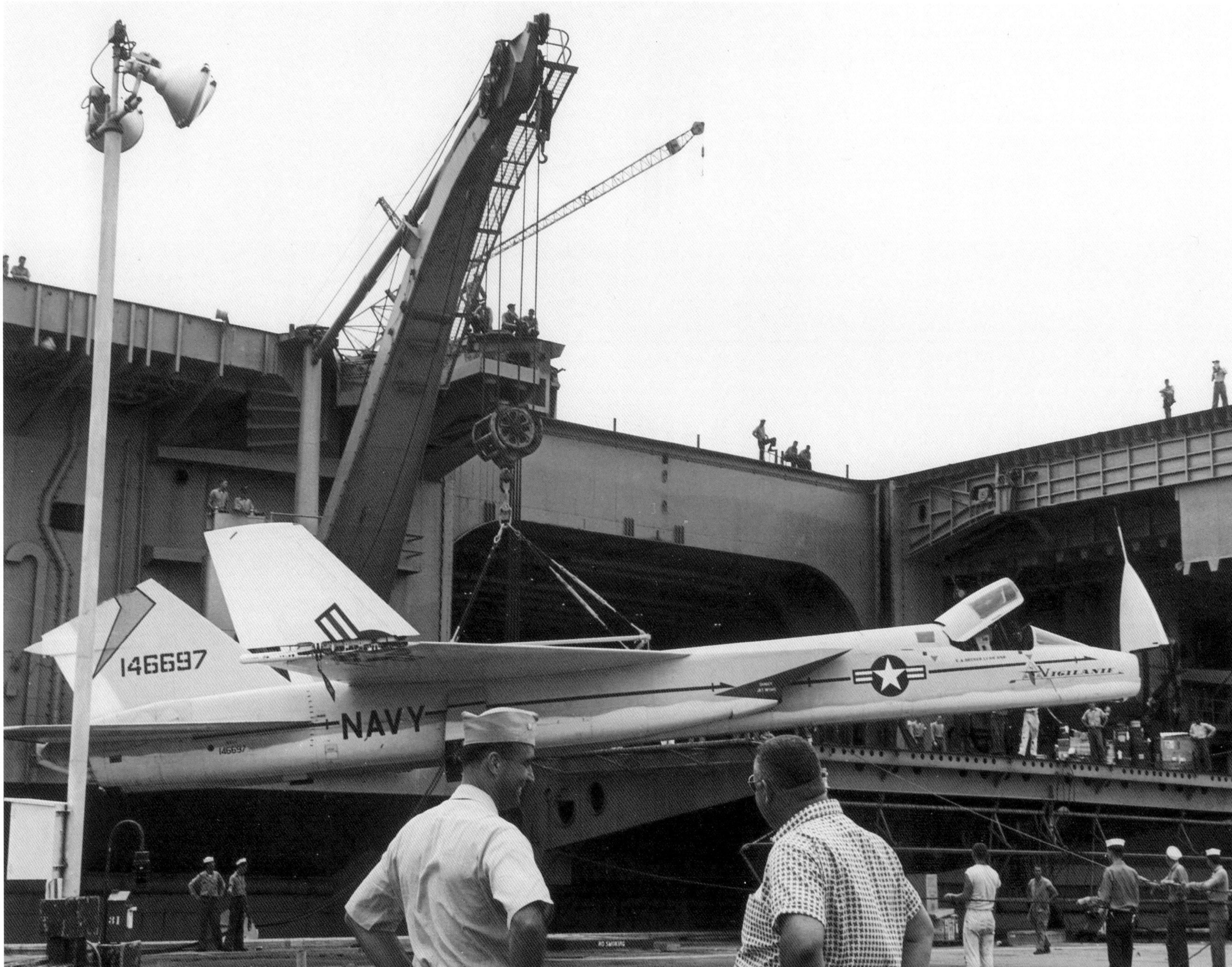
146697
NAVY
146697

to be expected with an aircraft of such complexity, the initial flight test program was very successful. The reliability of the REINS system was questionable in early tests, but as development progressed the system improved dramatically.

Dummy weapon drop-tests, begun in October at the Naval Weapons Test Center, China Lake, presented a variety of problems. The original plan was that the fuel tanks, attached to the Mk 28 weapon in a tandem arrangement, would be emptied during the early part of the mission. At the target release point the bombing system would jettison the package and the empty tanks would stabilize the released weapon. The system was not always predictable, and as used in conjunction with the REINS and VERDAN systems, initially had a mean time between failures (MTBF) of only 15 minutes. By 1961, the MTBF had been brought up to 240 hours, and would ultimately peak at 500 hours in service. The original design called for no external weapons to be carried, but production A3J-1s would have two under-wing pylons for weapons or additional fuel tanks. The A3J-1 and subsequent Vigilante variants could be equipped with a "buddy pack" reel-and-drogue that allowed the aircraft to refuel another A3J.

The second prototype was flying by mid-1959 and the Navy began carrier evaluations aboard the USS *Saratoga* in July 1960. On December 13, 1960, the plane made a zoom-climb flight from Palmdale, California, carrying a 1000-kilogram (2200-pound) payload to 91,451 feet, setting a new world record. The pilot was Commander LeRoy Heath and the navigator Lieutenant Larry Monroe.

Operational deployment of the A3J-1 began in June 1961 with four aircraft delivered to VAH-3 (Heavy Attack Squadron Three), a training squadron of Heavy Attack Wing One, based at NAS Sanford, Florida. The first operational unit to receive the Vigilante was VAH-7, commissioned on January 25, 1962, and later serving aboard the nuclear-powered carrier USS *Enterprise.* Deliveries to VAH-1 and VAH-3 followed, and service aboard *Forrestal*-class carriers and the

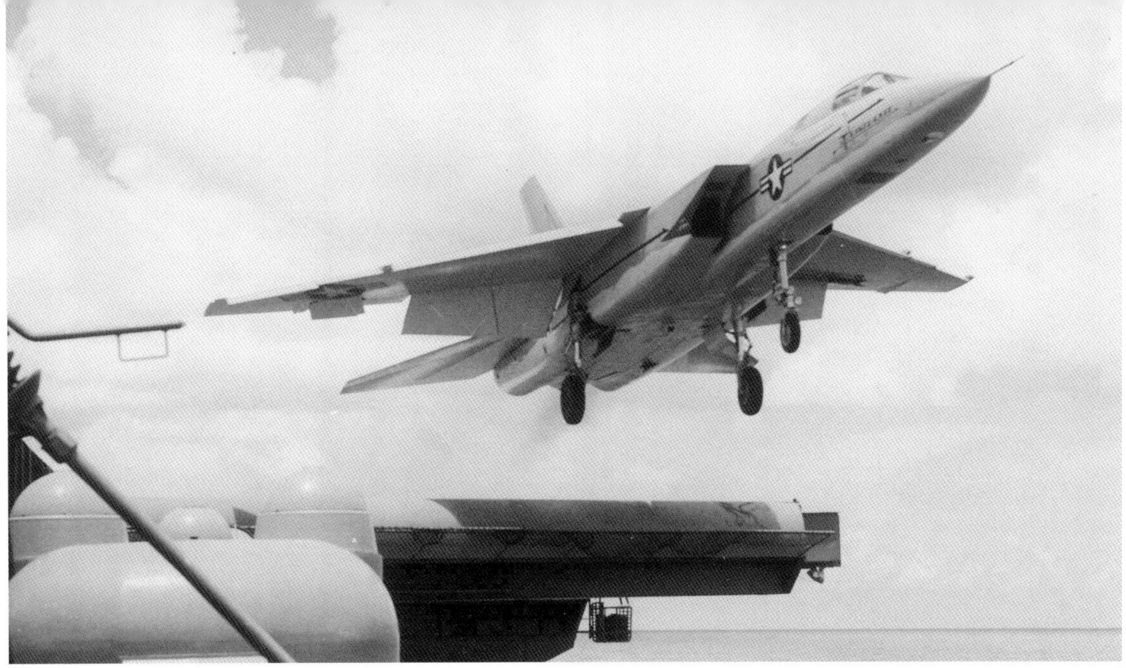

Being hoisted aboard the USS Saratoga, *on opposite page, and catapulted from the deck for the first time, above and on page 88, is A3J-1, BuAer 146697. Lieutenant Commander E.A. Decker was the pilot for the qualifications that began in July 1960.*

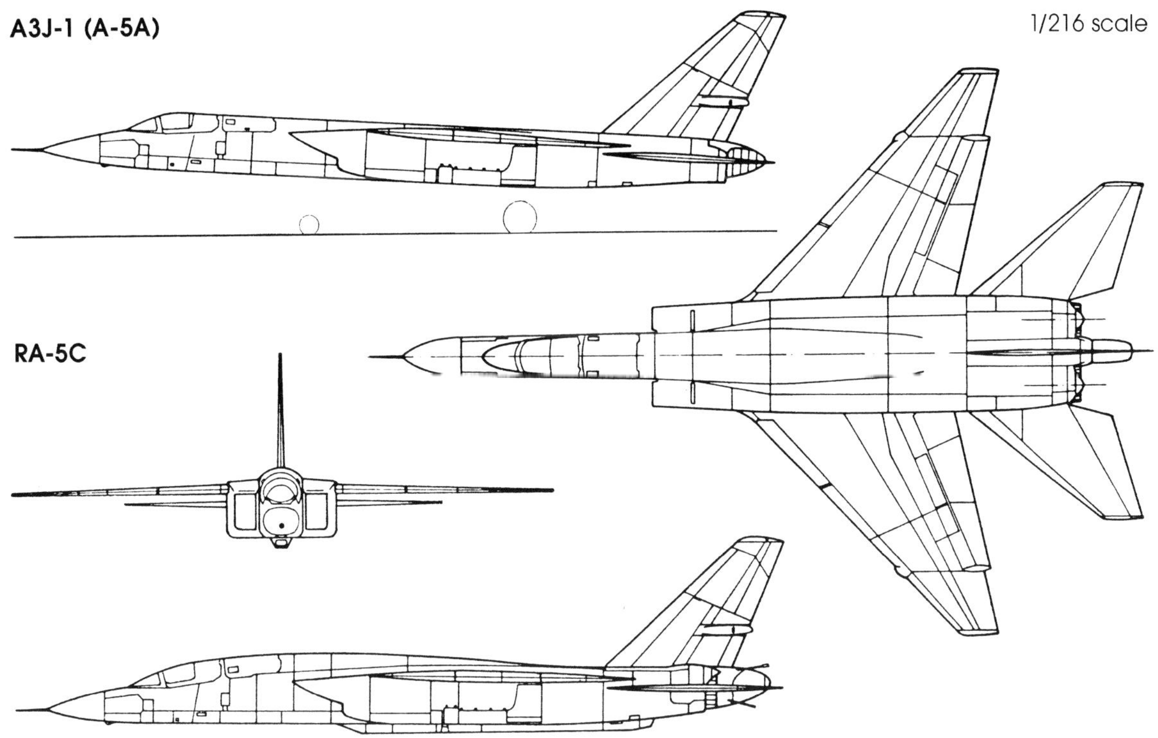

88

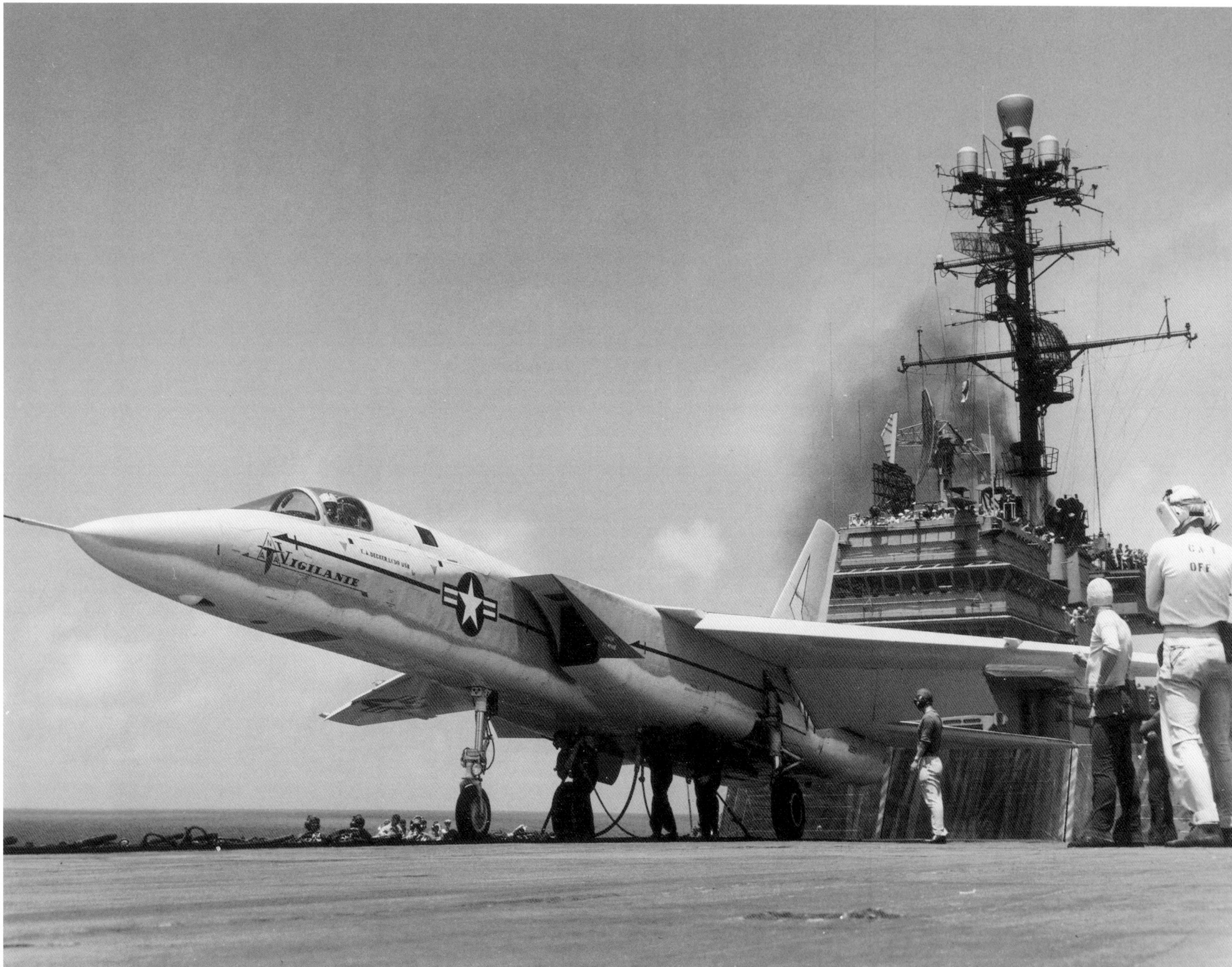

Enterprise began with a strength of twelve aircraft per ship. A total of 57 A3J-1s were built.

As the A3J-1s went into service, the engineers at Columbus made studies for a revised Vigilante. The availability of the more powerful J79-GE-8 engine, along with other improvements, allowed a 60-percent increase in overloaded gross weight (80,000 pounds). Changes to the configuration of the wing included the introduction of "blowing" to the leading-edge droop-section, deleting the blown flaps and increasing flap chord and span. With the additional weight, the fuselage was deepened from the cockpit aft to increase the capacity of the two center fuel tanks and allow for a new tank to be fitted over the wing center section. These changes increased the capacity by 500 gallons.

Other changes included the addition of two more wing pylons, improvements in braking capacity and an increase in engine inlet-duct capture area to improve high-altitude performance. These revised NA-269 aircraft were designated A3J-2, and the last eighteen A3J-1s were also converted to this specification. The prototype A3J-2 first flew on April 29, 1962.

During 1962 the U.S. Government standardized military aircraft designations by implementing a tri-service system that did away with the traditional U.S. Navy manufacturer's codes. With the new system, Vigilantes would all become A-5s. The A3J-1 was changed to A-5A, and the A3J-2 became the A-5B.

NAA also developed a photo reconnaissance/attack version, and this variant was initially designated A3J-3 (NA-279/NA-283), later changed to YA-5C for prototype, and RA-5C for service versions. Using the same basic airframe as the A-5B, the RA-5C used a long "canoe" fairing under the fuselage to house multiple sensors and optical and electronic equipment. These included APD-7 SLR (side-looking radar), AAS-21 linescan infrared sensors and various electronic countermeasures (ECM) equipment. The RA-5C made its first flight on June 30, 1962, and was put into production during 1963 with wing fillets that improved low-speed handling and the J79-

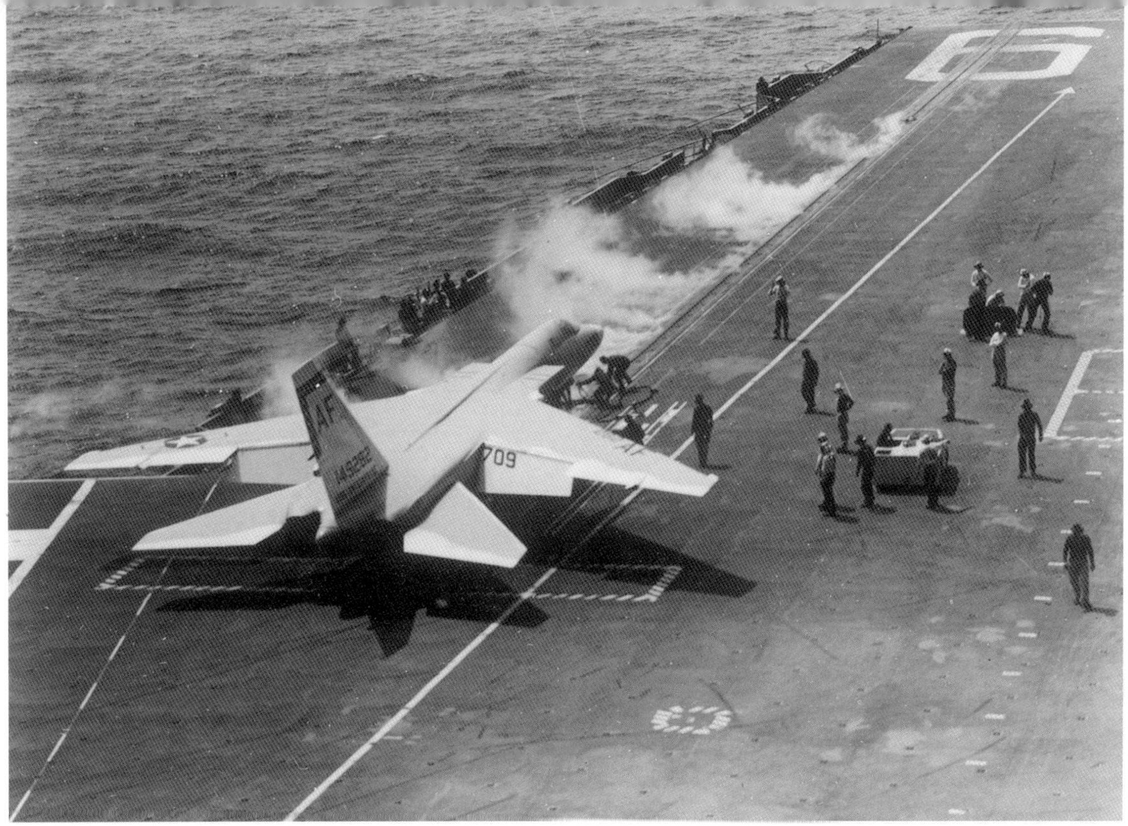

Above, A3J-2, 149282, of VAH-7, about to be catapulted from CVAN-65, the USS Enterprise. *Below, tufting for in-flight aerodynamic tests of A3J-1, 146696, serving as the first prototype for the extensively modified RA-5C reconnaissance/attack version.*

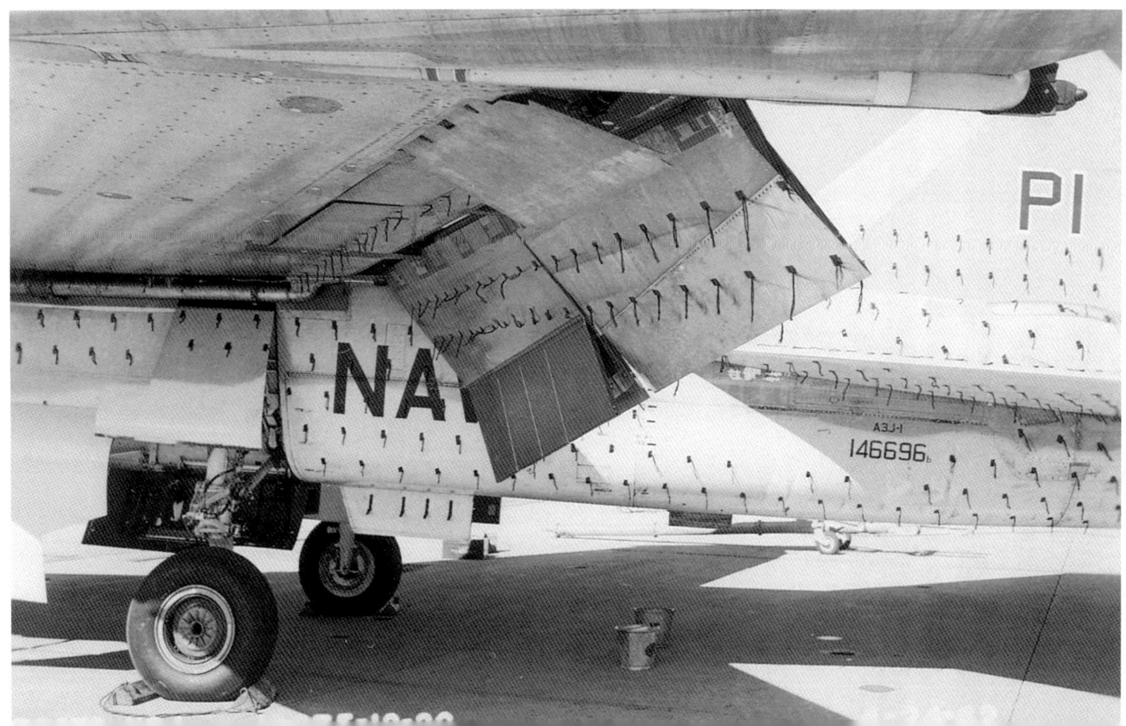

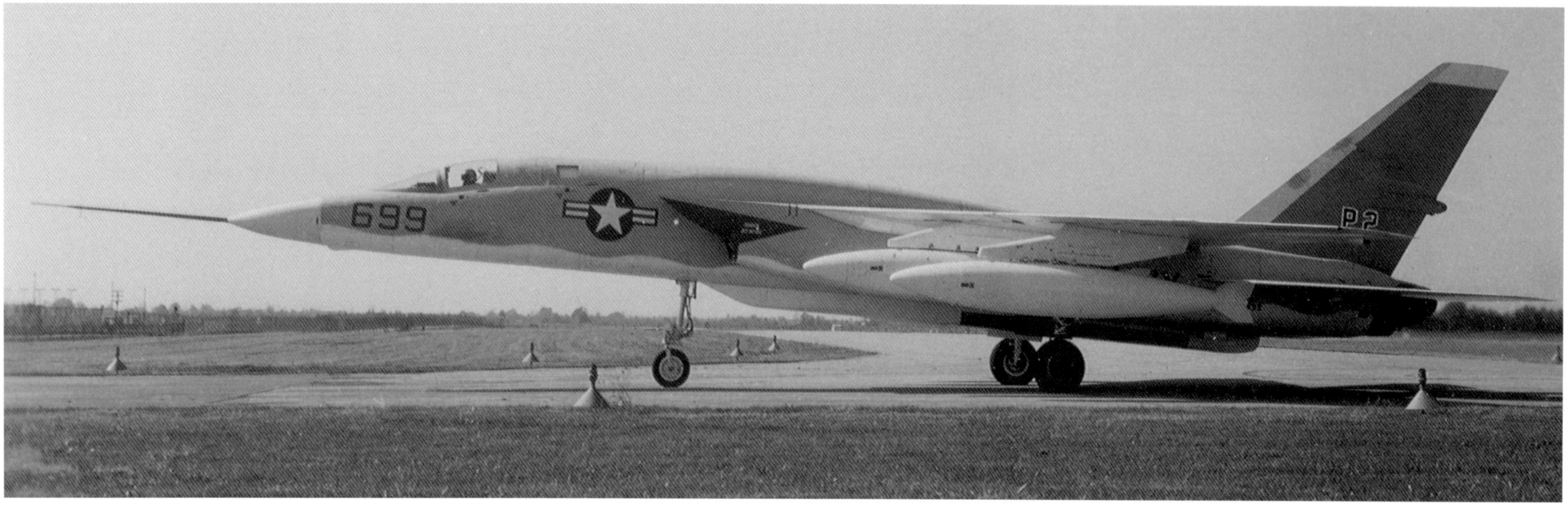
699
P2

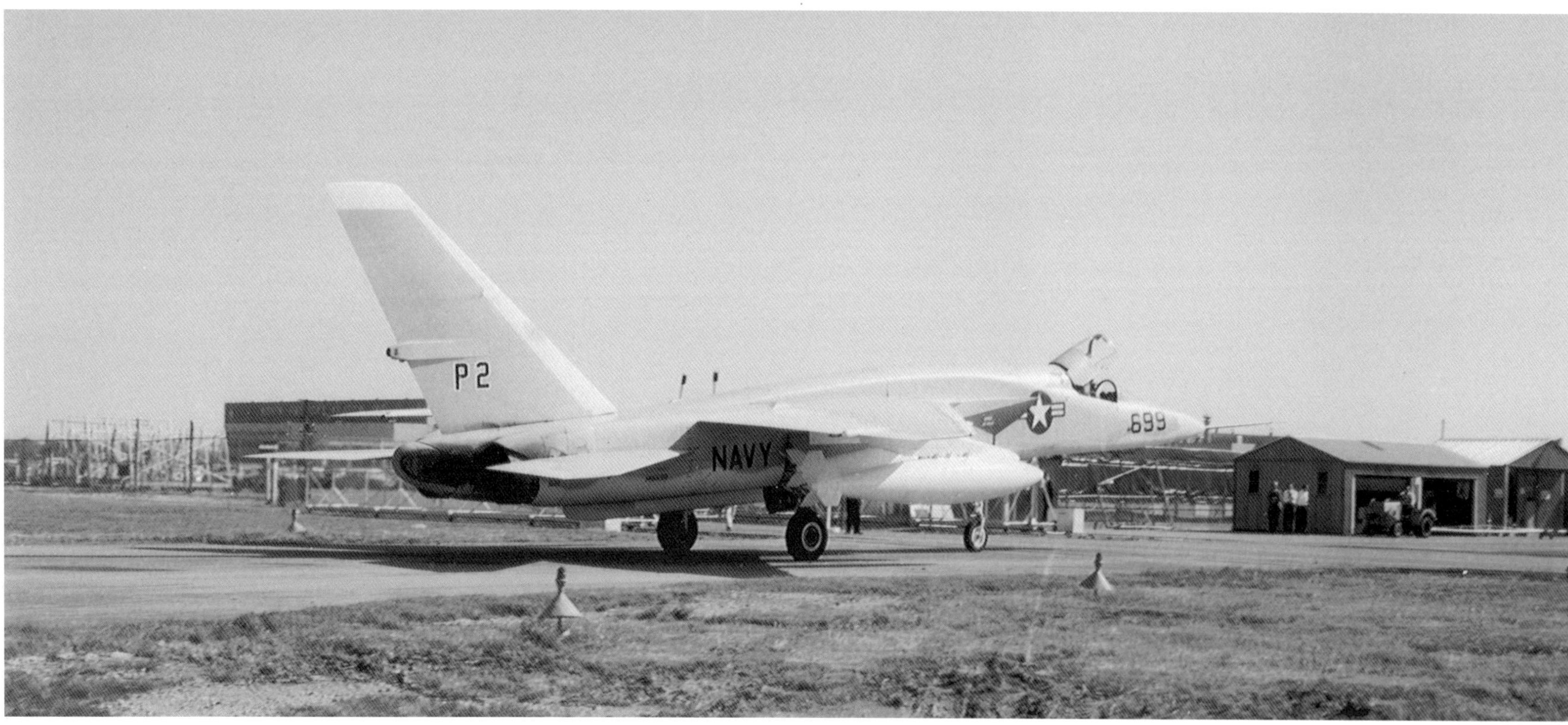
P2
NAVY
699

Opposite page and at right, three views of the second prototype for the RA-5C, originally an A3J-1, 146699. Note the characteristic arch of the fuselage, and the four wing pylons.

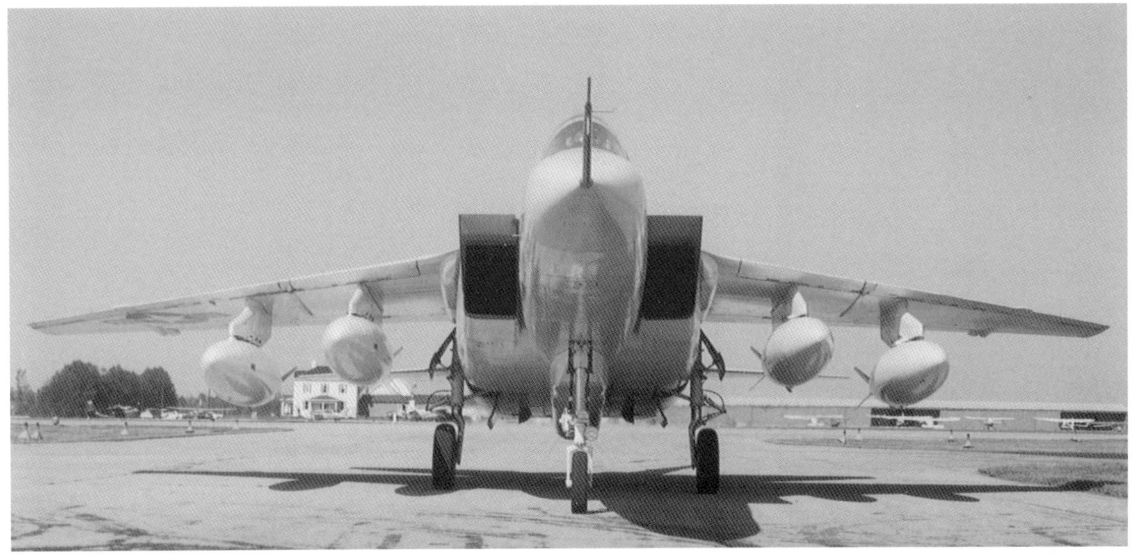

GE-8 engine of 17,900 pounds thrust. Later RA-5Cs had airframe changes allowing either -8 or -10 engines to be installed; the last 36 were manufactured with -10s. Deliveries to RVAH-3, a reconnaissance training squadron, and to nine operational RVAH squadrons began in July 1963.

The RA-5Cs performed duties as the airborne portion of the fleet-wide Integrated Operational Intelligence System (IOIS). Photographic, SLR, IR (Infra-red) and ELINT (Electronic Intelligence) duties were performed with automatic real-time secure feedback to an interpretation center in the carrier. The RA-5C could effectively perform these duties over a 1500-mile radius.

During the Vietnam War, RA-5Cs were a major factor in intelligence gathering and were flown round-the-clock from mid-1964 to the very last months of 1972. RVAH-13 Vigilantes routinely flew missions over the Hanoi-Haiphong area during early deployment, and some were painted in USAF-style camouflage for these duties, instead of the standard gray-white Navy scheme.

Although the A-5A had been used for pilot training before RA-5Cs were delivered, the A-5B never served operationally. The RA-5C was by far the most useful of the Vigilante variants. Approximately sixty aircraft saw duty in South East Asia, and by the end of U.S. involvement eighteen had been lost in combat missions. After the Vigilantes returned home, their days were numbered. Early RA-5Cs were put into storage in 1974 and the last operational squadron stood down in 1979.

The Vigilante was the only Mach 2 strike/reconnaissance aircraft used by the U.S. Navy and was the heaviest aircraft to serve regularly aboard U.S. aircraft carriers. It was advanced for its day and also very innovative; much of its technology set the precedent for military aircraft development during the next two decades.

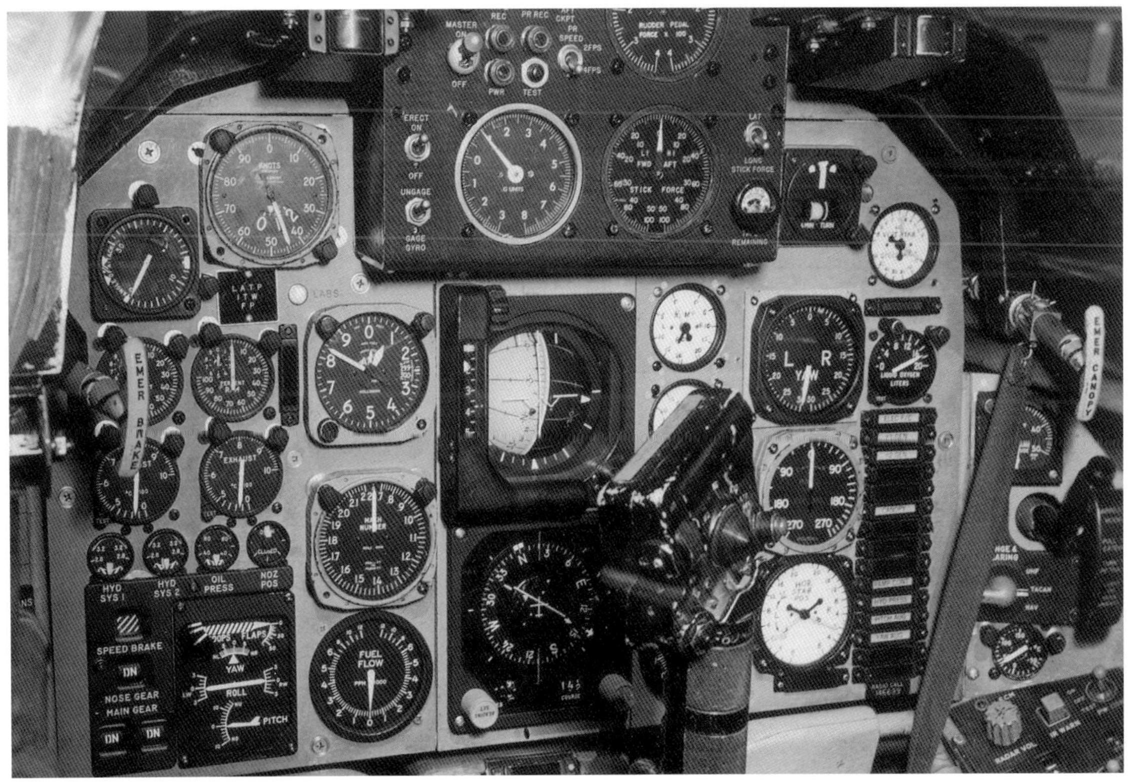

Right, the instrument panel of 146699, as modified to RA-5C configuration. Compare with the A3J-1 panel on page 85.

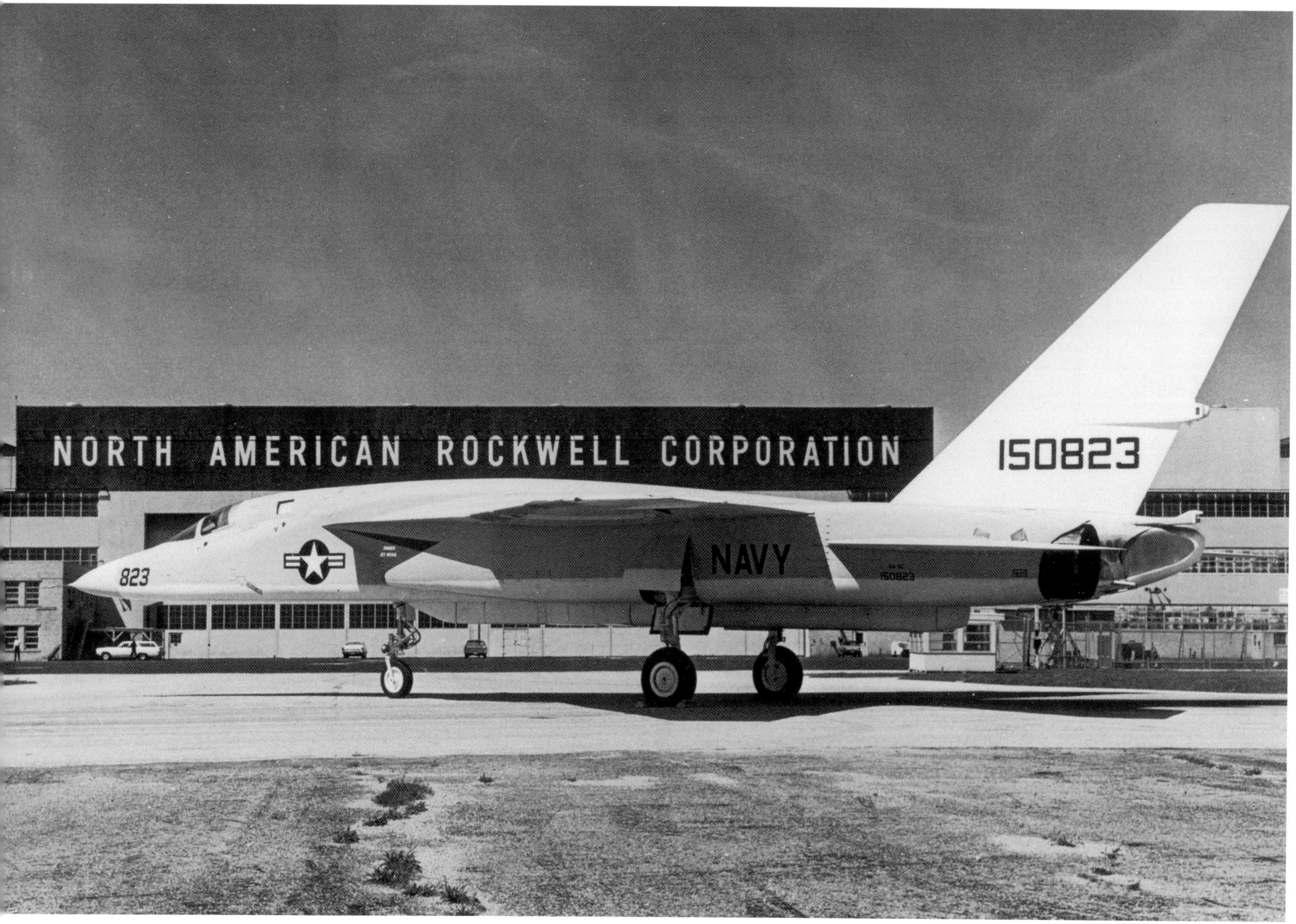
NORTH AMERICAN ROCKWELL CORPORATION
150823
823
NAVY
150823

Specifications: North American NA-233>298 A3J-1/-2 A-5A/A-5B / RA-5C Vigilante

Description: U.S. Navy heavy attack/reconnaissance bomber. Twin-jet high-wing monoplane of all-metal construction with retractable landing gear. Crew 2.

Engines: Two 10,800-lb-thrust General Electric J79-GE-2/-8/-10, 17,900-lb-thrust with afterburner.

Dimensions: Span 53 feet (42 feet 4 in folded), length 76 feet 6 in, height 19 feet 4.75 in, wing area 753.7 sq.ft.

Weights: 37,498 lb empty, 65,589 lb take-off with internal fuel, 79,588 lb maximum take-off. Fuel 3600 gal. plus four 400-gal. tanks on wing hardpoints.

Armament: B27, B28 and B43 nuclear bombs, Bullpup missiles, conventional 500-, 1000- and 2000-lb bombs.

Performance: Cruising speed 1254 mph, maximum 1385 mph, climb 27,900 ft/min, service ceiling 48,400 ft, combat radius 2000 miles.

Opposite, the first production RA-5C, BuAer 150823, outside the Columbus factory in 1963. Seventy-nine RA-5Cs were built new and another 43 converted from A3J-1s (A-5As).

Right, RA-5C of RVAH-5, aboard CVA-61, the USS Ranger, *Summer 1964, with McDonnell F-4s and Grumman S2F.*

Production: North American YA3J-1 / A3J-1/-2 (A-5A/B) / RA-5C Vigilante

NA-233	Design, engineering and mock-up	—
NA-247	YA3J-1 BuAer 145157/145158	2
	A3J-1 BuAer 146694/146702	9
NA-258	NAGPAW program, transferred to NA-247	—
NA-263	A3J-1 (A-5A) BuAer 147850/147863	14
NA-269	A3J-1 (A-5A) BuAer 147924/148933	10
	A3J-2 (A-5B) BuAer 149276/149317	42
NA-272	Long-lead-time effort, to NA-269	—
NA-279	RA-5C BuAer 150823/150842	20
NA-283	RA-5C BuAer 151615/151634, 151726/151728	23
NA-296	RA-5C (converted from A-5A)	(27)
NA-298	RA-5C (converted from A-5A)	(16)
NA-316	RA-5C BuAer 155608/156653 (last 10 canceled)	36
	Total, A3J/A-5 attack bombers, 1953-1963	156

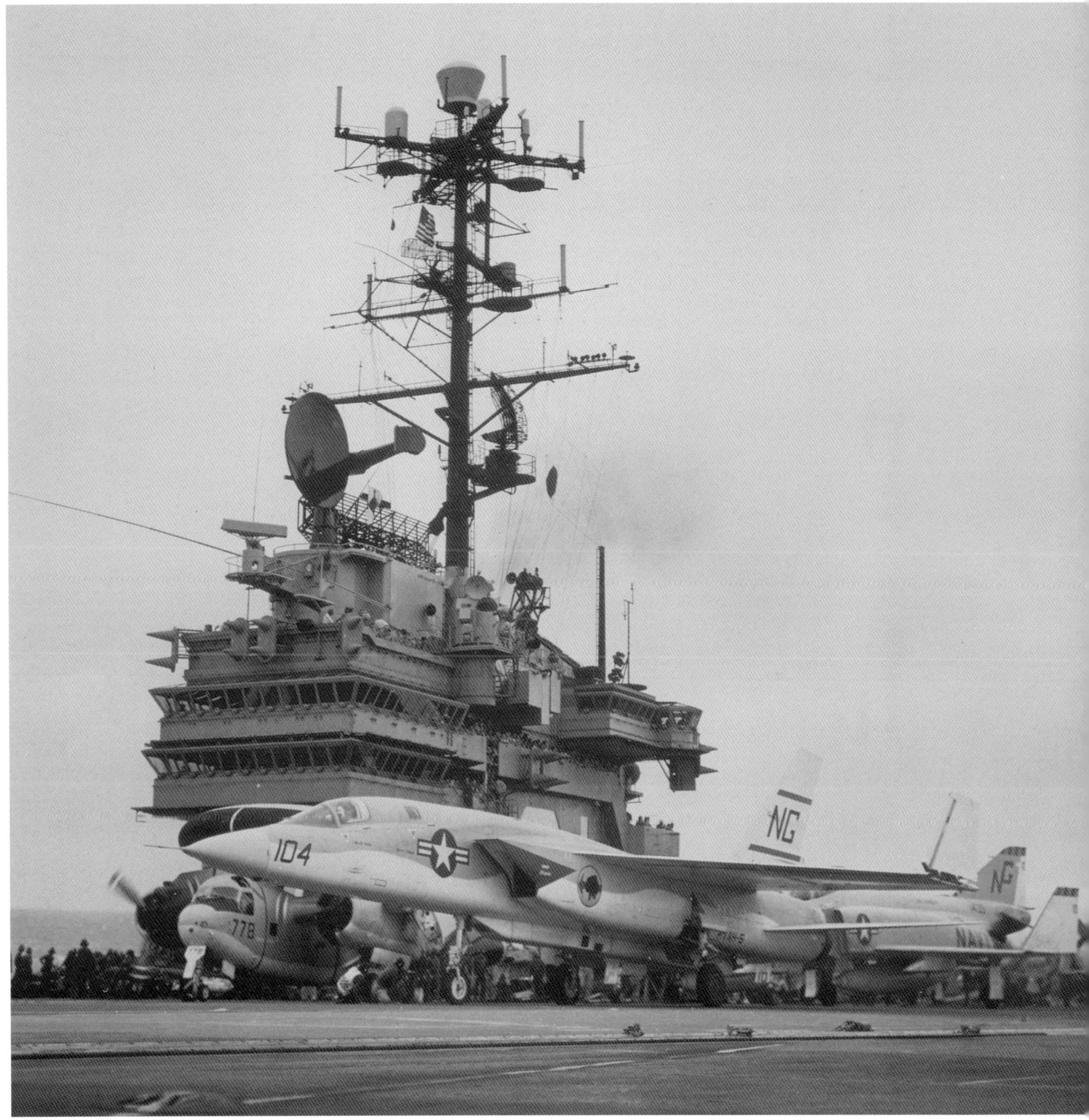

North American NA-240 X-15 / X-15A-2

AFTER MANNED SUPERSONIC FLIGHT became a reality in late 1947 attainable speeds and altitudes increased rapidly and during the early 1950s the U.S. Government and the aerospace industry began studying the possibility of manned hypersonic flight. On June 24, 1952, the National Advisory Committee for Aeronautics (NACA) passed a resolution to "increase its program dealing with the problems of unmanned and manned flight in the upper stratosphere at altitudes between 12 and 50 miles, and Mach numbers between 4 and 10." NACA's executive committee ratified the proposal the following month and subsequently asked officials at its Ames, Langley and Lewis laboratories for recommendations on implementing the program.

The several reports and design proposals submitted to NACA included studies by Robert Woods and Walter Dornberger of Bell Aircraft, Hubert Drake and Robert Carman of NACA's High Speed Flight Research Station (HSFRS) at Edwards AFB and David Stone of the NACA Piloted Aircraft Research Division (PARD). Although these proposals generally endorsed the feasibility of manned hypersonic and re-entry flight, some NACA officials felt that such testing should be applied primarily to missiles and that manned flight should be limited to Mach 4.5.

By 1954, progress made by the second-generation Bell X-1s and new X-2 research aircraft had heightened U.S. Air Force interest in a manned hypersonic and re-entry program. As with supersonic flight research, hypersonic flight held many unknowns, perhaps the greatest being structural heating. The proposed speeds and altitudes would allow research into the void of near-space and on the physiological effects of high-speed, high-altitude flight.

After intensified studies were conducted at

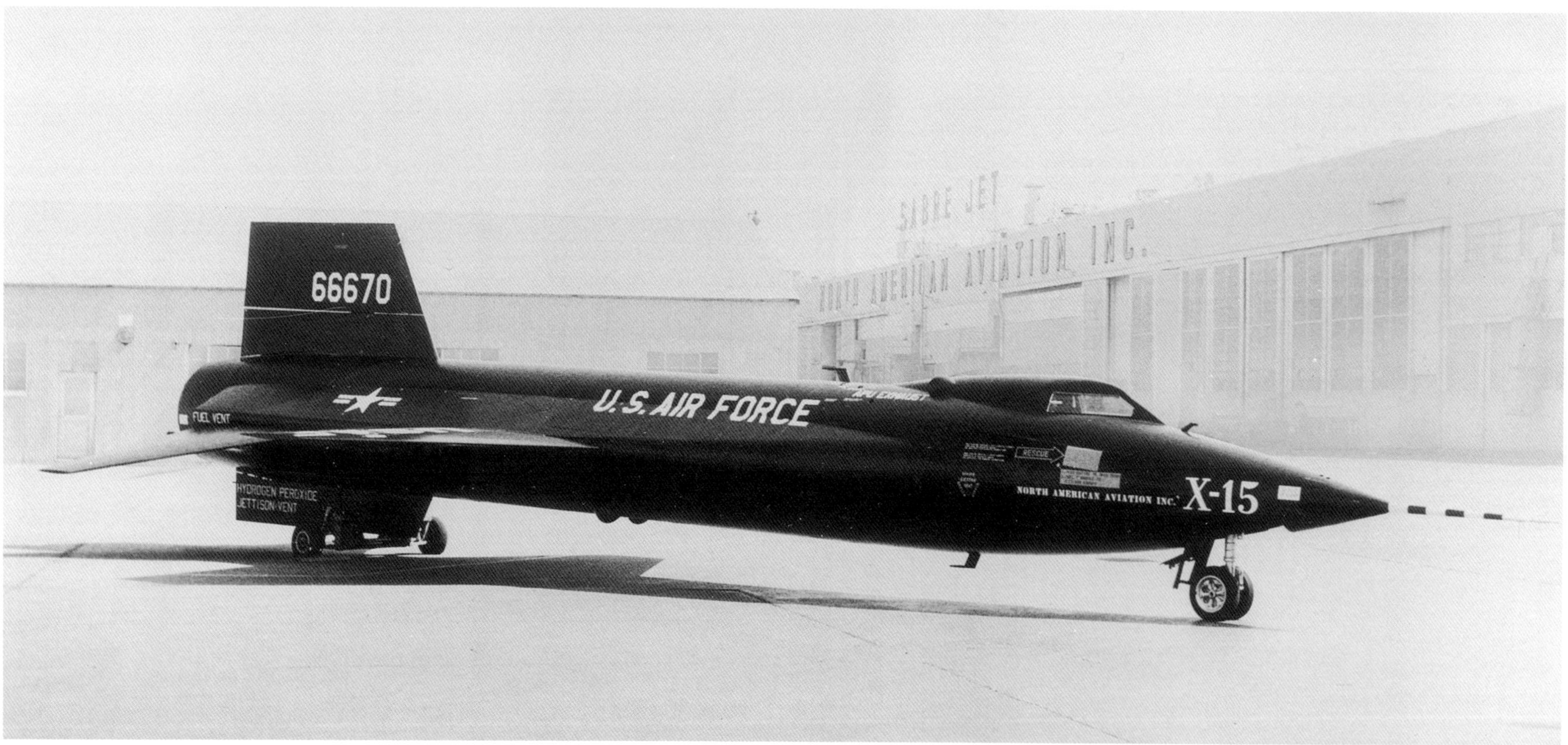

Langley through the end of 1954, NACA compiled a comprehensive report that set out the requirements and projected performance of the program, which would be known as Project 1226. Specifications called for an aircraft able to attain speeds up to Mach 6 and to withstand temperatures in the range of 1200 degrees Fahrenheit.

These reports were submitted to nine contractors at a bidder's briefing at Langley in January 1955. Between February and May five companies dropped out of the competition, leaving Bell, Douglas, North American and Republic to compete. By May 9, 1955, these four manufacturers had submitted design proposals to NACA and eight days later NACA and USAF evaluation groups reviewed these proposals.

Thanks to its detailed data concerning structural heating, the North American design was eventually chosen. Raymond H. Rice, then vice president and chief engineer at NAA, sent a letter to NACA explaining that, due to the man-hours spent on the F-107 and other projects, it would be necessary to extend the completion schedule by eight months. This was approved by the Air Force and formal contract ratification took place on September 30, 1955. On June 11, 1956, NAA's Los Angeles Division was given the go-ahead to produce three X-15 (Project 1226) aircraft.

The X-15 design team, headed by Harrison "Stormy" Storms and Charles Feltz and directly assisted by veteran test pilot Scott Crossfield, went to work on design NA-240, also known as ESO 7487. Much of the aircraft's structure utilized Inconel-X, a nickel-alloy capable of withstanding the expected 1200 degrees F. The fuselage core, made of titanium, housed the fuel and oxidizer tanks. The airframe featured a small, low-aspect-ratio wing, anhedral horizontal stabilizers and a wedge-shaped vertical stabilizer/rudder on the upper and lower rear fuselage.

Two Reaction Motors XLR11 rocket engines were chosen as the initial powerplant while the more powerful Thiokol XLR99 was still under development. To aid in trimming the aircraft at higher speeds, small ballistic control rockets were mounted in the nose. A three-stick control system was used: a conventional central stick to control pitch and roll during low-speed atmospheric flight, a side-mounted stick to be used during acceleration and re-entry, and another

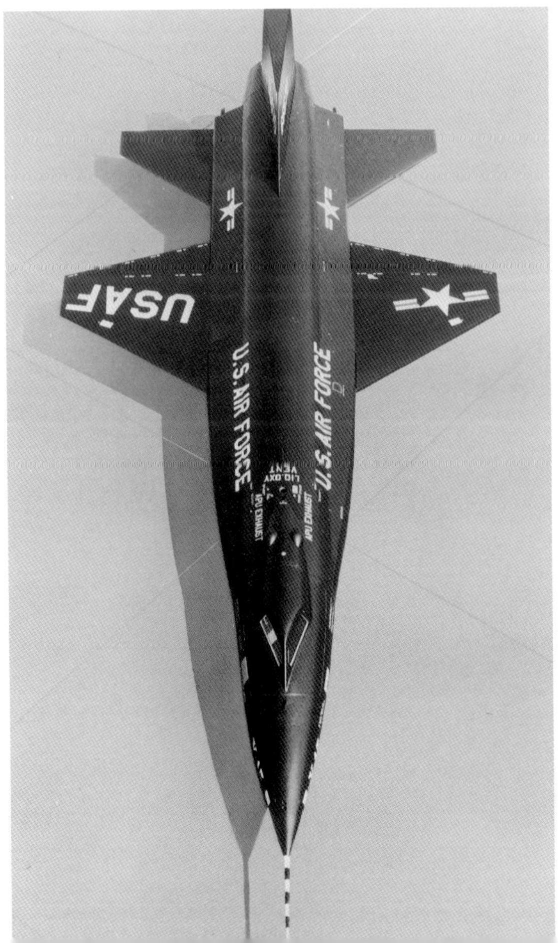

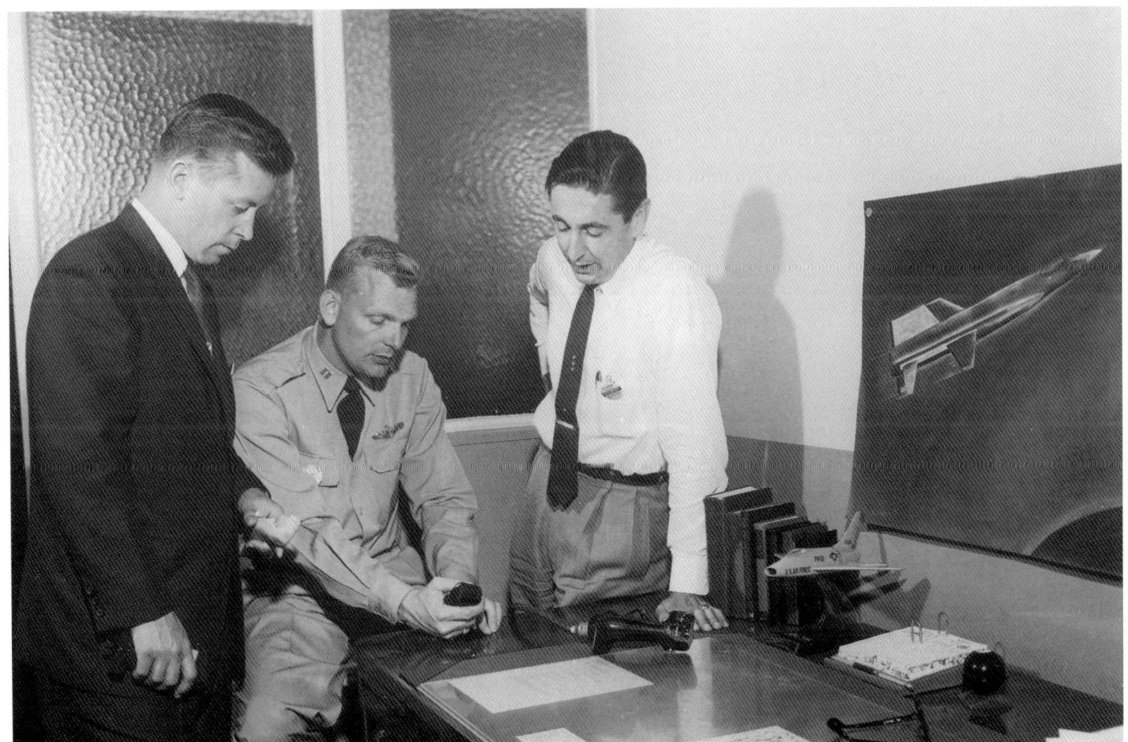

Left, brand-new, all black X-15, 56-6670, after roll-out at NAA's Los Angeles factory on October 15, 1958. Above, pilots Alvin White, USAF Captain Ivan Kincheloe and Scott Crossfield discuss the side-stick control system designed for the X-15.

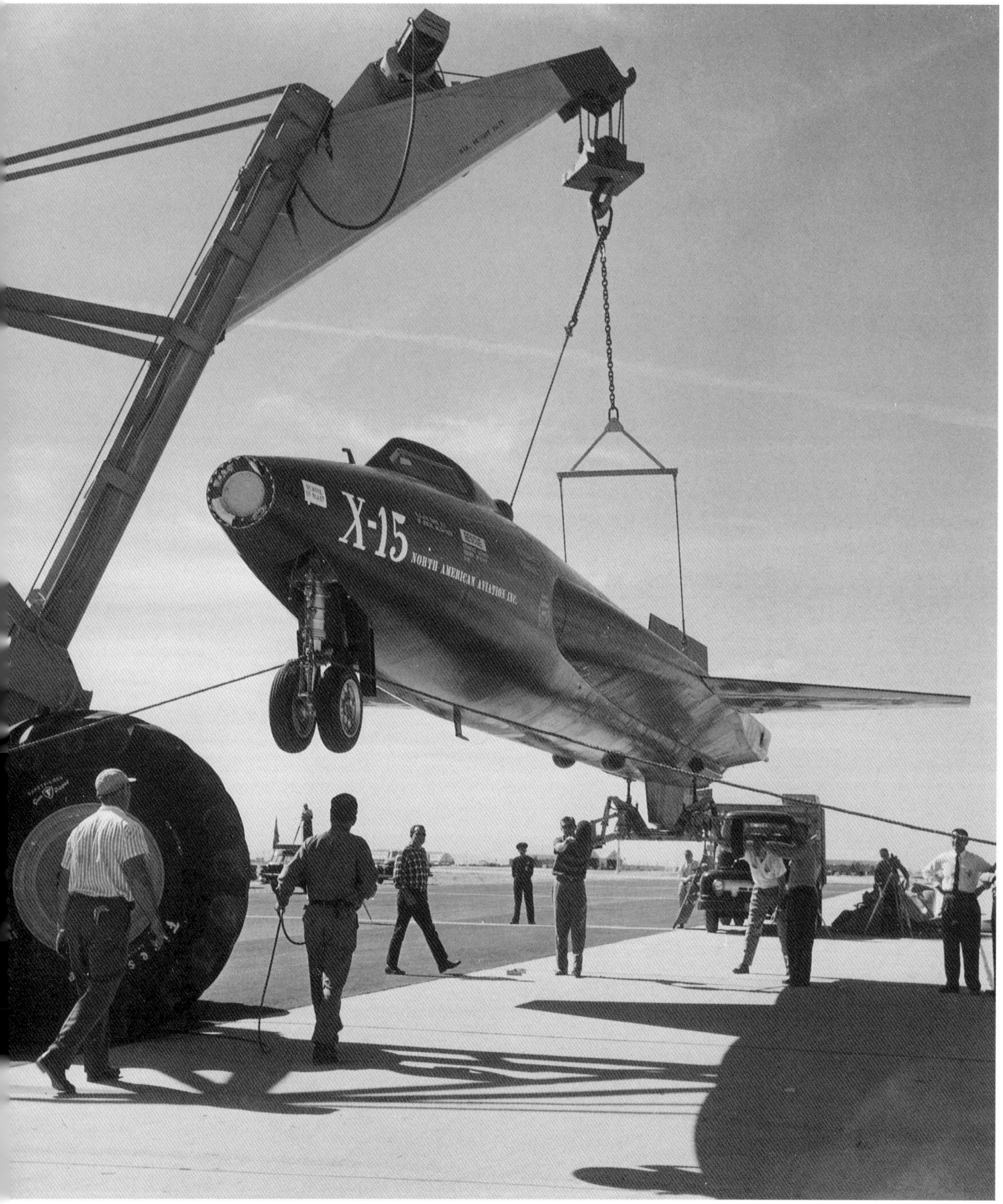

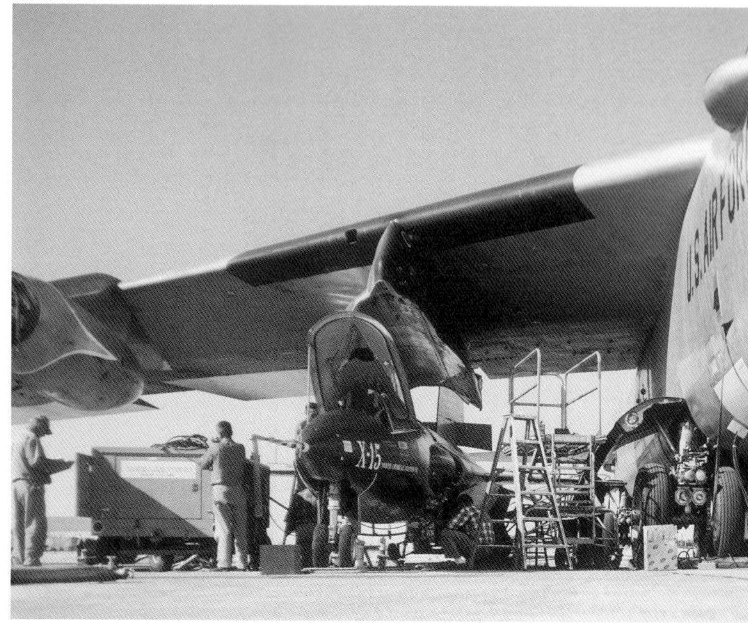

The X-15 was trucked to Edwards Air Force Base, reassembled and mounted below the right wing of NB-52A, 52-0003. As seen from the rear, the two XLR11 rocket engines take up only part of the space designed for the XLR99, still under development.

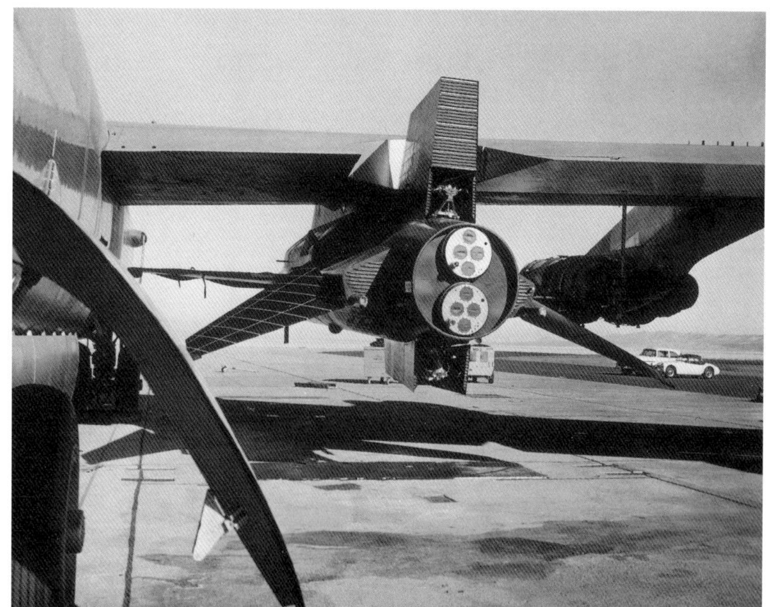

side stick to actuate reaction jets during exo-atmospheric flight. The X-15s would be carried aloft by "mother" aircraft, released for flight and land at approximately 200 miles per hour. A retractable nose gear and rear skids were fitted. The aircraft was equipped with an ejection seat engineered for speeds up to Mach 4.

The Boeing B-52 was chosen as the launch platform, and two (B-52A, USAF 52-0003, and B-52B, 52-0008) were modified with systems to carry the X-15 aloft, maintain fuel storage, communication and test pilot support, release the X-15 for glide or powered flights and monitor the flights in conjunction with ground personnel. These modified bombers were designated NB-52A and NB-52B, respectively.

The official roll-out ceremony took place on October 15, 1958, and on March 10, 1959, the first X-15, 56-6670, made its first captive flight beneath the starboard wing of the NB-52B from the AFFTC at Edwards AFB. Several captive flights were made, followed by the first glide flight on June 8. During this flight, Crossfield found that the control system was sensitive to pitch on landing, and had to use a very light touch to land the aircraft safely. The first powered flight was made on September 17, 1959, using the second aircraft, 56-6671. During this flight Crossfield reached Mach 2.11 and 52,341 feet.

During the fourth flight, on November 5, an inflight explosion and engine compartment fire forced Crossfield to make an emergency landing on Rosamond Dry Lake. His efforts to jettison propellants were incomplete, due to the steep descent angle, and the landing weight was heavier than normal. As the nosewheel touched the ground, the fuselage buckled just aft of the cockpit, dragging the rear fuselage on the ground. After repair to the fuselage and nose gear at NAA, the aircraft was ready to fly again in February 1960.

While 56-6671 was under repair, aircraft 56-6670 made its first powered flight. On this occasion, Crossfield set new speed and altitude marks for the program: 1669 miles per hour and 66,844 feet. Joe Walker made the first flight for NASA

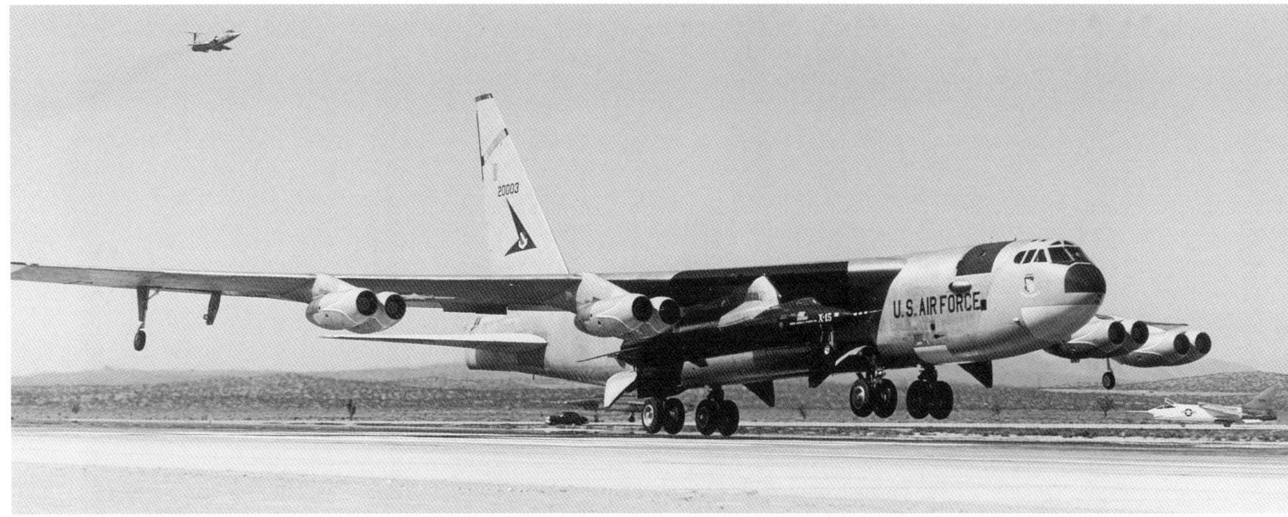

Above, with the X-15 underwing, an F-104 chase plane overhead and an F8U in the background, the NB-52A takes off at Edwards. Below, after landing, Crossfield heads for wagon and debriefing. Note dolly for X-15 ground transport beside wing.

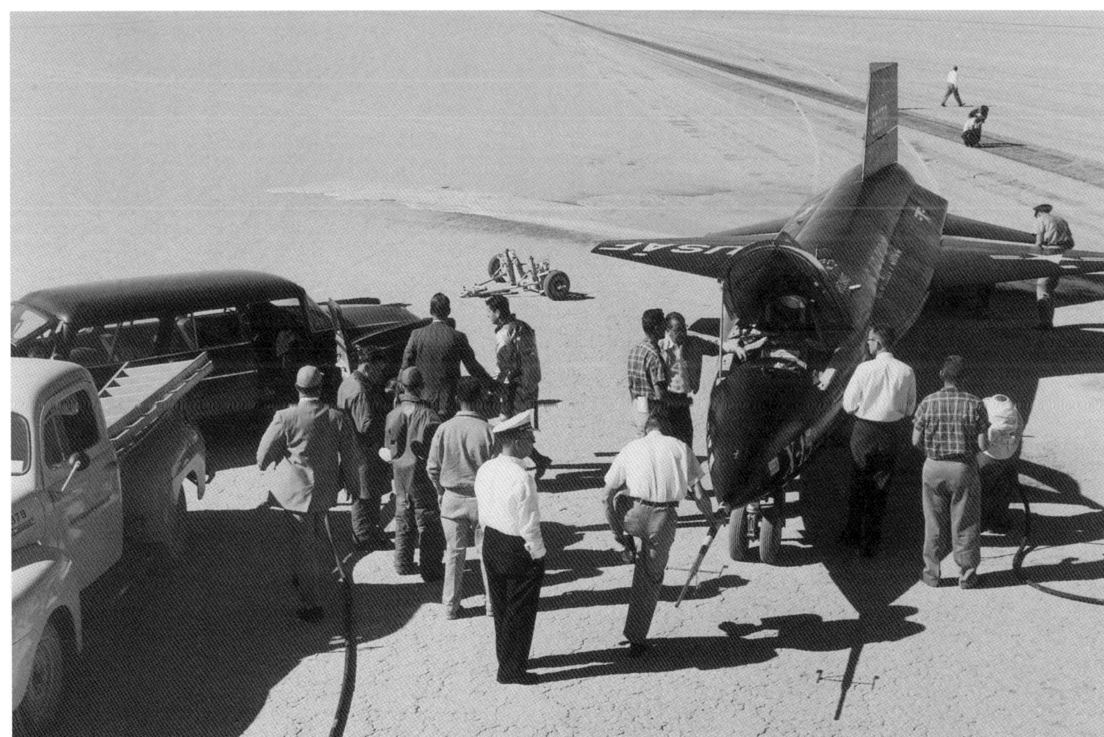

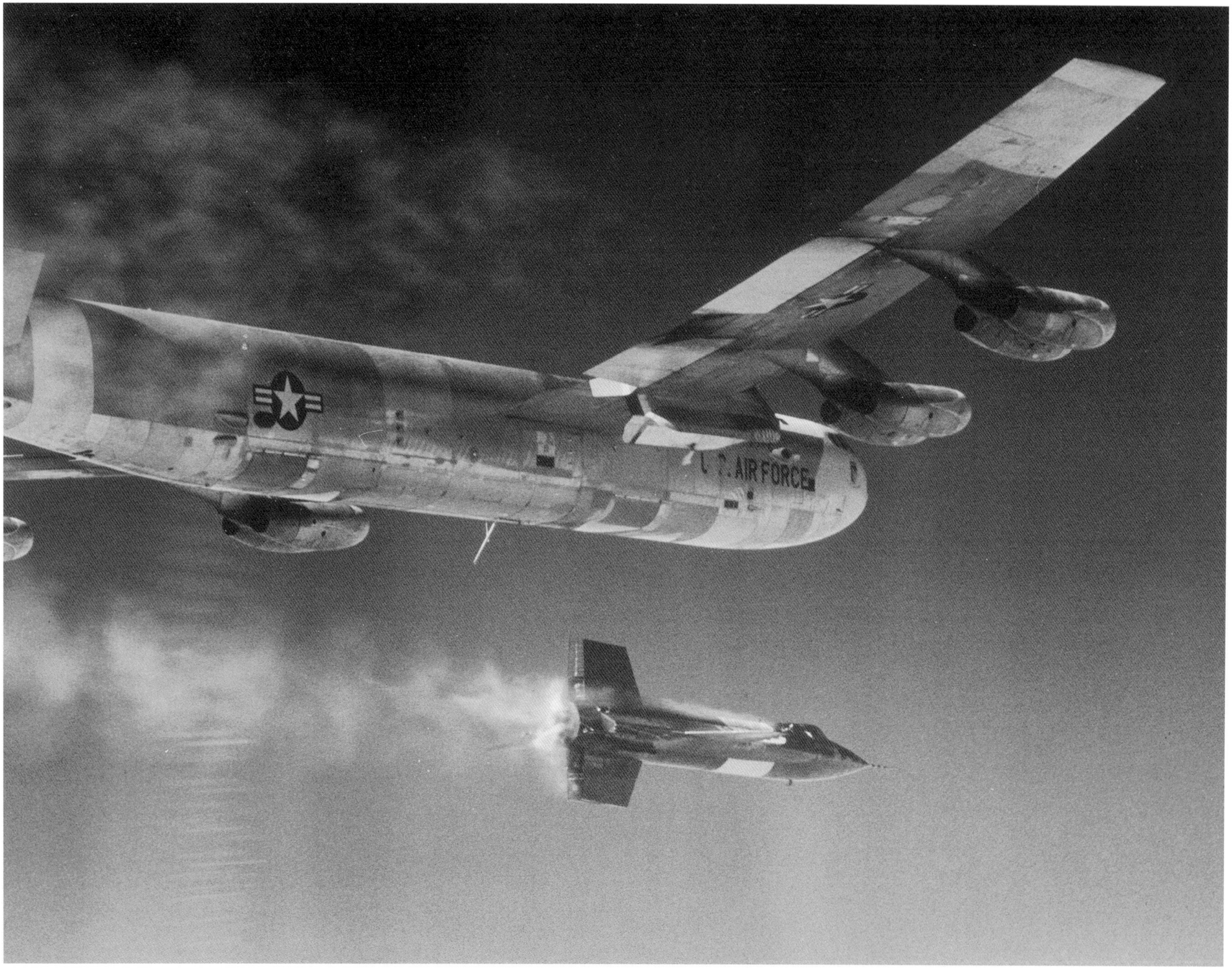
U.S. AIR FORCE

Opposite, XLR11s fire as X-15 drops from NB-52A. White band around center of fuselage is condensation from tank, not paint.

on March 20 and Captain Robert White was the first USAF pilot to fly the X-15, on April 13, 1960. By this time, six different pilots were assigned to the program: Crossfield (NAA), Joe Walker and Neil Armstrong (NASA), Captain White and Major Robert Rushworth (USAF), and Commander Forrest Peterson (USN). A total of twenty-nine flights were made with the XLR11 engines, and during this time White set new marks for speed and altitude: 2275 miles per hour (Mach 3.5) and 126,200 feet. Pilots added to the program included NASA's Milton Thompson and William Dana, and Captain Joe Engle, Captain William "Pete" Knight and Major Michael Adams of the USAF.

Bio-medical base-line data gathered during the early XLR11-powered flights contributed to the development of the David Clark Company MC-2 and A/P22S-2 full-pressure suits. The X-15 tests also provided more data on pilot weightlessness than any previous aircraft.

In March 1960 the new XLR99 engine was installed in the third aircraft, 56-6672, and a series of ground tests were begun. The new engine, fueled by a mixture of liquid oxygen and anhydrous ammonia, provided much more than three times the thrust of the twin XLR11s. It could be throttled between 30 and 100 percent power and had a burn time of 86 seconds on internal fuel. Although the new engine would allow the X-15 program to explore much wider performance envelopes, and although it was ultimately safer and more reliable, it experienced its own development difficulties. On June 8, a fire ignited by an accident during ground tests destroyed the new engine and severely damaged the aircraft.

The second XLR99 arrived at Edwards later that month and was installed in X-15 number 2

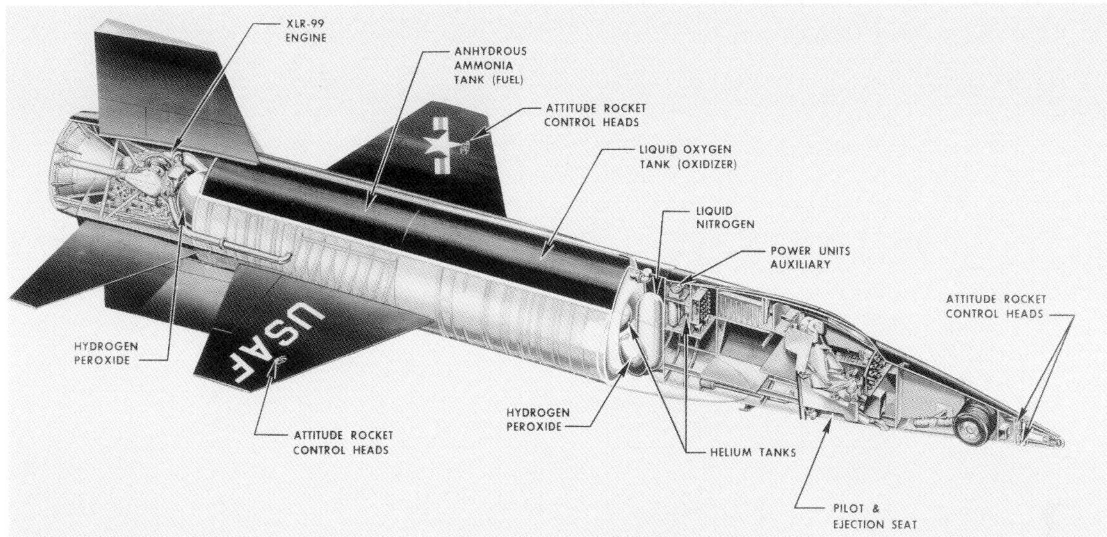

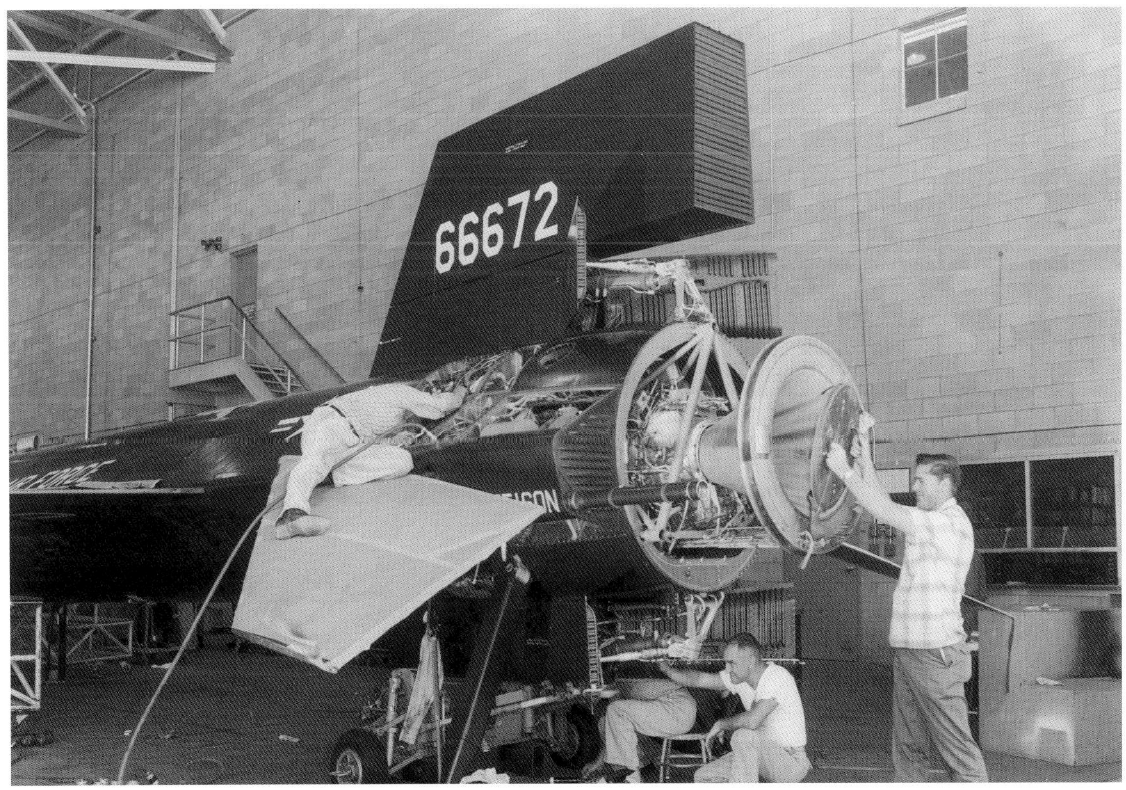

Above right, cutaway illustration of X-15 shows the much larger XLR99 engine mounted in tail. Right, installation of the engine by NAA technicians in the third X-15, 56-6672.

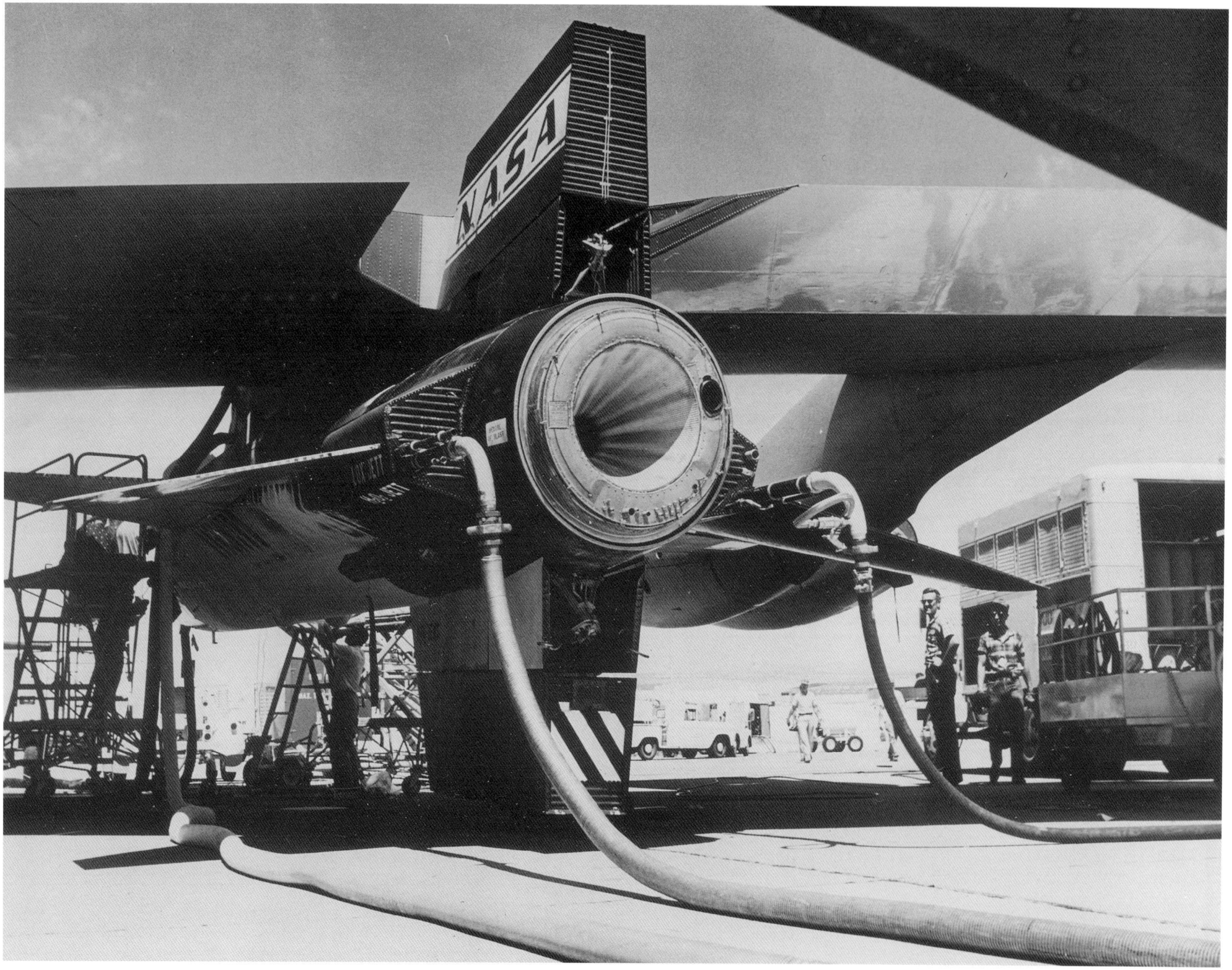
NASA

Opposite, the first X-15, with XLR99 engine and NASA band on tail. NASA pilots flew a total of 91 flights with the X-15s.

while number 3 was at NAA for repairs. Crossfield made the first XRL99 flight on November 15, 1960. Aircraft number 1 was also fitted with the new engine by June 1961. In the fall of that year, the X-15s exceeded their design Mach number by attaining 4093 miles per hour (Mach 6.04) and had flown as high as 217,000 feet.

In late October, number 3 was completely rebuilt following the ground-test engine fire and was fitted with a Minneapolis-Honeywell MH-96 self-adaptive flight control system, designed to simplify control of the X-15's integrated systems throughout the performance envelope. On April 30, 1962, Joe Walker achieved design altitude in X-15 number 1, reaching 246,700 feet. Neil Armstrong made the first flight of number 3 on December 20, 1962, and this aircraft later reached 314,750 feet with White at the controls. White was awarded astronaut wings for this accomplishment (attaining an altitude of 50 miles or greater), and thirteen more astronaut flights followed.

On November 9, 1962, the second aircraft, with John McKay at the controls, suffered extensive structural damage during an emergency landing at Mud Lake, Nevada. McKay survived, but sustained back and lung injuries. Returned to NAA for repairs, X-15 number 2 was extensively modified to become the X-15A-2. The changes included lengthening the fuselage 29 inches to provide room for a liquid hydrogen tank (to be used with an experimental ramjet known as a "scramjet"), reshaping the windshield glass from rectangular to oval, applying a special ablative coating to withstand local temperatures, and providing two large external fuel tanks that would increase engine burn time from 86 to 145 seconds. These tanks were jettisonable and recoverable/reusable.

These modifications allowed the X-15 to realize its full performance capabilities. On October 3, 1967, by which time the X-15 program was winding down, Captain Knight achieved 4520

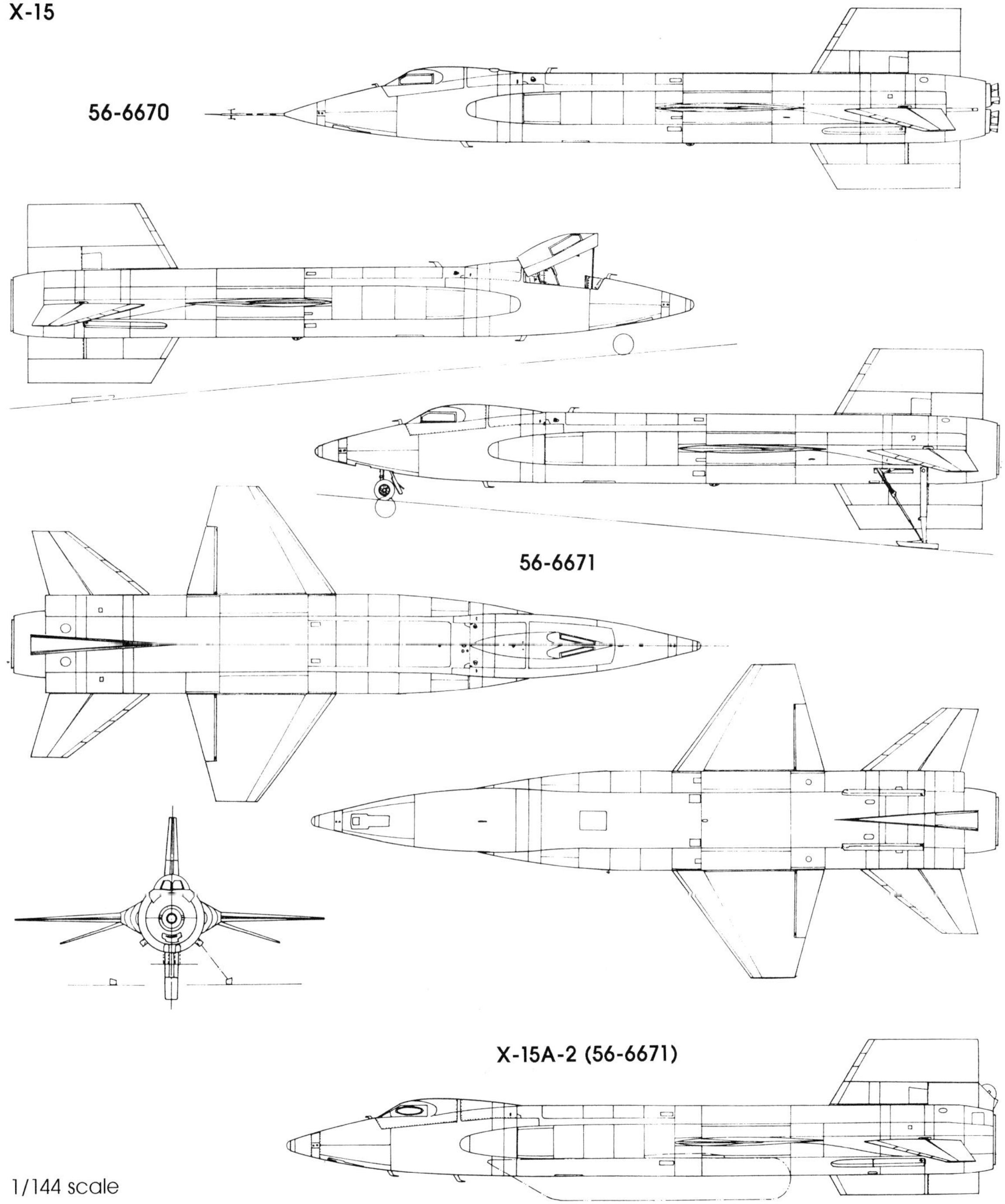

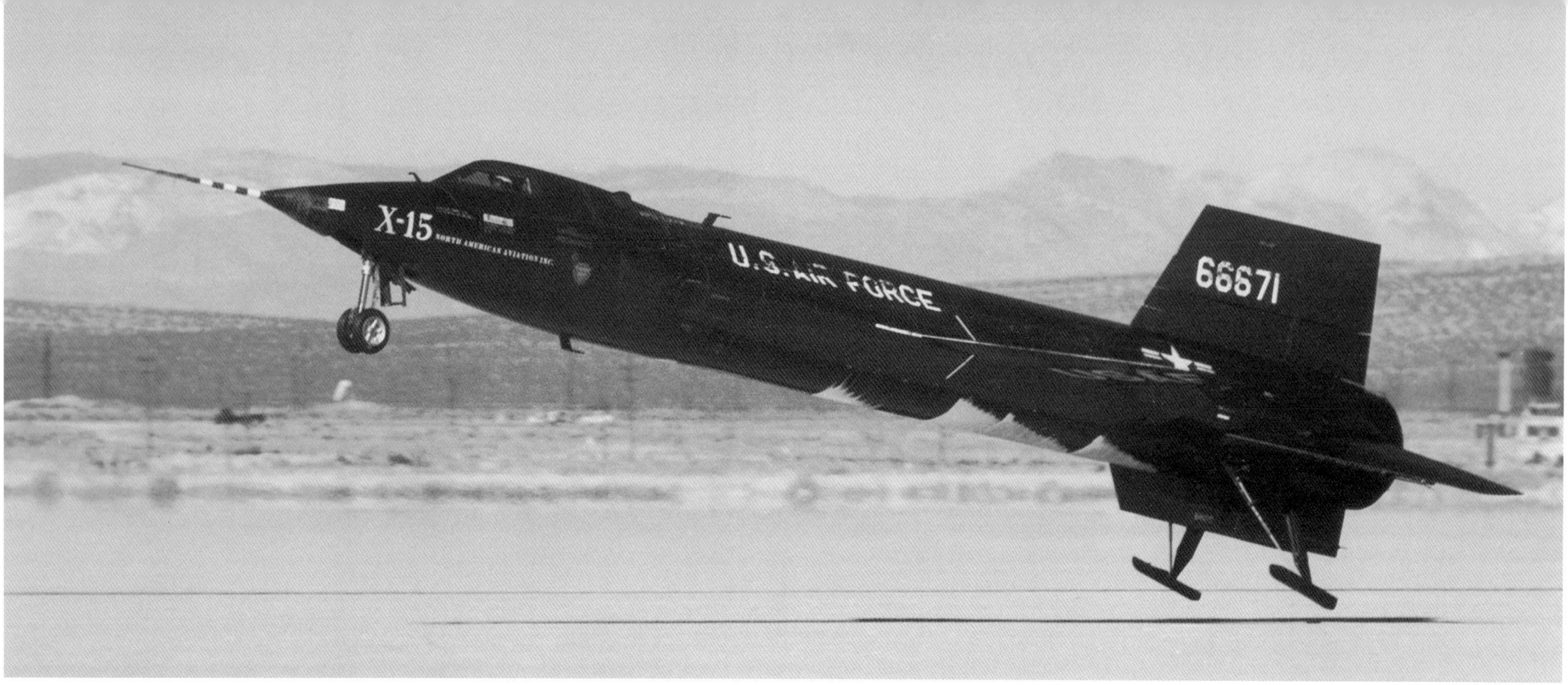

Typical X-15 landing was nose-high, with tail skids touching down first. Above is second X-15,56-6671, still fitted with a nose probe, during NAA evaluations. In the lower photo the plane, showing a later nose configuration and a different condensation pattern under the fuel tanks, lands at a lesser angle of attack. The tail band indicates a NASA pilot at the controls.

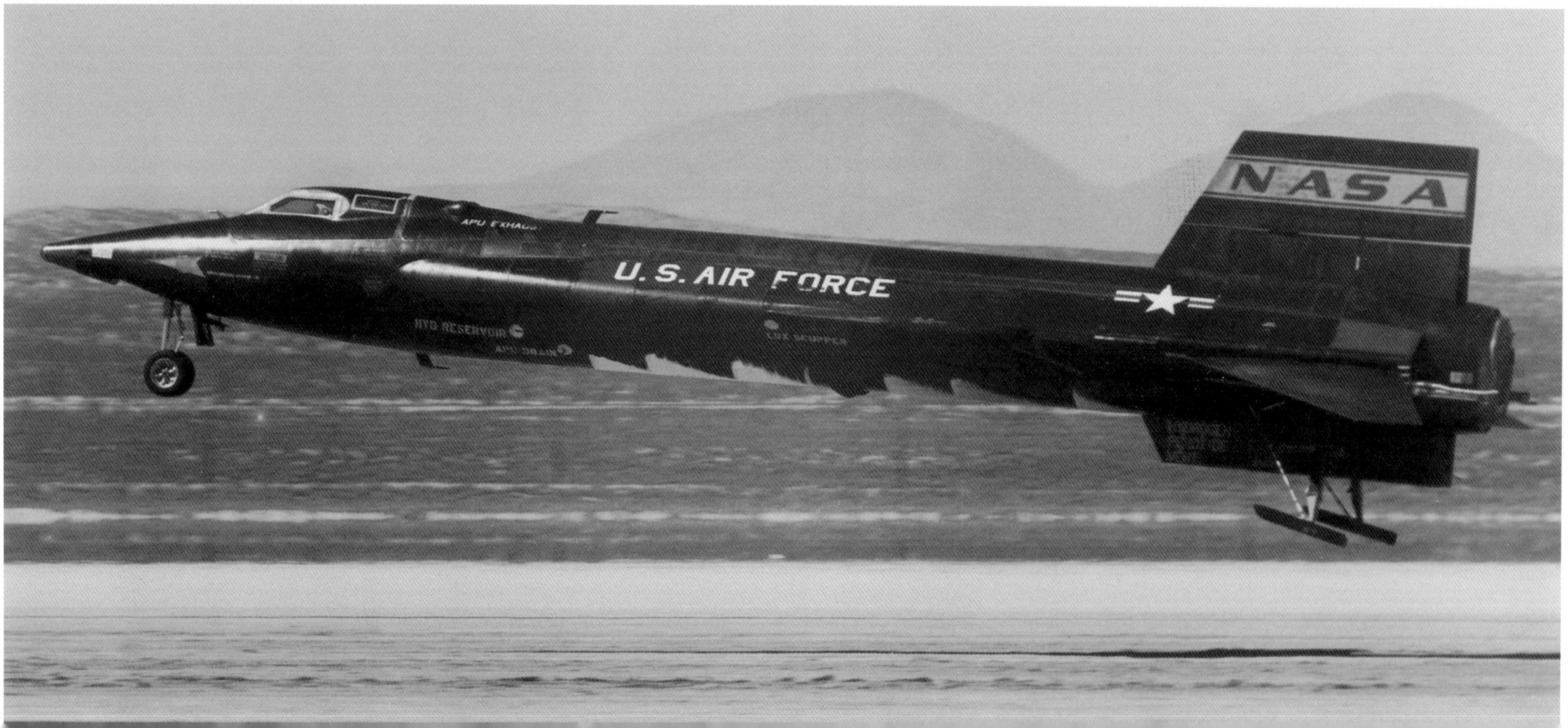

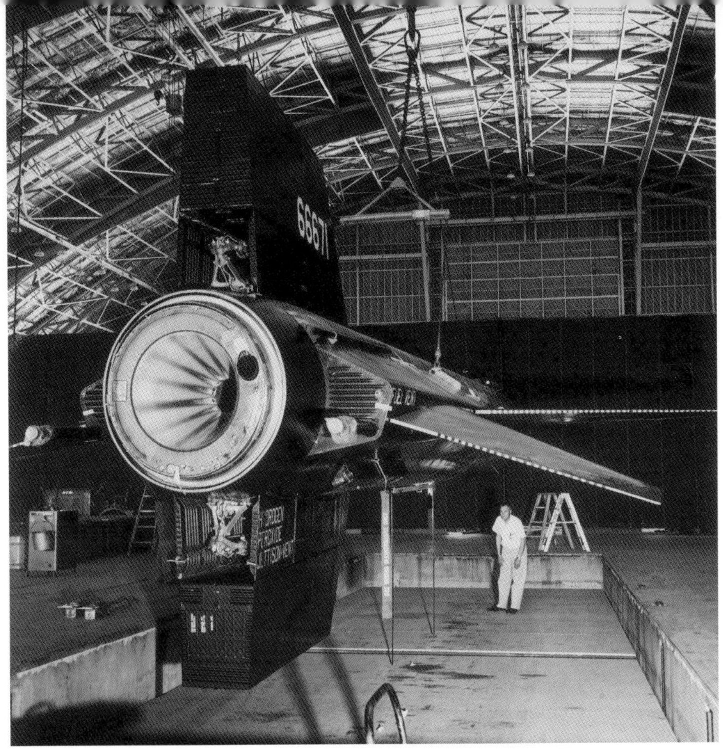

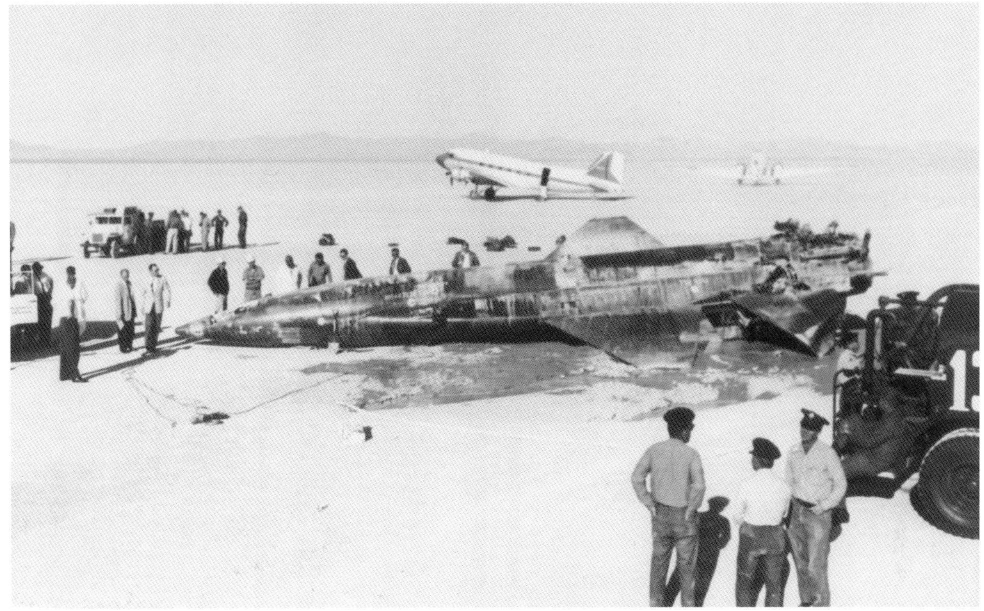

Above, a weight and balance check of X-15 number two. Right, crash site of 56-6671 at Mud Lake, Nevada, on November 9, 1962. The pilot, John McKay, survived despite serious injuries and the aircraft was rebuilt by North American as the X-15A-2, below. Among the extensive changes were huge external fuel tanks that nearly doubled the rocket's burn time before being jettisoned.

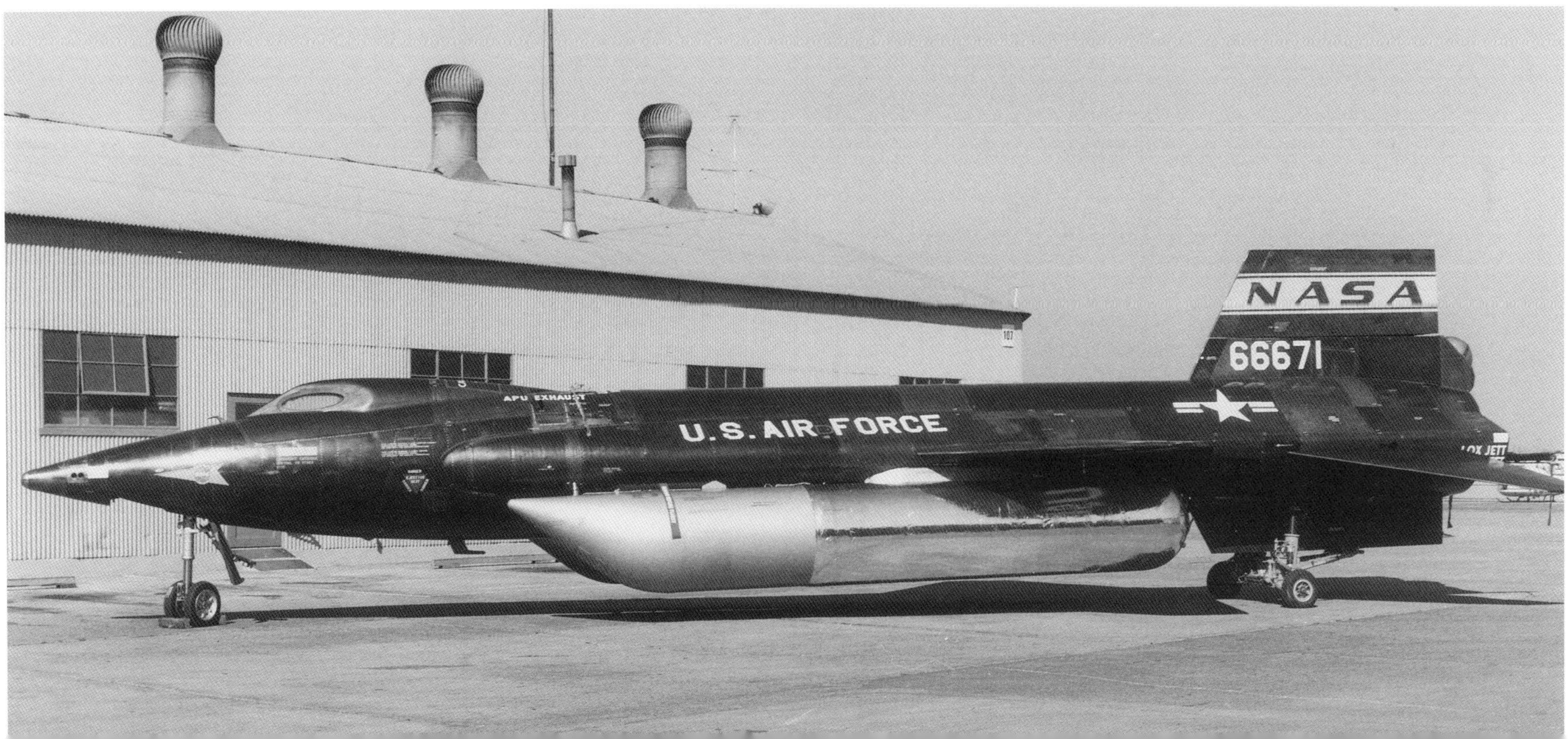

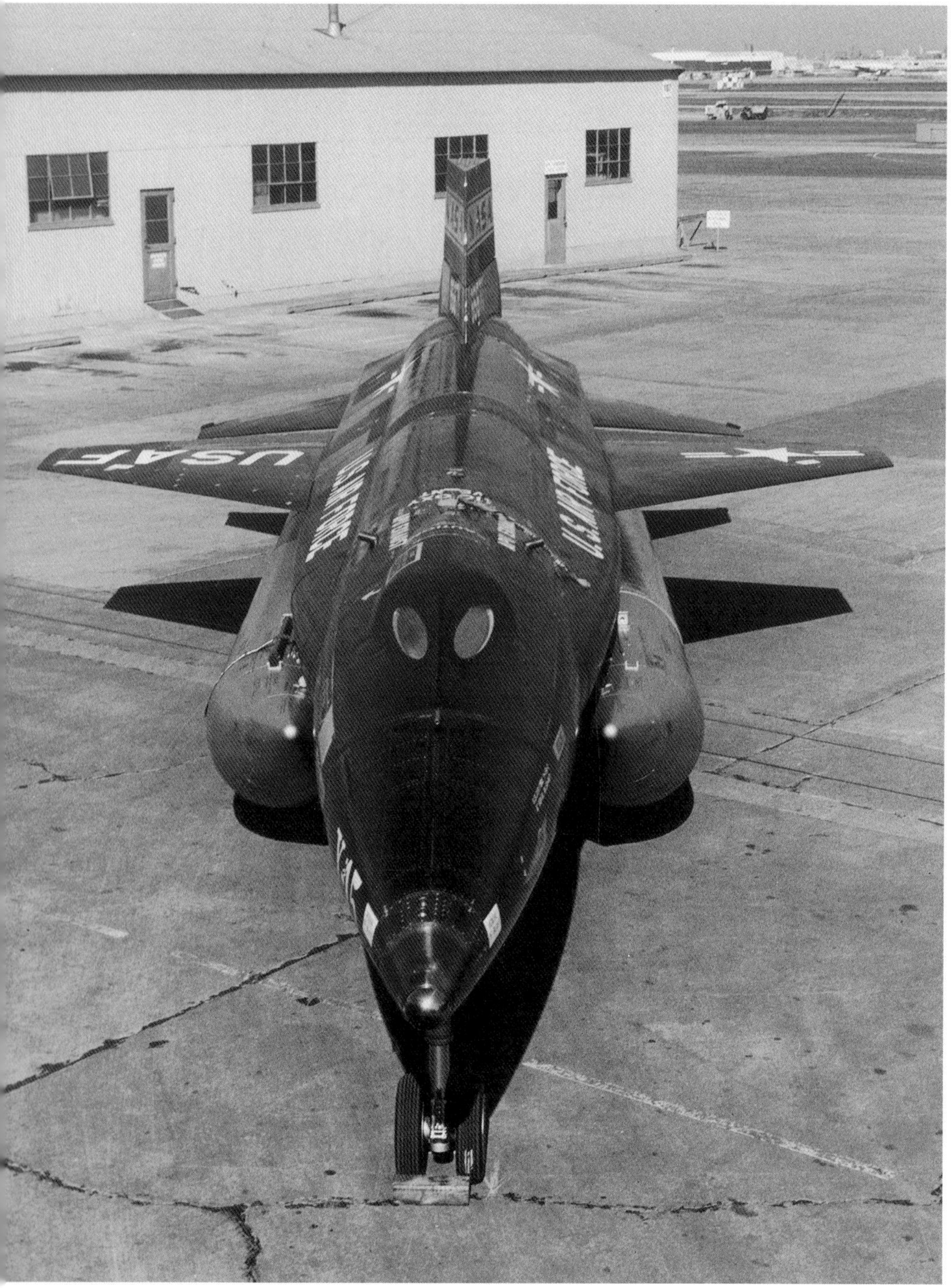

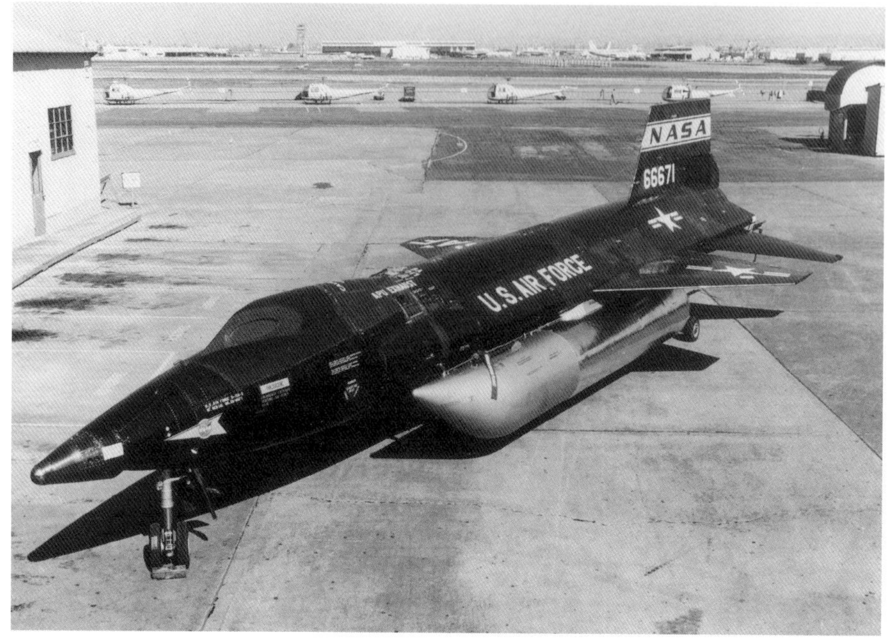

Left and above, the X-15A-2 as completed, showing the revised window configuration, and below, mounted beneath the wing of NB-52B, 52-0008 (note X-15 mission markings). Opposite page, with a white ablative coating, the just-launched X-15A-2 beginning its 42520-mph record flight.

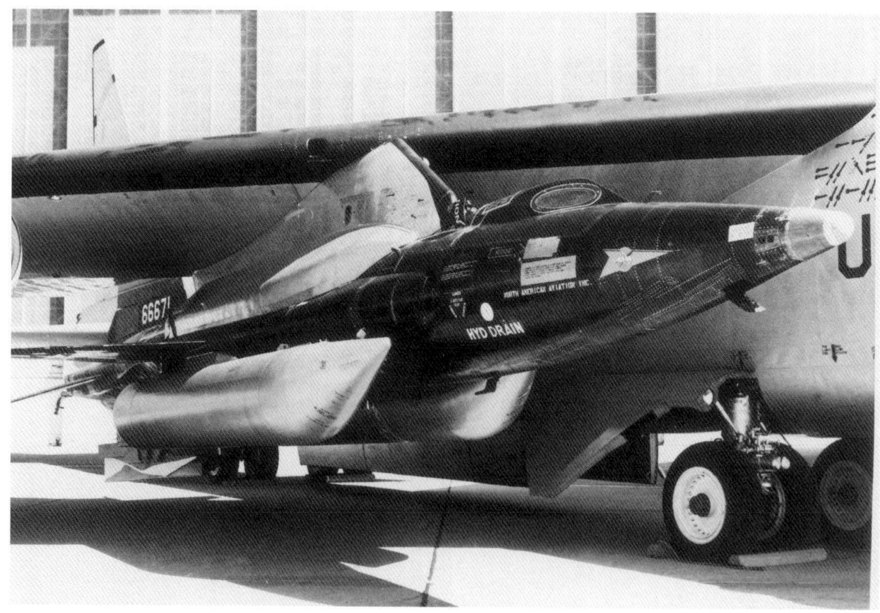

miles per hour (Mach 6.7), the absolute fastest a winged, piloted aircraft has ever flown. On November 15, 1967, the third X-15 disintegrated in flight, killing its pilot, Major Adams.

Overall, the X-15 program was a success, with a total of 199 flights completed. The aircraft broke new ground and explored speed and altitude regimes previously unattainable. Much of the data gathered would later be applied to the Saturn V Apollo missions and, more recently, the Space Shuttle program. Of the two remaining X-15s, number 1 is on display at the National Air and Space Museum in Washington, D.C., and number 2 resides at the USAF Museum at Wright-Patterson AFB in Dayton, Ohio.

Specifications: North American NA-240 X-15/X-15A-2

Description: USAF/NASA rocket-powered hypersonic research aircraft. Mid-wing monoplane of all-metal construction with retractable skid and nosewheel landing gear.

Engines: Two 8000-lb-thrust Reaction Motors XLR11 liquid-propellant rocket engines; later one 57,850 Thiokol XLR99-RM-2 liquid-propellant rocket engine.

Dimensions: Span 22 ft 4 in, length 50 ft, height 13 ft, wing area 200 sq ft. (A-2) Length 52 ft 5 in, height 14 ft.

Weights: 15,000 lb empty, 33,000 lb maximum. (A 2) 18,340 lb ompty, 56,130 lb maximum.

Performance: Maximum speed 4159 mph, maximum ceiling: 354,200 ft, typical range 275 miles. (A-2) Maximum speed 4520 mph.

Production: North American NA-240 X-15/X-15A-2

NA-240	X-15	AF 56-6670/56-6672	3
	X-15A-2	(56-6671 modified)	(1)
Total, X-15 research aircraft, 1955-1967:			3

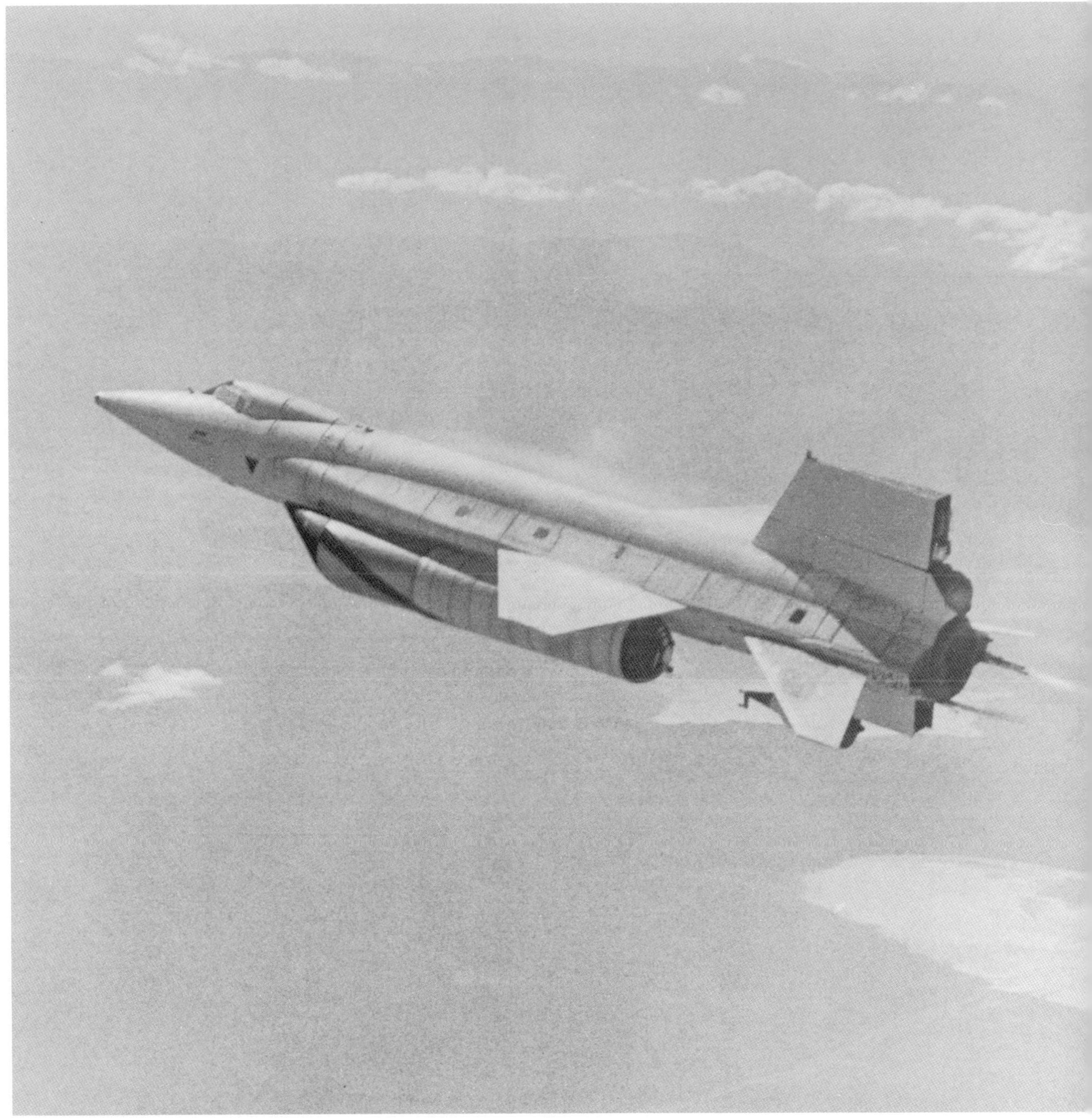

106

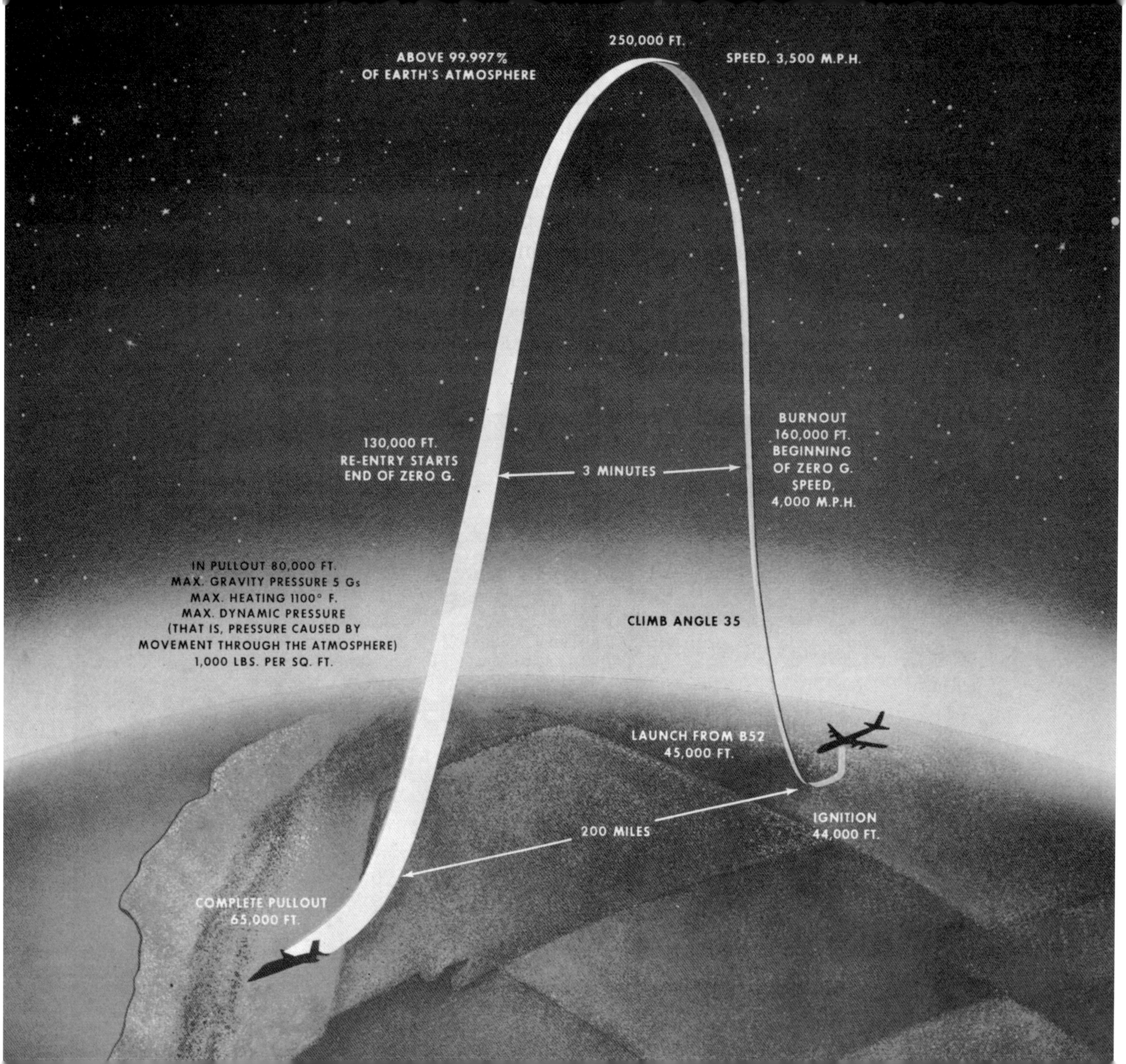
250,000 FT.
ABOVE 99.997%
OF EARTH'S ATMOSPHERE
SPEED, 3,500 M.P.H.
BURNOUT
160,000 FT.
BEGINNING
OF ZERO G.
SPEED,
4,000 M.P.H.
130,000 FT.
RE-ENTRY STARTS
END OF ZERO G.
3 MINUTES
IN PULLOUT 80,000 FT.
MAX. GRAVITY PRESSURE 5 Gs
MAX. HEATING 1100° F.
MAX. DYNAMIC PRESSURE
(THAT IS, PRESSURE CAUSED BY
MOVEMENT THROUGH THE ATMOSPHERE)
1,000 LBS. PER SQ. FT.
CLIMB ANGLE 35
LAUNCH FROM B52
45,000 FT.
IGNITION
44,000 FT.
200 MILES
COMPLETE PULLOUT
65,000 FT.

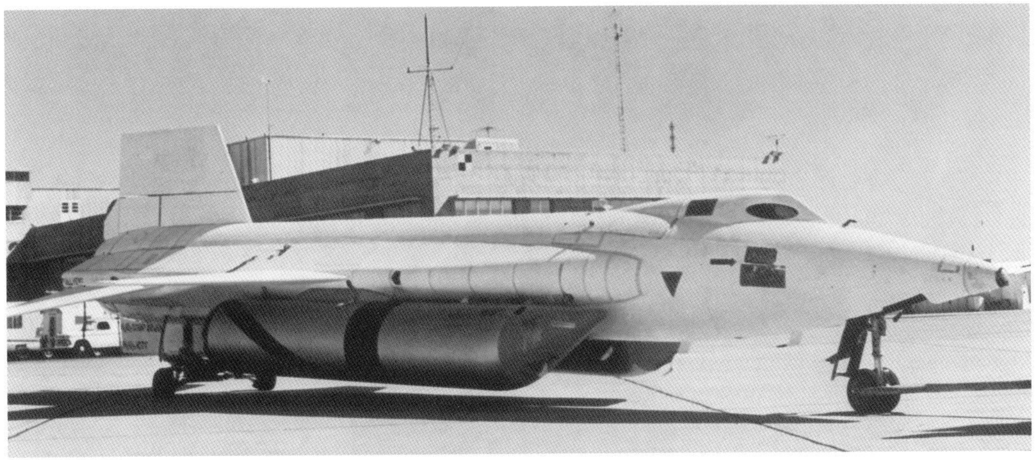

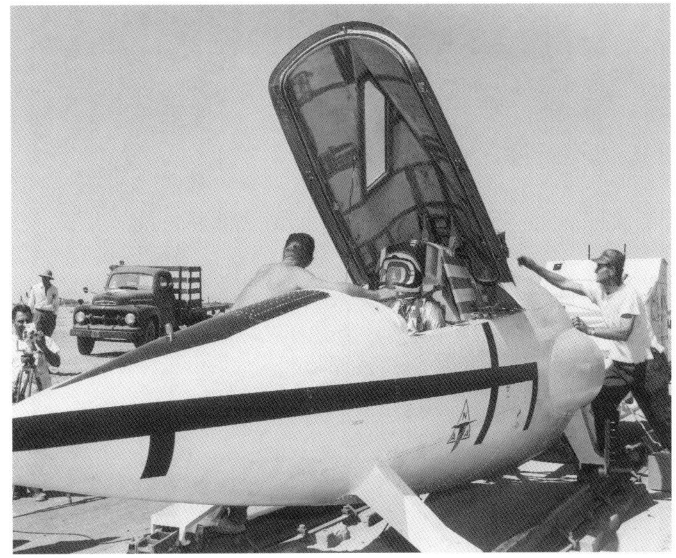

Opposite page, a mission profile for an X-15 flight. Above, the X-15A-2 with ablative covering; the external tanks, jettisoned before higher speeds were reached, were not coated. Right, an X-15 nose section was used on a rocket sled for ejection-seat tests. Below, 56-6671 displayed by NASA in company with the first XB-70, 62-0001.

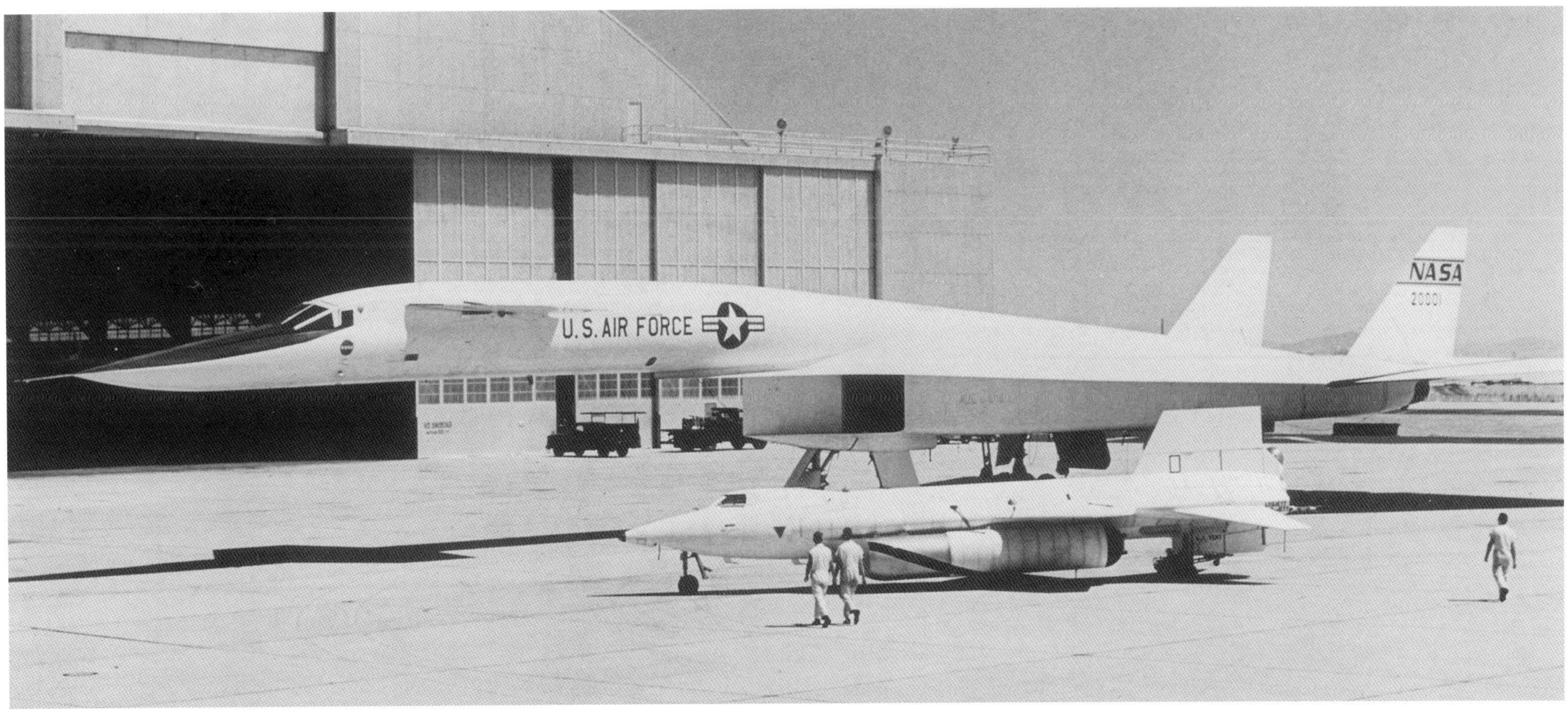

North American NA-241>NA-405 T2J-1/T-2A / T-2B/C/D/E Buckeye

THE U.S. NAVY submitted a specification for a multi-purpose jet trainer to the American aviation industry early in 1956. The requirements covered the entire jet-training program from transition and intermediate through carrier qualification, gunnery/weapons delivery and advanced training. For the aircraft to perform the variety of duties, the Navy specified a simple, rugged construction and an ease of maintenance that would greatly reduce downtime and increase flight time.

North American's success with previous training planes, the AT-6/SNJ and more recently the T-28, made the company the obvious choice for the contract, and on June 29, 1956, the Navy awarded the design and production of six YT2J-1 aircraft (NA-241) to NAA's Columbus, Ohio, division.

The T2J was a mid-wing design with a tandem arrangement for student and instructor front/rear in LS-1 ejection seats. Because simplicity was a major consideration, NAA engineers drew on lessons learned from earlier designs. The wing and main landing gear configuration was a refined version of that originally used on the FJ-1 Fury, and the basic control system of the T-28, with an added hydraulic-boost package, was chosen for the new aircraft.

By implementing many proven systems, the lead-time to test them was greatly reduced, simplifying both production and in-service maintenance. A large equipment bay, accessible from either side of the fuselage at waist height, allowed ground crews to service the electronic equipment and batteries. The engine bay beneath the central fuselage section was also made for easy servicing, the single 3400-pound-thrust

The sixth T2J-1, BuAer 146001, in carrier landing attitude with tailhook extended. Only 217 single-jet Buckeyes were built before more versatile twin-engine T-2Bs went into production in 1962.

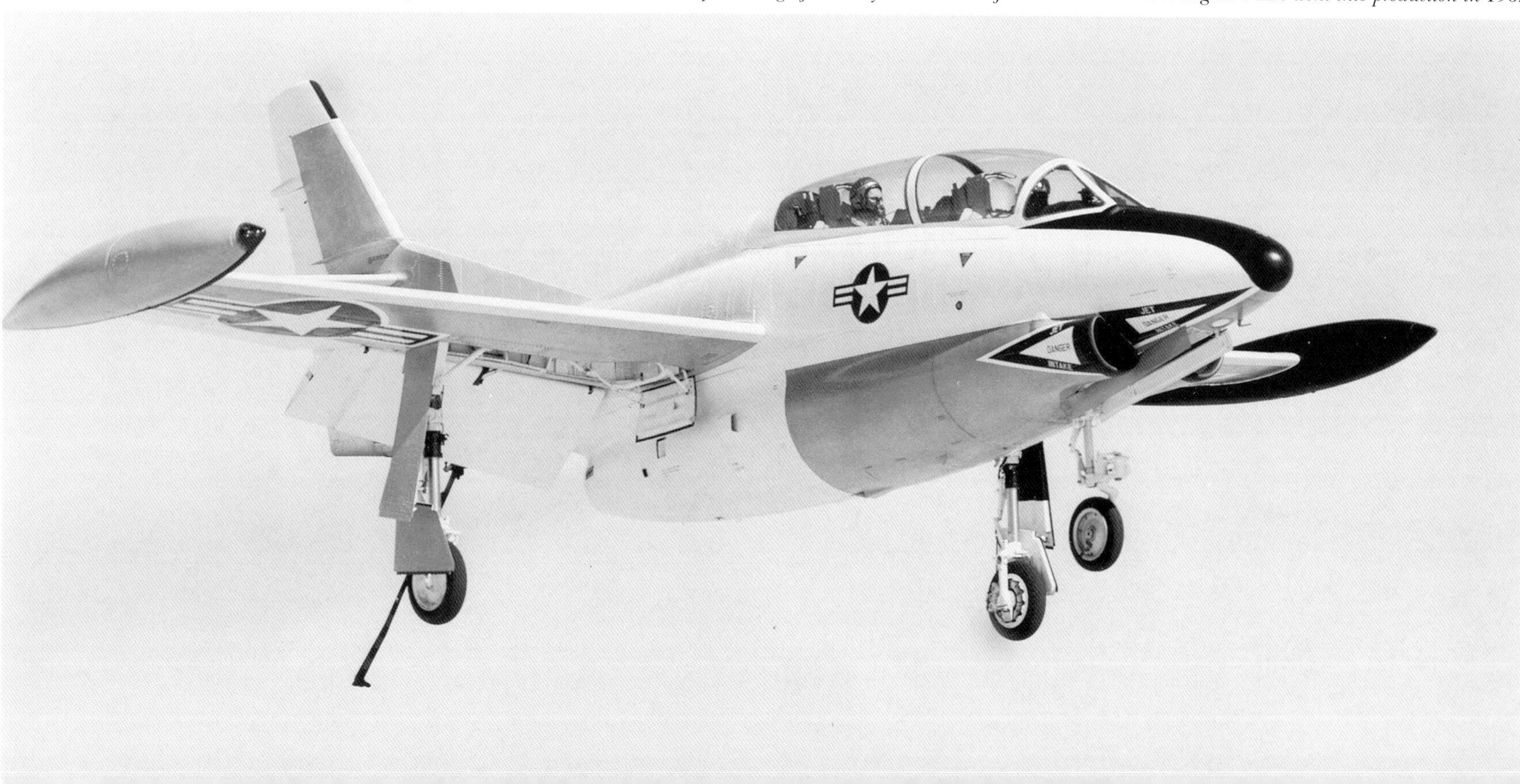

Westinghouse J34-WE-48 turbojet being accessible through three main doors. The rear fuselage section housed two hydraulically actuated airbrakes and a retractable arrestor hook for carrier operations. The Navy was sufficiently impressed with the program to place an order for an additional 121 T2J-1s (NA-253) in October 1956.

The first YT2J-1, BuAer 144217 rolled out at Columbus on December 27, 1957, and flew on January 31, 1958, with Dick Wenzell at the controls. Slight difficulties with the landing gear were encountered but quickly remedied. The first public demonstration (the third flight) took place on February 10 and the aircraft was subsequently tested and evaluated by Navy pilots at Palmdale, California. Three YT2J-1s underwent further evaluation at the Naval Air Test Center, NAS Patuxent River, Maryland. The Navy placed a third order, for an additional ninety T2J-1s, on February 17, 1959.

Carrier suitability tests were conducted aboard the USS *Antietam* (CVS-36) in May 1959, and the T2J-1 was cleared for introduction into the Training Command. A contest within Training Command to name the new aircraft was initiated, and the winning name of Buckeye, for the T-2's Ohio origins, was announced on June 3. The first T2J-1 was delivered to Training Squadron Four (VT-4) at NAS Pensacola, Florida, on July 9. Deliveries to VT-7 and VT-9 followed in 1960 and 1961 respectively.

As flight crews became familiar with it, the Buckeye earned a reputation as an excellent overall training platform with safe and predictable handling characteristics. It was underpowered, however. The J34 was an obsolete powerplant originally developed in 1944, chosen for the T2J because it was the only suitable engine available at the time.

By 1960 many new and efficient lightweight turbojets had become available, and on January 26, 1962, the Navy awarded NAA a $3.3 million contract to modify two T2J-1s to twin-engine configuration as T2J-2s. The selected powerplants were Pratt & Whitney J60-P-6s, each with 3000 pounds of thrust. The combined thrust of 6000 pounds was an 88-percent increase over the single J34, and this resulted in a safer aircraft. Because the engines were positioned beneath the central fuselage section, the modifications to accommodate them were engineered with minimal difficulty. The engine bay was widened to accommodate side-by-side installation and the air intakes were enlarged to provide greater airflow.

In 1962, under the new Tri-Service system,

The second T2J-1, BuAer 145997, modified as a YT-2B with two Pratt & Whitney J60-P-6 turbojets. With nearly double the power, the speed and climb rate were markedly improved.

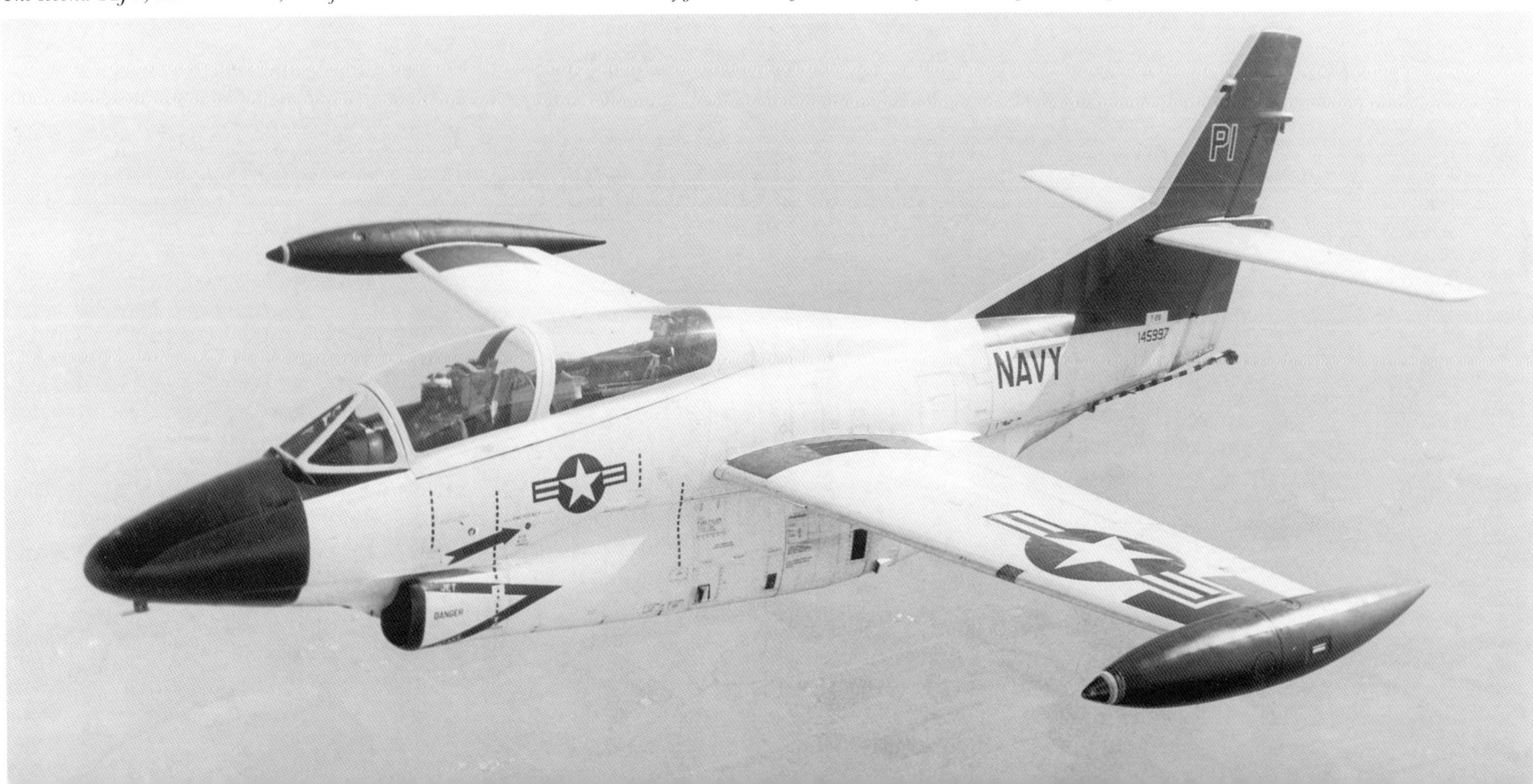

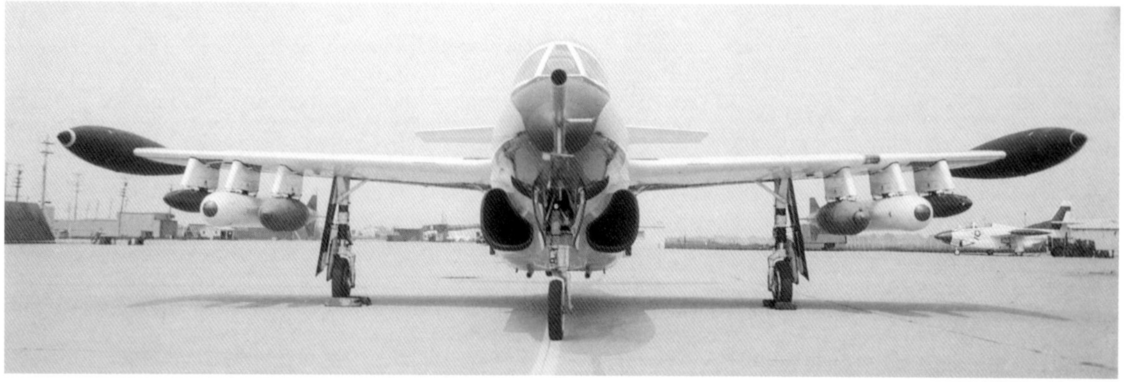

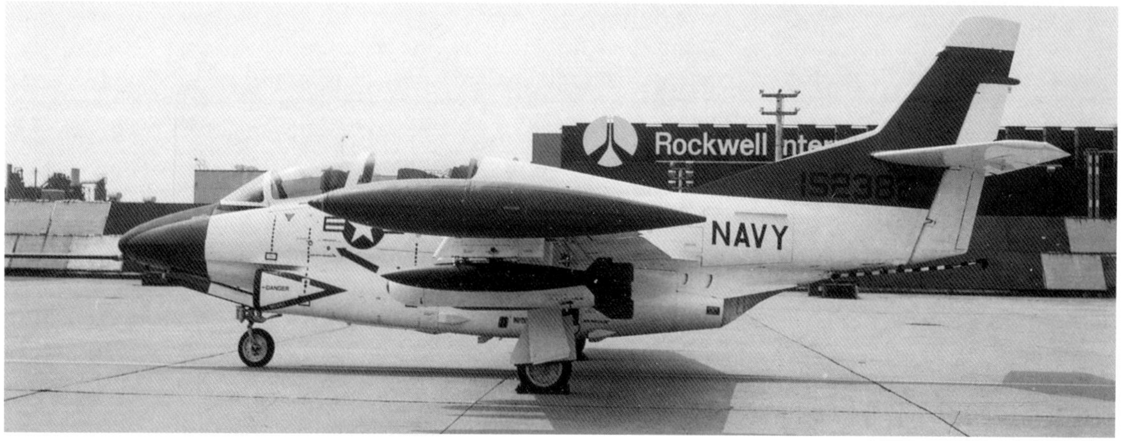

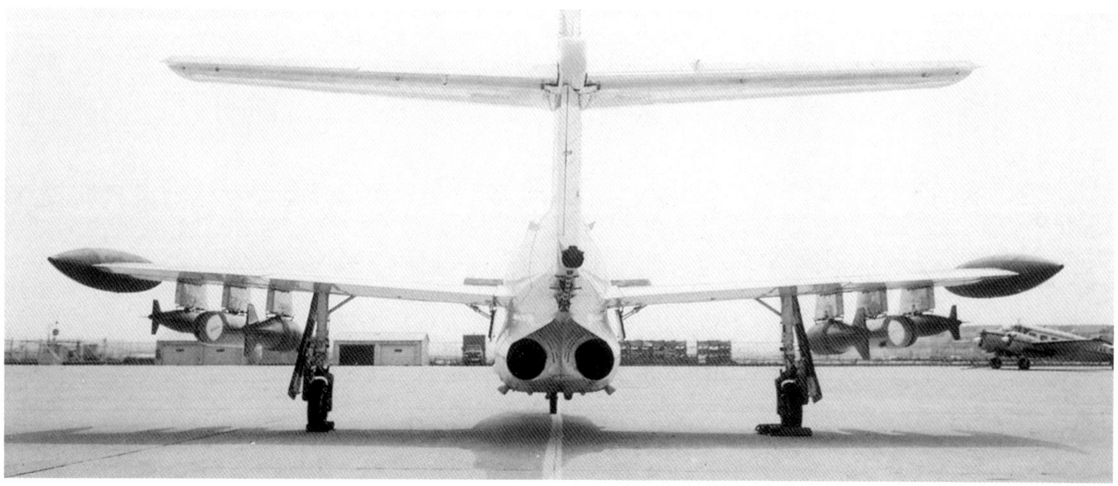

Three views of the first production T-2B, BuAer 152382, posed at North American Rockwell in Columbus in 1962.

the T2J was redesignated T-2. The original T2J-1 became the T-2A, and the new twin-engined version the T-2B. The YT-2B, BuAer 145997, first flew on August 30. Navy evaluation followed shortly thereafter, and an initial order for ten T-2Bs was placed on March 3, 1964, with subsequent orders for 87 additional aircraft through August 1, 1966.

Beginning with the 34th production T-2B an additional 50-gallon fuel tank was installed in each wing leading edge. The twin-engined Buckeyes required 1800 feet less take-off distance and had an increase in range of 165 nautical miles. They also featured improved avionics with the AN/APX-64(V) identification system that greatly enhanced instrument flying and long-distance flights. T-2B deliveries began in 1965, supplementing existing T-2A squadrons and equipping new training units.

In 1967 a newer, more cost-effective engine became available. The General Electric J85-GE-4 provided virtually the same thrust as the J60, allowing identical performance with a reduced cost-per-unit. The Navy placed a $2.2 million contract to have the first production T-2B fitted with these engines, this configuration designated T-2C. The YT-2C, converted from T-2B BuAer 152382, first flew on April 17, 1968.

The T-2C went into service in early 1969, and became the definitive Buckeye variant. Various improvements to the avionics systems, the use of two under-wing hardpoints and a target-towing mechanism allowed the T-2C to be a truly versatile training aircraft. With the weapons training package, a large variety of external ordnance could be carried on the under-wing hardpoints. These included 50-caliber gun packages, AA 37B-3 multiple bomb racks, Mk 86 and Mk 15 practice bombs, 2.75-inch rocket packages, Aero 1A or 1B banner target carriers, and Mk 1 target containers. An attack package with six under-wing stations yielded a maximum capacity of 3500 pounds of external stores, including GPU-3A gun pods, Mk 81 and Mk 82 general-purpose bombs,

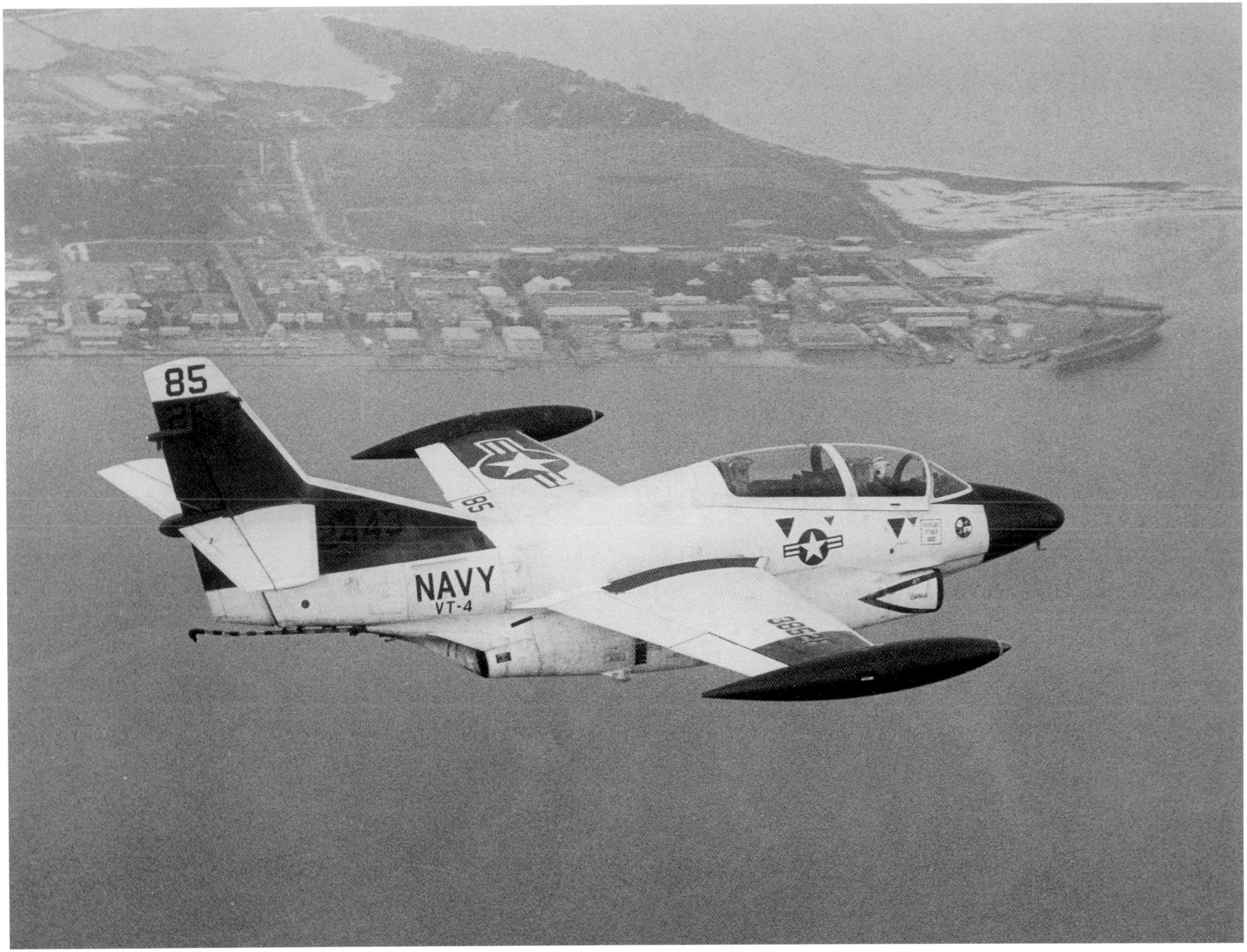

T-2B, BuAer 152443, serving with Training Squadron Four, in flight near NAS Pensacola in 1966. Note the carrier docked at right. VT-4 had begun training with T2J-1s in July 1959.

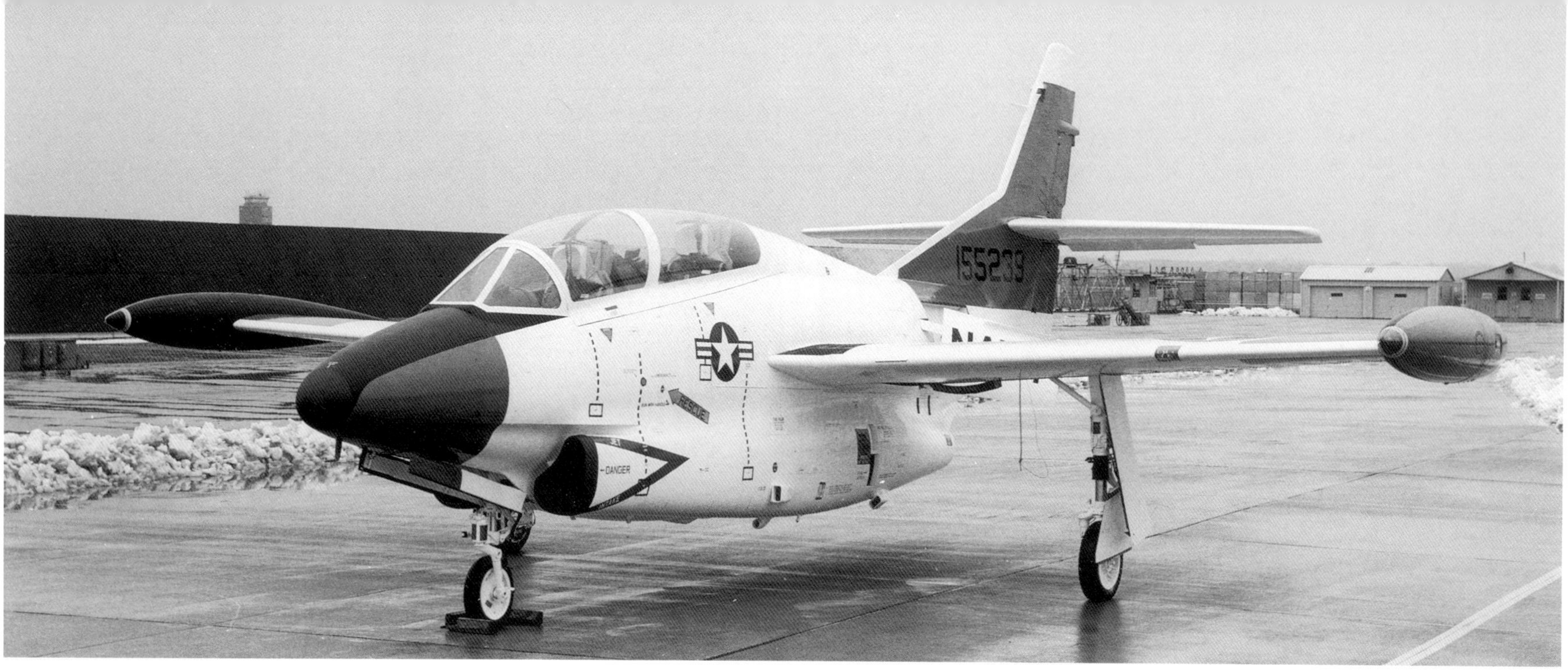
155239

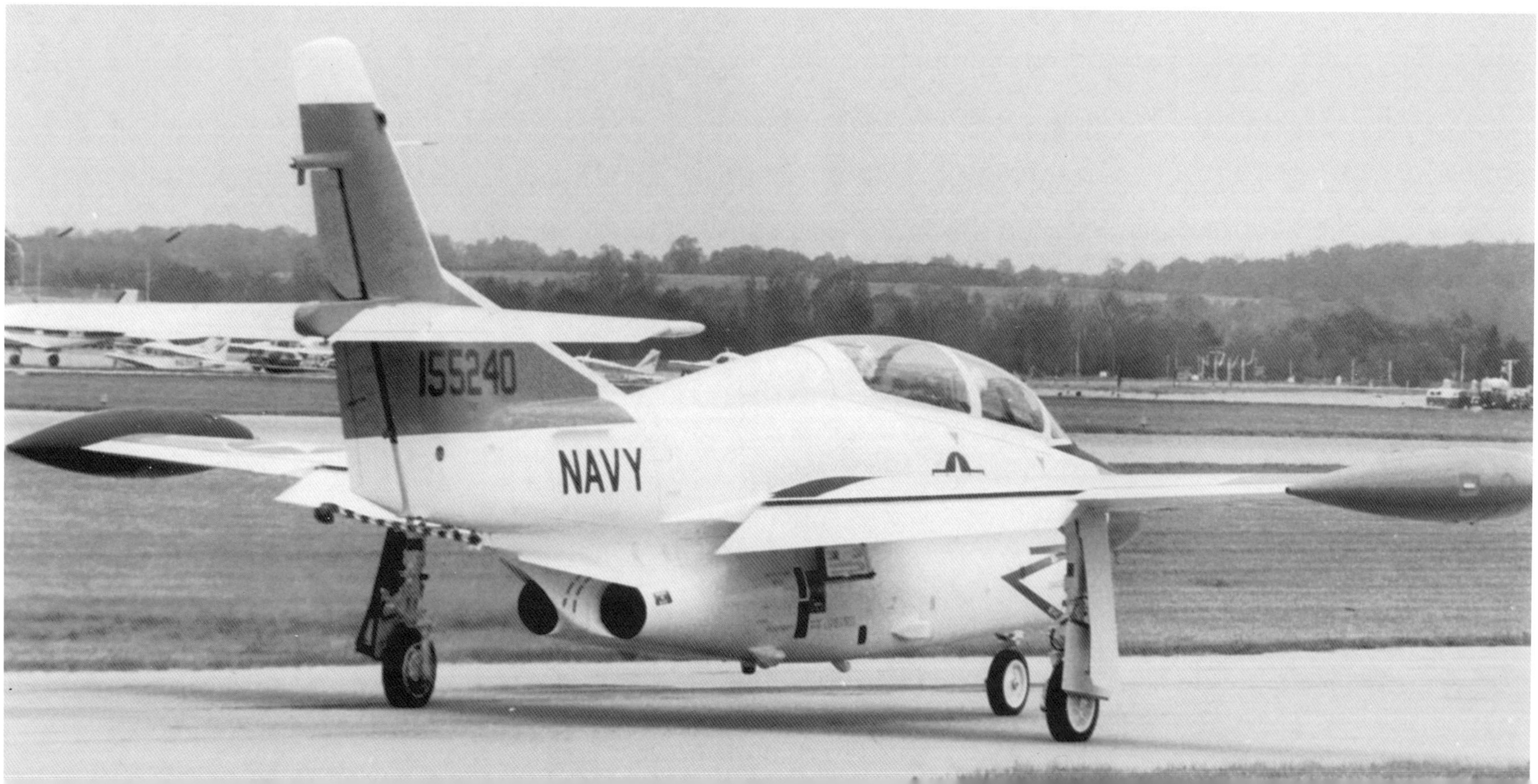
155240
NAVY

Opposite, the first two production T-2Cs, BuAer 155239 and 155240, at Columbus in 1969. T2C-1s used J85-GE-4 engines.

retarded "Snakeye" bombs, flare dispensers, various rocket packages and fire bombs. This package enabled the Buckeyes to serve as cost-effective light attack aircraft.

As originally planned, the Buckeye proved to be a simple aircraft to operate and service. All service operations could be performed at ground level. The engines had a built-in self-start capability, eliminating the need for additional ground power equipment, and engines could be replaced in less than three hours. Maintenance required 7.1 man-hours per flight hour, a very low figure among operational military planes in the early 1970s, with average aircraft availability of over 75 percent.

The T-2C eventually served in the following Navy (as well as some Marine Corps) training squadrons: VT-4, VT-7, VT-9, VT-10, VT-19, VT-23, VT-26, VT-43 and VT-126. The latter two squadrons were adversary training units using the Douglas A-4, Northrop F-5 and IAI Kfir (F-21) as adversary aircraft for air combat training. The T-2Cs serving in these units provided out-of-control or spin-recovery training. Early in the program, minor structural problems temporarily halted these maneuvers, but when they were remedied the training resumed. The safe and predictable manners of the Buckeye have made it an excellent aircraft for student pilots to learn spin recovery procedures.

In 1970, a NASA-Naval Air Systems Command joint venture explored the supercritical wing concept on a T-2C. The supercritical wing was designed to eliminate transonic drag, had a flatter upper surface and was much thicker than the production wing. By adding fiberglass and balsa wood to the wing and shaping the airfoil to the desired thickness and contour, a direct comparison of flight characteristics between the production and supercritical wings was made without having to alter the landing gear or other systems.

Another major change in configuration was a proposal for a four-seat multi-purpose trainer.

T-2C

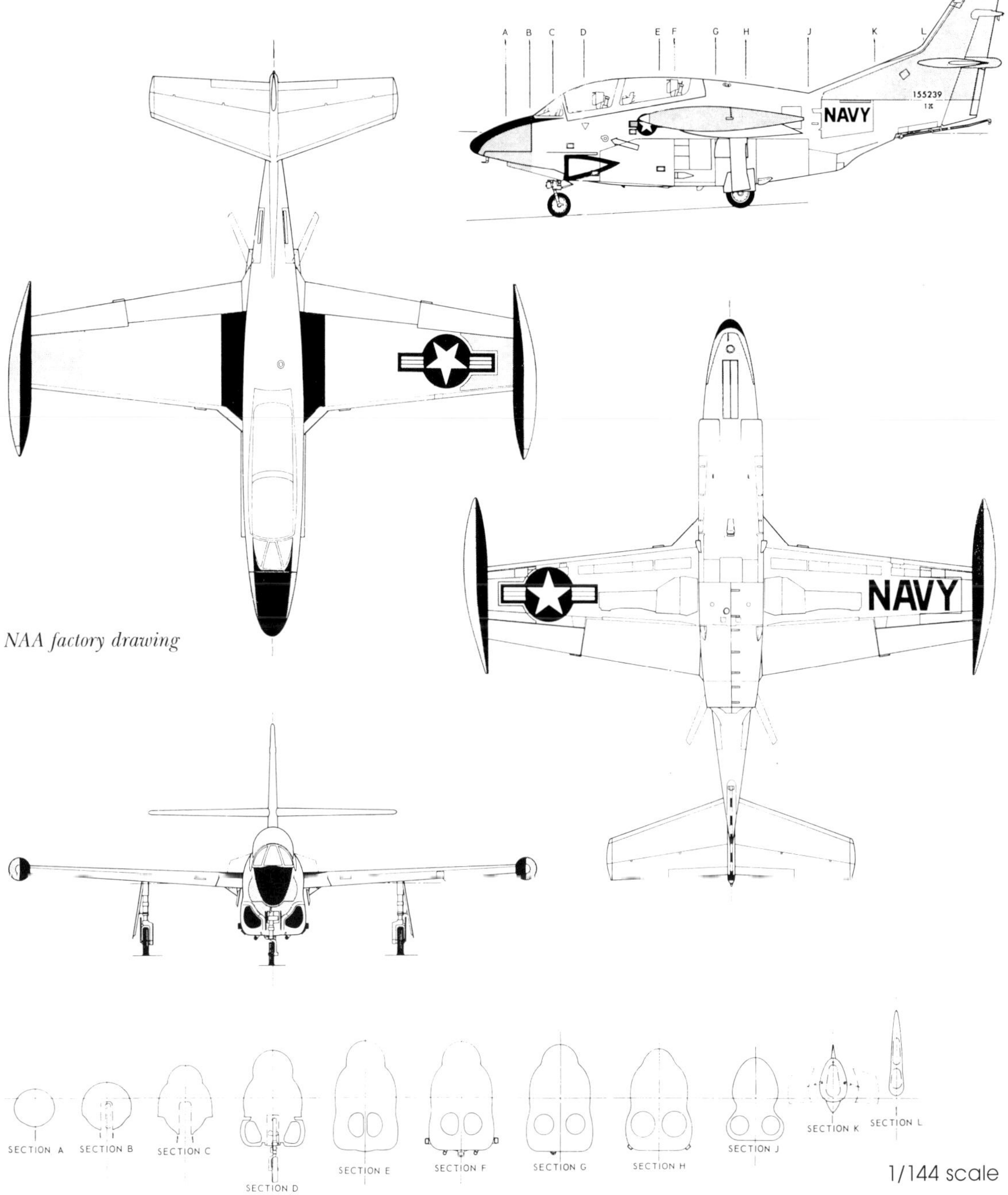

NAA factory drawing

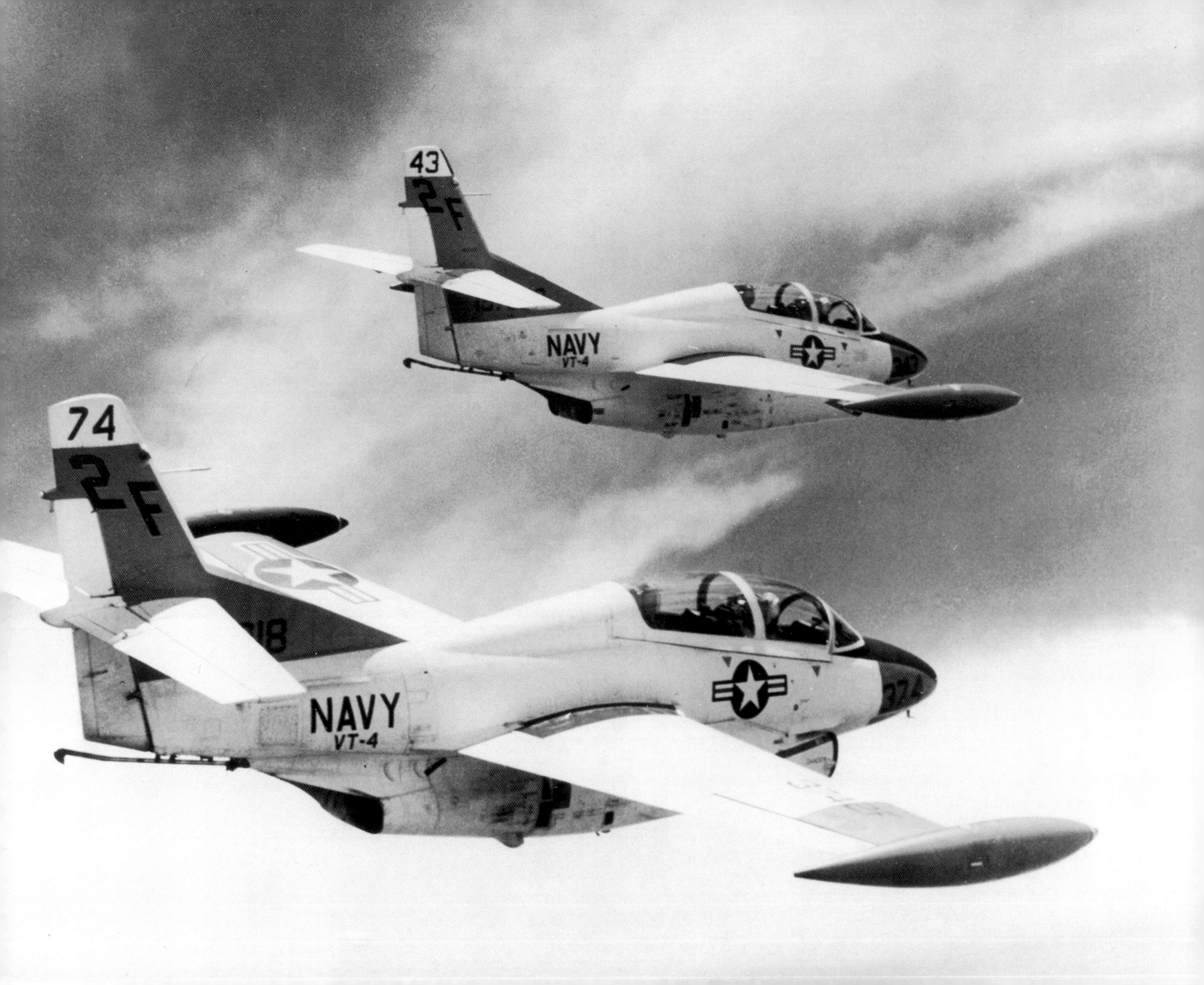
43
2F
NAVY
VT-4
74
2F
NAVY
VT-4

Opposite page, two T-2Cs flying with VT-4 of Training Air Wing Six in 1971.

As the Rockwell International T-2, the Buckeye trainer was offered to foreign air forces in essentially T-2C configuration. Besides its training role, it was also suitable for light attack duties. This cutaway illustration from a Rockwell marketing brochure shows the basic model without weapons pylons.

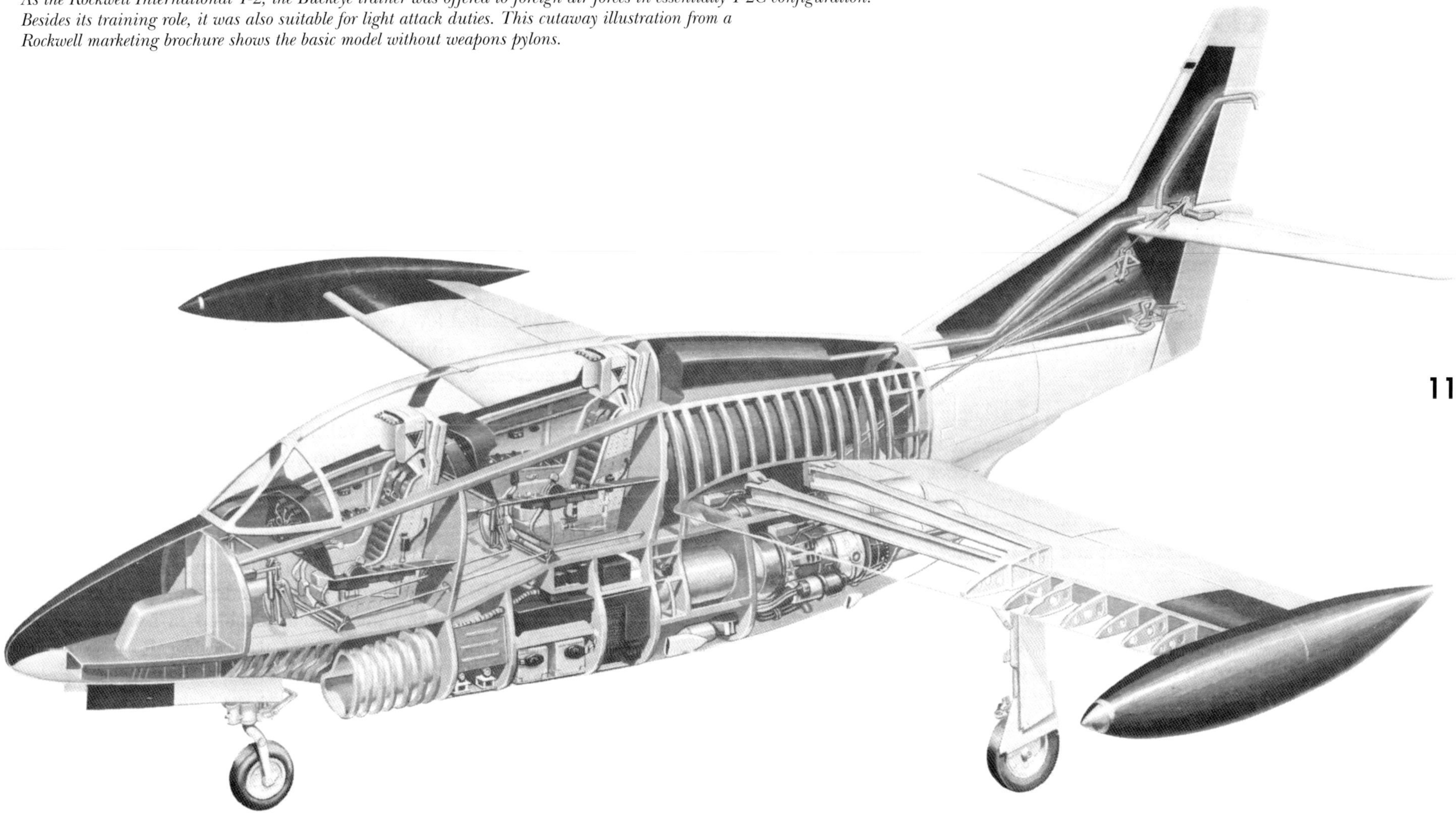

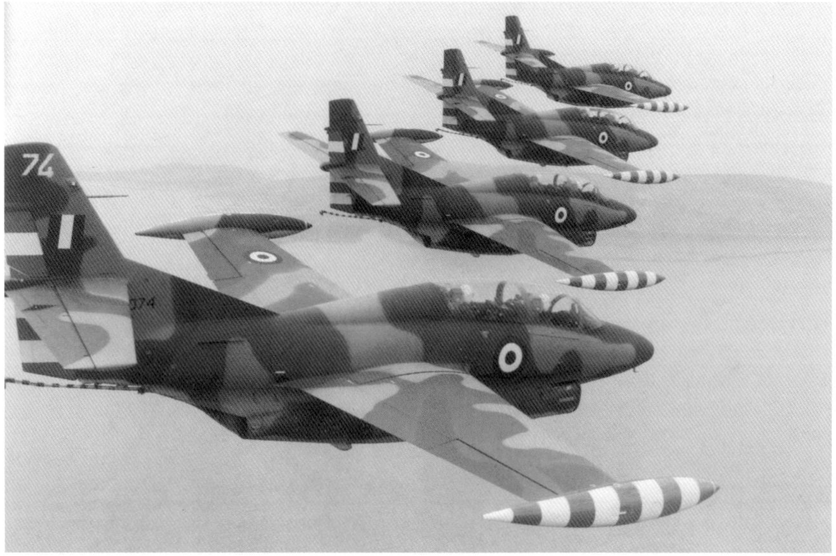

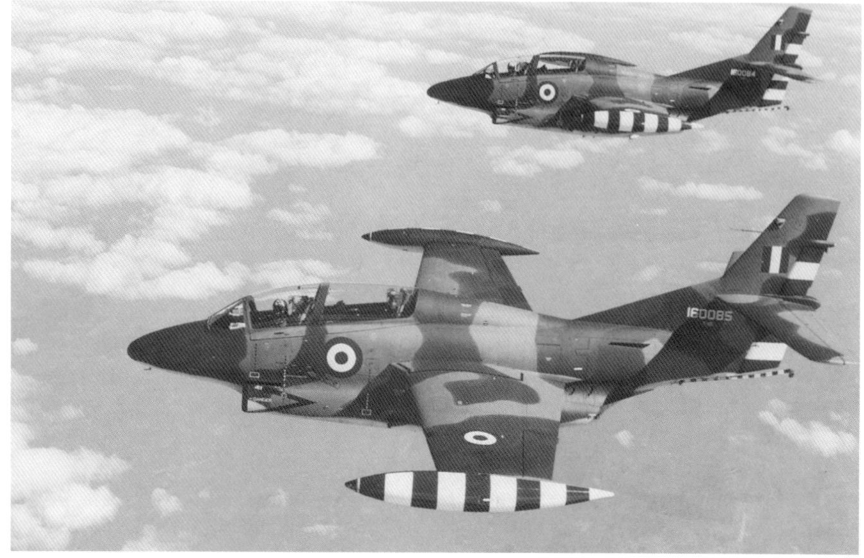

Rockwell delivered forty T-2Es to Greece for the Hellenic Air Force in 1976-1977. To implement the sale, these went through the U.S. Navy with the BuAer serial numbers 160059/160098.

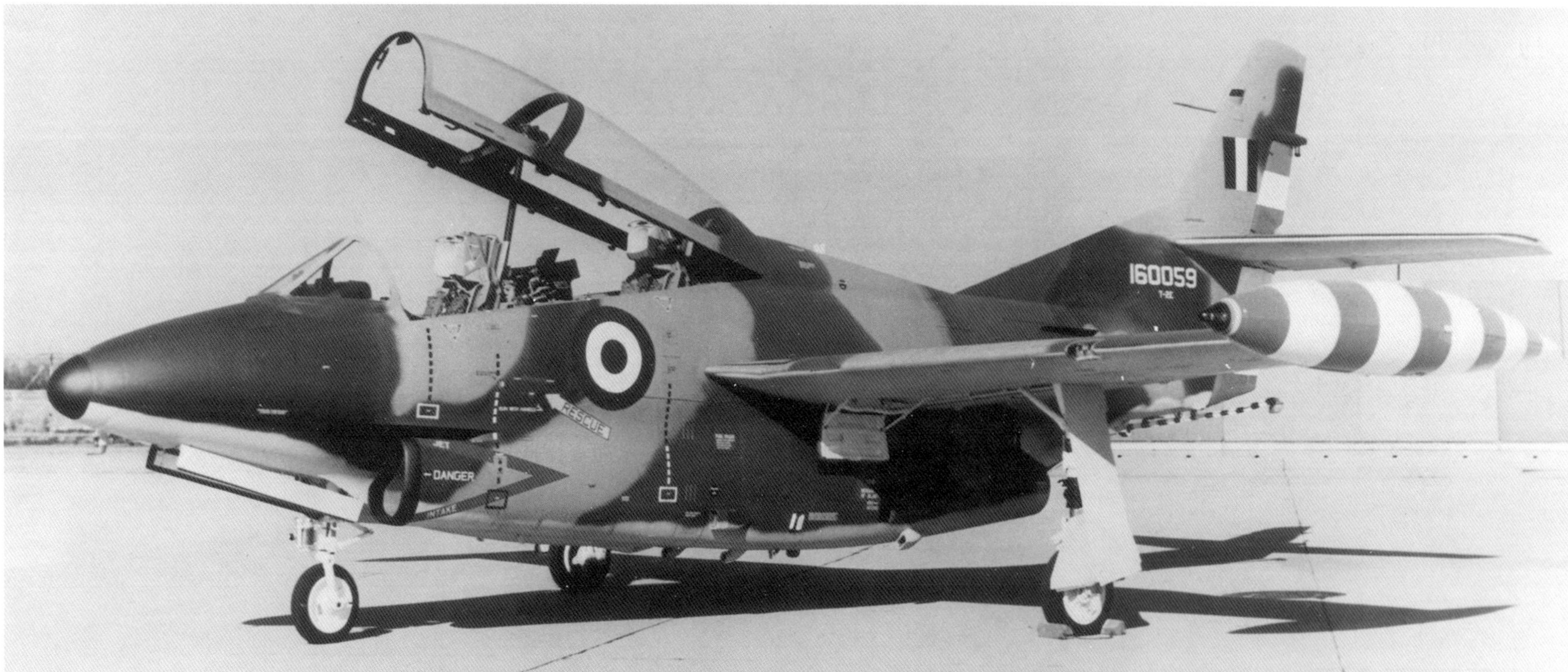

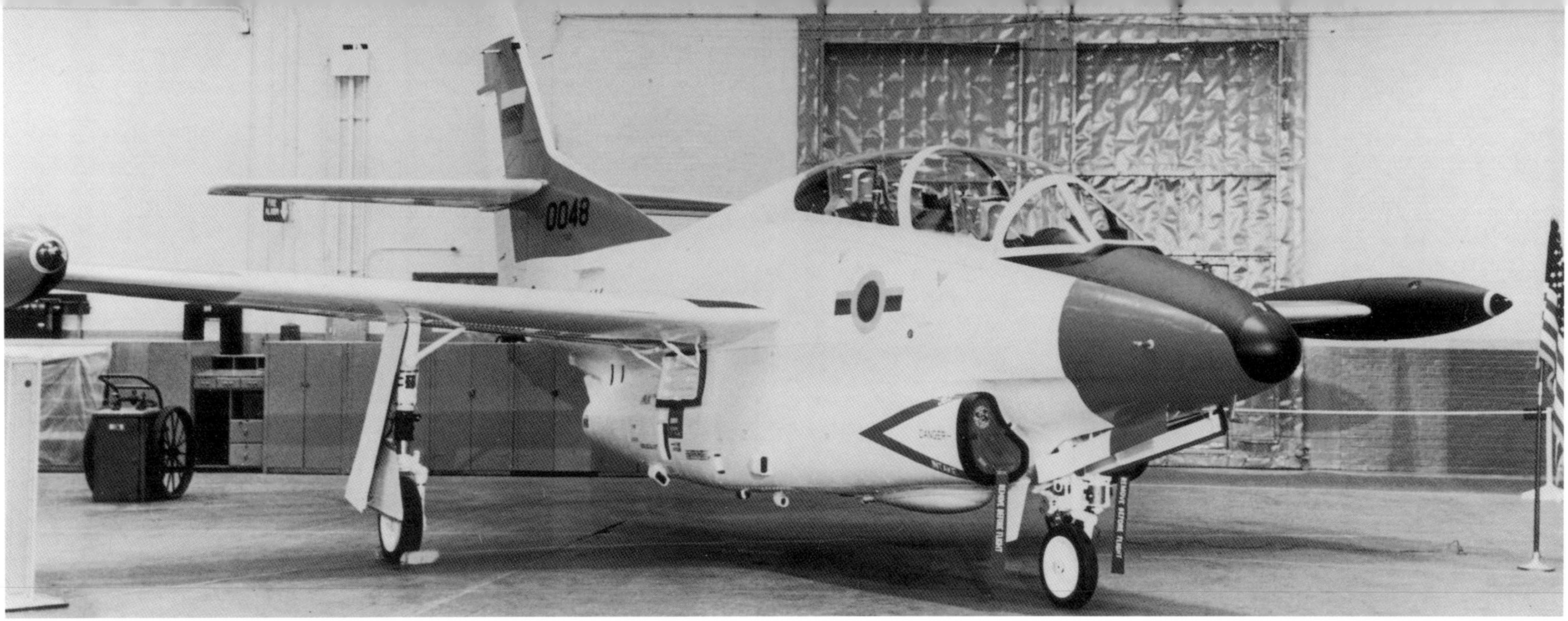

Above, wearing Venezuelan insignia and serial 0048, is the last of 24 T-2Ds delivered during 1973-1977. These also had BuAer numbers for contractual reasons. The second group of twelve was originally ordered by Morocco.

Specifications: North American NA-241>NA-405 T2J-1/-2 / T-2A/B/C/D/E Buckeye

Description: U.S. Navy multi-purpose trainer. Single-jet (T2J-1/T-2A) or twin-jet (T-2B/C/D/E) mid-wing monoplane of all-metal construction with retractable landing gear. Crew 2 (instructor and student).

Engines: One 3400-lb-thrust Westinghouse J34-WE-48 turbojet. (T-2B) Two 3000-lb-thrust Pratt & Whitney J60-P-6 turbojets. (T-2C/D/E) Two 2950-lb-thrust General Electric J85-GE-4 turbojets.

Dimensions: Span 38 feet 2 in, length 38 feet 8 in, height 14 feet 9 in, wing area 255 sq ft. (T-2B/C/D/E) Height 15 feet.

Weights: 8006 lb empty, 11,498 lb maximum take-off. Fuel 3842 lb. (T-2B) 7630 lb empty. (T-2C/D/E) 8115 lb empty, 13,180 lb maximum take-off. Fuel 4492 lb.

Performance: Maximum speed 488 mph, climb 5000 ft/min, service ceiling 40,000 ft, range 964 miles. (T-2B/E) Maximum 521 mph, climb 6560 ft/min, range 908 miles.

This was engineered to allow a wider range of training programs, including electronic warfare systems for transition to such aircraft as the Grumman EA-6 Prowler. The four-seat arrangement went only as far as the mock-up stage, built around T-2A BuAer 144218.

In February 1972 the Venezuelan Air Force ordered twelve T-2Cs from NAA. These aircraft had minor changes in avionics and the attack package with six hardpoints fitted, but lacked carrier capability. The Venezuelan Buckeyes were designated T-2Ds, and a total of twenty-four were delivered between 1973 and 1977. The Hellenic Air Force also ordered Buckeyes, forty of which were designated T-2E and delivered to Greece in 1976 and 1977. The T-2Es were similarly equipped to the Venezuelan aircraft, with the exception of retaining the tailhook. Both the T-2D and T-2E, delivered by way of U.S. Navy contracts and serial numbers, had reinforced wingtip tanks for protection from small arms fire.

The Buckeyes were due to be phased out in 1987 and replaced by the McDonnell Douglas T-45 Hawk, but as of 1999 a few T-2s remain in service with both the Navy and Marine Corps, testimony to the versatility and durability of these excellent multi-purpose training aircraft.

Production: North American NA-241>405 / T2J-1/T-2A / T-2B/C/D/E Buckeye

NA-241	T2J-1	Engineering	—
NA-249	YT2J-1	BuAer 144217/144222	6
NA-253	T2J-1	BuAer 145996/146015, 147430/147530	121
NA-266	T2J-1	BuAer 148150/148239	90
NA-280	YT2J-2/YT-2B	(BuAer 144218, 145997)	(2)
NA-288	T-2B	BuAer 152382/152391	10
NA-291	T-2B	BuAer 152440/152475	36
NA-294	T-2B	BuAer 153538/153555	18
	YT-2C	(Converted T-2B BuAer 152382)	(1)
NA-307	(T-2C production transfered to NA-310)		—
NA-310	T-2B	BuAer 155206/155238	33
	T-2C	BuAer 155239/155241	3
NA-318	T-2C	BuAer 156686/156733	48
NA-332	T-2C	BuAer 157030/157065	36
		(BuAer 157066/157101 canceled)	(36)
NA-340	T-2C	BuAer 158310/158333	24
NA-346	T-2C	BuAer 158575/158610	36
NA-352	T-2C	BuAer 158876/158911	36
NA-358	T-2D (Venezuela)	BuAer 159330/159341	12
NA-367	T-2C	BuAer 159150/159173, BuAer 159704/159727	48
NA-380	(T-2C long-lead materials)		—
NA-396	T-2E (Greece)	BuAer 160059/160098	40
NA-398	T-2D (Venezuela)	BuAer160228/160239	12
NA-405	(T-2 production for Turkey—canceled)		—
Total, T2J-1 and T-2 trainers, 1956-1977			609

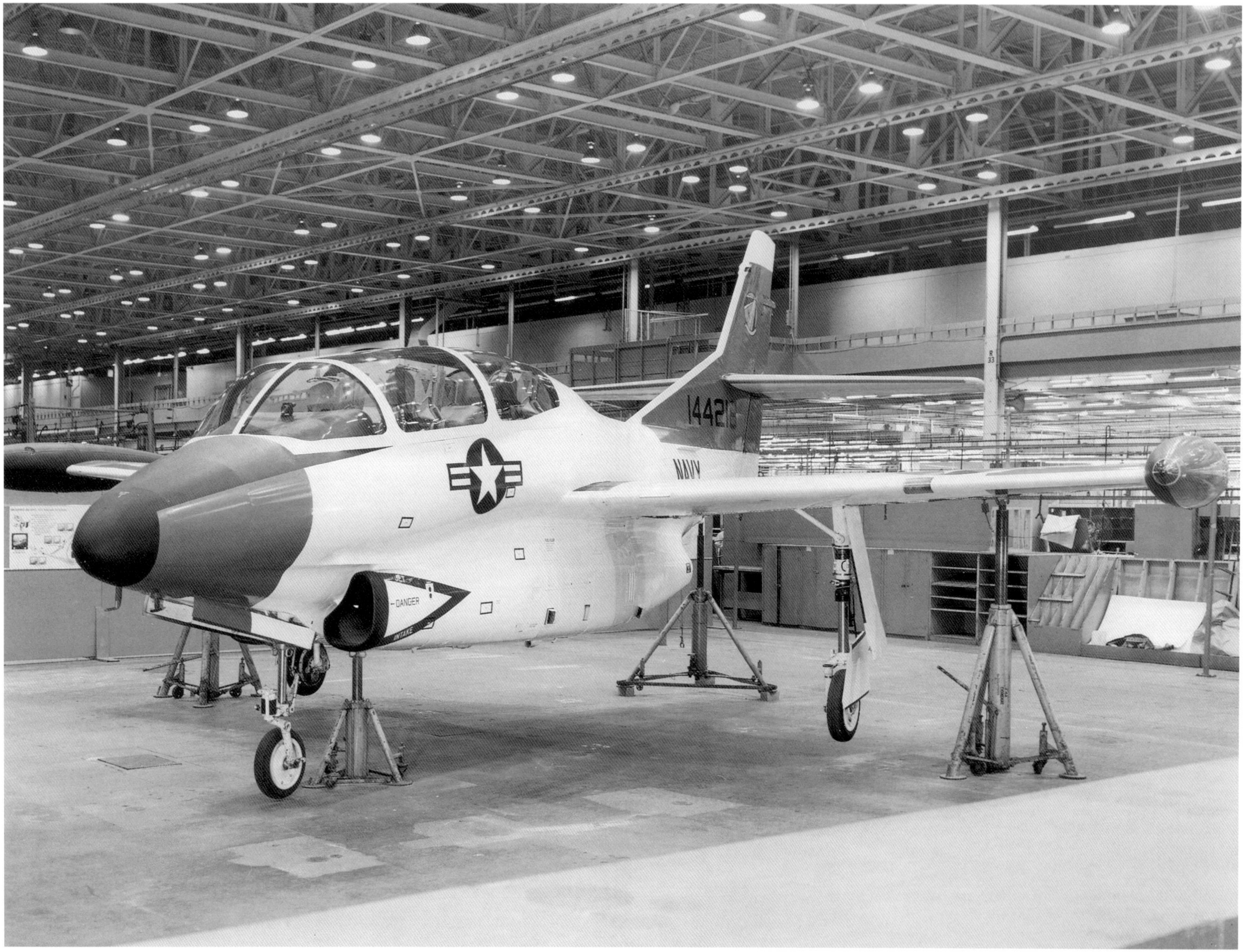

Rockwell proposed a 4-seat multi-purpose trainer to the Navy to facilitate a wider range of instruction, including electronic warfare systems. This mock-up was built on YT2J-1/T-2A, 144218.

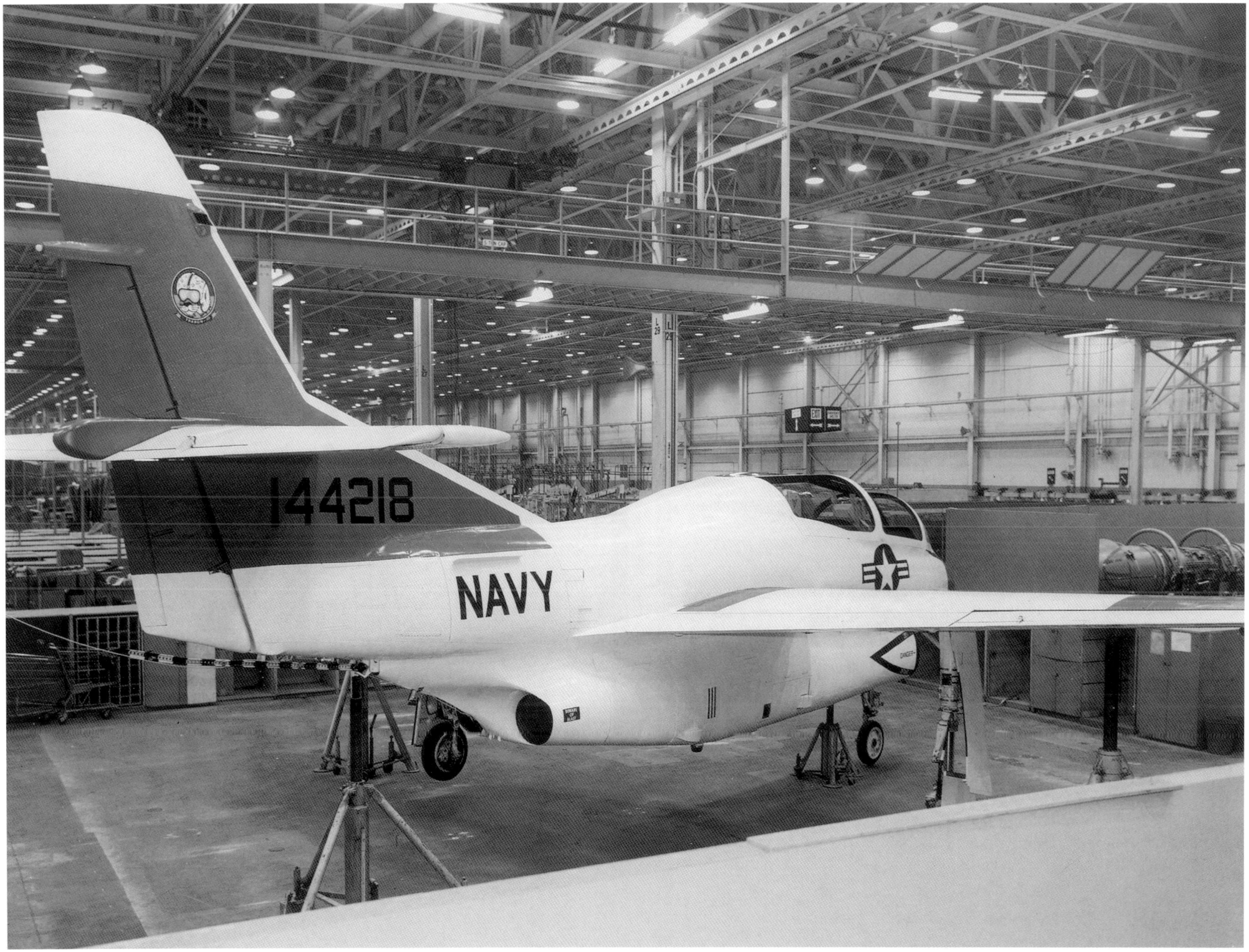

In addition to the 4-seat cockpit, the mock-up was brought up to twin-engine configuration, as appropriate for a maximum weight of 16,000 pounds. Tiger-in-helmet insigne says "TRARON-10."

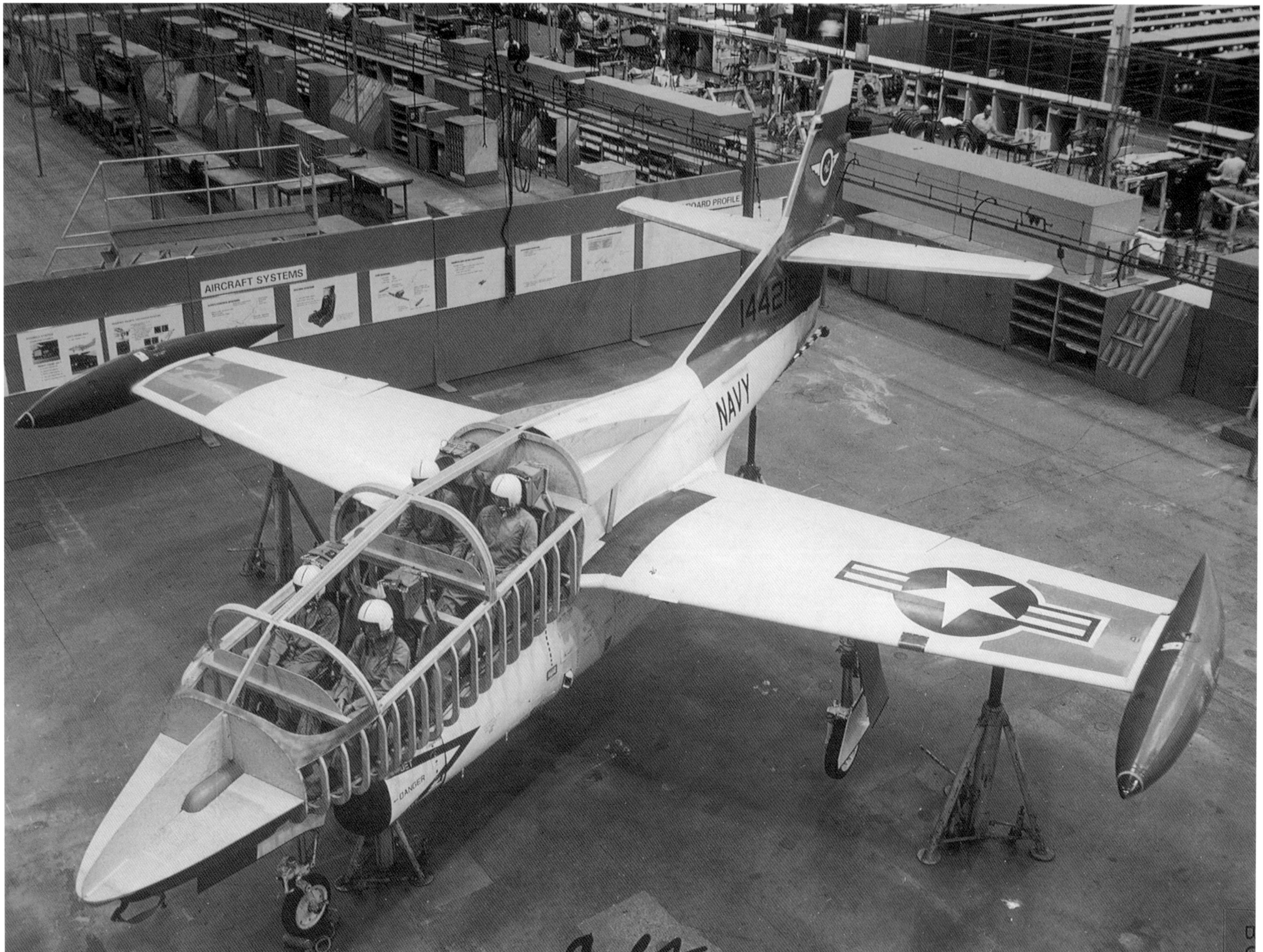

A four-seat T-2 would have been a flying classroom: "Pilot to co-pilot, electronics officer and co-electronics officer, we're ready for take-off." The instructional needs were better met by T-39Ds.

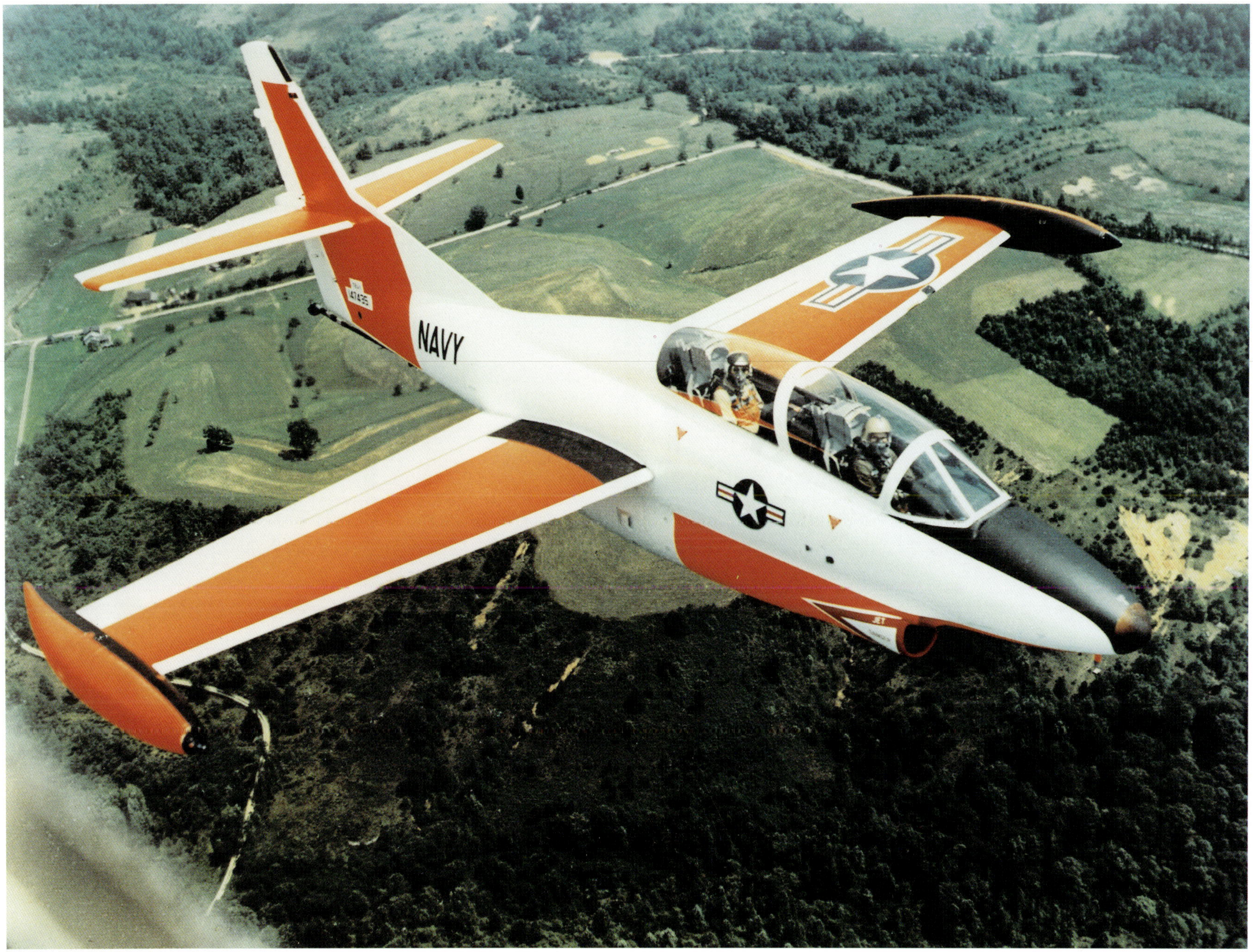

T2J-1 BuAer 147435, before assignment to a training squadron. The high-visibility da-glo orange and white scheme was used, with several variations in pattern, on almost all Buckeye trainers.

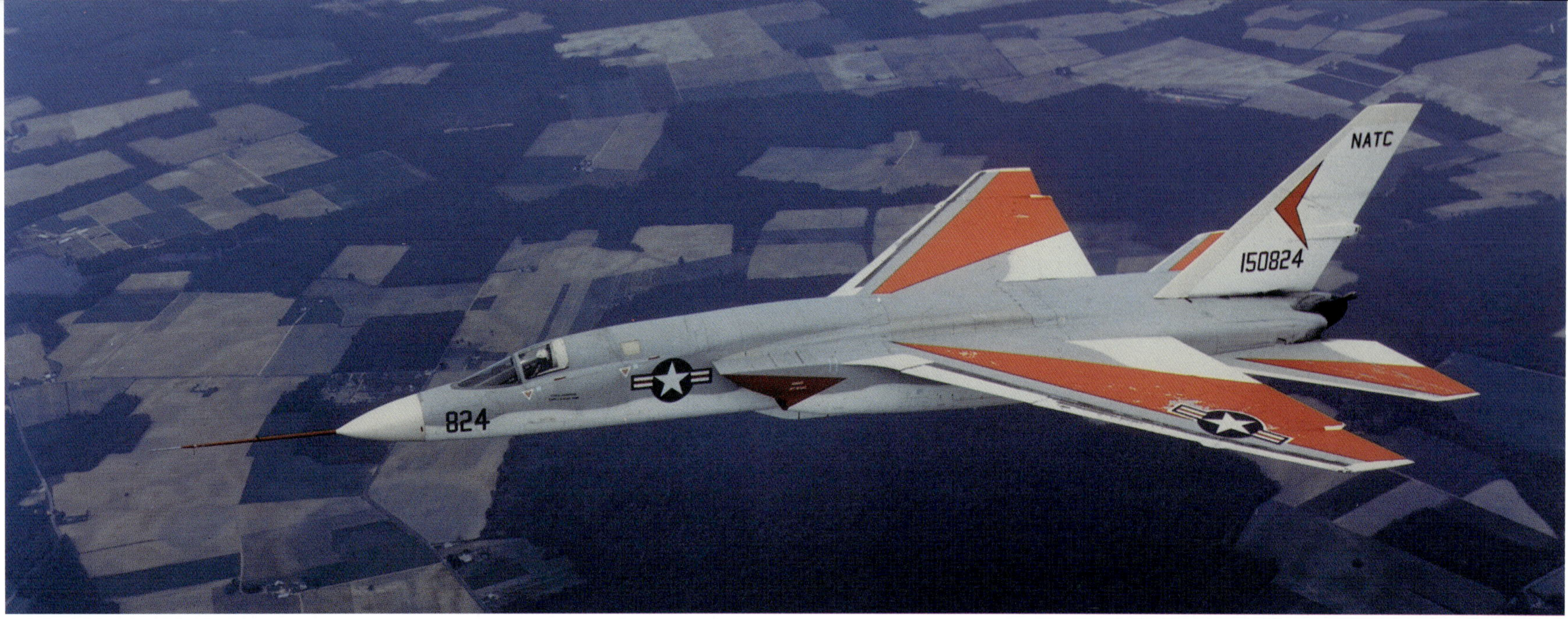

RA-5C, BuAer 150824, during NATC tests. Below, RA-5Cs serving with RVAH-3. Opposite, XB-70A, 62-0207, landing at Edwards with drag chutes deployed. Note the XB-58 chase plane.

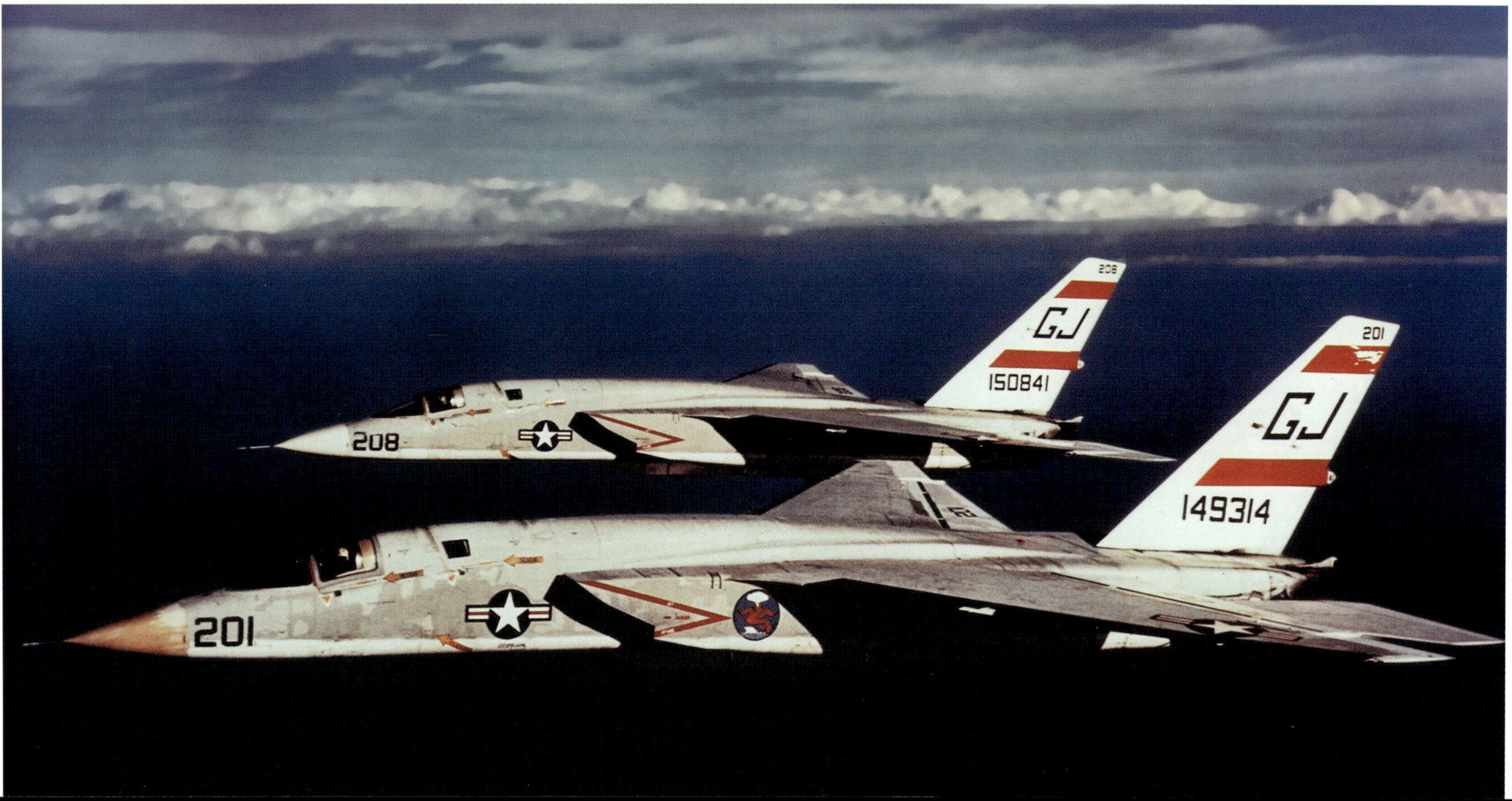

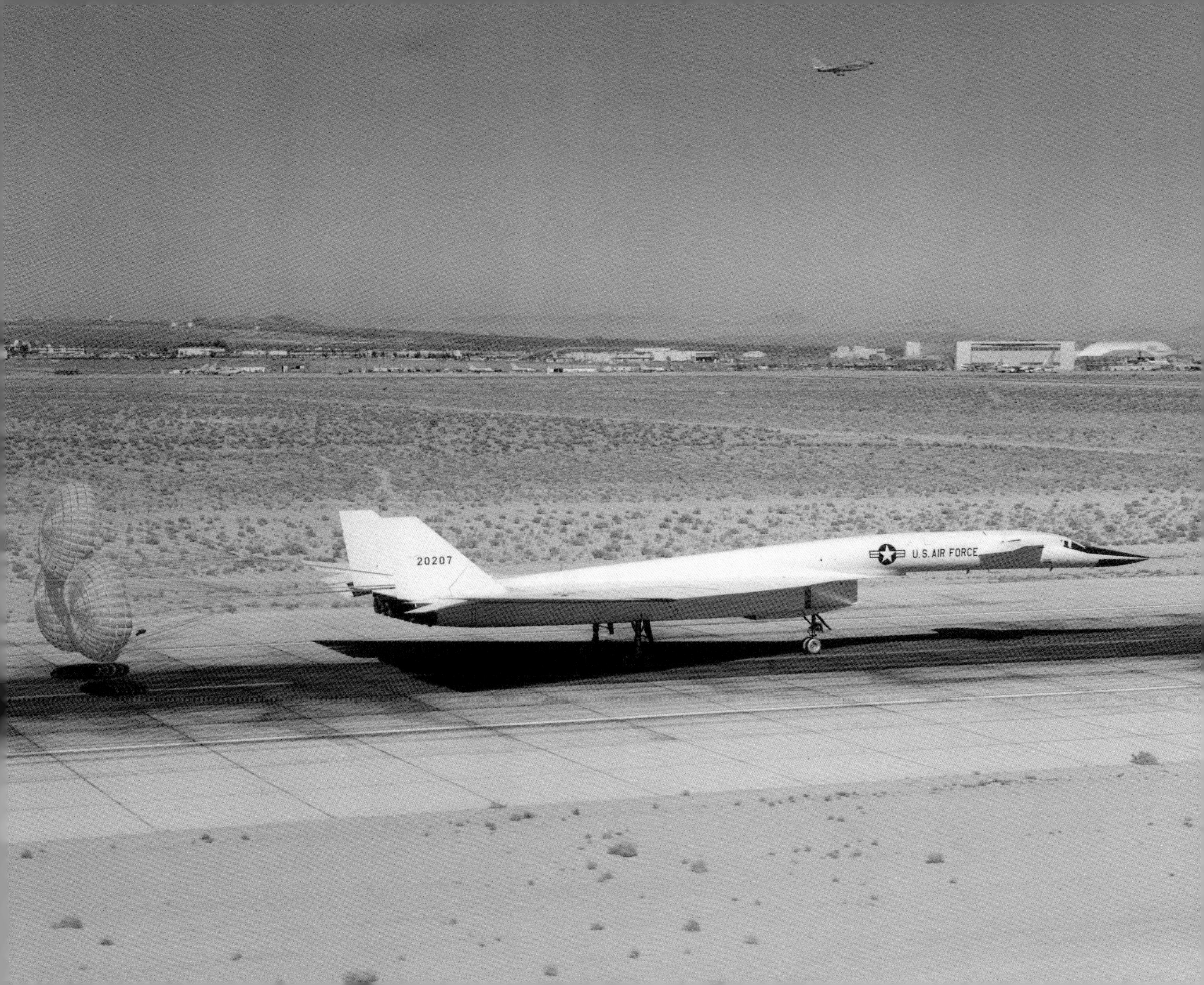
20207
U.S. AIR FORCE

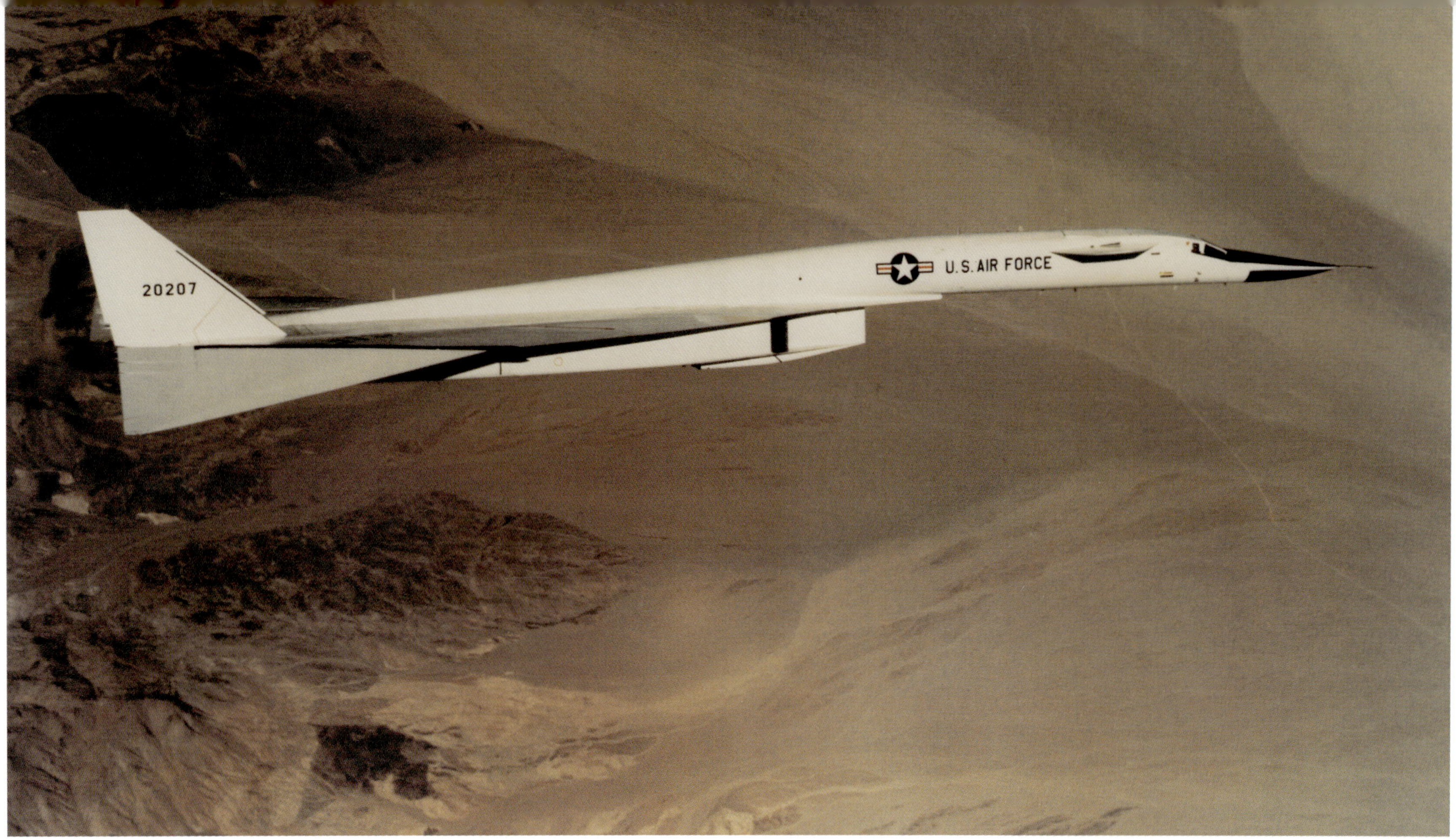

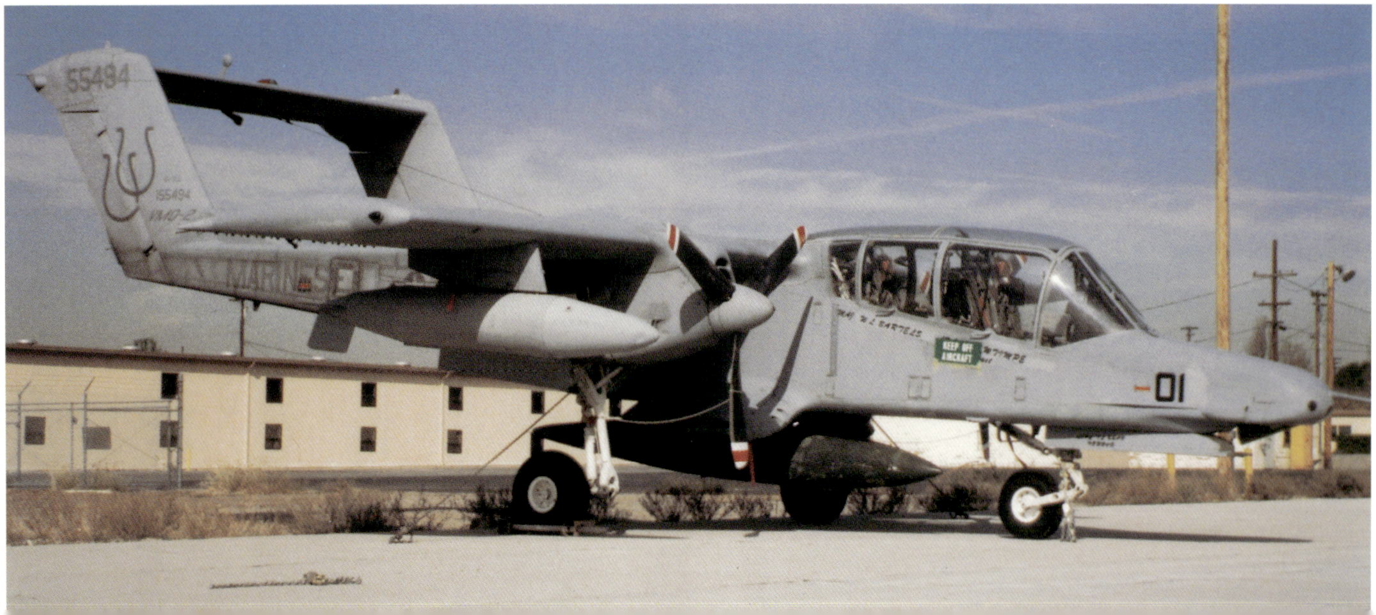

Above, XB-70A, 62-0207, with its wingtips 25-degrees down, over the Mojave desert in November1965.

Opposite above, Marine YOV-10D, 155395, in flight during NOGS tests, with an FLIR turret in the nose and a 20-mm gun turret below the fuselage.

Left, Marine OV-10, 155494, in VMO-2 markings on display at MCAS El Toro in January 1994.

Opposite, the second B-1A, 74-0159, modified for the B-1B test program (as indicated by the tail graphics), takes off at Edwards in March 1983.

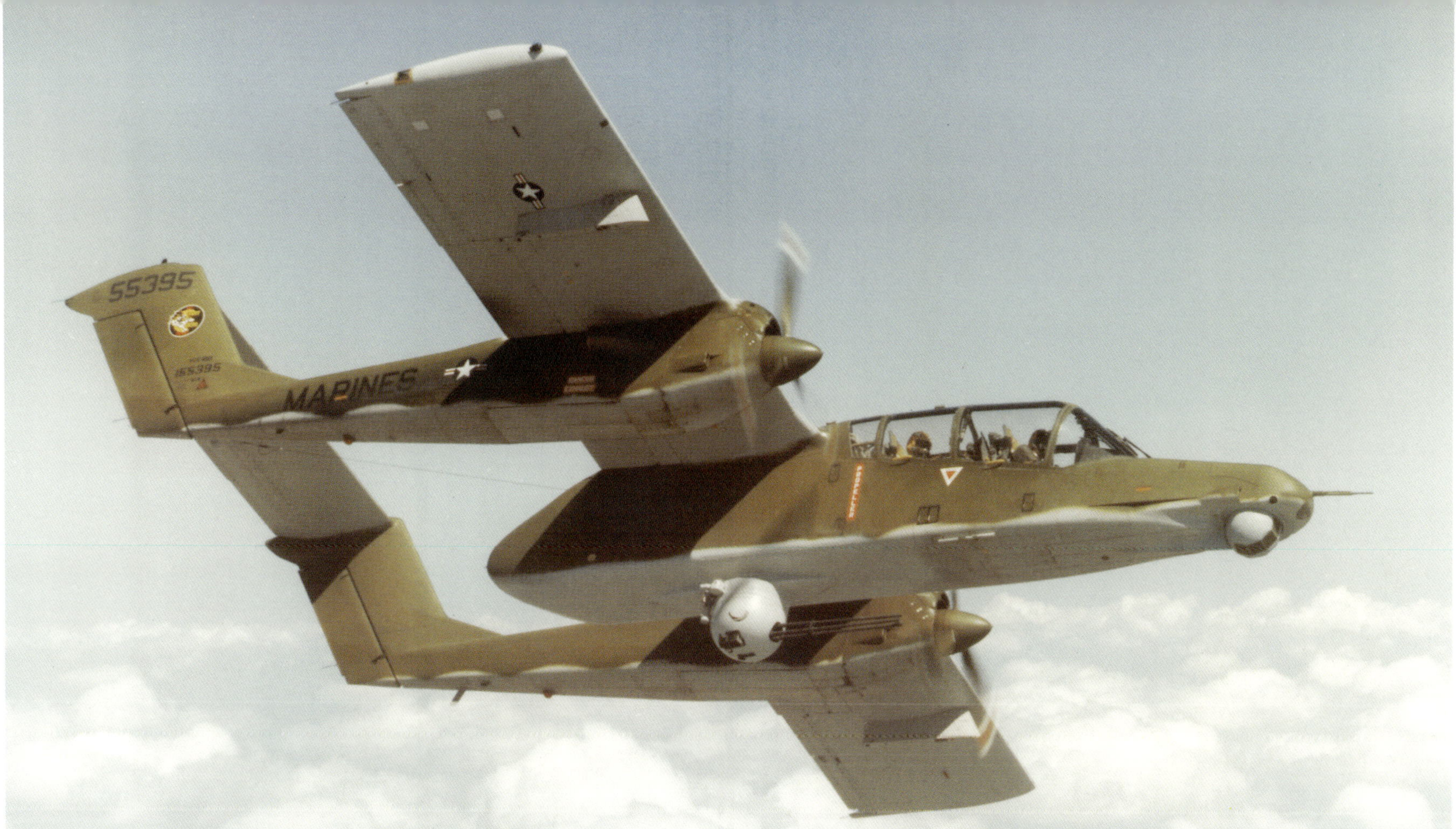
55395
MARINES
155395

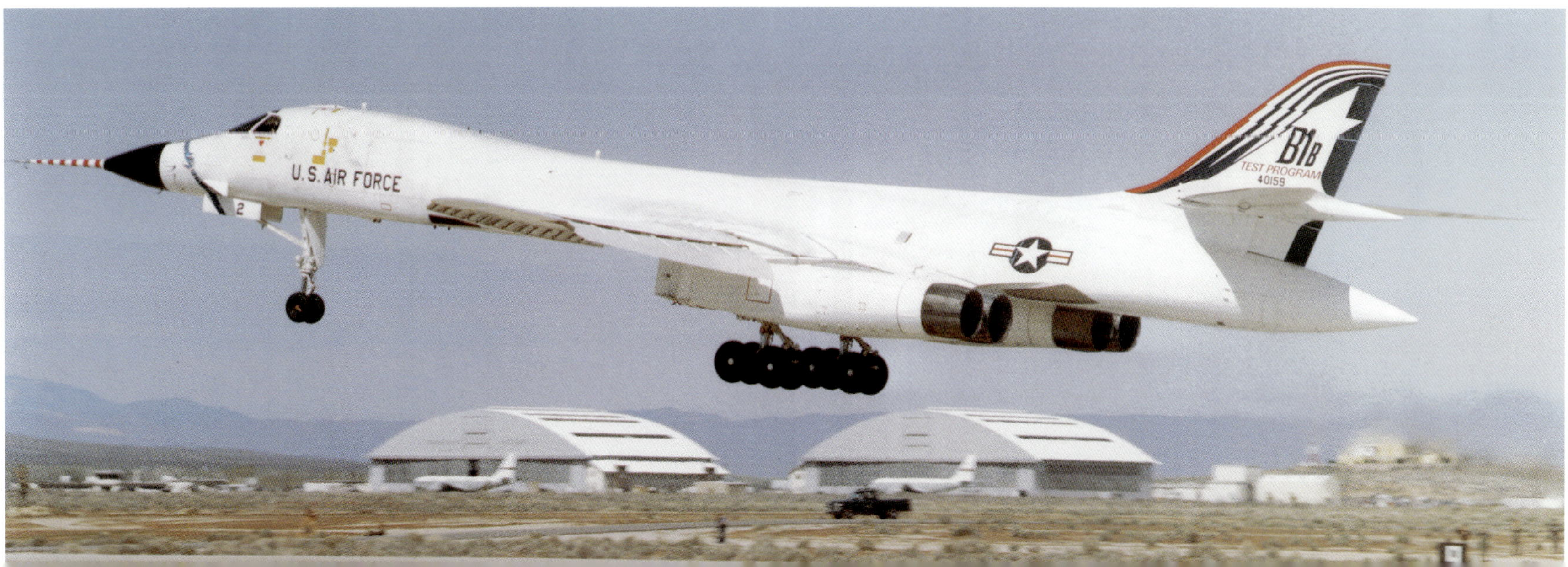
U.S. AIR FORCE
B1B
TEST PROGRAM
40159

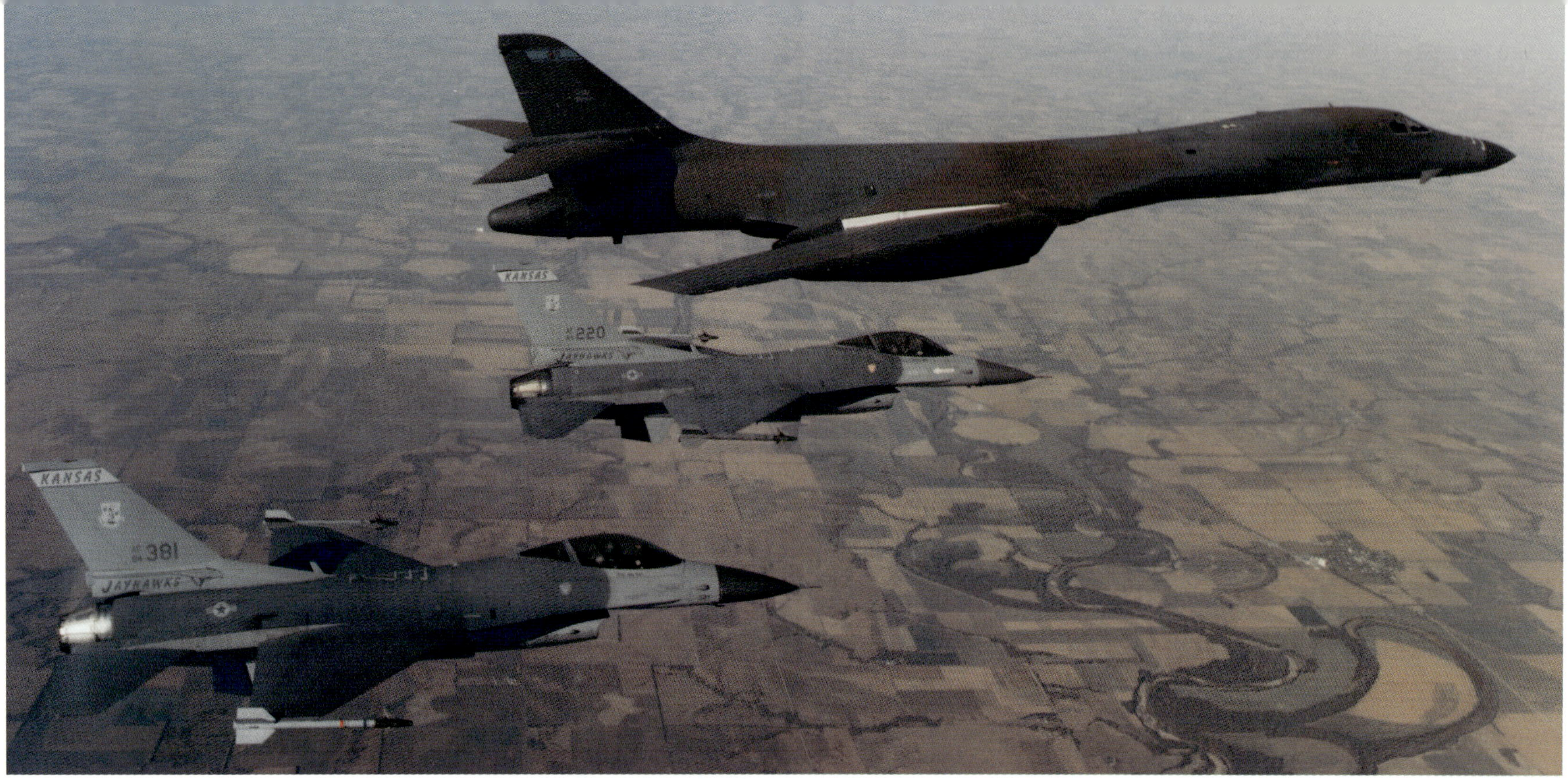

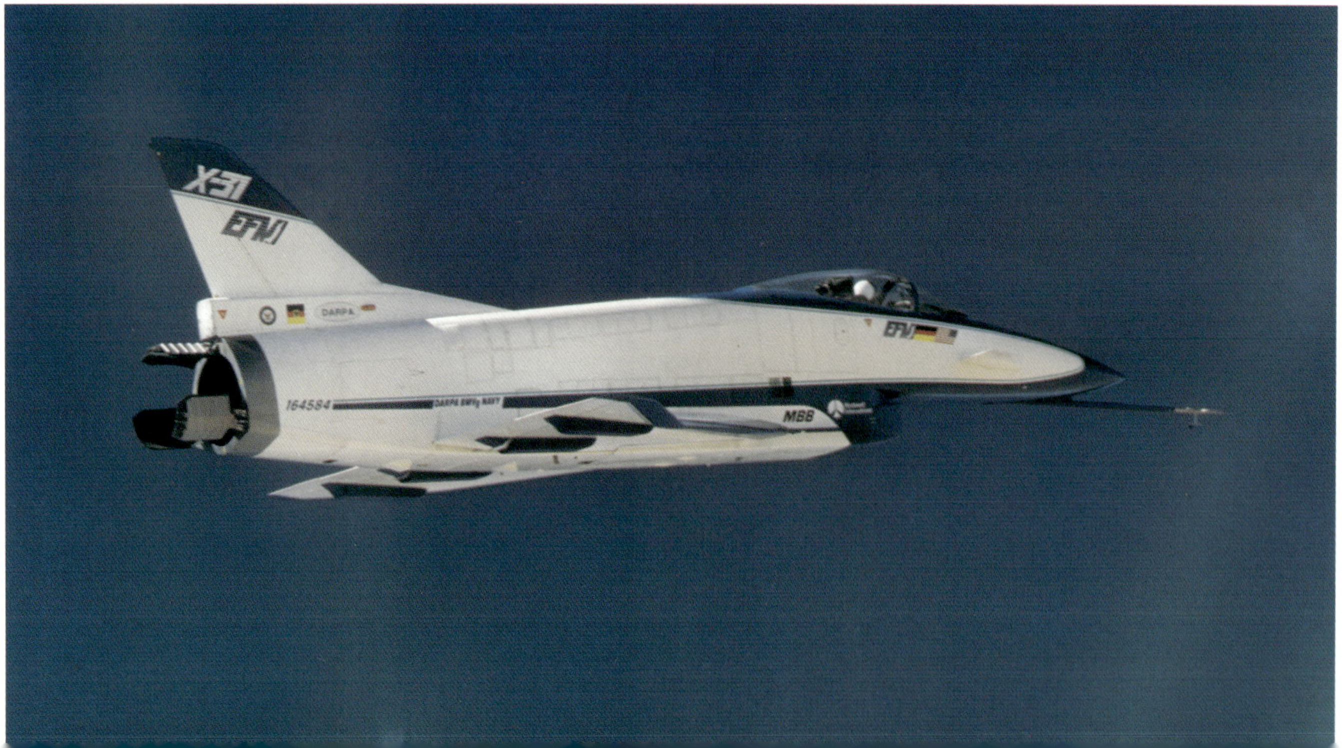

Above, with extremely drab camouflage compromised only at the wing pivot, and tiny low-visibility insignia, the last B-1B, 86-0140, sneaks along with Kansas ANG "Jayhawks" F-16As.

Right, the first of two Rockwell/MBB X-31A Enhanced Fighter Maneuverability Demonstrators, BuAer 164584, in level flight over the Dryden Flight Research Center at Edwards AFB.

Opposite above, the same X-31A, banking with its gear down for the camera, 15 miles north of Edwards, with California City in the background.

Far right, a dramatic night shot of a U.S. Marines CT-39G tactical support transport, BuAer 159364, taken at Cherry Point MCAS in 1974.

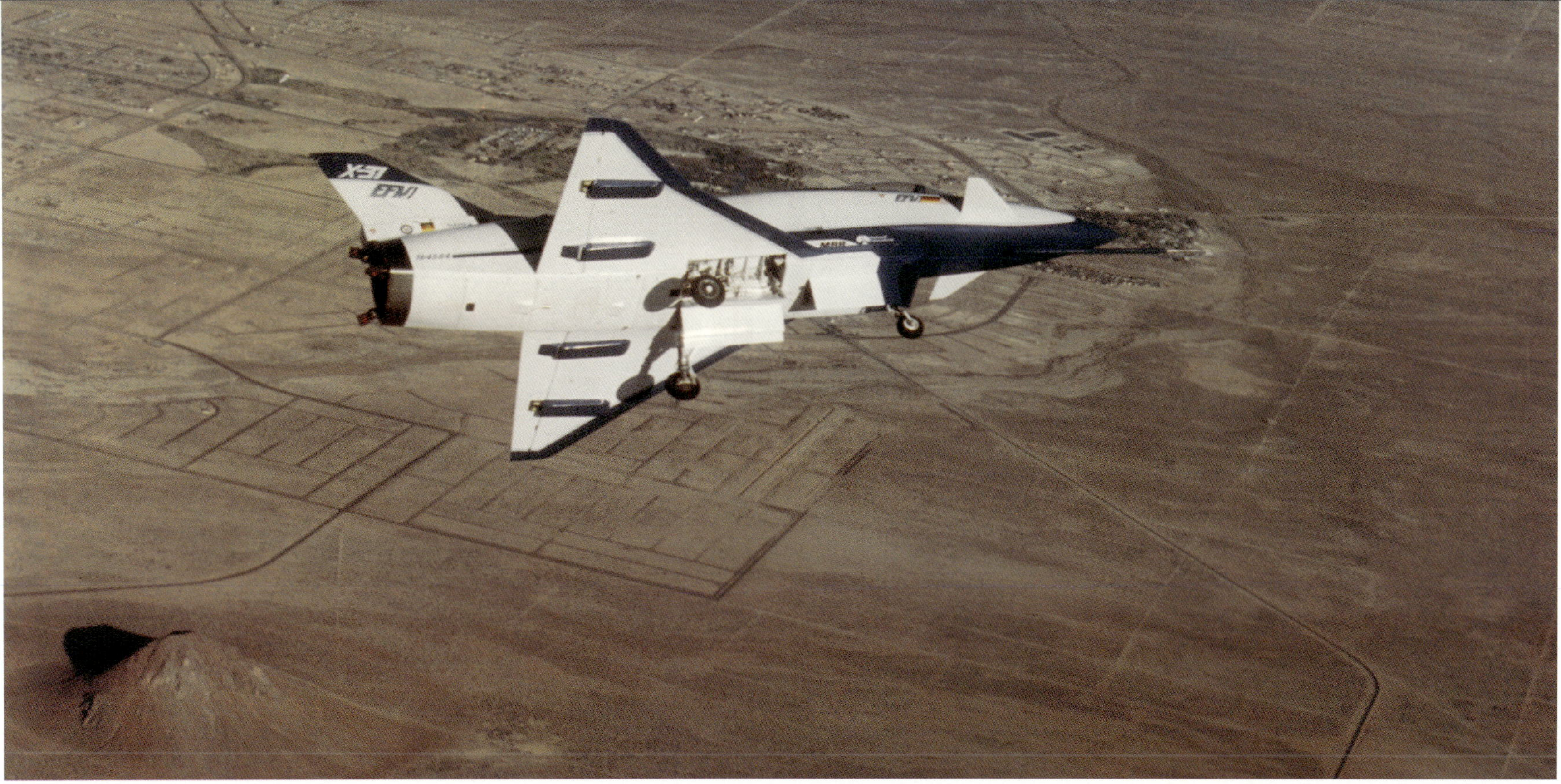

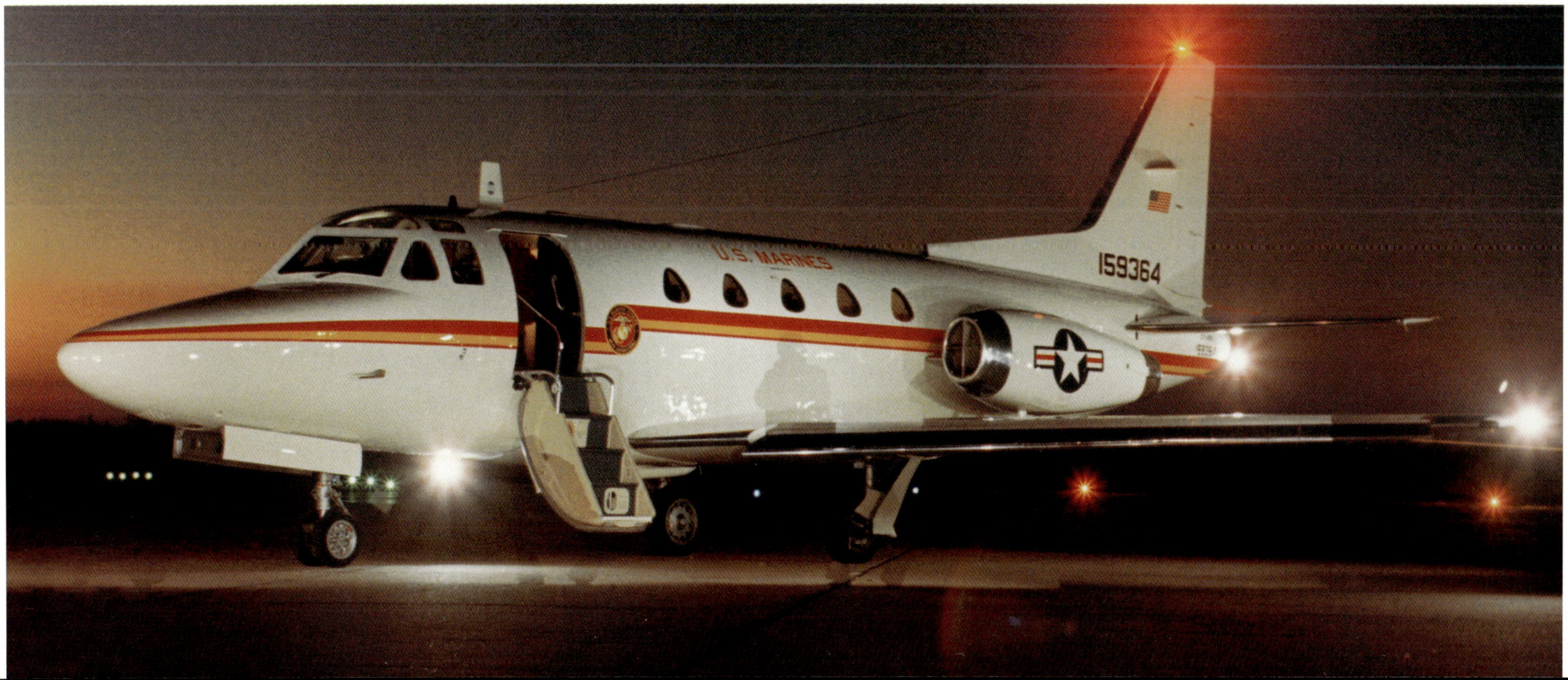

127

128

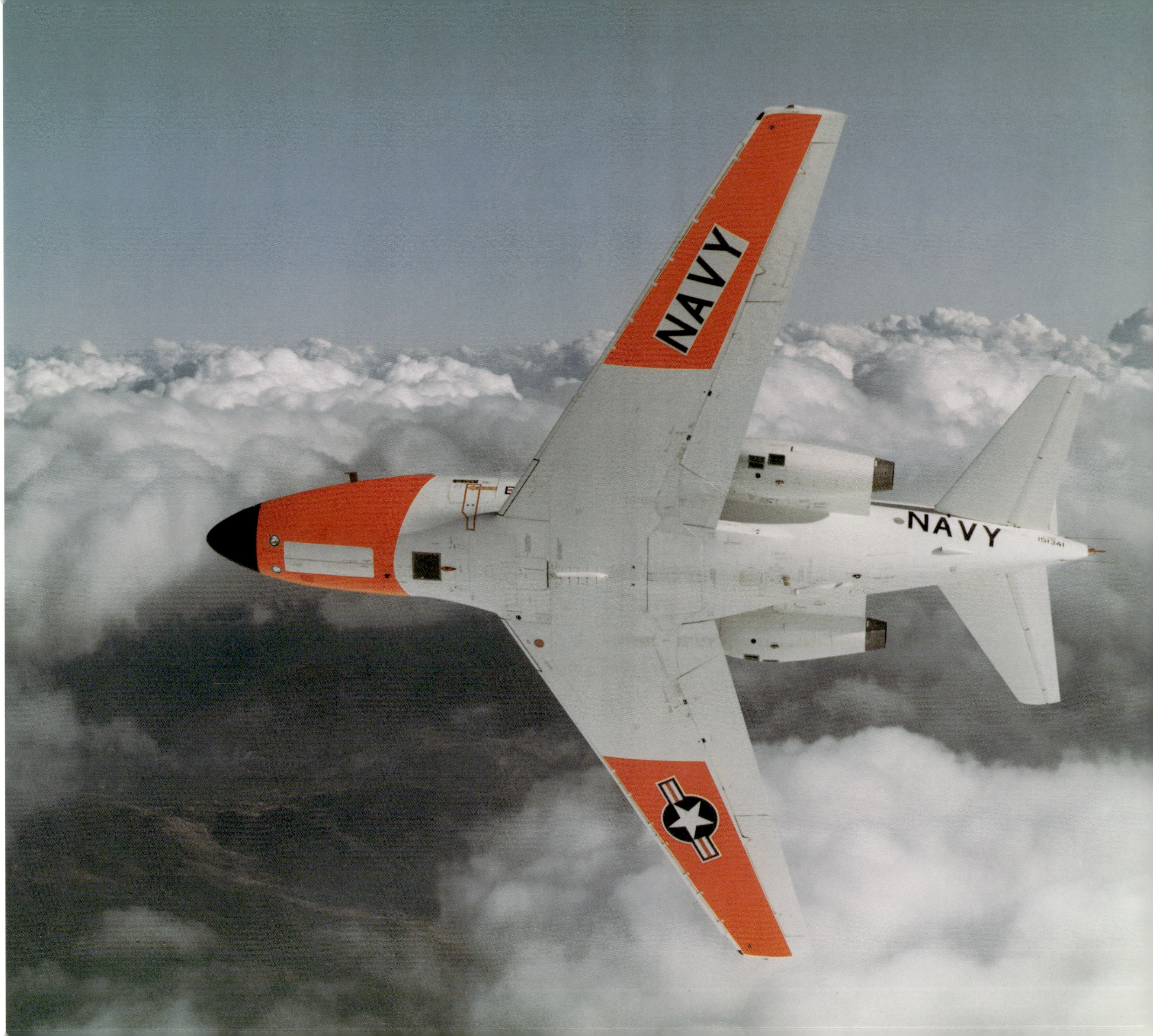

NORTH AMERICAN NA-246>NA-403 UTX / T-39/CT-39/T3J Sabreliner / Sabre 75/80

NORTH AMERICAN BEGAN DESIGN and marketing studies of a small twin-jet aircraft, for both military and civilian use, early in 1952. The initial configuration featured a swept low-wing passing beneath the fuselage structure. Both jet and turboprop engines were considered, and initially two General Electric J85 turbojets with 2500 pounds of thrust were chosen, mounted at the wing/fuselage joints. The design study was subsequently shelved, however, and remained only on paper for nearly four years.

In March 1956 the U.S. Air Force announced plans to acquire small and medium jet trainer/transports from the aircraft industry. In August, Major General David H. Baker, director of procurement and production of the USAF Air Materiel Command, sent letters to twenty-eight aircraft manufacturers expressing interest in acquiring two categories of jet trainer/transports. The estimated quantities were 1500 of the smaller aircraft and 300 of the medium type, at costs-per-unit ranging from $200,000-400,000, although these estimates were overly optimistic. Two General Design Specification documents outlined a light-weight twin-jet utility trainer (UTX) and a medium-weight four-jet utility transport (UCX).

Because the USAF believed there would also be a commercial market for these aircraft and that development costs were well within the capability of private industry, manufacturers were requested to design and build prototypes free of charge to the U.S. Government. Full commercial certification would therefore be required for any type selected by the USAF under this specification.

The project was a gamble and only three manufacturers expressed interest in the program. North American was the only company to commit itself, announcing its intention to proceed with design and engineering on August 27, 1956. NAA's original studies corresponded most closely with the UTX requirements, so development of the smaller type went forward at the Los Angeles Division. A six-seat aircraft with a crew of two and tandem controls was specified, and its performance requirements included a range of 1725 miles and a cruising speed of Mach 0.76 at 45,000 feet.

The original 1952 design was re-configured and on July 1, 1957, a revised study was initiated. The layout was a low-wing aircraft with a 28.5-degree sweep, leading edge slats and swept control surfaces. Tricycle landing gear and a pressurized cabin were featured. The small side-by-side cockpit included tandem controls, necessary for the training mission, and the cabin had seats for four passengers. Changes were made to the wing/fuselage layout, and the engines were positioned alongside the rear fuselage section to optimize wing slat aerodynamics and to facilitate engine maintenance. The effectiveness of these changes was later confirmed in wind-tunnel tests and engineering mock-ups. The design was designated NA-246, and in the tradition of the F-86 and F-100, the aircraft was named Sabreliner.

The design changes delayed the program to the point where the original first flight target of early 1958 was impossible to meet, and further delays with the J85 engines set the program back an additional four months. The completed UTX

In flight, left, the thirtieth T-39D, BuAer 151341, in 1964.

The NA-246-1 UTX , the prototype for the T-39, was given no military designation, just the civil registration N4060K.

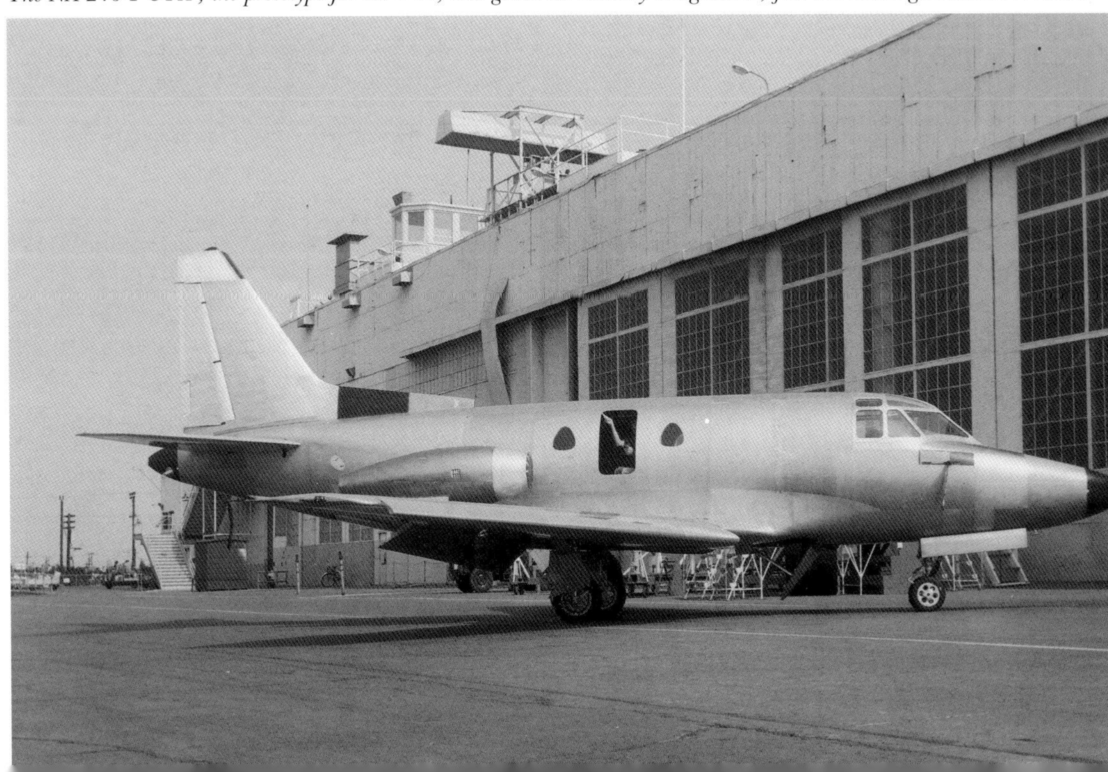

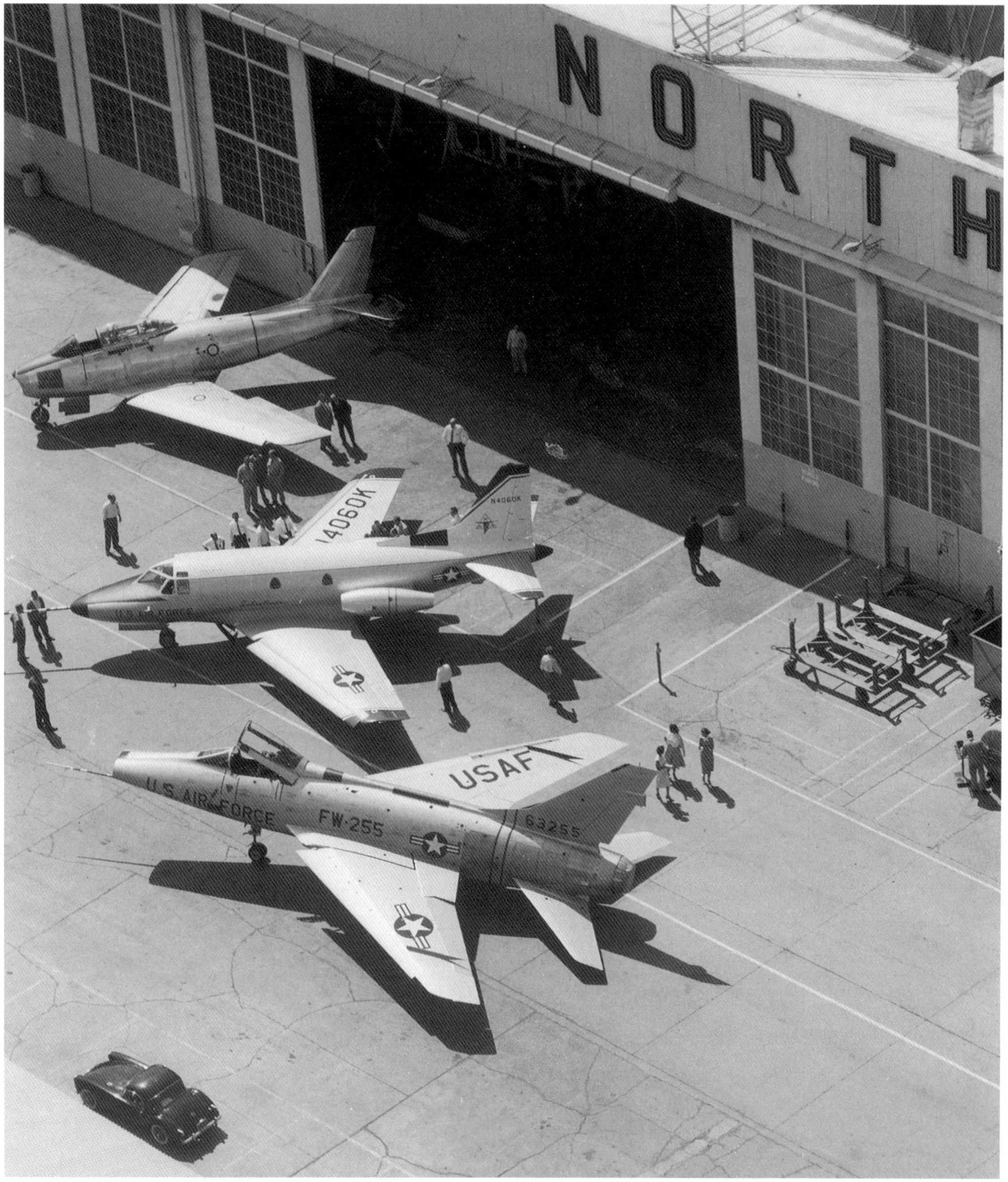

NAA Sabres: Flanked by an F-86 Sabre and an F-100 Super Sabre, the new UTX Sabreliner poses at the factory in May 1958.

prototype, with the civil registration N4060K, rolled out of the Inglewood factory on May 8, 1958. It flew for the first time on September 16 at NAA's Palmdale facility, piloted by J. O. Roberts and Gage Mace.

The first flights were very encouraging, and Roberts commented that the Sabreliner performed as predicted, handling well and demonstrating a good rate of acceleration. After further flight testing by NAA pilots, the Sabreliner was handed over to the Air Force for evaluation at Edwards AFB.

The USAF made twenty-seven flights logging 31 hours, and its evaluation, completed in December 1959, was also very favorable. By then NAA had already won the UTX competition and production was scheduled for an initial batch of seven NA-265 aircraft to be tested by NAA, the USAF and the Federal Aviation Administration. These aircraft, designated T-39A by the USAF, used Pratt & Whitney J60-P-3 engines with 3000 pounds of thrust instead of the 2500-pound/thrust J85s of the UTX prototype. Assembly of the first T-39A started in late-summer 1959.

During this time the USAF Tactical Air Command was developing the Republic F-105D Thunderchief all-weather supersonic fighter-bomber and needed an economical training platform for its Search and Ranging Radar system (NASARR, developed by NAA's Autonetics Division). The T-39 airframe was found to be well suited to installation of the system. NAA received the go-ahead from the USAF to develop this version, designated T-39B, and the sixth flight-test T-39A was converted to this specification. Before the end of the year NAA received an order for an additional thirty-five T-39s; four of these were T-39Bs, given initial priority and entering USAF service before the T-39A.

The first T-39A flight took place on June 30, 1960, and shortly thereafter the extensive NAA/USAF/FAA test program began. By March 1961 the evaluation was completed and the Sabreliner reached full commercial FAA certification on March 23, 1962. T-39A deliveries to the USAF began on June 4, 1961, and 143 were built,

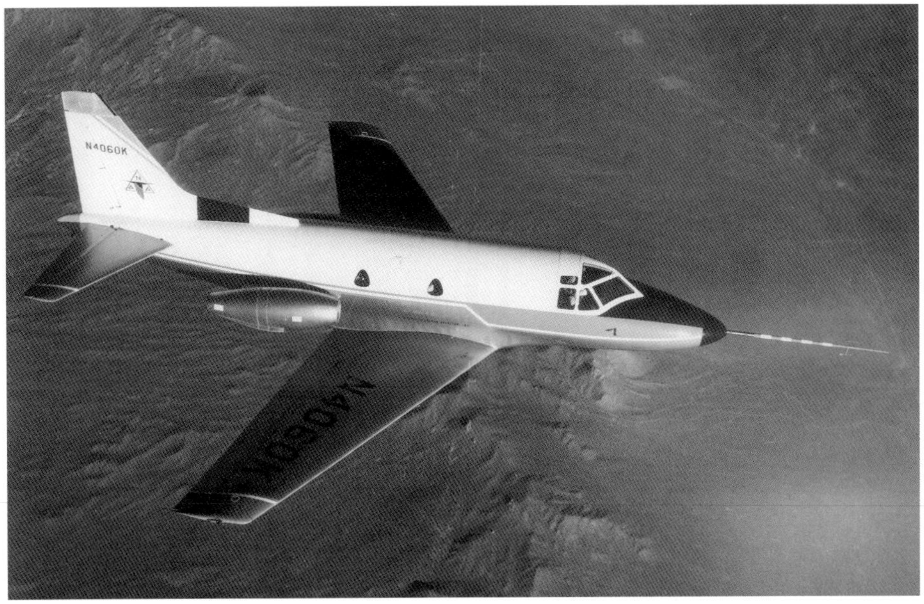

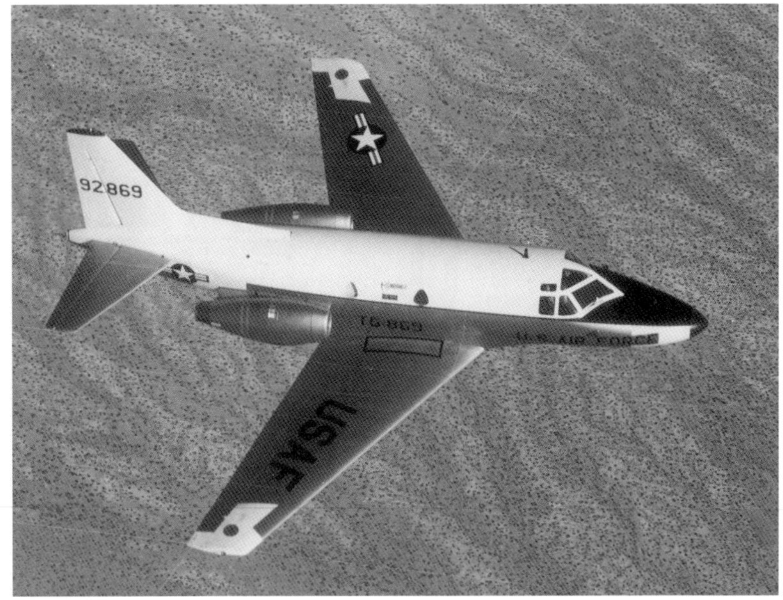

Without the USAF markings seen on page 130, the UTX first flew in September 1958. At right, also flying above the desert near Palmdale, is the second T-39A. Below, the first T-39A, 59-2868.

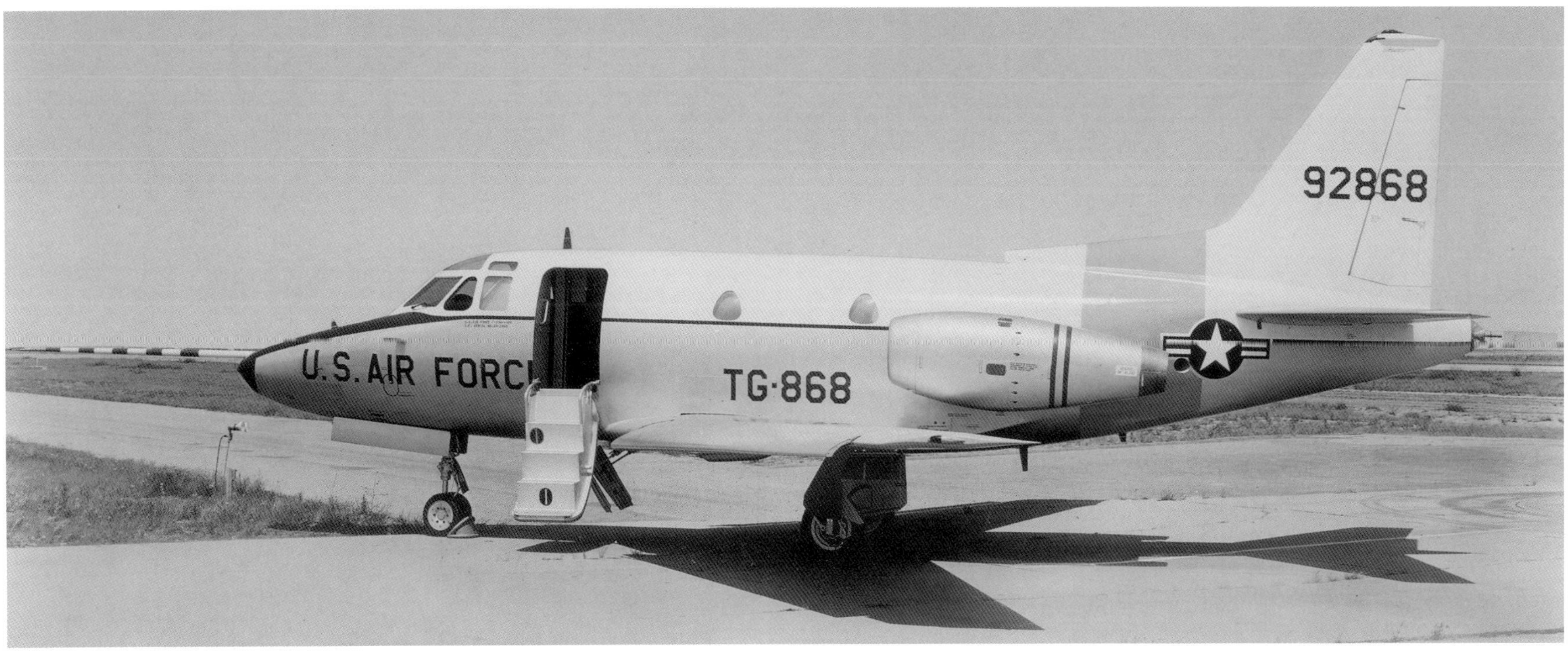

UTX (T-39)

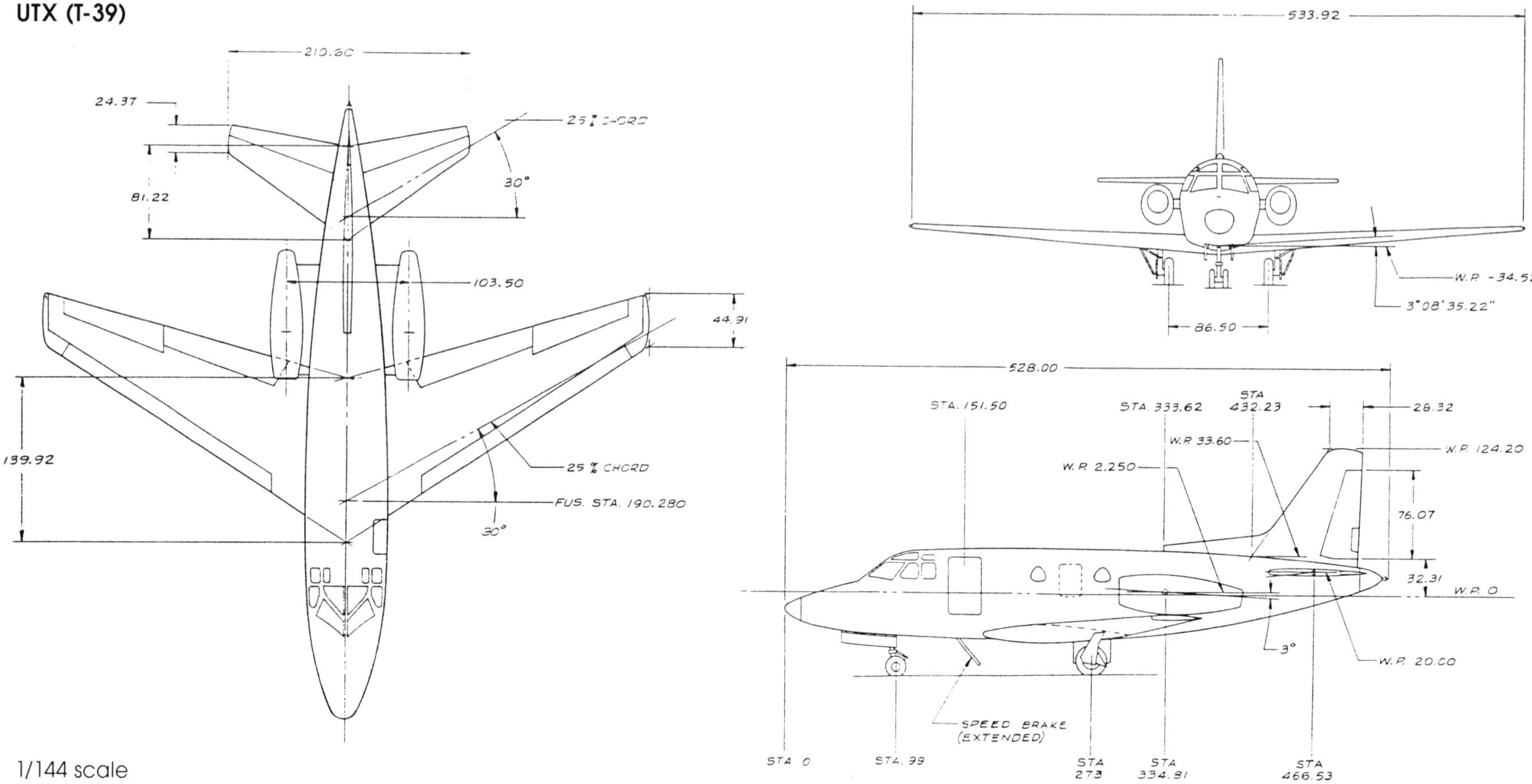

1/144 scale

the last being delivered in late-1963. Despite the trainer designation, T-39As were used primarily as utility aircraft, transporting military personnel and high-priority cargo. Some T-39As were used for training, however, including the USAF Instrument Pilot Instructors needed to train subsequent T-39 pilots.

Several T-39As were modified by NAA to seven-passenger configuration, the first of which was delivered to the USAF in June 1967. These aircraft had strengthened landing gear to accommodate higher take-off weights, as well as an improved cabin environment. In 1975 direct control of all USAF transport aircraft was transferred to the Military Airlift Command and the T-39As came under control of the 89th Military Airlift Wing until MAC later became the Air Mobility Command in the 1990s.

The T-39B flew for the first time in November 1960 and the first deliveries arrived two months later. The T-39B was externally very similar to the T-39A, except for a larger nose radome. The inside, configured for the F-105's NASARR radar system, was considerably different. The baggage compartment was converted into an avionics bay, containing the training radar and other equipment moved from the nose section to accommodate the R-14 NASARR antenna dish. Consoles for radar and navigation systems were provided in the cabin for three F-105 student pilots.

All T-39Bs were originally assigned to the 4520th Combat Crew Training Wing at Nellis AFB, Nevada, and later transferred to the 4519th CCTS at McConnell AFB, Kansas. In the early 1970s, F-105 training was taken over by the Kansas Air National Guard and the T-39Bs were delivered to the 127th TFTS (Kansas ANG).

During development of the McDonnell F-101B

Voodoo a similar radar platform was planned, to be designated T-39C. This was cancelled before production began, due to a reduction in the scope of the F-101B program.

In 1968 the USAF converted three T-39As as special-purpose training aircraft for the Electronic Countermeasures (ECM) system fitted to the F-105G "Wild Weasel" aircraft used in Vietnam. This ECM system was used in conjunction with Shrike and ARM missiles to locate, jam and destroy the North Vietnamese radar sites that directed Surface-to-Air Missiles (SAMs). These trainers, designated T-39Fs, were fitted with three "Wild Weasel" system consoles, one each for a trainee pilot, trainee Electronic Warfare Officer and an instructor. The T-39Fs, nicknamed "Teeny Weeny Weasels," were assigned to the 66th Fighter Weapons Squadron at Nellis AFB.

In 1961 the U.S. Navy had become interested in a radar trainer similar to the T-39B, but fitted with the Magnavox APQ-94 radar system. Originally designated T3J-1, it became the T-39D under the tri-service system initiated in 1962. The first of these aircraft was delivered to the Naval Air Training Command at NAS Pensacola, Florida, in August 1963. Forty-two T-39Ds were produced, the last completed in November 1964. T-39Ds have been used to train Navy pilots and Radar Intercept Officers, as Basic Jet Navigation trainers, and to flight-test missile-seeking and guidance systems.

Some Navy and Marine Corps Sabreliners were also used as personnel transports. Five T-39Ds were redesignated CT-39Ds for the purpose and the designations CT-39E and CT-39G were given to "off-the-shelf" civilian Sabreliners purchased by the Navy. The USAF, Navy and Marine Corps phased out most of their T-39s by 1986 and only a few remain in active service.

As a commercial aircraft, the Sabreliner was quite successful and was upgraded several times. The original NA-265 aircraft, the T-39A, B and D, were certified by the FAA as NA-265, -265-20 and -265-30 respectively, and the first civilian version, certified as NA-265-40, was marketed as the Sabreliner 40. Because NAA had little experi-

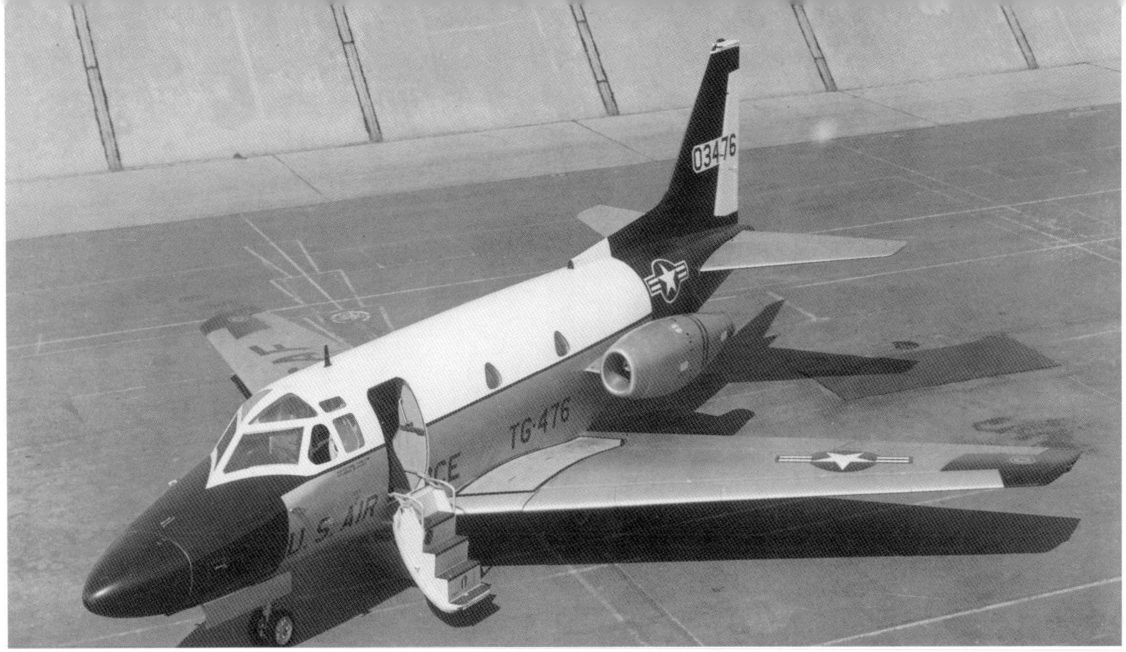

Above, the fifth T-39B, 60-3476. The USAF used this version, which had a larger nose radome, for NASARR systems training.

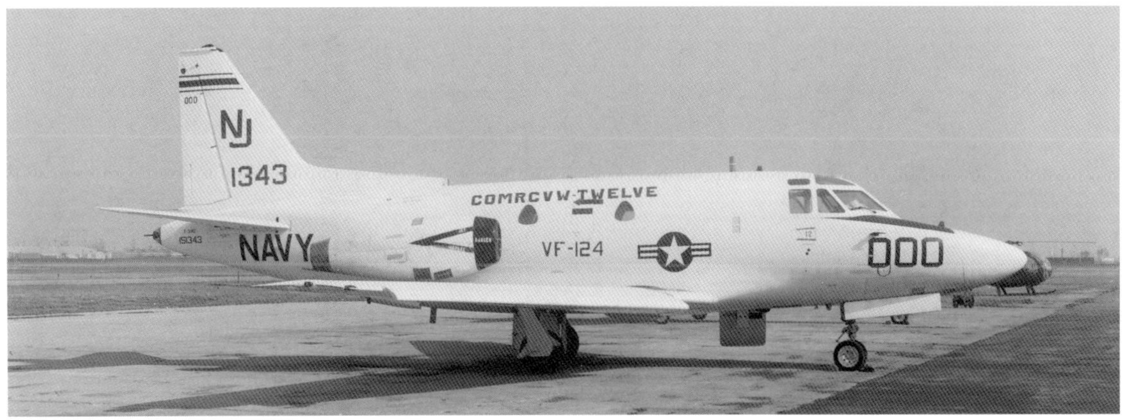

Above, the last Navy T-39D, 151343, of VF-124, CVW-12. Below, a CT-39G tactical support transport, one of thirteen commercial NA-306 Sabreliner 60s acquired by the USN/USMC. The tail number, not BuAer, comes from the NAA series.

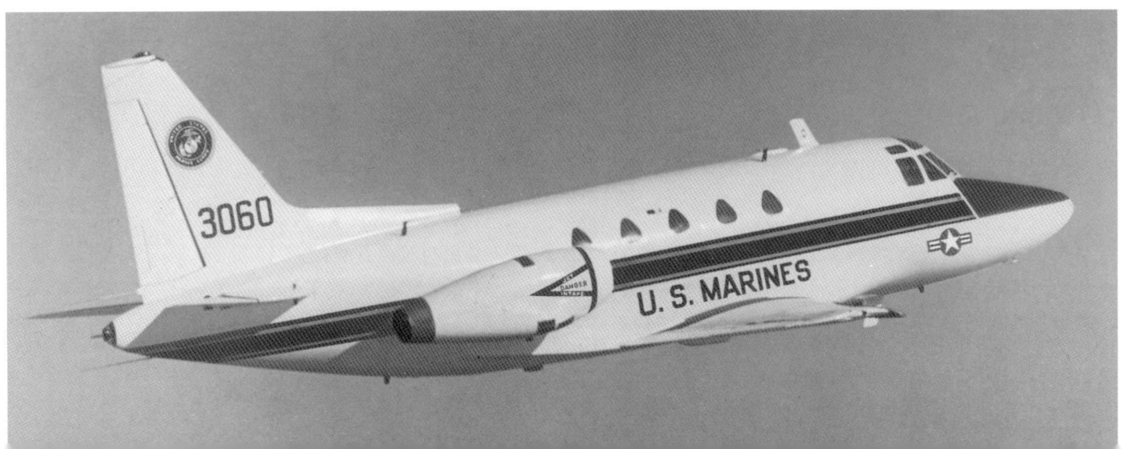

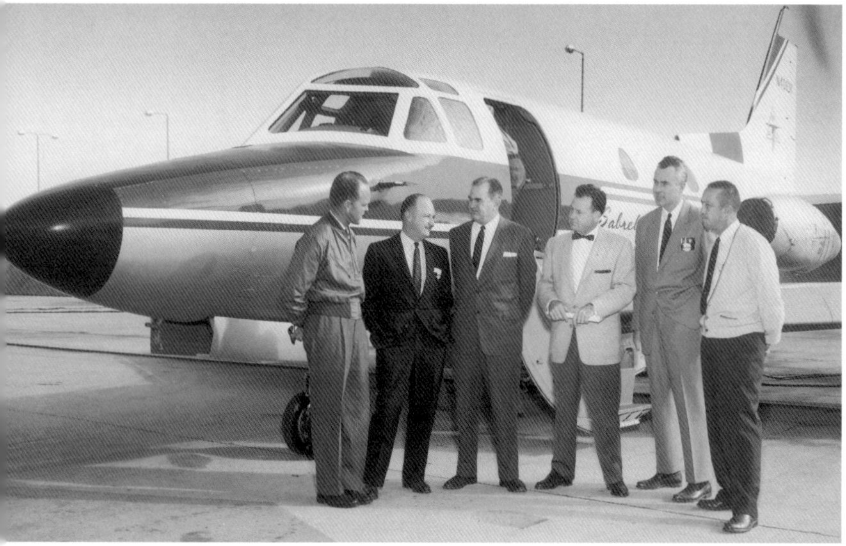

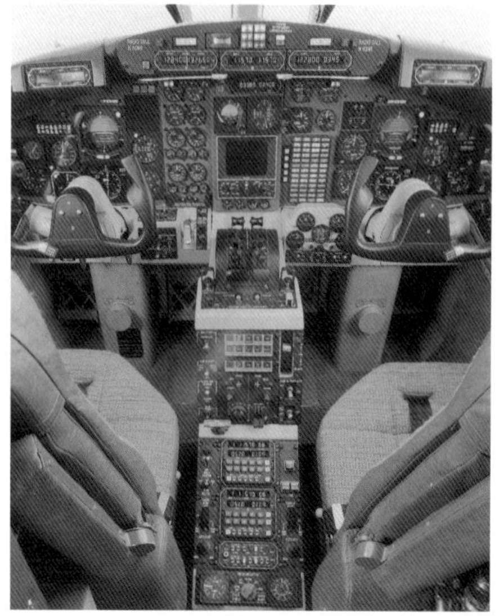

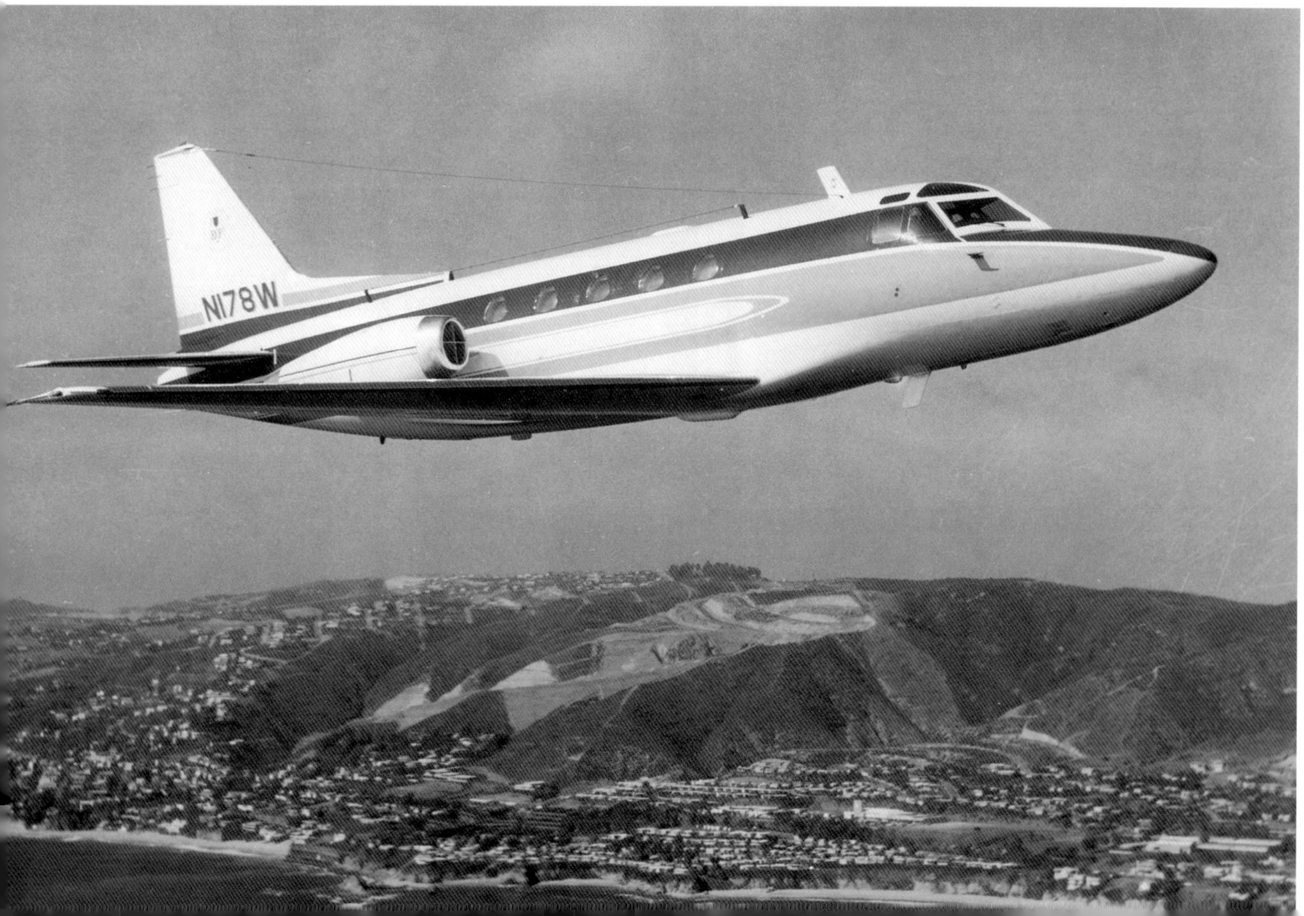

ence marketing commercial aircraft, it made an agreement with Remmert-Werner in 1962 for the sole distribution of the type. NAA delivered the Sabreliners in green zinc chromate primer, to be custom-finished by Remmert-Werner according to customer requirements. The Sabreliner 40 was similar to the T-39A, but used the commercially approved Pratt & Whitney JT12A-6A engines with essentially the same thrust rating of the military J60. Later Sabreliner 40s with 3300-pounds/thrust JT12A-8 engines featured a third side window on the fuselage.

In 1964 a single Sabreliner 50 was built with a longer fuselage for use as an electronics systems test bed. This led to a stretched version of the Sabreliner 40 in 1966. The 3-foot, 2-inch increase allowed six-passenger seating and a toilet at the rear of the cabin, and the new version with five side windows, certified as the Sabreliner 60, went into production alongside the model 40 in mid-1967.

After the merging of NAA and Rockwell resulted in the North American Rockwell Corporation, a new Sabreliner appeared, with deepened fuselage and improved headroom. First flown as the Sabreliner 70 on December 4, 1969, the aircraft was redesignated Sabre 75 in 1971. It was identifiable by square side windows, replacing the curved triangular windows of the previous models, and was powered by General Electric CF700-2D-2B turbofans with 4135 pounds of thrust each. In 1974 an improved Sabre 75A was introduced, accommodating up to 10 passengers. The Sabre 80 with AiResearch ATF 3 turbofans was introduced in 1974.

A development of the Sabreliner 60 with two 3700-pound/thrust Garrett AiResearch TFE turbofans was projected in 1978 as the Sabreliner 65, with vast improvements in fuel economy and range.

The Sabreliner series achieved just what it was projected to do, providing a versatile train-

Above left, test pilots and engineers Roberts, Ferren, Virgin, Brewer, O'Tool and Wells with N4060K as a civil Sabreliner Series 40 prototype. Above right, cockpit of N80NR. Left, a Series 60, N178W, which accommodated ten passengers.

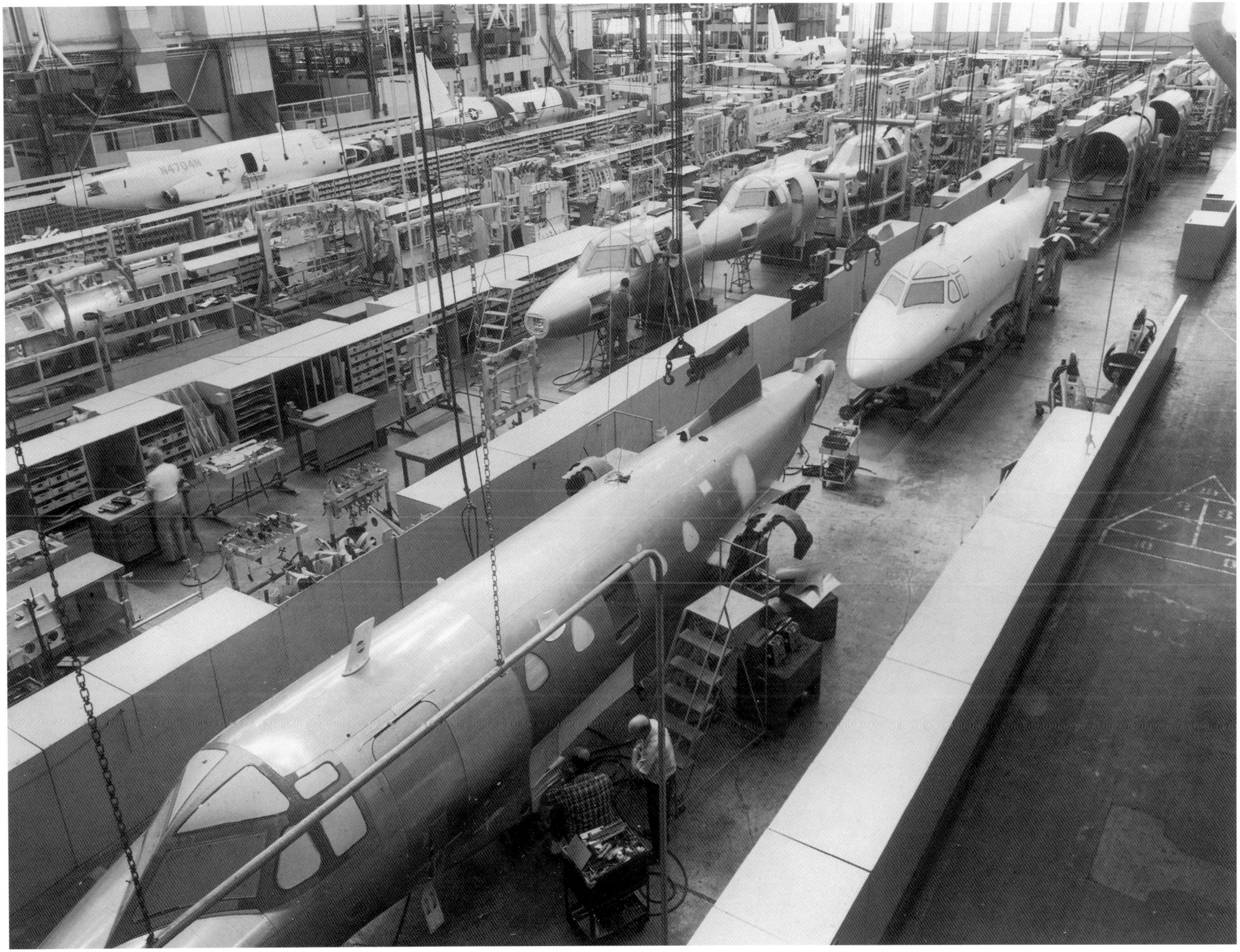

Components in various stages of manufacture for sixteen Sabreliners for both civil and military customers. A gamble at first, the venture proved profitable for NAA's Los Angeles Division.

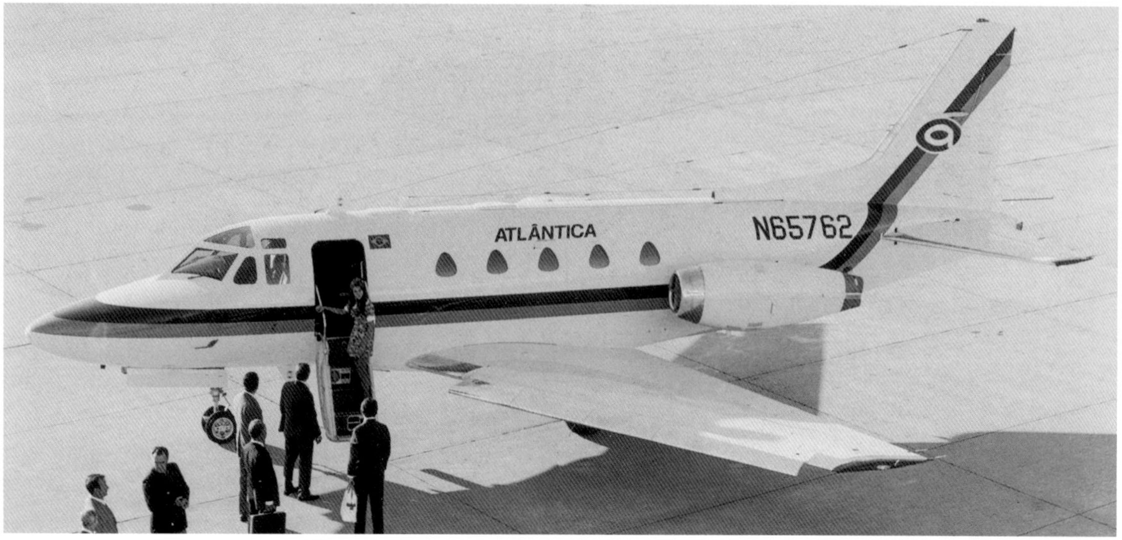

Above, a Sabreliner Series 65 Taxi Aereo for the Brazilian Atlantica line. Below, the Series 75 prototype, N7590N, easily identified by its rectangular windows. With 4135-lb/thrust GE CF700 turbofans, the 75A was the best-performing Sabreliner.

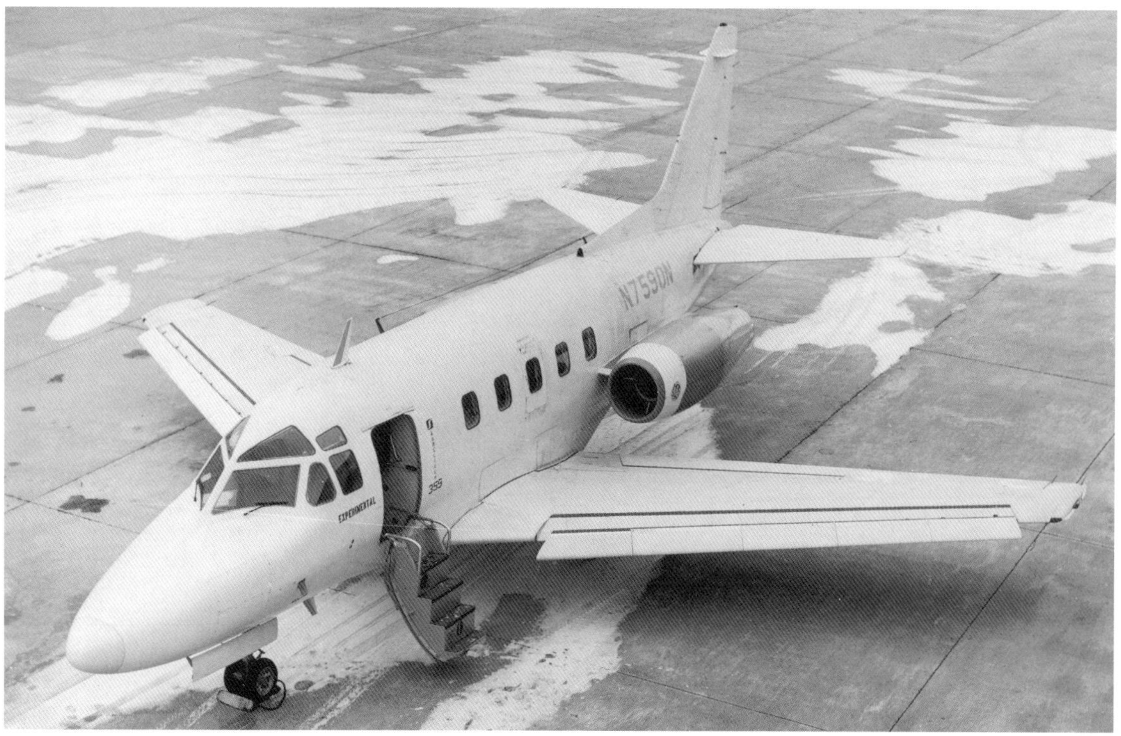

ing and transport aircraft for the military while earning a sizeable share of the civilian business-jet market. Nearly 600 military and civil Sabreliners were delivered; the program, initially considered a major risk, ultimately proved to be one of North American's most successful.

Specifications: NA-246>NA-403
North American UTX/T-39/Sabreliner

Description: U.S. Air Force and Navy light transport and systems training aircraft. Twin-jet low-wing monoplane of all-metal construction with retractable landing gear. Crew 2, passengers 4-7. (Sabreliner 40-Sabre 80) Light commercial transport and business aircraft.

Engines: Two 2500-lb/thrust General Electric J85 turbojets. (T-39) 3000-lb/thrust Pratt & Whitney J60 turbojets. (Sabreliner 40-60) 3300-lb/thrust Pratt & Whitney JT12A-8 turbojets. (Sabreliner 65) 3700-lb/thrust Garrett AiResearch TFE 731-3-100S turbofans. (Sabre 70-75A) 4135-lb/thrust General Electric CF700-2D-2B turbofans.

Dimensions: Span 44 feet 5 in, length 43 feet 9 in, height 16 feet, wing area 342.05 sq ft. (Sabreliner 60) Span 44 feet 8 in, length 46 feet 11 in. (Sabreliner 65) Span 50 feet 5 in, length 46 feet 11 in, wing area 380 sq ft. (Sabreliner 70/75) Span 44 feet 8 in, length 47 feet, height 17 feet 3 in.

Weights: 9257/9895 lb empty, 17,760/18,650 lb maximum take-off. (Sabreliner 60/65) 11,250/13,730 lb empty, 20,172/23,800 lb maximum take-off. (Sabreliner 75A) 13,220 lb empty, 23,300 lb max take-off.

Performance: Cruising speed 502 mph, climb 4700 ft/min, service ceiling 44,000 ft, range 1950 miles. (Sabreliner 40) Cruising 500 mph, ceiling 45,000 ft, range 2100 miles. (Sabreliner 60) Cruising 562 mph, range 2023 miles. (Sabreliner 65) Climb 3540 ft/min, range 3315 miles. (Sabre 75A) Climb 4500 ft/min, range 1956 miles.

Production: North American UTX / T-39/CT-39 / T3J / Sabreliner 40/60/65/70 / Sabre 75/80

Number	Type	Details	Qty
NA-246	UTX	246-1, N4060K	1
NA-265	T-39A	USAF 59-2868/2872	5
		USAF 60-3478/3508	31
		USAF 61-634/685	52
NA-270	T-39B	USAF 59-2873/2874,	2
		60-3474/3477	4
NA-271	(T-39A for FAA Type Certificate)		—
NA-276	T-39A	USAF 62-4448/4502	55
---	T-39C	ECM/radar trainer, not developed	—
NA-277	T3J-1/T-39D	BuAer 150542/150551	10
NA-282	Sabreliner	282-1/35 (Remmert-Werner)	35
NA-285	T3J-1/T-39D	BuAer 150969/150992, 151336/151343	32
		(151344/151347 canceled)	—
NA-287	Sabreliner	Autonetics test bed	1
NA-290	Sabreliner	Series 40, 282-36/49 (R-W)	14
NA-292	Sabreliner	Series 40, 282-50/65 (R-W)	16
NA-293	Sabreliner	Series 40, 282-66/98 (R-W)	33
NA-297	Sabreliner	Series 40, 282-99/123 (R-W)	25
—	CT-39E	Converted NA-282 Sabreliner	(7)
—	T-39F	ECM converted from T-39A	(3)
NA-306	Sabreliner	Series 60, 306-1/2 (R-W)	2
NA-308	Sabreliner	Series 60, 306-3/37 (R-W)	35
NA-320	Sabreliner	Series 60, 306-38/55	18
NA-327	Sabreliner	Series 60, 327-1/26	26
NA-336	Sabreliner	Series 70	25
NA-343	Sabreliner	Redesignated NA-327	(26)
NA-344	Sabreliner	Redesignated NA-336	(25)
NA-345	Sabreliner	Redesignated NA-320	(18)
NA-369	Sabreliner	Series 40	16
NA-370	Sabreliner	Series 70	9
		Series 80	2
NA-371	Sabreliner	Series 60	18
NA-372	CT-39G	BuAer 158843/158844, 159361/159365, 160053/160058	13
NA-373	T-39	Sierra TACAN sets	(6)
NA-374	Sea Sabre	Proposal for USN/USCG	(40)
NA-375	Sabreliner	Series 80	25
NA-376	Sabre 75A	conversion drawings from 75	—
NA-377	Sabreliner	Series 60	10
NA-378	Sabre	Series 75A	24
NA-385	Sabre	Series 85 (canceled)	—
NA-386	Sabreliner	Series 60	18
NA-390	Sabreliner	(40/60/80 parts, assemblies)	—
NA-391	Sabreliner	(40/60/80 parts, assemblies)	—
NA-392	Sabreliner	(40/60/80 parts, assemblies)	—
NA-402	Sabreliner	Series 40A	26
NA-403	Sabreliner	Series 40A	15
Rockwell	Sabreliner	Series 40R	1
		Series 40R converted from 40A	(6)
Total UTX, T-39 and Sabreliner transports and trainers, 1958-1981			599

(Military T-39 versions 204; civil Sabreliners 395)

Rockwell's prototype Sabreliner 65, N65R, was a Series 60 development tested in 1978; with two 3700-lb/thrust Garrett AiResearch TFE turbofans, it had excellent economy and rate of climb.

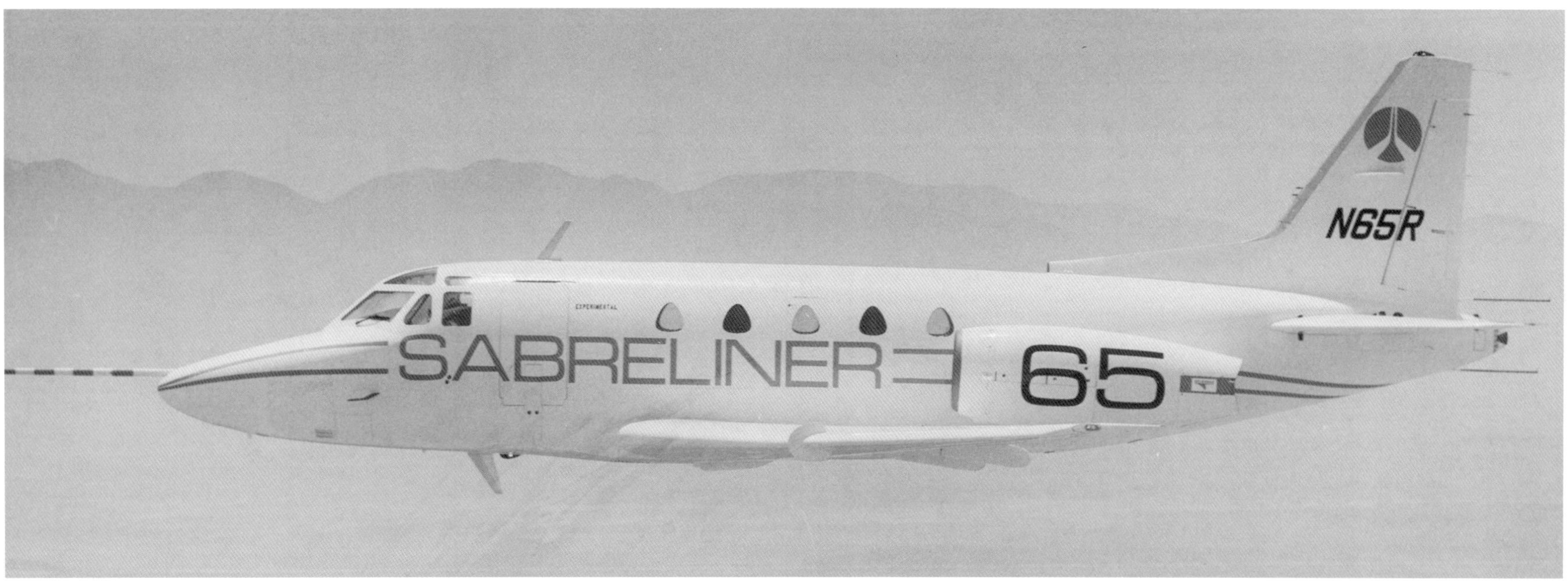

NORTH AMERICAN NA-259>NA-315 XB-70A Valkyrie

CALLING FOR A MANNED intercontinental bomber to replace the Boeing B-52 Stratofortess by 1965, the U.S. Air Force issued General Operational Requirement No. 38 in October 1954. By March 1955 this specification had been modified to General Operational Requirements 81 and 82, which outlined a nuclear-powered strategic bomber with a range of 11,000 miles, Mach 2 capability for 1000 miles at altitudes above 60,000 feet and a payload of 20,000 pounds of nuclear weapons. These specifications later became known as Weapon System 110A.

In November 1955, Boeing and North American were awarded contracts for initial studies of WS-110 Phase 1 design. After two years, during which time several changes to the original requirements were made, the USAF eventually chose the North American entry. On January 2, 1958, the Air Force awarded contract number AF-36599 to NAA for Phase 1 development, followed in December by the Phase 2 contract.

Between 1958 and 1961, re-evaluation of the program by NAA, changes in policy within the Air Force and budget constraints drastically altered the WS-110 program. The USAF dropped the nuclear-powered concept for reasons of cost, and NAA explored a more conventional means of achieving the projected performance.

Although the original specification called for an aircraft capable of supersonic dashes for brief periods, the requirements were later revised to a sustained cruise of Mach 3 (2000 mph) at 70,000 feet, a range of 7500 miles and the ability to use facilities originally designed for the B-52 without modifications.

NAA engineers found it feasible to develop a large aircraft capable of triple-sonic cruise by use of the "compression lift" theory that had been outlined in a previous NACA study. Compression lift enabled a supersonic aircraft to ride its own shock wave using the lift generated by it, greatly reducing aerodynamic drag and increasing range. With six General Electric YJ93 afterburning engines each delivering up to 30,000 pounds of thrust and a gross weight of around 500,000 pounds, the strategic bomber would be the fastest large aircraft in the world, and certainly the heaviest.

The USAF assigned the designation B-70, and following a contest within the Strategic Air Command the bomber was named Valkyrie (warrior maiden from Norse mythology). Considering the size, complexity and projected performance of the aircraft, this was to become the most ambitious aircraft project North American would ever undertake. The problems associated with triple-sonic flight would require the use of

Construction of the first XB-70A, 62-0001, at NAA's Palmdale facility was completed in May 1964. The techniques learned allowed improvements to be made before the second was begun.

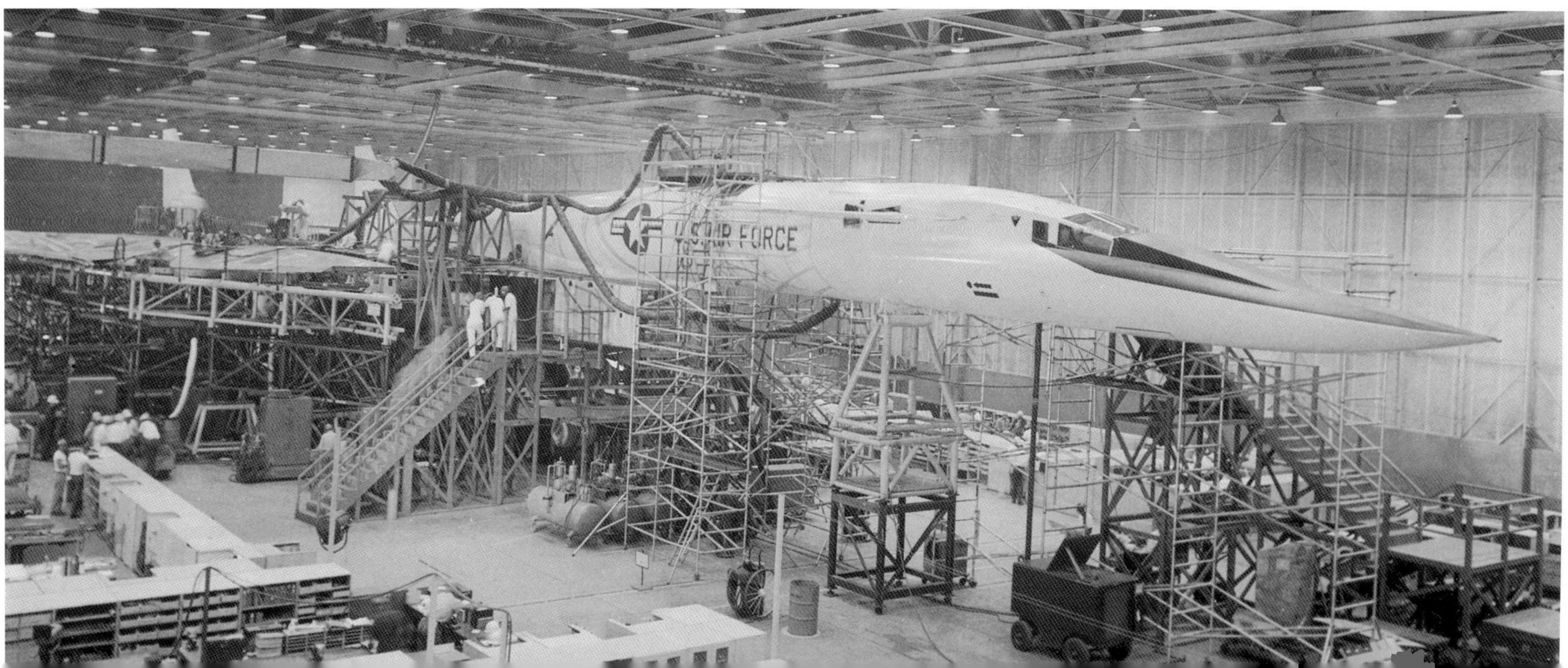

new technologies, construction techniques and materials, and the compression-lift theory had yet to be put into practice.

By 1961 the USAF had abandoned its long-range strategic bomber theory. Because of recent advances made in the Soviet air-defense system, Intercontinental Ballistic Missiles (ICBMs) were now considered to be a safer and more cost-effective method of delivering nuclear weapons. The Air Force subsequently cancelled the B-70 as a weapons system, but kept the program alive under Contract AF-42058O, issued on April 10, 1961. It gave NAA's Los Angeles Division the go-ahead to produce three XB-70A prototypes on charge number NA-278; no longer intended as bombers, the Valkyries would instead be high-speed research vehicles.

The Valkyrie was unique in both design and construction. Because the XB-70 design originated as an intercontinental bomber, it was a very large aircraft. Using a delta/canard layout, it had a wing area greater than 6000 square feet and the fuselage length was more than 190 feet.

To put the compression-lift theory into practice, the outer wing panels were ingeniously hinged to fold at anhedral angles to trap the shock wave, generating high-pressure air directly beneath the main wing surface. These outer panels, the largest movable aerodynamic devices ever fitted to an aircraft, folded down 25 degrees between 300 knots and Mach 1.4, and 65 degrees from Mach 1.4 to Mach 3. When folded, these outer panels streamlined the aircraft at high speed and greatly enhanced directional stability. Two vertical stabilizers were mounted at the extreme rear of the upper wing surface.

Beneath the center of the aircraft was a large cone-shaped box section housing the six engines and a multi-shock variable-area inlet system. Because it was necessary for intake air to remain subsonic at all times, a series of movable ramps was designed into the inlet system to optimize airflow at varying Mach numbers. This system was one of many innovations incorporated into the Valkyrie to solve the problems of triple-sonic flight.

The forward fuselage measured a massive 84

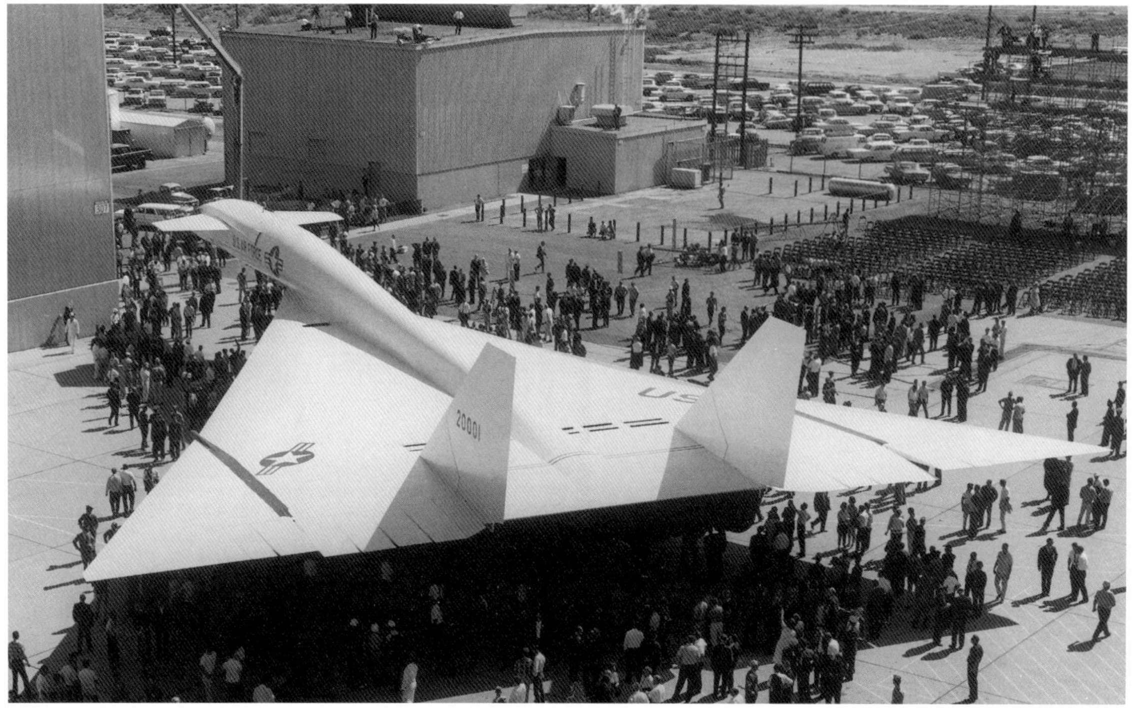

Above, the roll-out ceremony for the XB-70A, and below, the aircraft lifting off the runway at Edwards Air Force Base.

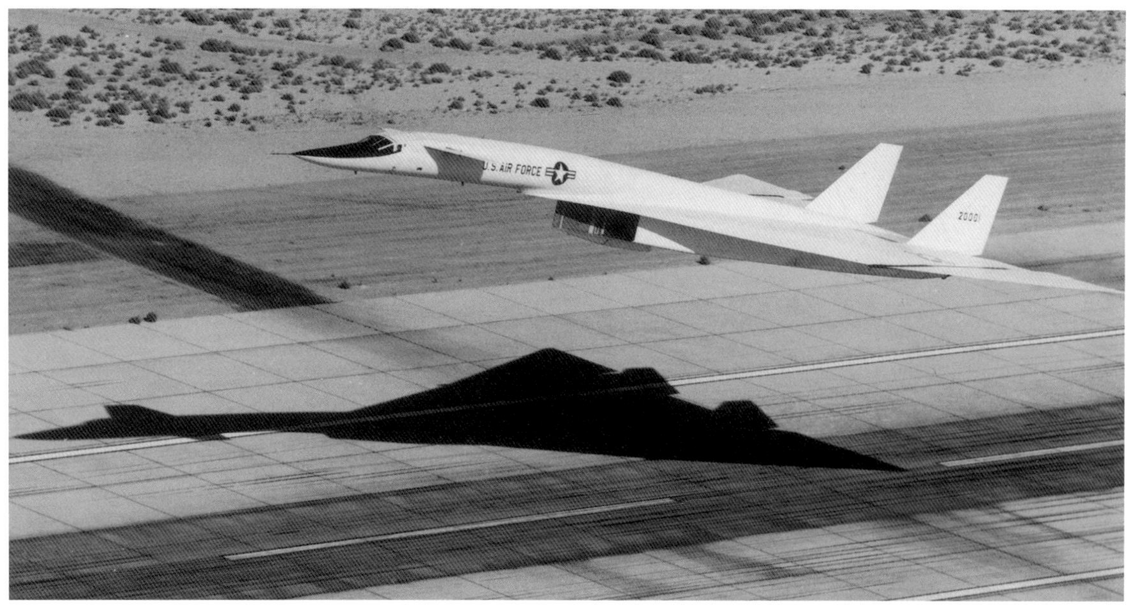

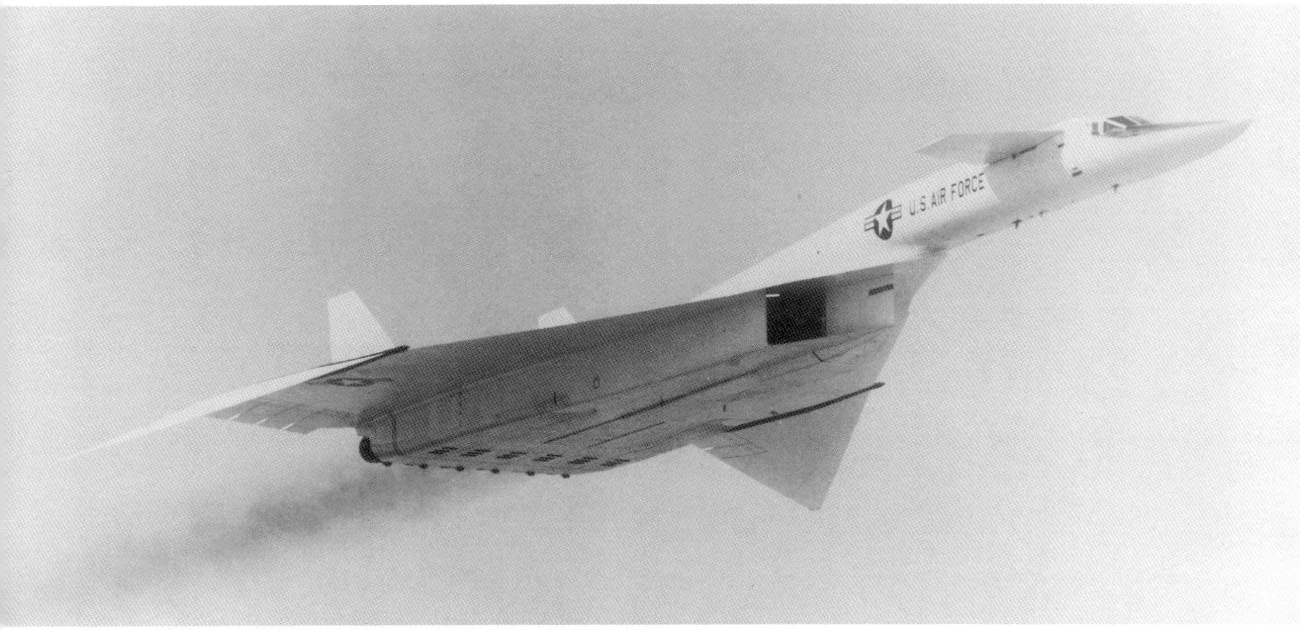

XB-70A wingtips drooped 25 degrees at speeds between 300 knots and Mach 1.4. Below, three drag chutes slowed the landing roll.

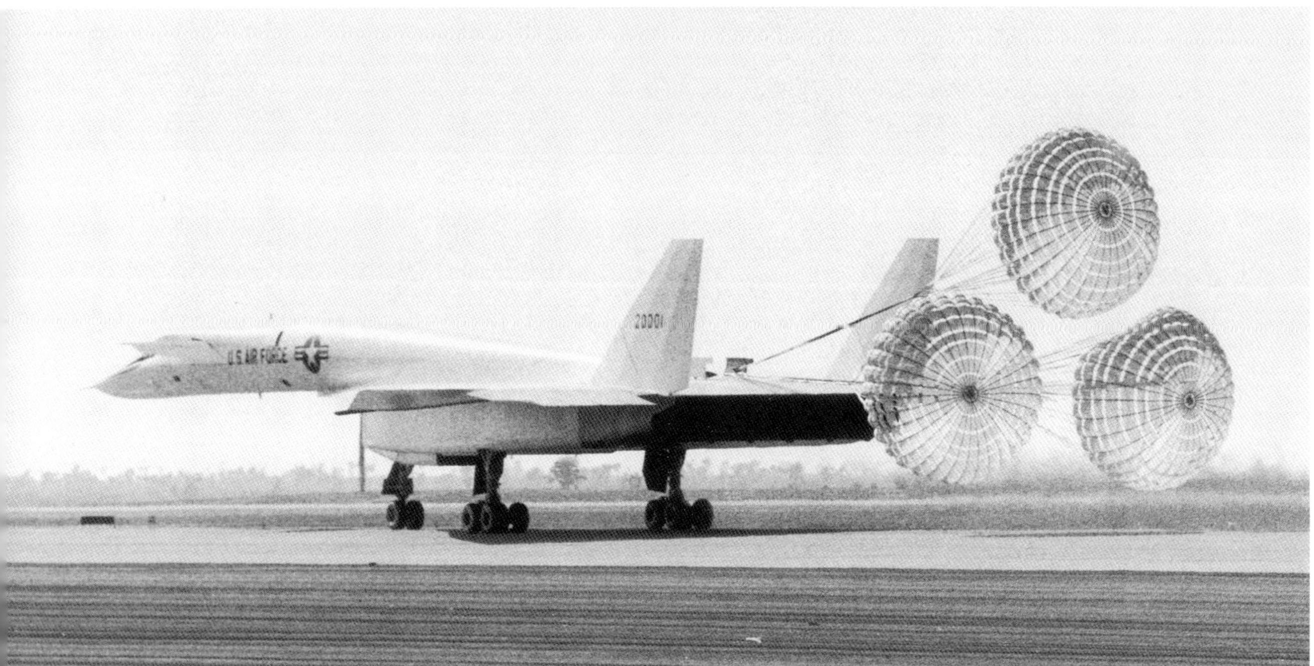

feet from the inlet opening to the nose. It housed the crew compartment and escape modules, canard horizontal stabilizers and variable-ramp windshield arrangement (for aerodynamic efficiency in high-speed flight). The cockpit was more than 20 feet above the ground and nearly 110 feet ahead of the main landing gear. Pivoting and jackknifing into its retracted position, the main gear was a very complex piece of engineering. The layout enabled the gear to fit into minimal space, providing more room for hydraulic systems, fuel storage and flight-data recording equipment that would soon be monitoring the aircraft's systems.

Mach 3 flight also presented a tremendous increase in skin temperature over conventional supersonic aircraft. To handle the expected temperatures of up to 620 degrees Fahrenheit, a portion of the XB-70 nose section utilized titanium (about 9 percent of the total structure), but the material was difficult to fabricate and very expensive.

To reduce production costs, NAA engineers developed a stainless steel honeycomb construction, another innovation, which proved to be strong as well as efficient at dissipating heat. The honeycombs were formed of material as thin as twenty-thousandths of an inch (.020"), and the outer surface skin was as thin as seven-thousandths (.007") in some areas. Because welding such thin material was impossible, a brazing technique was developed for the honeycomb structures.

On March 3, 1964, the Air Force cancelled the third prototype for budget reasons, and the XB-70 program was reduced to two aircraft. Lessons learned in the development and manufacture of the first Valkyrie were to be applied to the second aircraft, which by this time was already under construction.

On May 11, 1964, the first aircraft, XB-70A-1, AF 62-0001, was rolled out of its hangar at Air Force Plant 42 in Palmdale, California. The focus of a large ceremony attended by members of NAA, the Air Force, some members of Congress and the public, the large, all-white aircraft was both graceful and awe-inspiring. USAF Brigadier General Fred J. Ascani, deputy commander for the B-70 program, introduced and described the

aircraft as it emerged from the hangar.

On September 21 the Valkyrie made its first flight from Palmdale with Al White and Air Force Colonel Joe Cotton at the controls. The original flight plan included retracting the landing gear and accelerating the aircraft to Mach 1, then landing the Valkyrie at nearby Edwards AFB, where all subsequent flight testing would be conducted. Although the first flight was hampered by landing gear problems (a hydraulic system malfunction prevented gear retraction and a locked brake on the left main gear caused a fire and minor damage), White, Cotton and NAA Los Angeles Division President Warren Swanson all expressed enthusiasm at the post-flight press conference.

After repairs to the landing gear and hydraulic system were made, White and Cotton made the second flight on October 5. The landing gear was retracted, lowered and retracted again without problems. As the pilots prepared to accelerate to Mach 1, the primary hydraulic system showed a pressure drop, and the gear was lowered using the emergency electrical system. The landing at Edwards was uneventful, but again the hydraulic system needed to be sorted out.

The third flight was made one week later and the XB-70 successfully reached Mach 1.1, accelerating and decelerating in and out of the sound barrier to check transonic stability. The fourth flight occurred on October 24, and on this occasion the wingtips were folded down for the first time as the aircraft reached a speed of Mach 1.4 and set a record for sustained supersonic flight of over 40 minutes.

After this flight, which completed Phase 1 testing, the aircraft was structurally analyzed and proof-loaded at Palmdale. It also received a fresh coat of paint, as some had flaked off during the fourth flight. The Valkyrie returned to testing on February 16, 1965; on its fifth flight the wingtips were lowered to 65 degrees and Mach 1.6 was achieved. On the sixth and seventh flights, Lieutenant Colonel Fitzhugh Fulton was co-pilot with White and on the seventh flight, on March 4, the XB-70 achieved Mach 1.8, sustaining supersonic flight for over one hour.

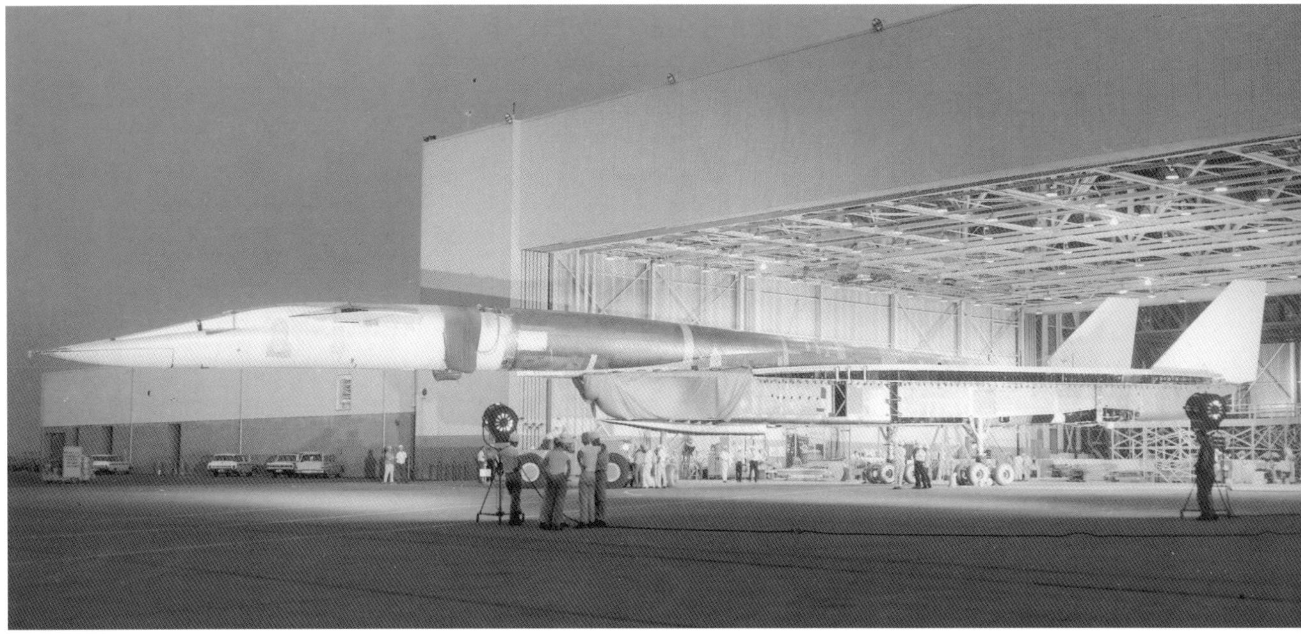

Above, before painting, the XB-70A-1 in a dramatic night shot. Below, systems tests on the completed second plane.

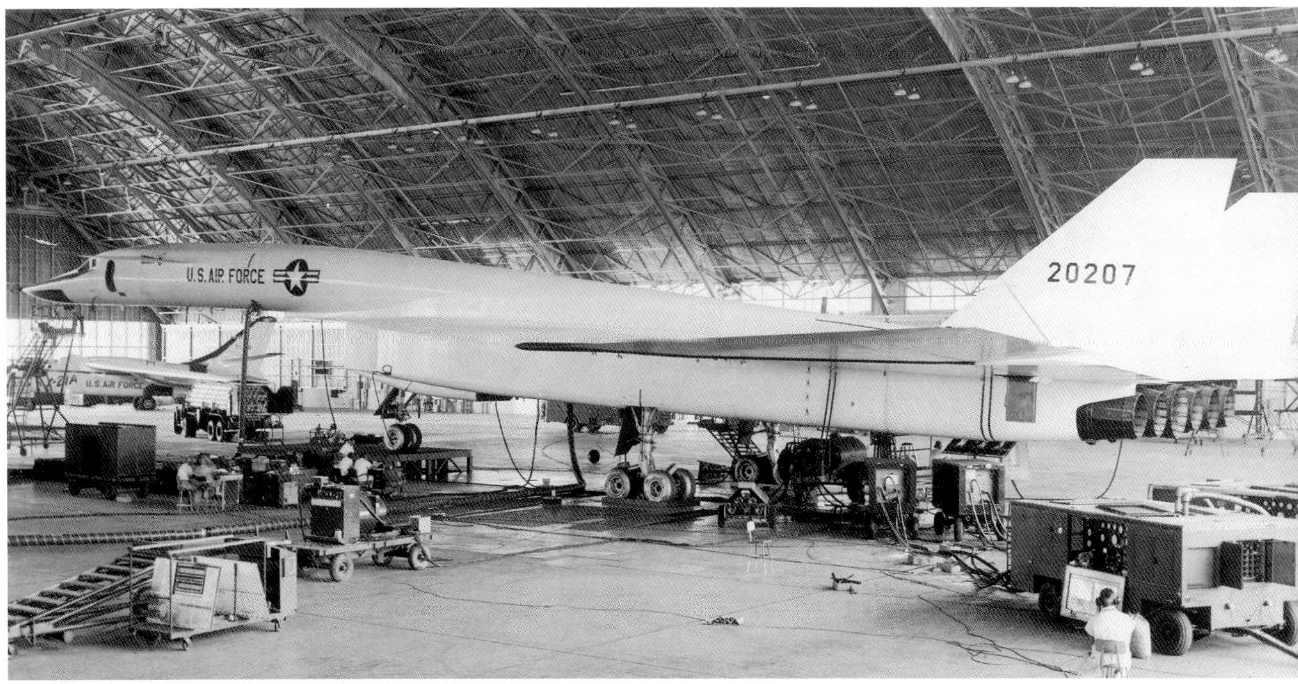

1/288 scale

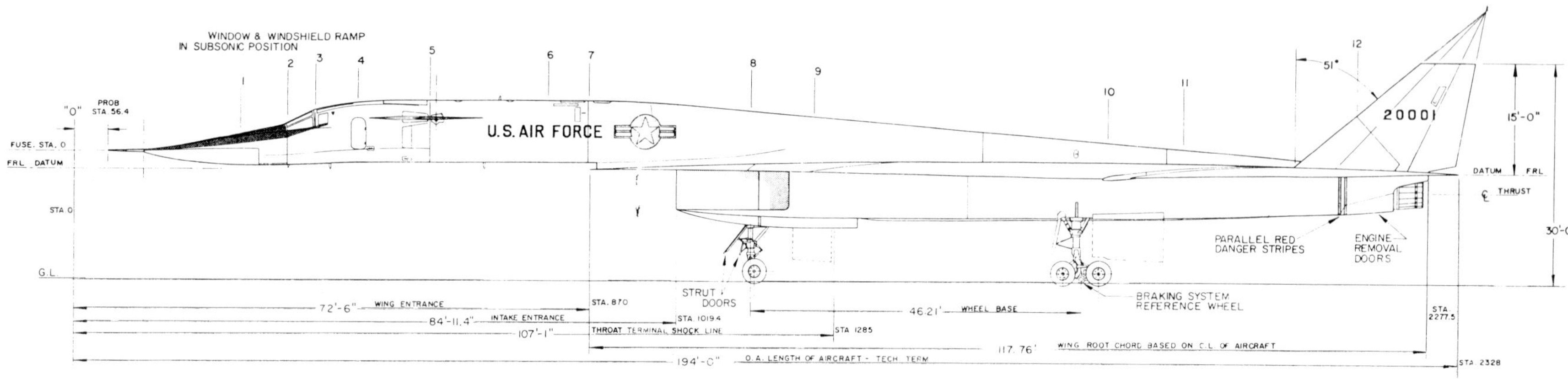

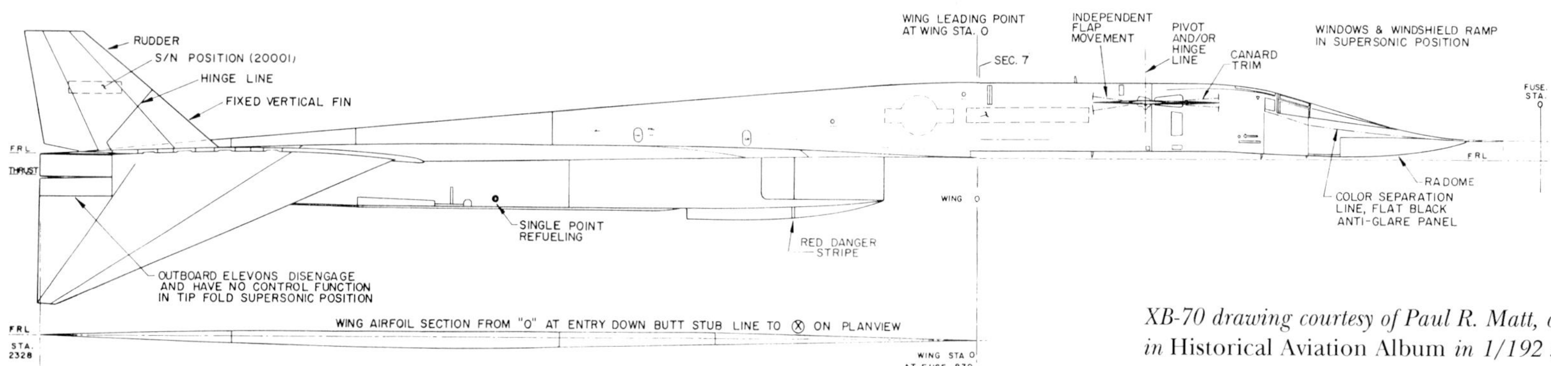

142

XB-70 drawing courtesy of Paul R. Matt, originally published in Historical Aviation Album *in 1/192 scale*

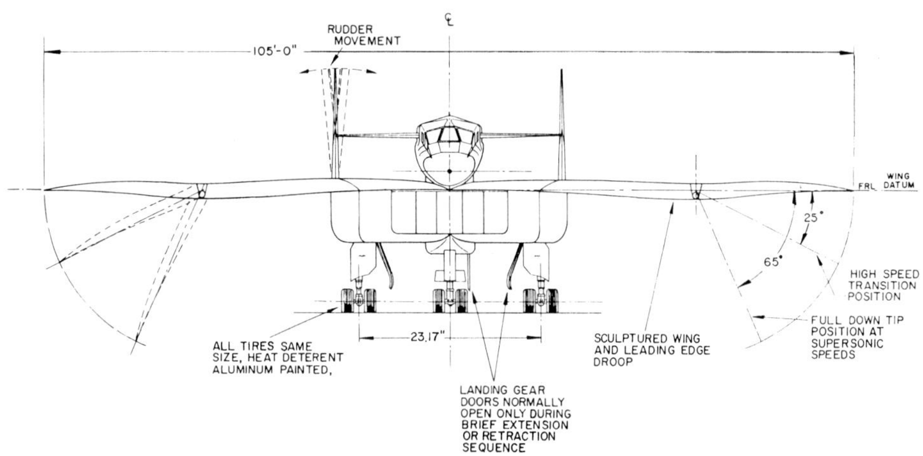

COLORING NOTES

ENTIRE AIRCRAFT - GLOSS WHITE
ENGINE BAYS AREA - NATURAL STEEL FINISH
TIRES - ALUMINUM PAINTED
ANTI-GLARE PANEL - FLAT BLACK
TURBO JET ENGINES - BLACK
ALL LETTERING - BLACK
WING TIP FOLD FAIRING - GLOSS BLACK
NATIONAL INSIGNIA - RED, WHITE & BLUE

GENERAL SPECIFICATIONS AND DATA

MAXIMUM GROSS WEIGHT - 550,000 lbs.
TOP SPEED - MACH 3 (2,000 m.p.h.) AT 70,000 FEET
POWER - SIX GENERAL ELECTRIC YJ-93 TURBO JETS, 30,000 lbs. THRUST EACH
FUEL - 11 TANKS, 47,000 gals. TOTAL CAPACITY
TAKEOFF AT 500,000 lbs., 200 knots IN 7,000 feet
AT 400,000 lbs., 180 knots IN 4,000 feet
LANDING APPROACH SPEED - 270 knots
AREAS,
TOTAL WING (INCLUDES 2,482.34 sq.ft. COVERED BY FUSELAGE BUT NOT 33.53 sq.ft. OF WING RAMP AREA) - 6,297.15 sq.ft.
CANARD (INCLUDES 150.31 sq.ft. COVERED BY THE FUSELAGE) - 415.59 sq.ft.
VERTICAL TAIL SURFACES (INCLUDES 8.96 sq.ft. BLANKETED AREA) - 233.96 sq.ft.
ENGINEERING AND DESIGN COST, REPORTED FIGURE - 1.3 BILLION DOLLARS
CONSTRUCTION COSTS, APPROX., REPORTED - 750 MILLION DOLLARS
ACCURATE AND TRUE TOTALS WILL PROBABLY NEVER BE KNOWN

XB-70 drawing courtesy of Paul R. Matt, originally published in Historical Aviation Album *in 1/192 scale*

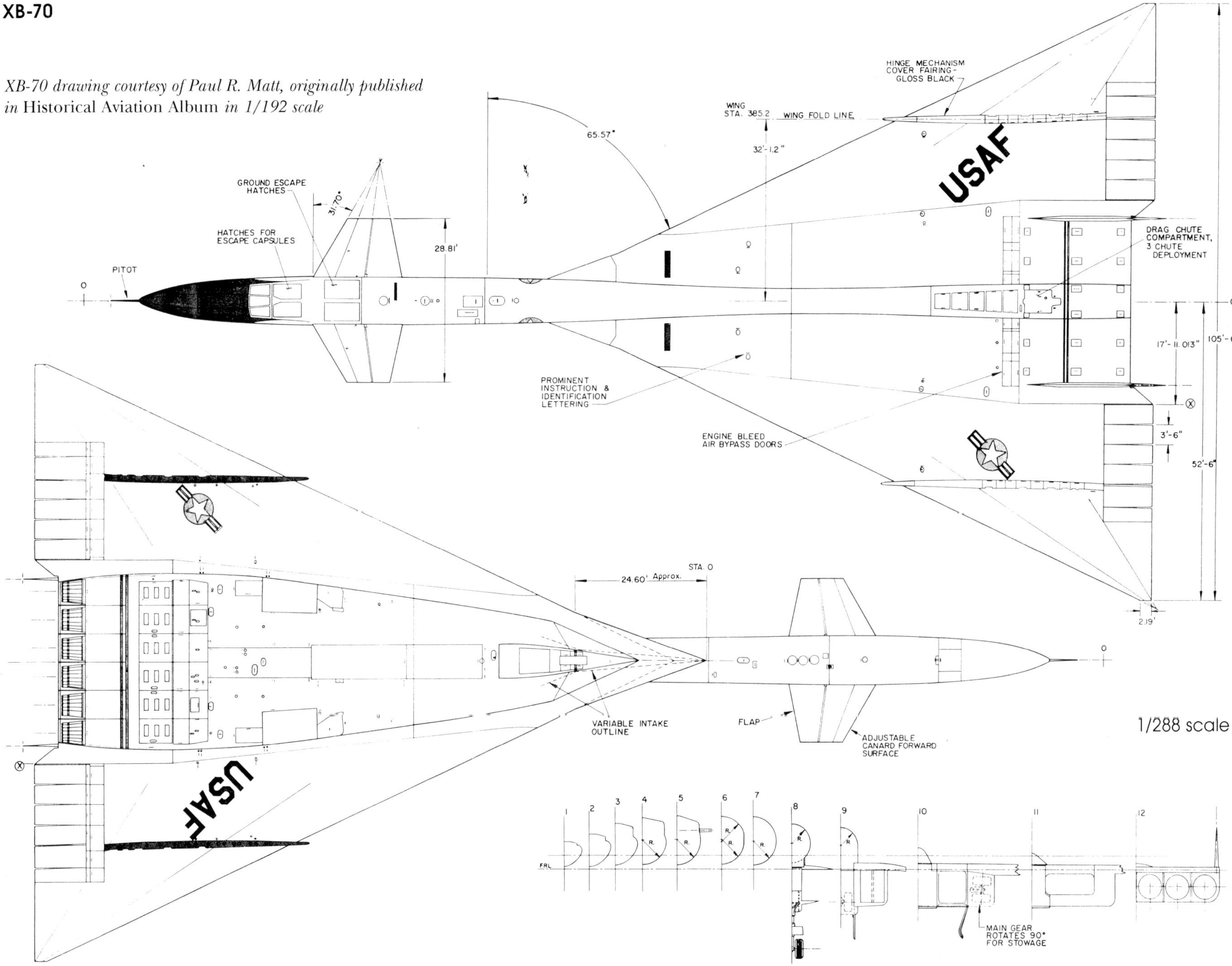

1/288 scale

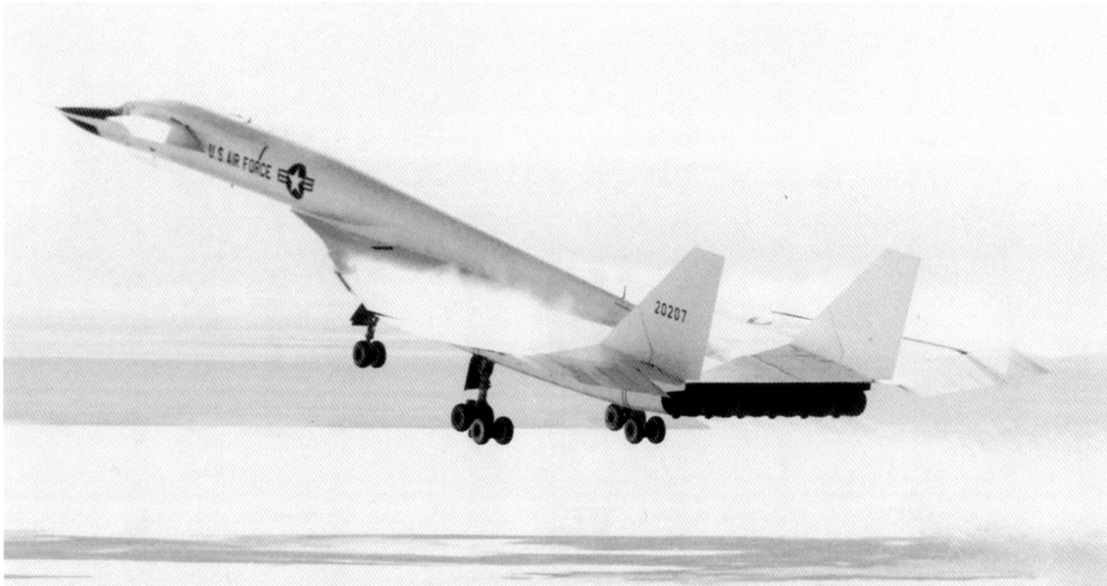

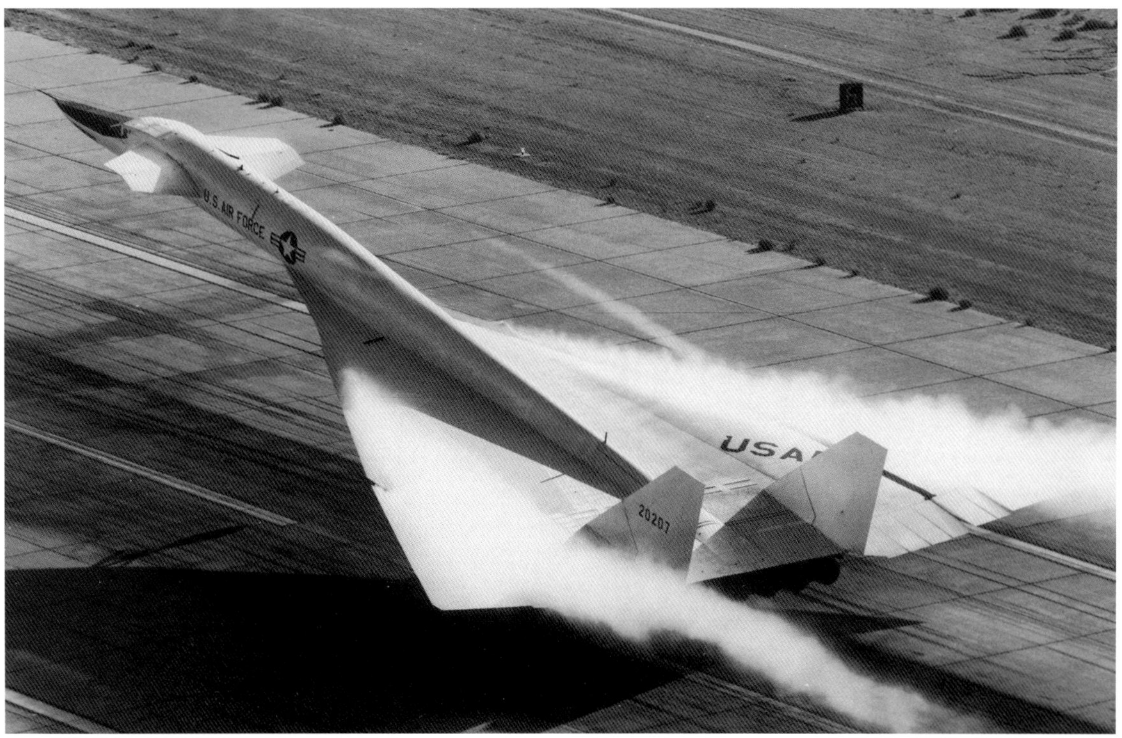

Tremendous condensation trails from huge delta as second XB-70A rotates and quickly climbs during take-off at Edwards. Compared to the original, the new wing had 5 degrees of dihedral.

NAA test pilot Van Shepard made the eighth flight, during which the XB-70 exceeded Mach 2. On the tenth flight, the aircraft sustained supersonic speed for 74 minutes, 50 of which were spent above Mach 2. The test program was progressing well, although hydraulic troubles would continue to plague the first Valkyrie throughout its life. Many of the faults were being eliminated in the construction of the second aircraft.

On July 17, 1965, the second plane, XB-70A-2, AF 62-0207, took to the air. An improved aircraft featuring 5 degrees of wing dihedral, it hit Mach 1.4 and lowered the wingtips 65 degrees on its first flight. The second Valkyrie had improved hydraulics, better fuel tanks and an automatic control system for the inlet ramp adjustment, as opposed to the manual system on the first aircraft. The XB-70A-2 accelerated through its test program more rapidly and with fewer problems.

During the mid-1960s U.S. aerospace companies were studying the possibilities of a supersonic transport (SST) aircraft for the airline industry, and much of the data gathered by the B-70 program would determine the viability of such a program. Sonic boom testing was conducted, as well as heat-soaking and several other speed-related studies for large aircraft.

On October 14, 1965, the XB-70A-1 reached its design goals of Mach 3 at 70,000 feet, but this was the only triple-sonic flight that this aircraft would ever make. Due to the improved structure of the second Valkyrie, all subsequent Mach 3 research would be done using that aircraft. On May 19, 1966, White and Cotton flew XB-70A-2 at Mach 3 for 33 minutes, covering eight western states and dashing from Salt Lake City, Utah, to the Blythe, California, in 18 minutes.

Two pilots were added to the XB-70 program in early 1966: Major Carl Cross and NASA's Joe Walker, who had earned Astronaut status in the X-15 program. Cross was co-pilot with White on June 8, 1966, during the 46th flight of the second

Valkyrie, making a low-performance, instrument-calibration flight followed by a publicity photo shoot for General Electric. A formation of GE-powered aircraft accompanied the XB-70; Walker was on the right wing in a Lockheed F-104 Starfighter. The photo shoot was nearly complete when the F-104 suddenly collided with the Valkyrie. The F-104 burst into flames and disintegrated, killing Walker instantly. The XB-70's vertical stabilizers were destroyed in the collision and the giant aircraft went into an uncontrollable spin, crashing into the desert northwest of Barstow, California.

White ejected but was seriously injured when his ejection module landed. Cross, probably incapacitated by the g-forces of the spin, did not eject and was killed on impact. It is likely that Walker's F-104 ventured too close to the Valkyrie's wing and the vortex generated by the larger aircraft's wingtip pulled the small fighter into it. Two test pilots and a multi-million dollar research aircraft were lost in just over one minute.

The XB-70A-1 continued flying with NASA through the end of 1968, during which time it was piloted by Emil "Ted" Sturmthal, USAF, and NASA's Don Mallick, who was later involved with the Lockheed YF-12A and Apollo Lunar Module programs. The sole remaining Valkyrie, with Fulton and Sturmthal at the controls, made its last flight when it was handed over to the USAF Museum at Wright-Patterson AFB, Dayton, Ohio, on February 4, 1969. It is currently on display inside a newly constructed building.

Although the XB-70 program consumed more than $1.3 billion, it proved very successful overall. The Valkyrie was the most aerodynamically efficient piloted aircraft ever produced, an engineering masterpiece in terms of performance and size. Although only two were built, they completed a combined 128 flights and the first plane was still collecting valuable data during its final flight.

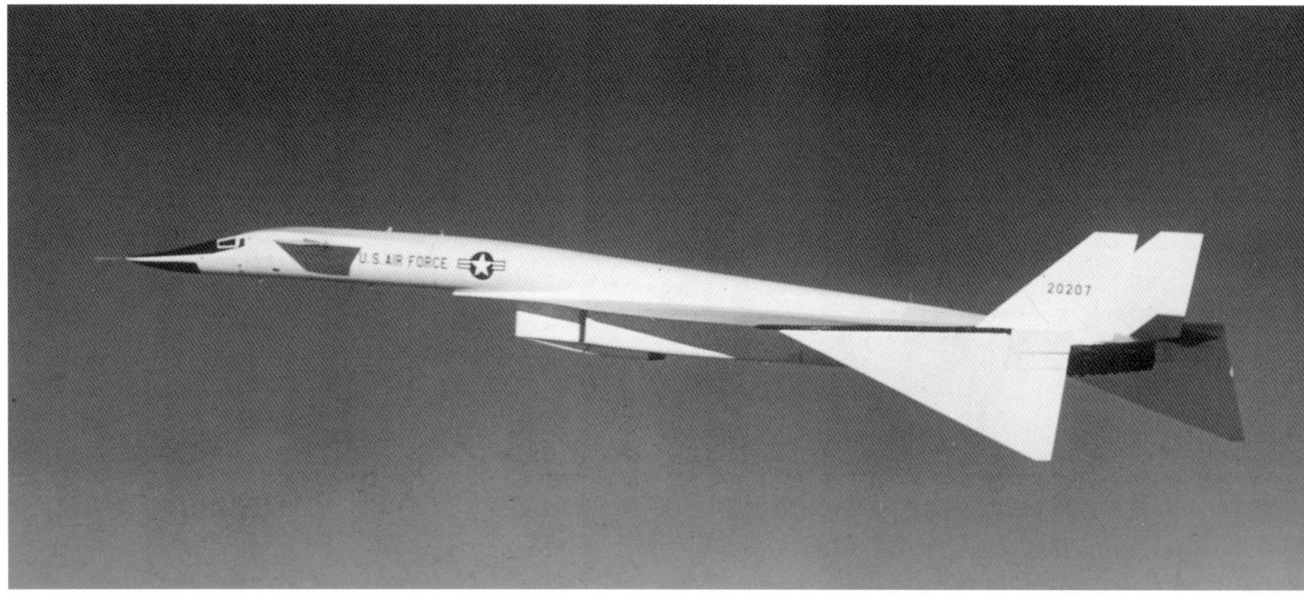

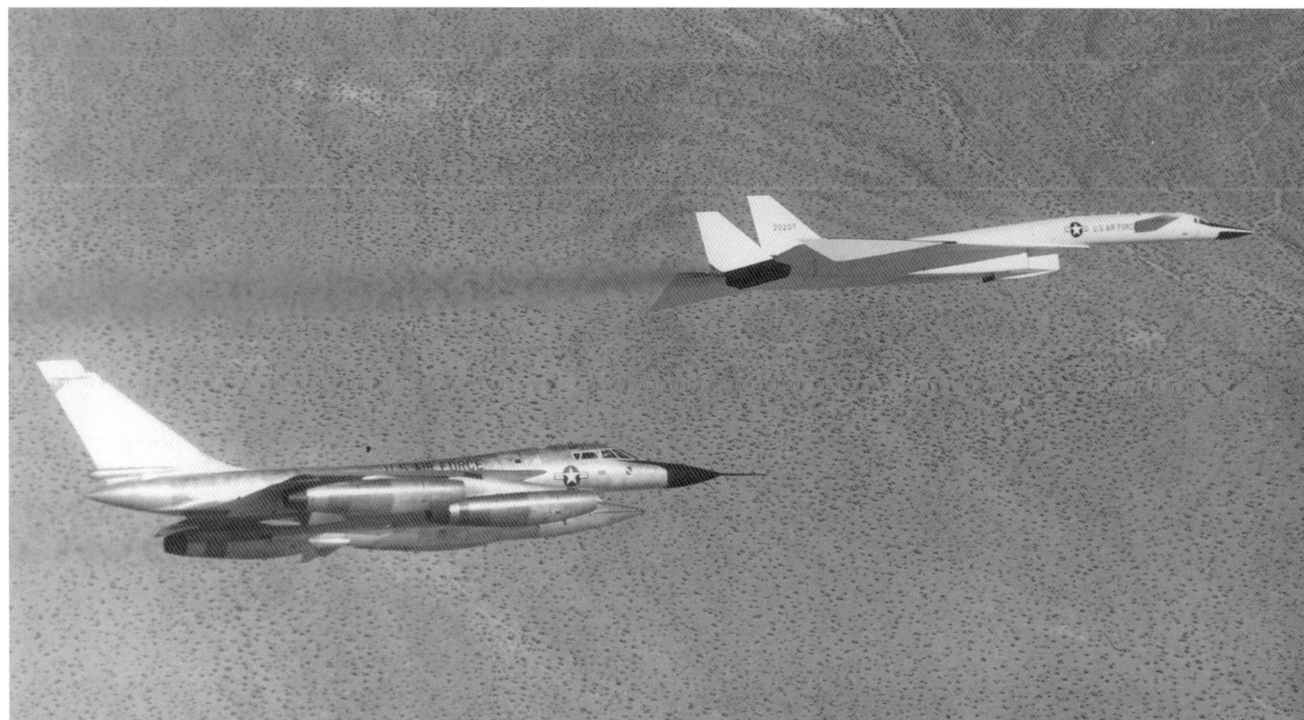

In top photo, wingtips are folded 65 degrees, improving stability for Mach 1.4 to 3.0. At right, nearer to camera, the B-58 chase plane looks big but is actually only half the XB-70A's length.

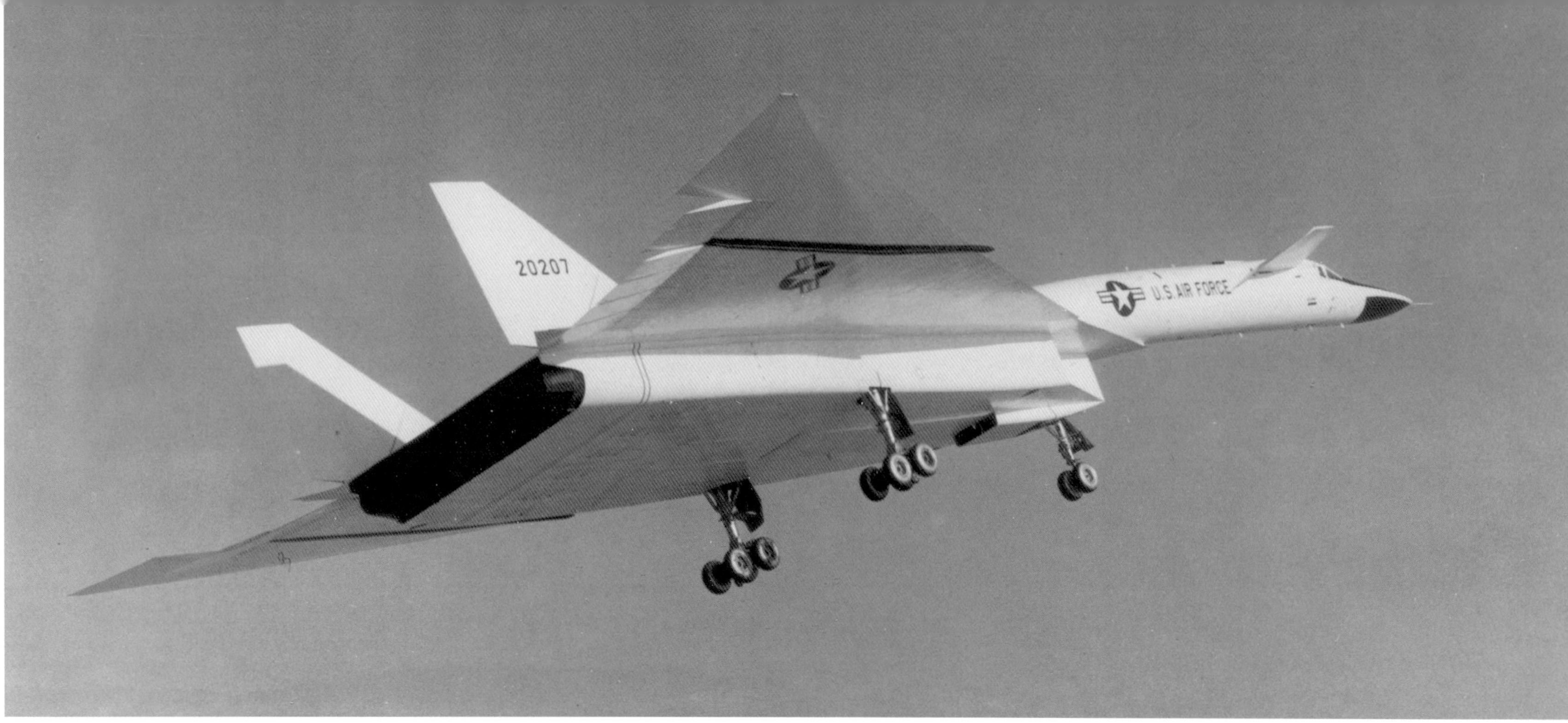

The six General Electric YJ93-GE-3 engines, producing a total of 168,000 lbs/thrust, dominate the underside of the XB-70A, on approach for Edwards with ten-wheel landing gear extended.

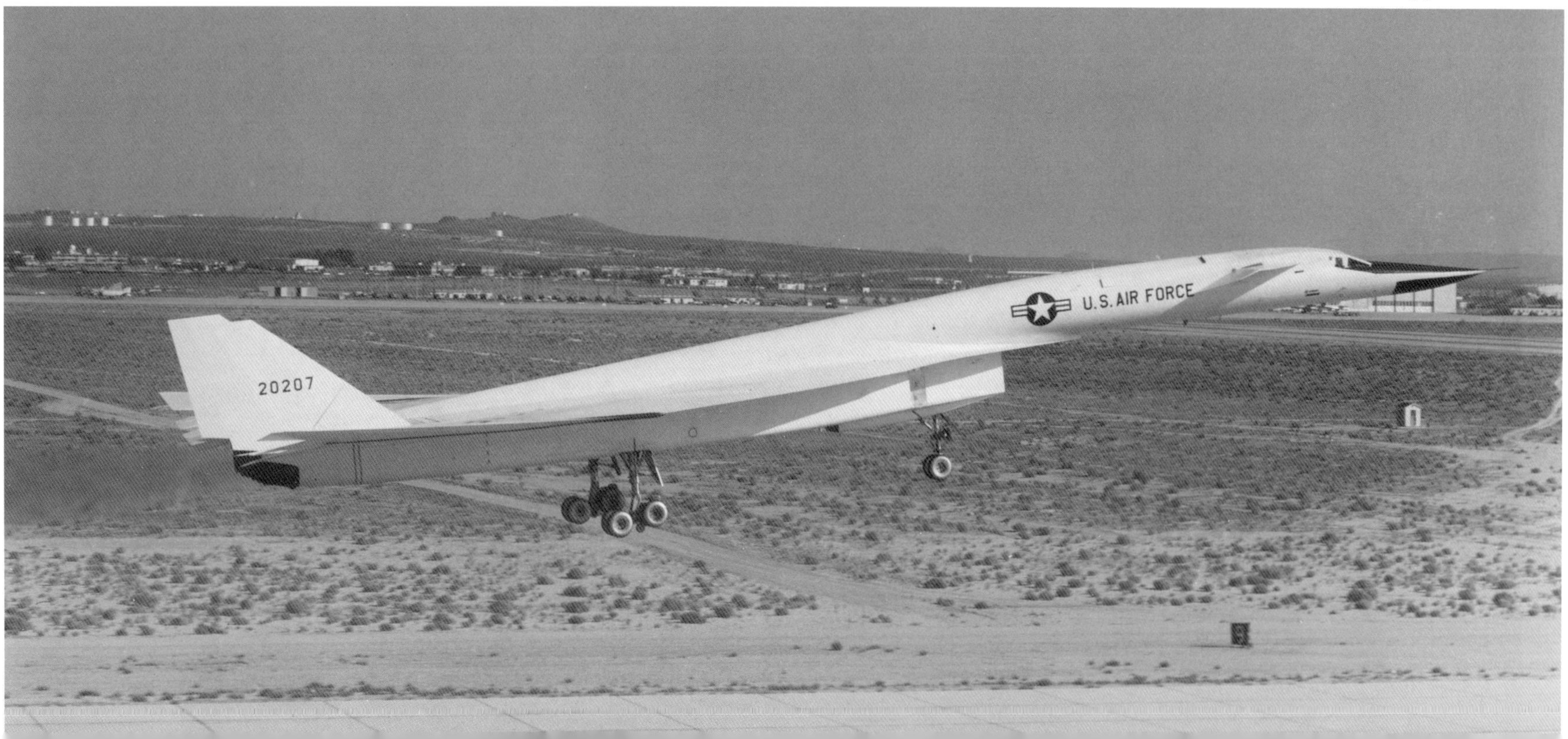

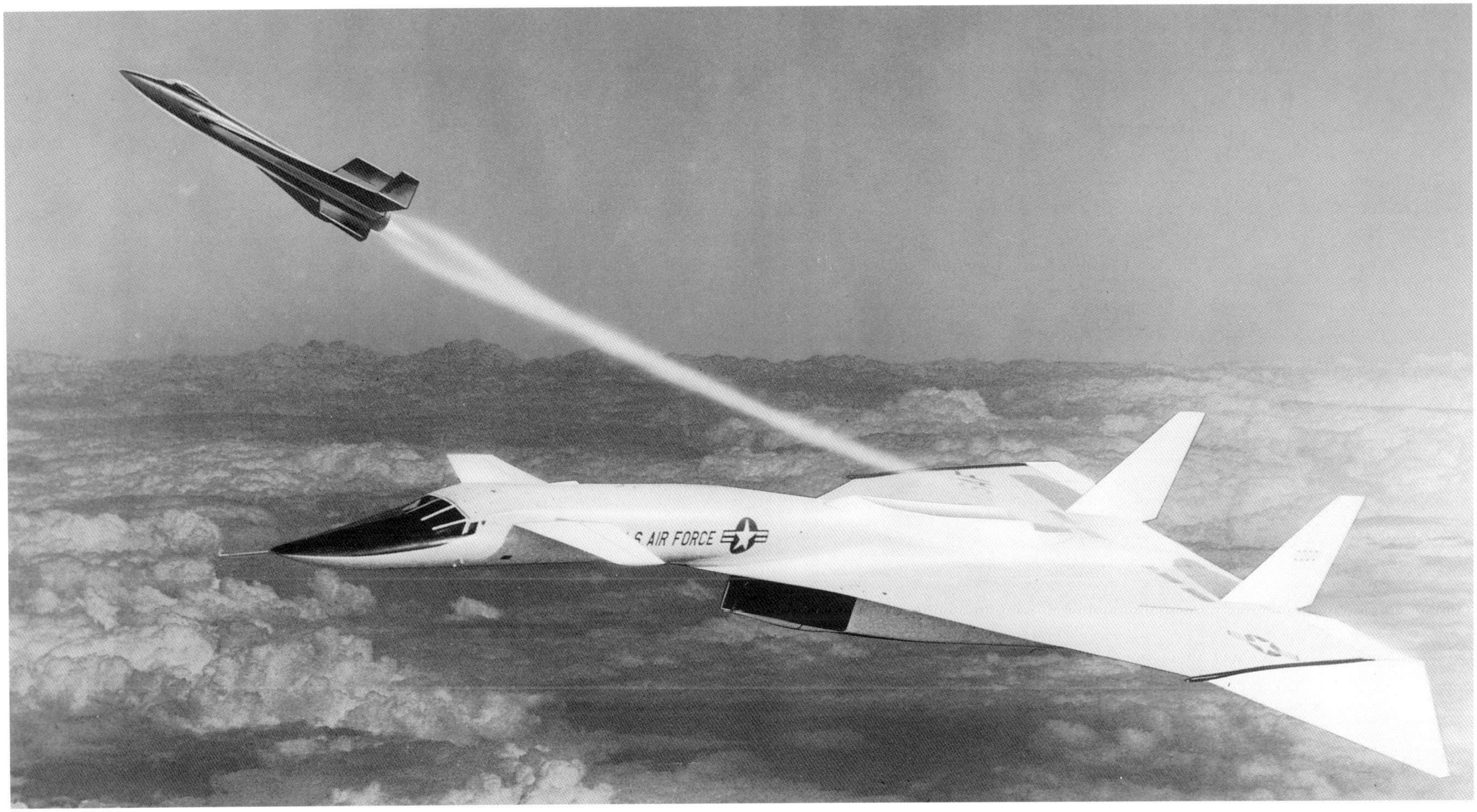

To reduce the amount of fuel the X-15 needed to attain higher altitudes, the XB-70A was considered as a launch plane. This composite rendering depicts a modified X-15, with vertical winglets, zooming away from the Valkyrie at 70,000 feet.

Specifications: North American NA-259>NA-315 XB-70A-1/A-2 Valkyrie

Description: U.S. Air Force long-range strategic bomber, completed as experimental research aircraft. Six-jet mid-wing monoplane of all-metal construction with retractable landing gear. Crew 4. (Research aircraft) 2.

Engines: Six 28,000-lb/thrust General Electric YJ93-GE-3 turbojets with afterburner.

Dimensions: Span 105 feet, length 196 feet 6 in, height 30 feet 3 in, wing area 6297.15 sq ft.)

Weights: 205,000 lb empty, 550,000 lb max take-off.

Armament: (Bomber) 14 nuclear bombs or various conventional bombs. (Research aircraft) None.

Performance: Cruising speed 1982 mph, maximum 2035 mph (Mach 3.08), service ceiling 75,500 ft, range approx. 5000 miles.

Production: North American NA-259>NA-315 XB-70/ XB-70A-1/A-2 Valkyrie

NA-259	B-70 engineering, Phase I	—
NA-264	B-70 engineering, initial Phase II	—
NA-267	B-70 engineering, Phase II, Part 2	—
NA-274	YB-70 weapon system development	—
NA-278	XB-70A-1 and -2 AF 62-0001, 62-0207	2
NA-281	XB-70A flight test program, AF 62-0001	—
NA-286	XB-70B termination (AF 62-0208)	(1)
NA-303	XB-70A flight test AF 62-0001, 62-0207	—
NA-315	XB-70A flight test AF 62-0001, 62-0207	—
Total, XB-70 research prototypes, 1955-1968		2

North American / Rockwell NA-300>NA-404
YOV-10/OV-10A/F Bronco

DURING WORLD WAR II—and even more so in Korea—U.S. ground forces relied heavily on close air support. Over the years, several types of light aircraft were given the duty of making low-level pinpoint strikes against opposing forces, to minimize U.S. casualties and to contain enemy troops, although no aircraft had been designed specifically for that purpose.

Lessons learned in the Korean War led to a U.S. Marine Corps study in 1959 for a new aircraft that could perform both observation and light attack duties. The study specified a relatively inexpensive aircraft, designated LLVMA (Light, Light Marine Attack Aircraft), that could operate from forward airfields, requiring minimal maintenance and able to deliver a wide variety of ordnance in support of Marine helicopters during air assaults.

Early in 1963, Secretary of Defense Robert McNamara formed a Tri-Service Committee by which Army, Navy, Marine Corps and Air Force representatives formulated requirements for an aircraft that could meet the needs of each military branch with little modification to the basic design. By the end of 1963, a specification for a twin-engine, two-man Light Armed Reconnaissance Aircraft (LARA) was issued, and several U.S. manufacturers submitted design proposals. The LARA parameters were very stringent: this versatile aircraft would have to carry 2400 pounds of ordnance, loiter over a target at least fifty miles from base for one hour and have the capacity to carry 2000 pounds of cargo, six paratroopers or several wounded troops on stretchers in the rear fuselage compartment.

Performance expectations included a top speed of 350 miles per hour, the ability to operate from aircraft carriers, a takeoff run of 800 feet and easy adaptability into an amphibian.

The first of seven YOV-10As evaluated under the Tri-Service program, carriying Navy BuAer number 152879. The YOV-10As were notable for their very short, 30-foot 4-inch wingspans.

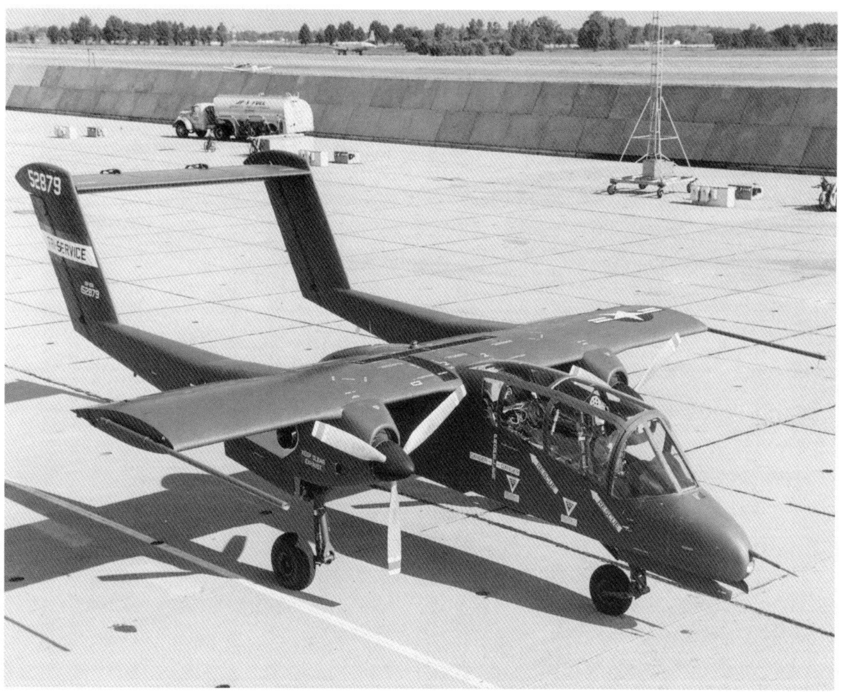

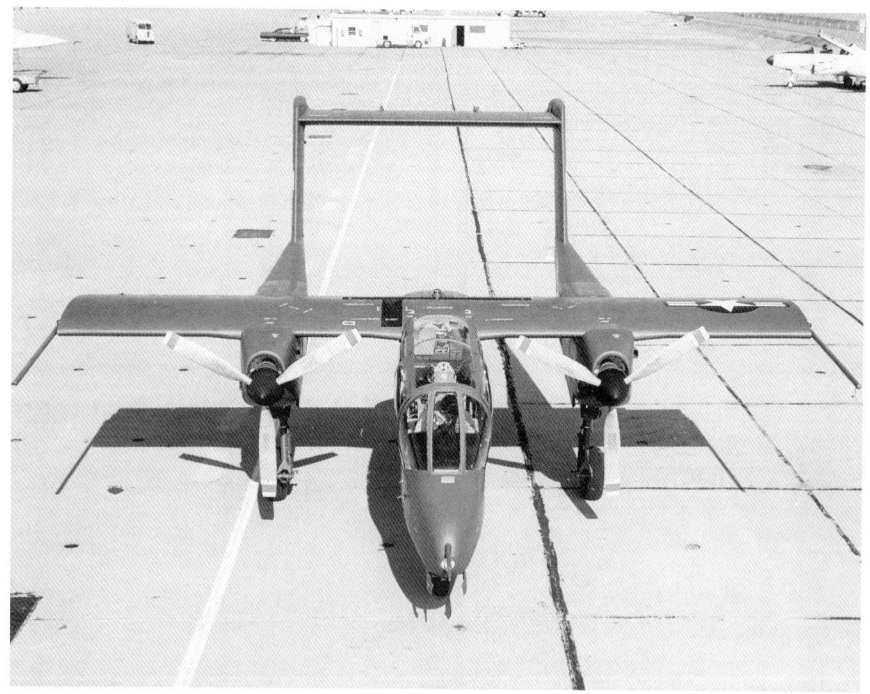

Specified armament was four 7.62-mm machine guns with 500 rounds each, a 20-mm gun pod and hardpoints to carry a variety of weapons, including Sidewinder air-to-air missiles.

Of the original eleven proposals originally submitted, seven were examined more closely. After evaluating proposals from Beechcraft, Douglas, General Dynamics, Helio, Lockheed, Martin and North American, the committee chose NAA's design in August 1964. The NA-300 was a product of the Columbus Division, which received a Navy-directed initial contract for seven prototypes. General Dynamics persisted in developing a prototype, hoping to win in a fly-off competition, but the Tri-Service committee would not change its mind and NAA retained the contract.

The general layout of the NA-300 was a twin-boom aircraft with a central fuselage pod beneath a rectangular high wing. A large single horizontal stabilizer joined the vertical fins at the very top, allowing easy loading and unloading at the rear fuselage compartment door.

The large segmented canopy bulged beyond the width of the fuselage, allowing excellent overall visibility for both pilot and observer, seated in tandem ejection seats. Horizontal sponsons were fitted beneath the central fuselage for installation of four 7.62-mm machine guns and underside ordnance. Conventional aluminum monocoque construction was used throughout. Power was supplied by two Garrett AiResearch T76-G6/8 turboprop engines of 660 shaft horsepower each, with the propellers turning in opposite directions to counteract torque and pitch-reversible for short-field operations, further improved by the employment of heavy-duty tricycle landing gear.

The mock-up of the NAA "Counter-Insurgency" (COIN) aircraft passed its inspection by the Tri-Service Committee in February 1965, and the first prototype, YOV-10A BuAer 152879, flew on July 16, four months ahead of schedule, as NAA had accelerated the initial development program.

The remaining six prototypes, BuAer 152880 through 152885, were delivered by October 1966 and underwent an intensified test program with NAA pilots, as well as pilots from all four services. By this time, much information had been gathered through combat reports from South East Asia, leading to new requirements and extensive modifications to the production aircraft. The use of self-sealing fuel tanks, 300 pounds of armor protection and more-comprehensive communications equipment were some of the changes that were eventually incorporated into series production, although these changes caused a reduction in performance.

The most obvious change was a 10-foot in-

Below, a mock-up of a YOV-10A seaplane with twin Edo floats. None was actually built. At right, the second YOV-10A, BuAer 152880, carrying four bombs on a horizontal airfoil.

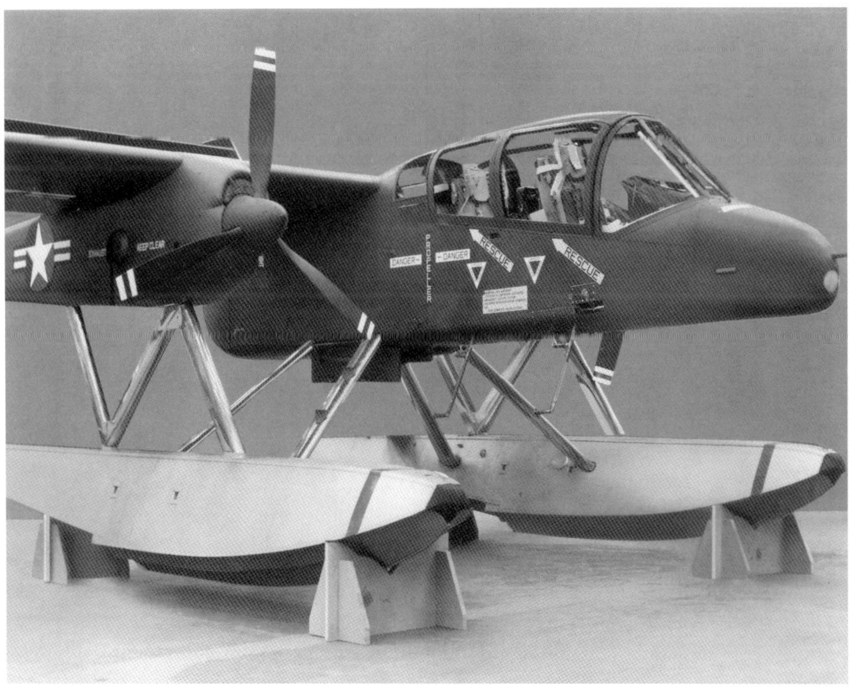

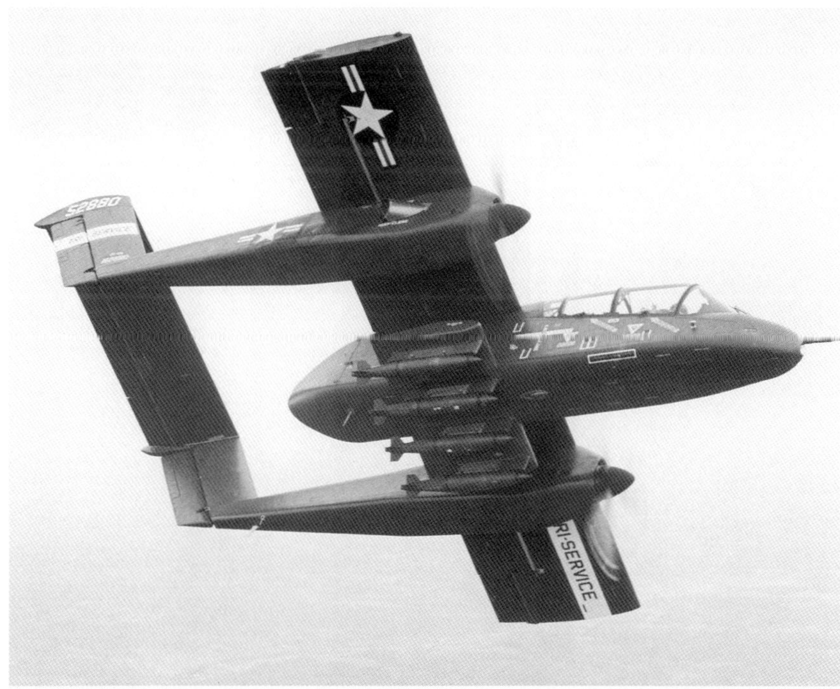

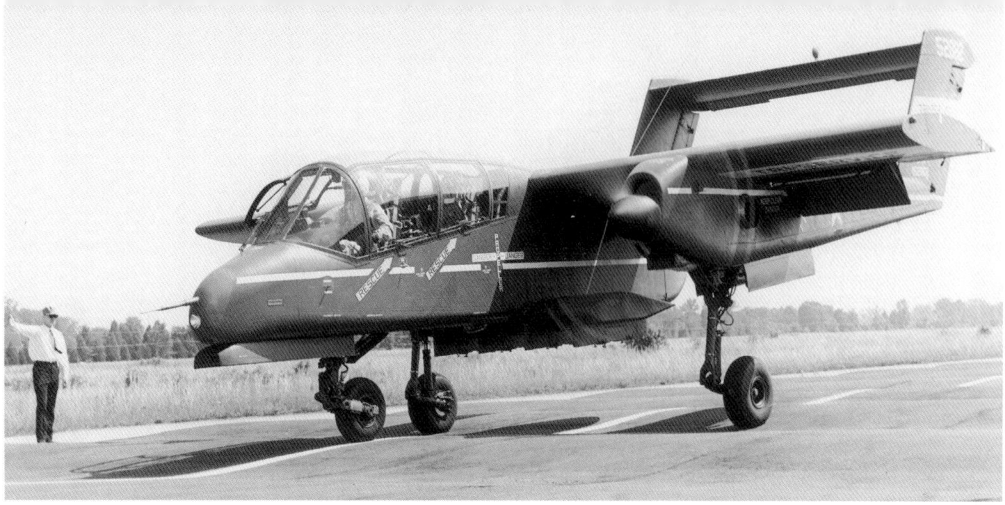

YOV-10A number four, BuAer 152882, testing the Bronco's extremely resilient landing gear over "speedbumps."

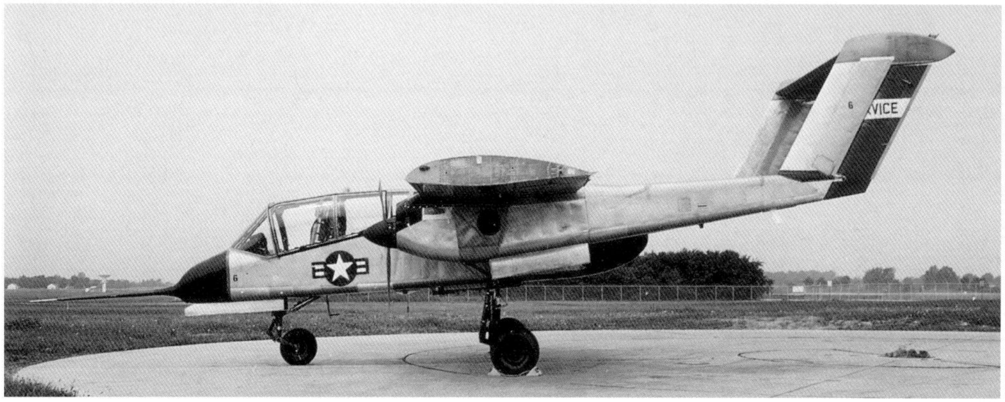

Above, unpainted YOV-10A number 6, 152884, and below, the same ship in flight, carrying rocket launchers.

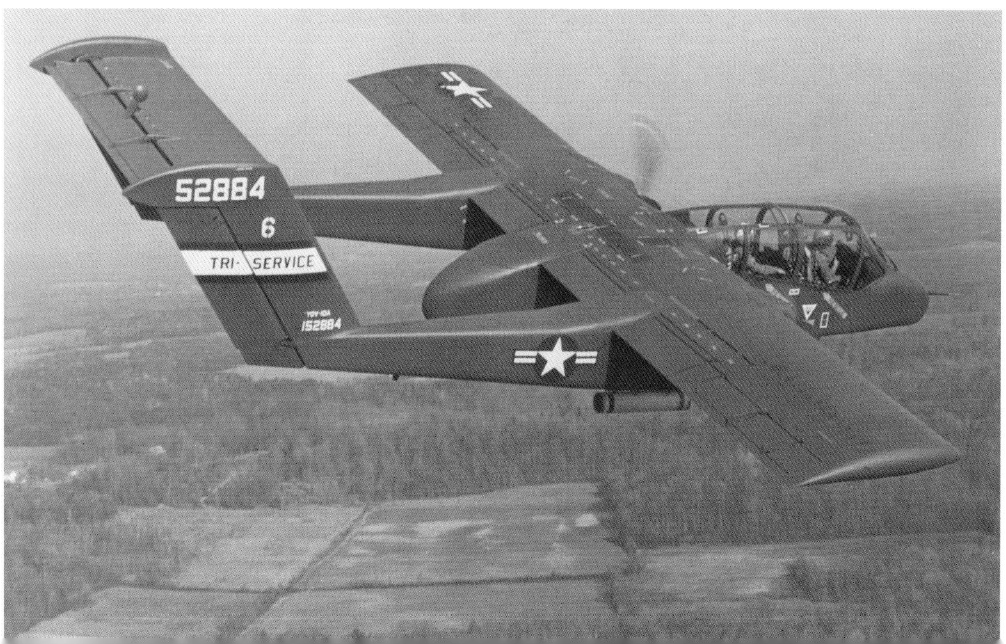

YOV-10A

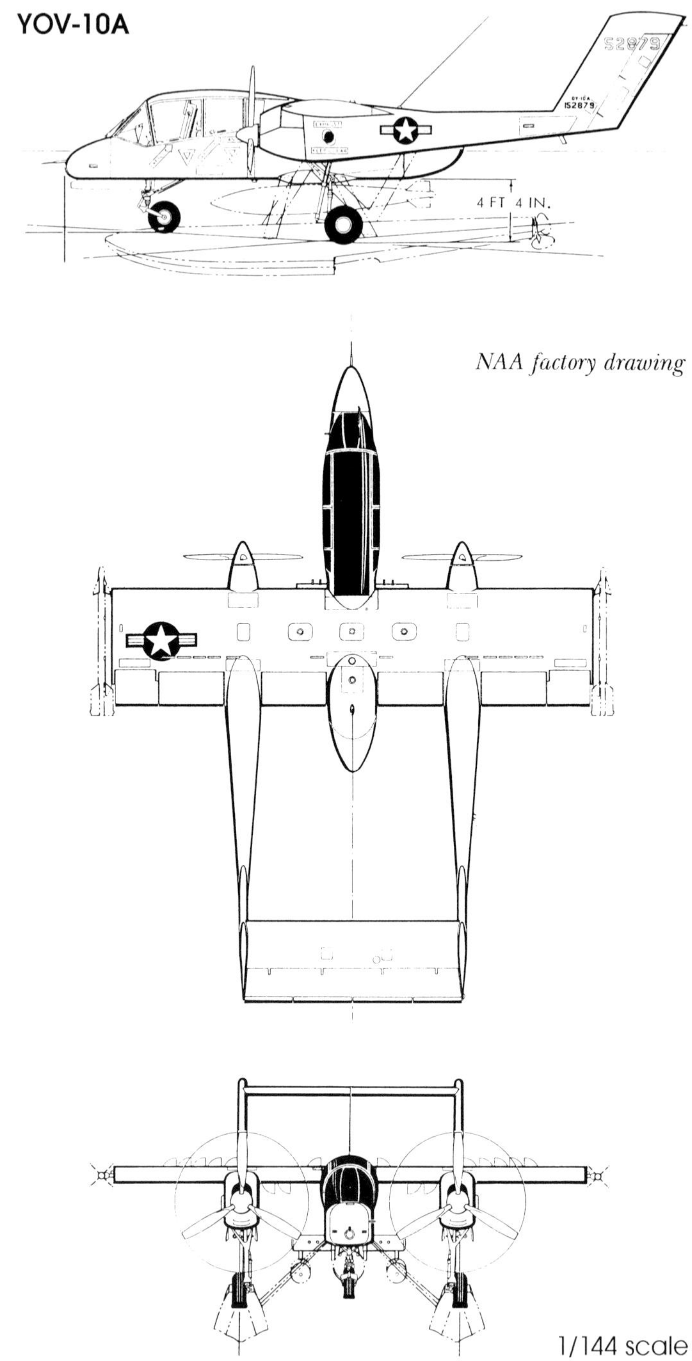

NAA factory drawing

1/144 scale

crease in wingspan, with the repositioning of the engine nacelle booms six inches further outward, primarily to reduce cockpit noise levels. With the added weight, more-powerful T76-G10/12 engines of 715 horsepower were used. The sponsons were now angled downward for cleaner weapons separation and a dorsal fillet was integrated into the base of each vertical fin to improve lateral stability. All of these changes were made to the prototypes for evaluation, and by early 1968 the problems were effectively resolved. Production of the OV-10A Bronco began shortly thereafter, the first deliveries to both USAF and USMC units taking place on February 23.

A total of 271 OV-10As were built, 157 of these entering service with the USAF and 114 with the USMC. The Navy eventually used a batch of 19 aircraft, borrowed from the Marine supply. It was anticipated that the U.S. Army would also be equipped with Broncos, but no orders ever materialized.

The USAF 504th Tactical Air Support Group pressed the Broncos into combat in Southeast Asia, employing them in Forward Air Control duties beginning in July 1968. The aircraft immediately became popular with its pilots, its most valuable assets being the outstanding all-around view, maneuverability and versatility in close-support warfare. The Broncos had an excellent response time, were very well suited for target designation and had the necessary communications equipment for both FAC and search-and-rescue duties.

Shortly after the initial OV-10A combat deployment, the USAF became interested in improving the aircraft's ability to operate at night and in poor weather. The ability to direct more precise air strikes was also studied, and LTV Electrosystems modified fifteen USAF Broncos to OV-10A Pave Nail specification in 1970. This consisted of adding a stabilized night periscope sight, combination laser range finder and target illuminator, LORAN navigation receiver and a Lear Siegler LORAN coordinator. Known as Pave Spot, the combination allowed target illumination with a laser beam and gave an accom-

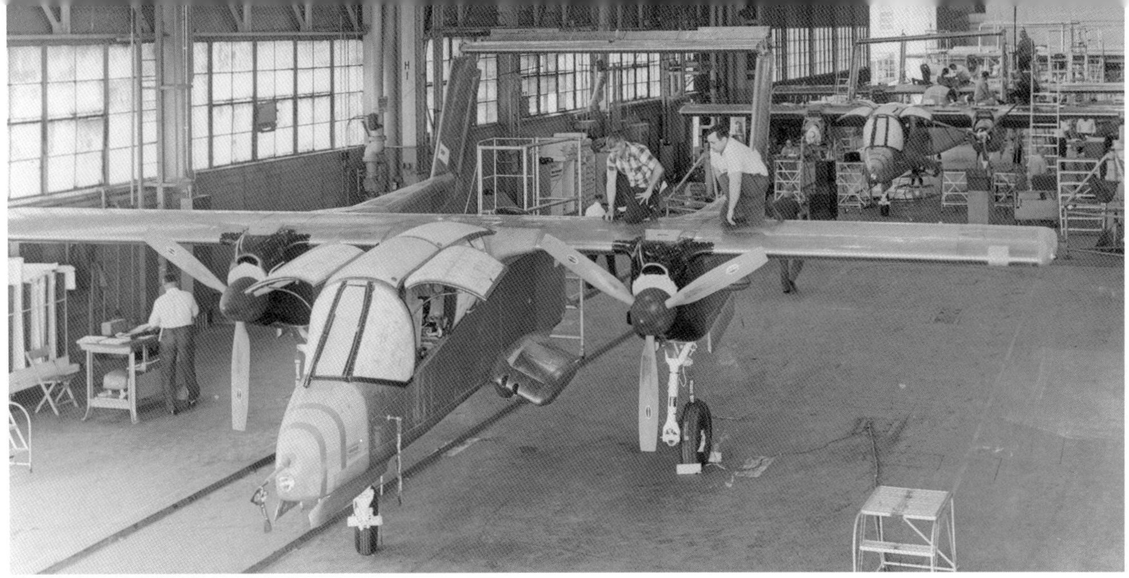

OV-10A production line, showing the longer 40-foot wing and angled weapons sponson. Below, the fifth Marine Bronco, 155394.

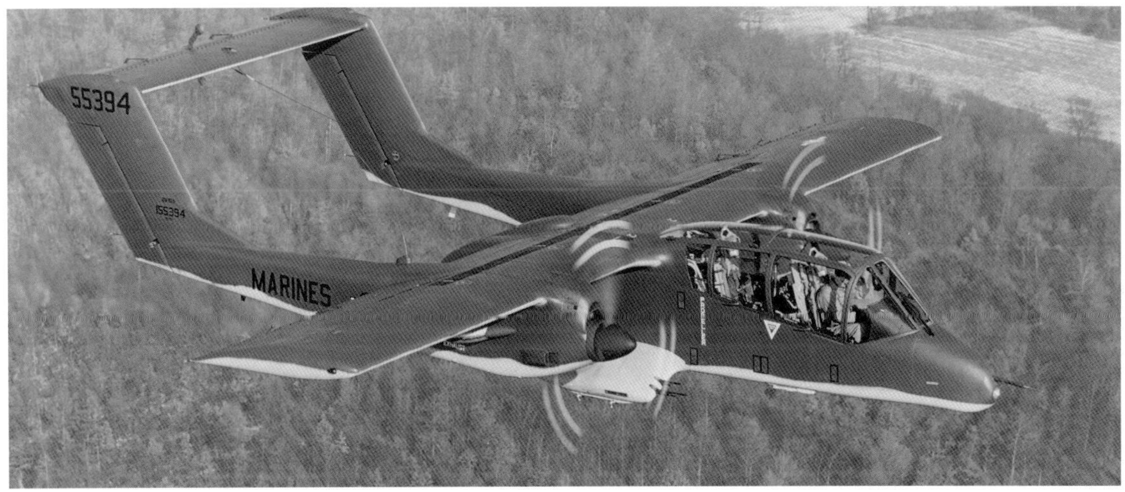

Another visible production change was the addition of fillets to the vertical tailpanes, as on the Air Force Bronco, AF66-13554, below.

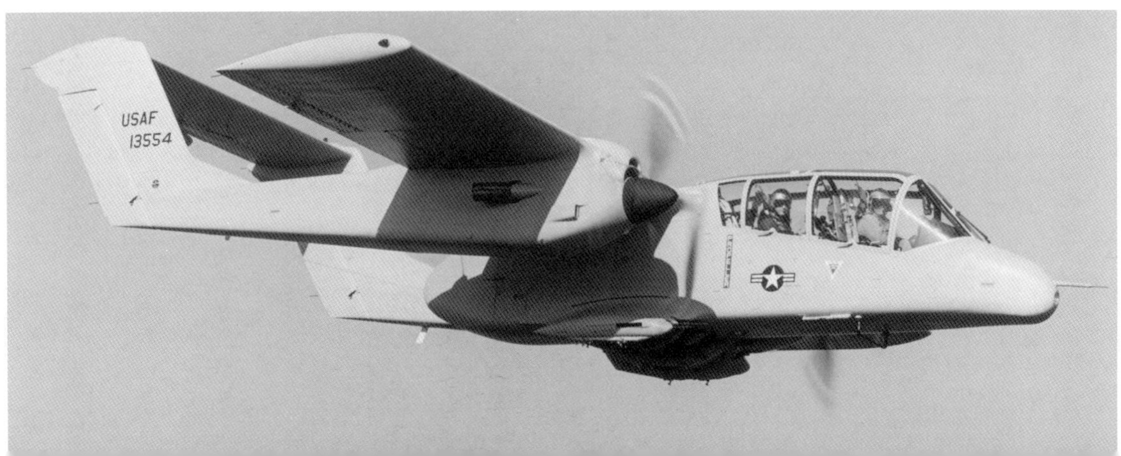

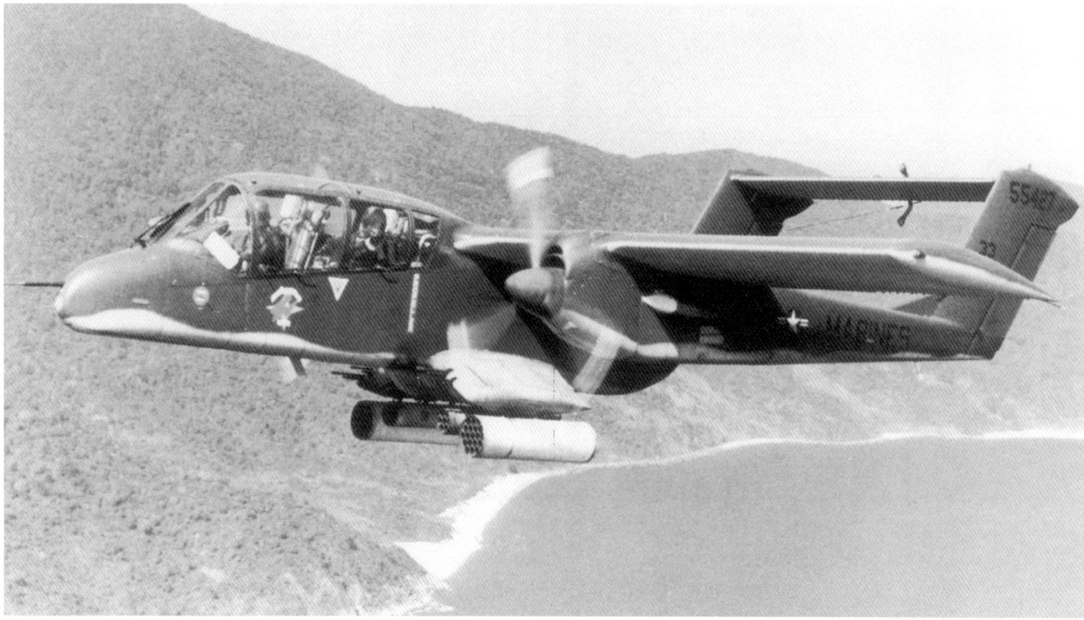

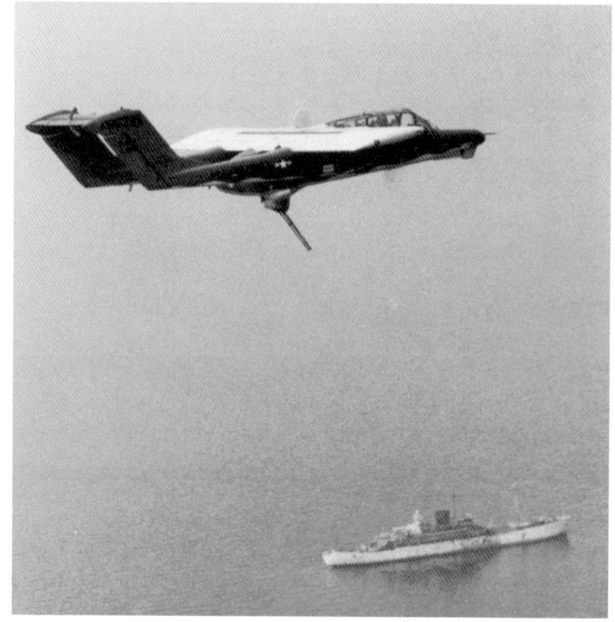

Above, an OV-10A of VMO-2 on patrol south of Danang, carrying four rocket launchers. Right, a Marine YOV-10D NOGS with ventral turret, and below, an OV-10D NOS with FLIR nose.

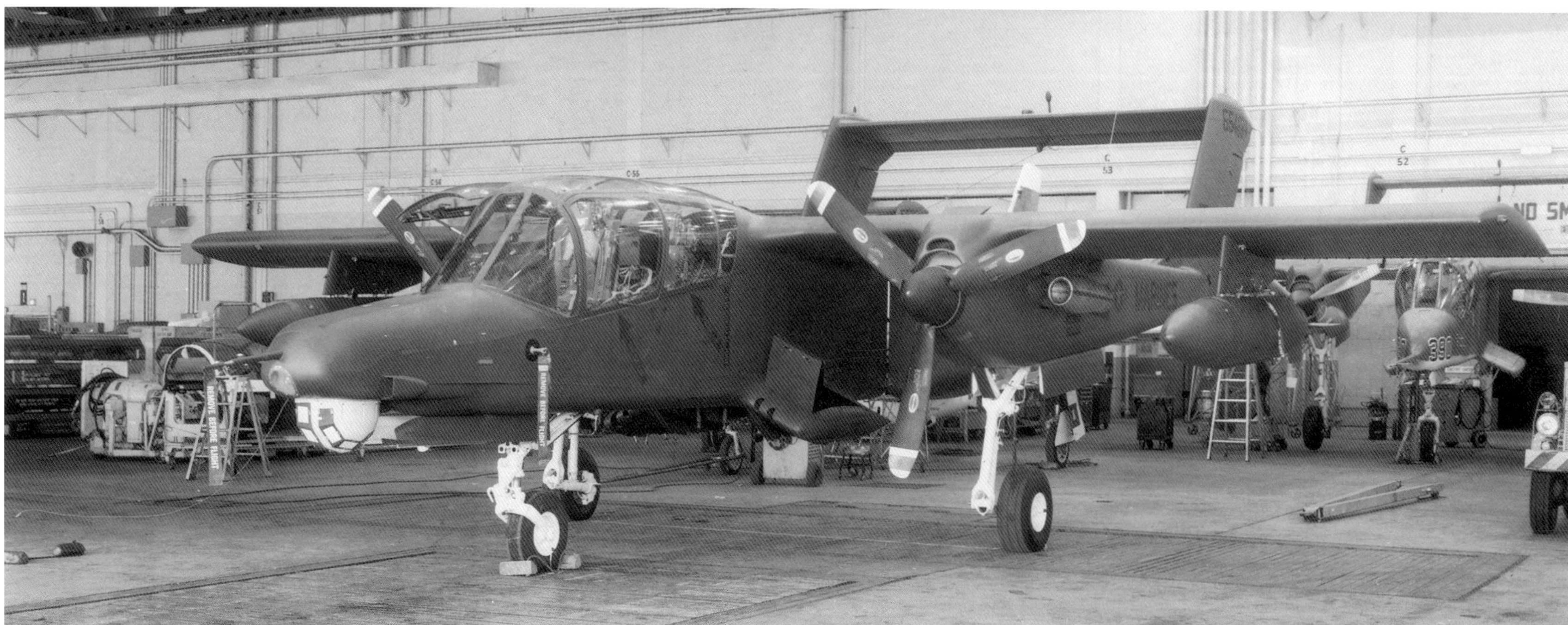

panying aircraft the ability to make an attack using an offset vector. The high-flying aircraft directed incoming strikes while the low aircraft designated targets and surveyed the battlefield. During the latter stages of the Vietnam war the system was most useful in locating downed pilots during search-and-rescue operations, performed by the 23rd Tactical Air Support Squadron, usually in pairs. After the war Pave Nail Broncos were converted back to standard OV-10A specification.

Navy and Marine Corps Broncos also saw action in Vietnam. The only naval squadron was VAL-4 (Light Attack Squadron Four), formed to provide air cover for naval forces operating in the Mekong Delta. Known as the Black Ponies, VAL-4 was based at Binh Thuy Air Base and flew numerous missions against Viet Cong targets in the Delta through late 1971. Marine Corps squadron VMO-2 was actually the first USMC unit to see action in South East Asia, on July 6, 1968, followed by the arrival of VMO-6 in November.

The OV-10A used a variety of ordnance throughout its career, including 2.75-inch and 5-inch rocket launchers, AIM-9 Sidewinder air-to-air missiles, Pave Nail laser designator pod, GPU-2 20-mm gun pods, SUU-11 mini-gun pods, Mk-6 float flares (for night target marking) and four different capacities of fuel tanks. The centerline hardpoint could accommodate 150-, 230- and 300-gallon tanks, and on later models 100-gallon tanks were often fitted to the wing hardpoints.

In the late 1960s the West German government showed an interest in the Bronco for target-towing. This version was designated OV-10B and first flew on April 3, 1970. The OV-10B differed from the OV-10A in that all armament, including sponsons, was removed and the rear cargo door was replaced with a transparent glass dome for rearward viewing. Due to an apparent lack of power, a 2950-pound/thrust J85-GE-4 engine was tested atop one plane, given the designation OV-10B(Z). The jet engine cut the takeoff run in half and tripled the climb rate. Six OV-10Bs were built and all were delivered to West Germany, operating there until the early 1990s.

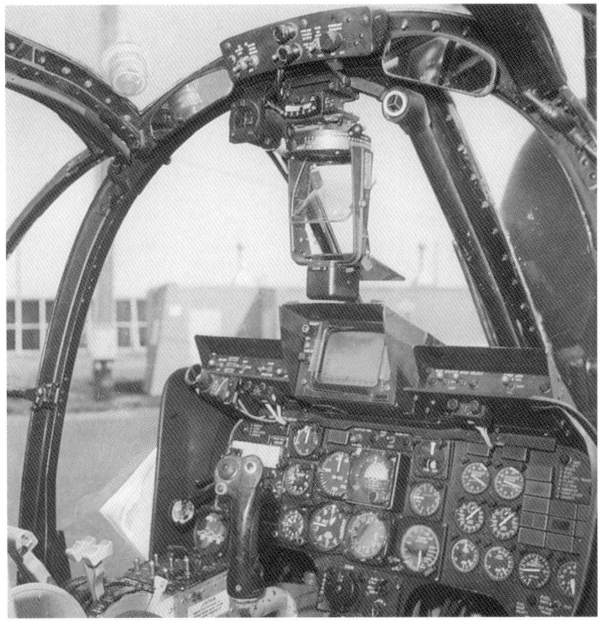

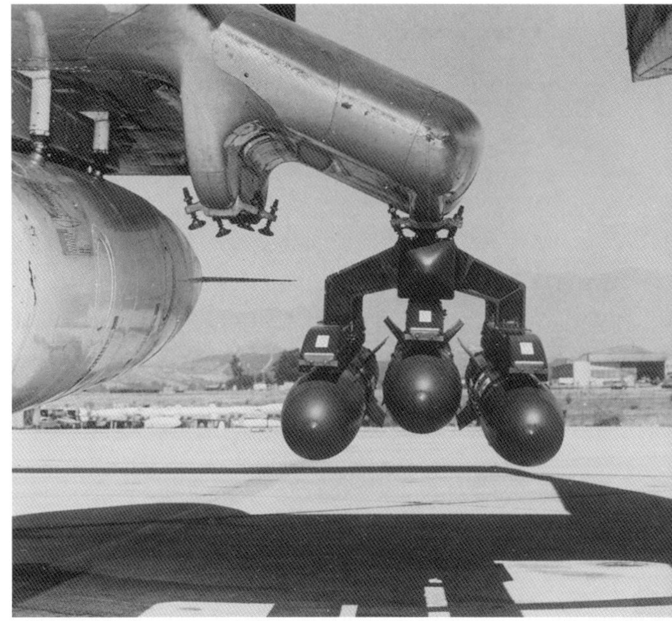

Above, OV-10A cockpit, and right, an Army weapons rack on Marine OV-10A. Below, the OV-10B(Z) with dorsal J85 jet.

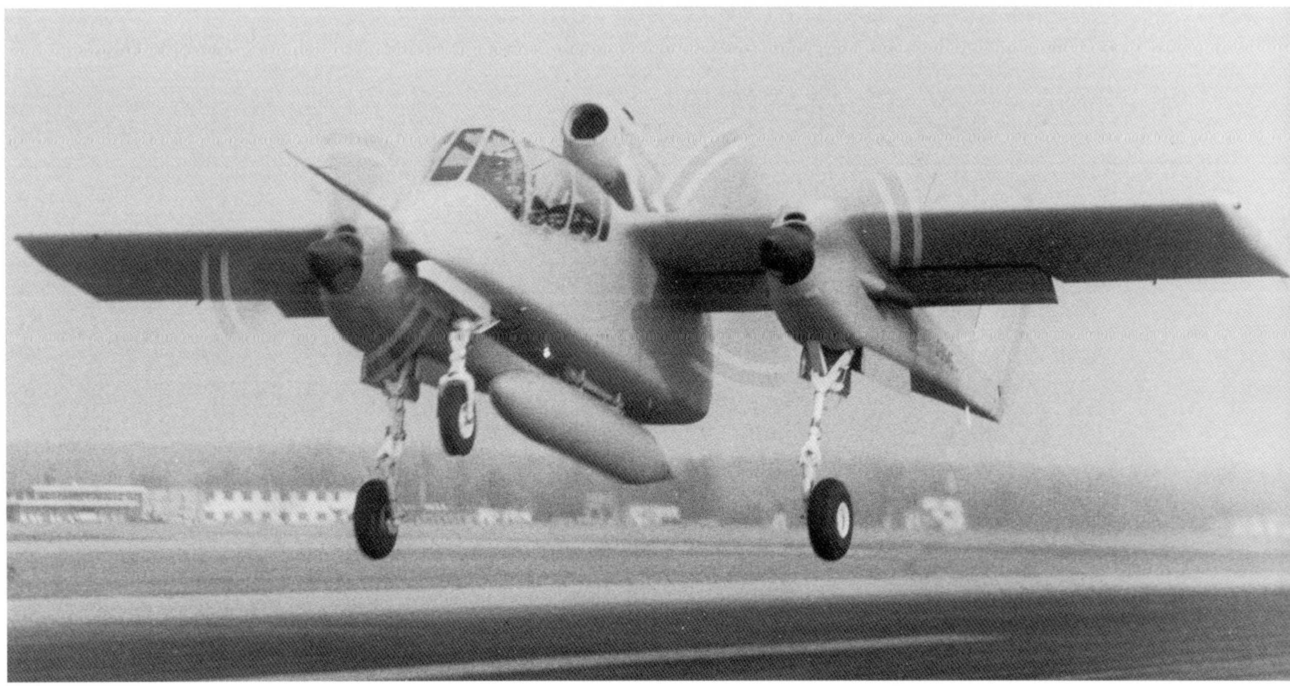

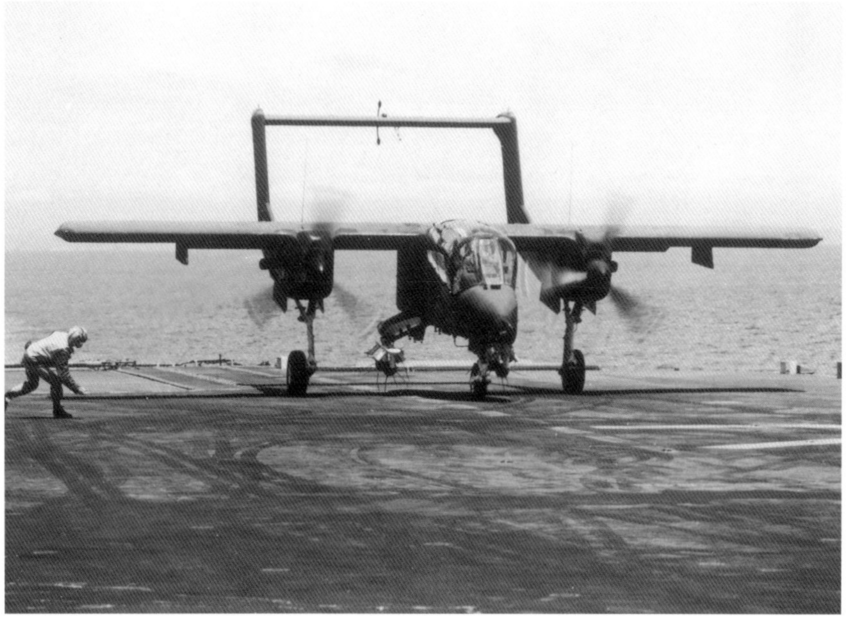

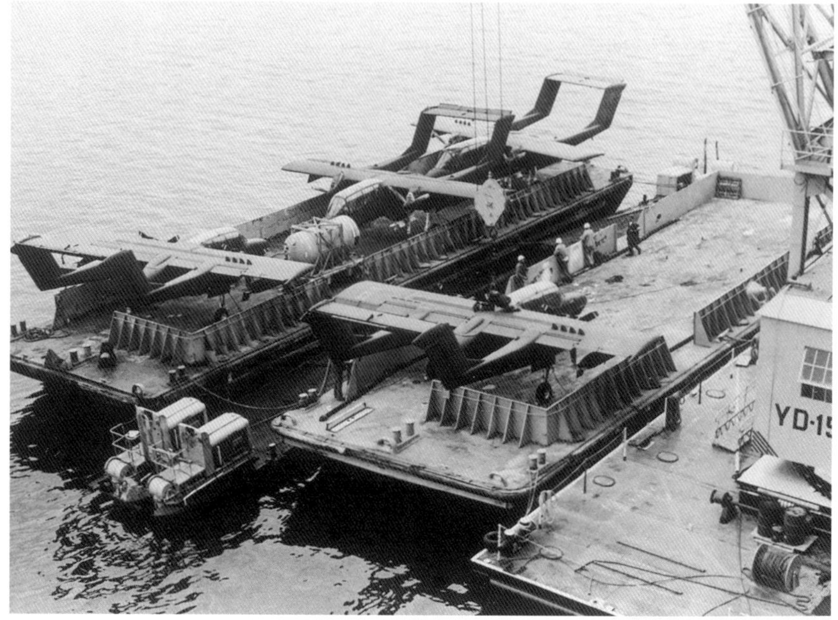

Above, a Marine OV-10A begins take-off from LPH-3, the USS Okinawa, *a Navy amphibious warfare ship. Without arrestor gear, the Broncos had to be transported to the carriers on barges.*

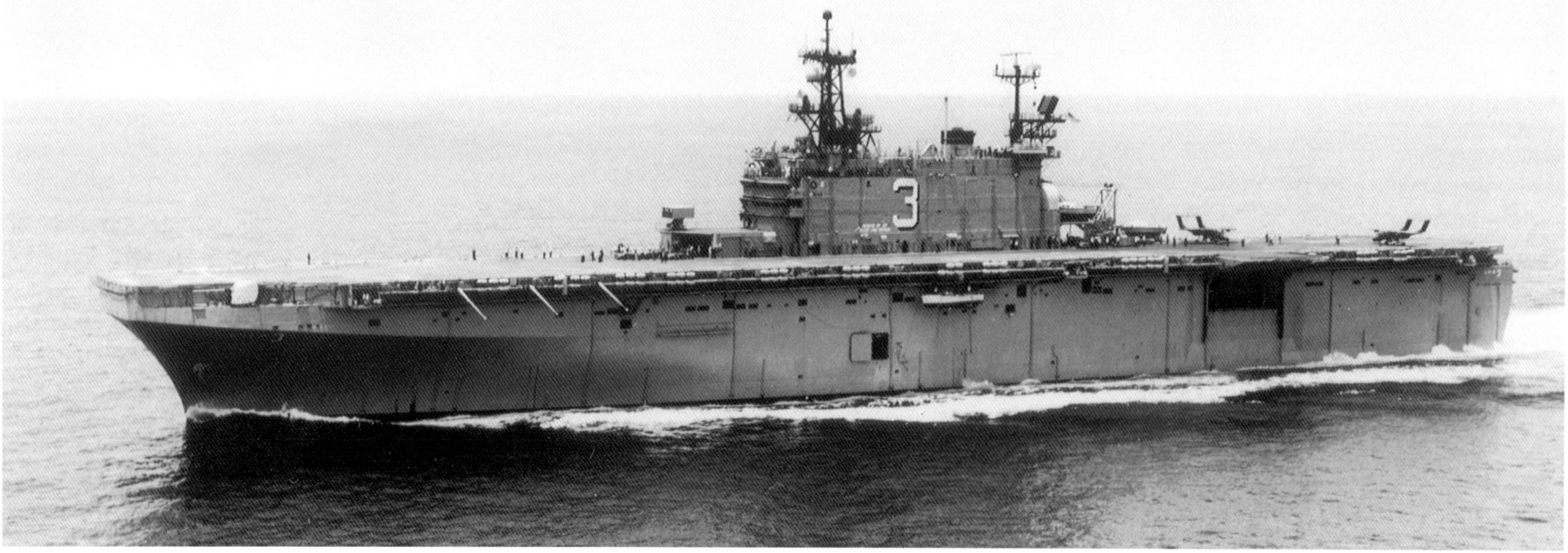

The Marine Corps wanted to develop a cost-effective night interdiction aircraft to monitor North Vietnamese Army (NVA) and Viet Cong supply movements. Two OV-10As were modified with an M-197 20-mm gun turret in the rear fuselage and a Forward Looking Infrared turret in the nose. The sponsons were removed to accommodate the rear turret and the nose was lengthened thirty inches. Designated YOV-10D Night Observation Gunship Systems, the two Broncos were tested at the Naval Weapons Test Center at China Lake, California, in 1970. The two NOGS aircraft were sent to Vietnam, serving with VAL-4, and flew more than 200 successful missions between June 5 and August 13, 1971. Both aircraft returned to the U.S. immediately afterward.

The added weight of the turrets was later overcome by the installation of 1040-horsepower Garrett T76-G-420/421 engines. With the new systems, including the Texas Instruments AN/AAS-37 FLR and Laser Rangefinder, the OV-10D weighed more than 14,000 pounds and no longer had para-drop mission capability. The OV-10Ds were also fitted with infrared suppressive exhaust stacks.

The program was approved in 1978, and eight OV-10Ds were delivered to USMC squadrons VMO-1 at MCAS New River, North Carolina, and nine to VMO-21 at MCAF Camp Pendleton, California. In 1985 a Service Life Extension Program was initiated to continue OV-10D service past the year 2000. Fourteen were upgraded, and 23 OV-10As were also converted to the SLEP standard. These Broncos, modified at NAS Cherry Point, North Carolina, were redesignated OV-10D+. They were among the last U.S. OV-10s to see combat, serving alongside OV-10As with the USMC during Operation Desert Storm in the Persian Gulf during 1991, deployed from carriers and later attacking Iraqi targets in Kuwait and Iraq. Two were lost to enemy ground fire during Desert Storm. By mid-1994, all Broncos were phased out of U.S. service.

In addition to the West German OV-10Bs, Broncos served in the air forces of several foreign nations. Most of these were similar in speci-

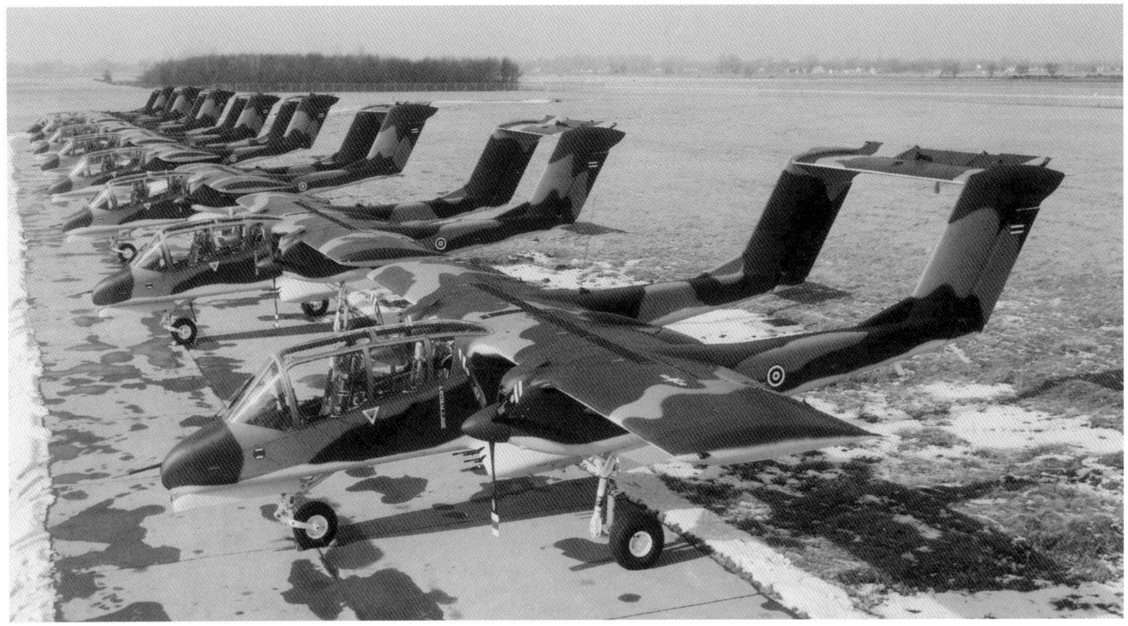

Above, eight Thai OV-10Cs (note U.S. wing insigne on the nearest), and below, six Venezuelan OV-10Es in a similar photo.

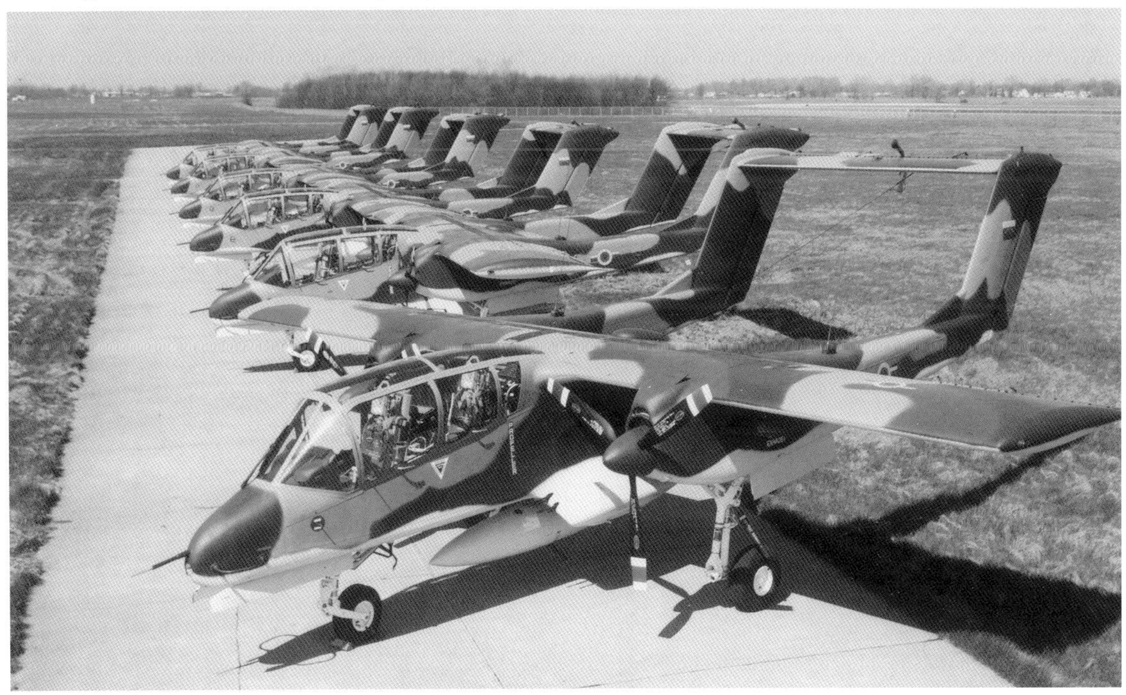

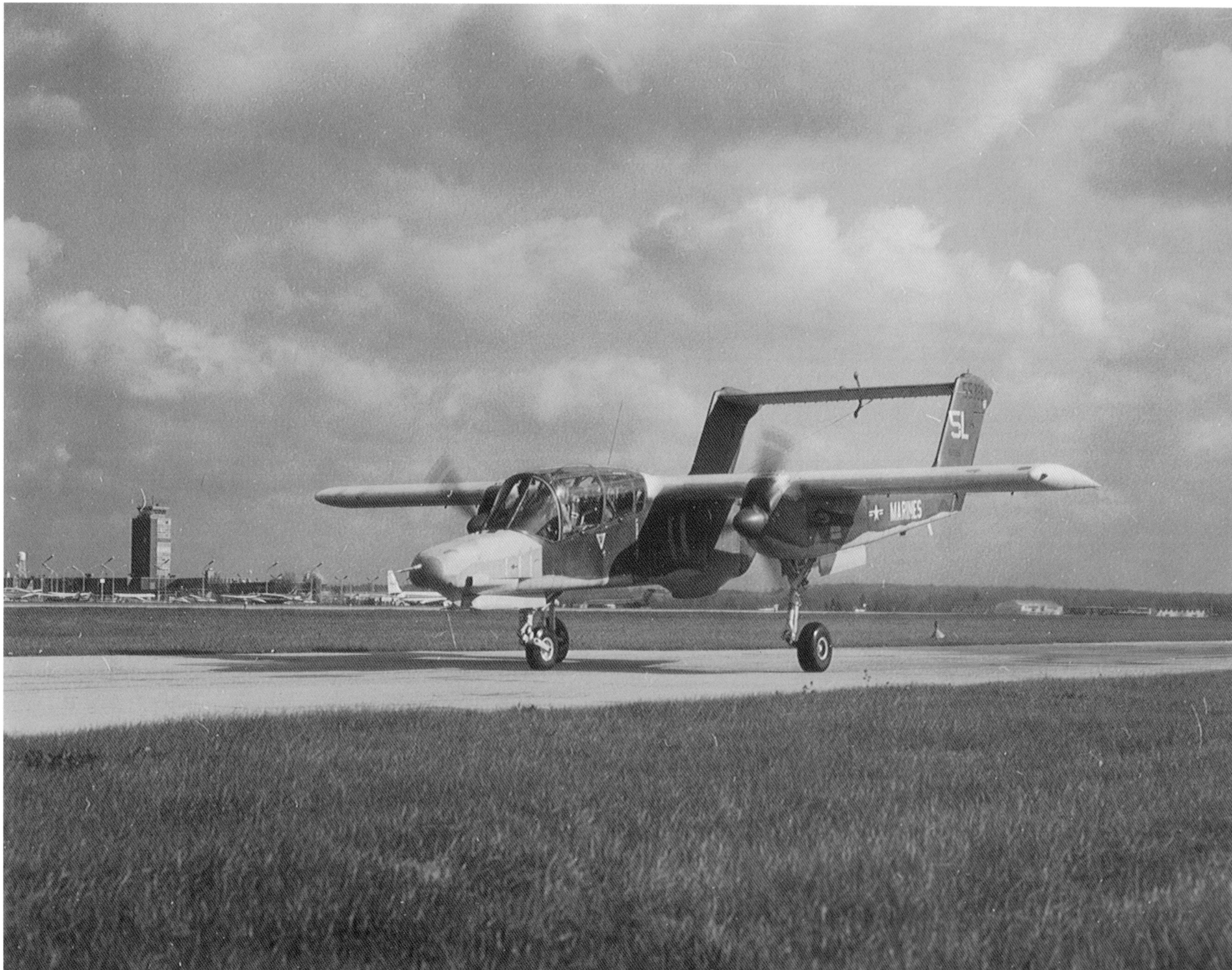
SL
MARINES

fication to the OV-10A, but had different designations. Thailand received OV-10Cs in 1971, and Broncos sold to Venezuela in 1973 were known as OV-10Es. A few of the latter saw combat in December 1993 during a coup against the Venezuelan government. The designation OV-10F was applied to sixteen Bronos acquired by Indonesia in 1976 to replace its F-51D Mustangs.

Six ex-USMC OV-10As were purchased by the Royal Moroccan Air Force in 1981 and these Broncos saw action in the Western Sahara against Polisaro insurgents. In the Philippines, after the demise of the Marcos regime, the U.S. sent an unspecified quantity of Broncos to the aid of the Corazon Aquino government, which was still using T-28s for the counterinsurgency mission as late as 1987.

A small number of Broncos are currently used by civilian organizations, including NASA to acquire air data, the California Department of Fire as a spotter aircraft, and the Bureau of Land Management to perform surveys.

The Bronco served its purpose well, becoming far more versatile than originally envisioned. It gave a good account of itself during twenty-five years of U.S. military service and continues in use with foreign countries around the world.

Specifications: North American/Rockwell NA-300>NA-404 YOV-10/OV-10A/OV-10D Bronco

Description: USAF, USMC and USN light attack and observation aircraft. Twin-engine high-wing aircraft of all-metal construction with retractable landing gear. Crew 2.

Engines: Two 660-hp Garrett AiResearch T76-G turboprops. (YOV-10A) T76-G6/8. (OV-10A) 715-hp T76-G10/12. (OV-10D) 1040-hp T76-G420/421.

Dimensions: Span 30 feet 4 in, length 40 feet, height 15 feet, wing area 218 sq ft. (OV-10A) Span 40 ft, length 41 ft 7 in, height 15 ft 2 in, wing area 291 sq ft. (OV-10D) As OV-10A except: length 44 ft.

Weights: 5257 lb empty, 10,170 lb maximum take-off. (OV-10A/D) 6893 lb empty, 14,444 lb maximum take-off.

Armament: Four 7.62-mm M60C machine guns with 500 rounds each, two AIM-9 Sidewinder air-to-air missiles, 2.75-in. and 5-in. rockets, GPU-2 20-mm gun pod, SUU-11 mini-gun pod, Mk 82 Snakeye and general-purpose bombs.

Performance: Maximum speed 288 mph, climb 3020 ft/min, service ceiling 30,000 ft, combat radius 228 miles, maximum range 576 miles.

Production: North American YOV-10/OV-10A/B/C/D/E/F Bronco

NA-300	YOV-10A	300-1/7	BuAer 152879/152885	7
NA-301	(OV-10A transport—canceled)			—
NA-302	(OV-10A reconnaissance-canceled)			—
NA-305	OV-10A	305-1/114	BuAer 155390/155503	114
	OV-10A-1	305-1/8	AF66-13552/13559	8
	OV-10A-5	-9/11	AF66-13560/13562,	3
		-12/21	AF67-14604/14613	10
	OV-10A-10	-22/31	AF67-14614/14623	10
	OV-10A-15	-32/41	AF67-14624/14633	10
	OV-10A-20	-42/51	AF67-14634/14643	10
	OV-10A-25	-52/61	AF67-14644/14653	10
	OV-10A-30	-62/71	AF67-14654/14663	10
	OV-10A-35	-72/83	AF67-14664/14675	12
	OV-10A	-84/109	AF67-14676/14701	26
NA-311	(OV-10A training, technical services)			—
NA-312	(OV-10A spares and repair parts)			—
NA-313	(OV-10A special support equipment)			—
NA-317	(OV-10D night forward control program)			—
	YOV-10D NOGS (BuAer 155395/155396)			(2)
	OV-10D NOS (modified OV-10A)			(18)
NA-321	OV-10A	321-110/157	AF68-3784/3831	48
NA-338	OV-10B	Germany	BuAer 158292/158297	6
	OV-10B(Z)	Germany	BuAer 158298/158309	12
NA-342	OV-10C	Thailand	BuAer 158396/158411	16
NA-348	(OV-10 Fixed Wing Ship Aircraft)			—
NA-354	OV-10E	Venezuela	BuAer 159057/159064	8
NA-360	OV-10C	Thailand	BuAer 159134/159149	16
NA-397	OV-10F	Indonesia	BuAer 160216/160227,	
			BuAer 160295/160298	16
NA-404	OV-10E	Venezuela	BuAer 159065/159072	8
Total, OV-10 Bronco attack aircraft, 1963-1977				360

On opposite page, OV-10A, BuAer 155396, one of two converted at North American Rockwell's Columbus Division to the YOV-10D NOGS configuration (Night Observation Gunship System). The ventral turret is not in place here but its housing can be seen. Below, two gray USAF OV-10A-1s lined up outside factory next to two of 114 olive drab Marine OV-10As.

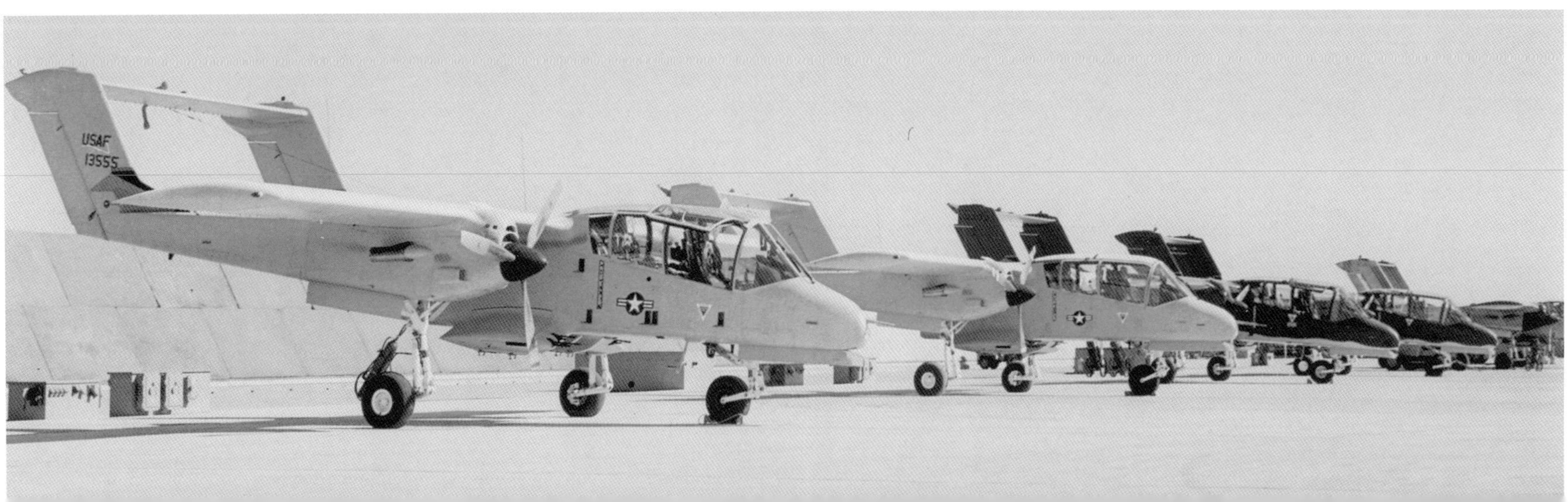

North American Rockwell D481 B-1A/B Lancer

AS THE FIFTIES DREW TO A CLOSE, top U.S. government officials began to re-evaluate their positions on the nuclear deterrent issue. At that time, the general opinion concerning nuclear strike capability favored the use of Intercontinental Ballistic Missiles, with the manned bomber considered inefficient and obsolete. This philosophy led to the cancellation of the B-70 as a weapons system in April 1961, several years before the first aircraft was even produced.

In the 1960s a new approach to America's ability to deliver nuclear weapons surfaced in the form of the Triad concept. The idea was that the U.S. should have three different means of delivering nuclear weapons at its disposal: manned bombers, land-based ICBMs and submarine-launched ballistic missiles (SLBMs).

As the Triad idea gained acceptance, serious thought was given once again to the development of a manned bomber to replace the Boeing B-52 Stratofortress, which was nearing the end of its first decade of service. At least one arm of the Triad should be flexible and capable of recall, if necessary. Although the B-70 had been intended as a B-52 replacement, it proved to be one-dimensional in its mission profile, and with the downing of Gary Powers' Lockheed U-2B by Soviet SA-2 missiles in 1960, the high-altitude triple-sonic bomber program was abandoned.

During 1960 and 1961 the Strategic Air Command examined a series of different advanced bomber concepts. Four studies grew out of these requirements: a Subsonic Low Altitude Bomber in 1961, an Extended Range Strike Aircraft in 1963, an Advanced Manned Precision Strike System in 1964 and an Advanced Manned Strategic Aircraft in 1965. The AMSA study was eventually chosen and in 1969 divided into five sub-studies concerning propulsion, alternate ar-

Unpainted B-1A, AF74-0158, with diverse exterior surfaces, looks almost camouflaged in this 1974 night shot at Plant 42 in Palmdale, compared to the all-white scheme it would receive.

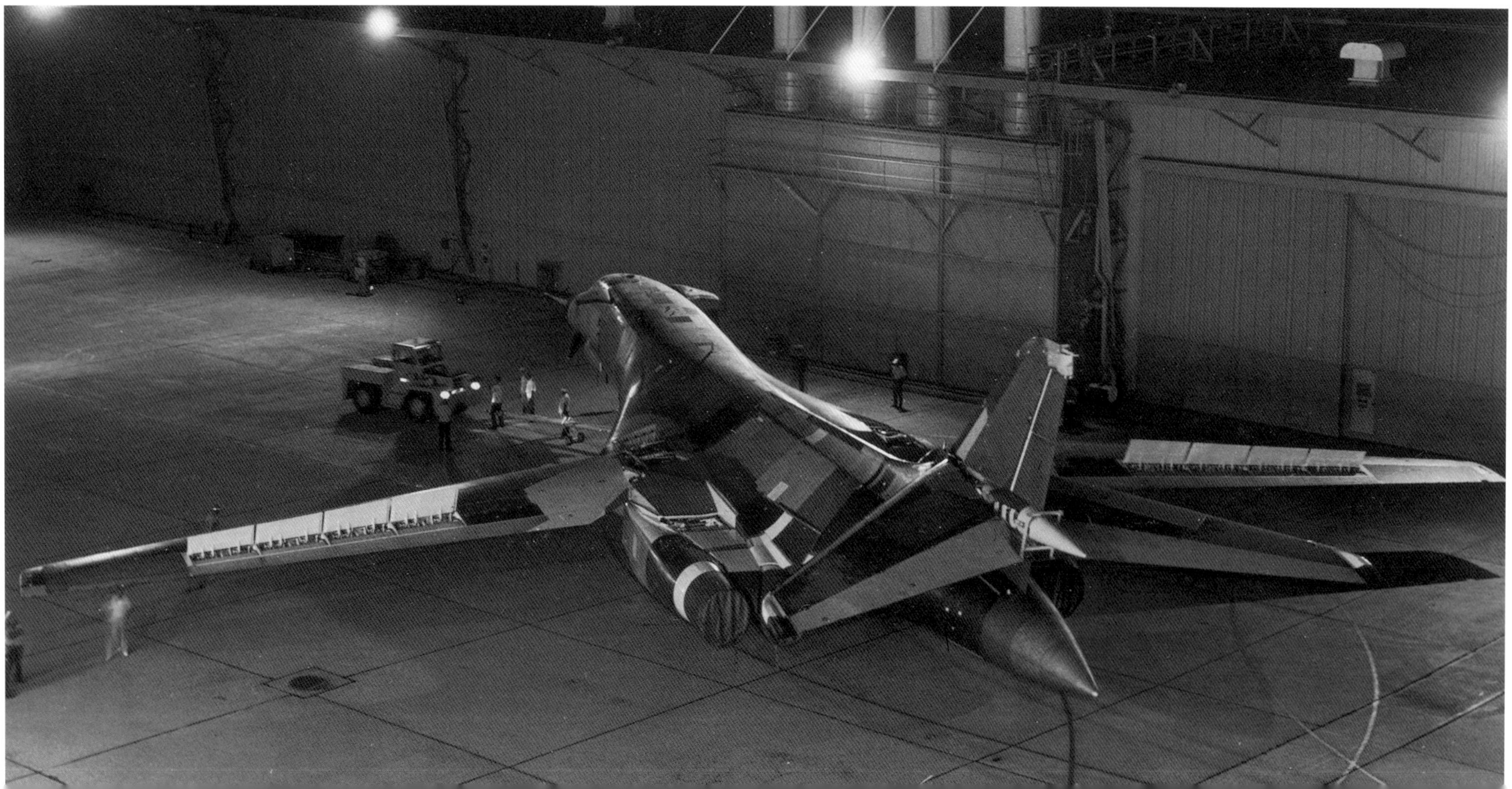

mament loading, reliability, titanium cost and maintenance.

As these studies were undertaken, Air Force officials were intent on developing a versatile aircraft that could perform intermediate- and low-altitude strikes effectively. By this time, the high-altitude penetration philosophy had been abandoned, and emphasis was put on the ability to operate at or near sea level.

On November 3, 1969, the official B-1 Request for Proposal was issued, and General Dynamics and North American Rockwell (as NAA had become two years before) were the only serious potential contractors. Both companies had explored various design studies concerning B-52 replacements. Pratt & Whitney and General Electric, the two major U.S. powerplant manufacturers, also submitted proposals.

On June 5, 1970, Rockwell's Los Angeles division received contracts for research, development, test and evaluation of the B-1 airframe. General Electric was chosen as the powerplant provider, to conduct similar studies with the new F101 turbofan engine. The original contracts called for five flight-test and two structural-test aircraft and 40 test and evaluation engines. Budget constraints later reduced the number of flight-test aircraft to three (although this was later increased to four in the FY 1976 budget), one structural-test airframe and 27 engines.

The Air Force ordered 244 production aircraft, with the first to be delivered by early 1975,and the last in 1981. Between January and September 1971, a full-scale, 45,000-pound mock-up was assembled, utilizing the combined efforts of 150 Rockwell employees. The Air Force reviewed the mock-up during October, and as many as 297 alterations were requested, the majority of which were quickly resolved.

Assembly of the first B-1A prototype began on March 15, 1972, at Air Force Plant 42 at Palmdale, California, with many sub-assemblies arriving from Rockwell's Los Angeles plant, as well as from various subcontractors nationwide. A wind-tunnel program was initiated (resulting in more than 22,000 hours of wind tunnel time), in addi-

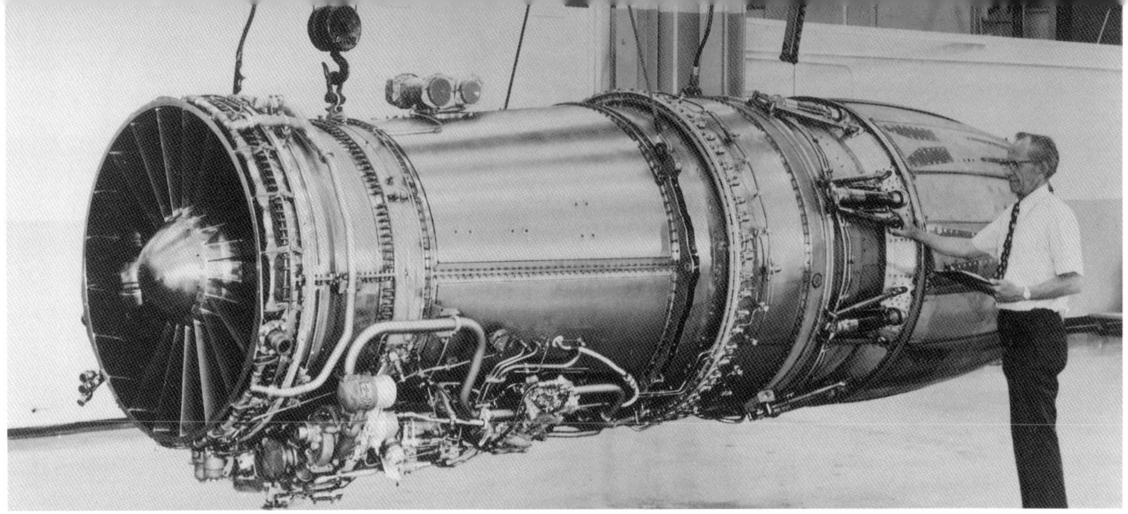

Above, GE press photo shows one of the B-1A's four huge 29,850-lb/thrust General Electric YF101-GE-100 turbofan engines.

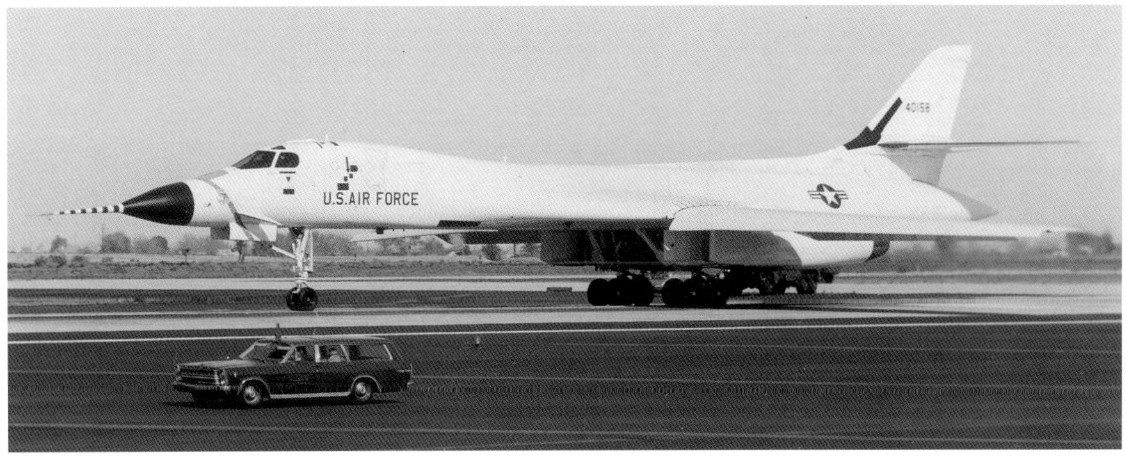

The first B-1A, AF74-0158, taxies out for its first flight. Below, the take-off, with wings forward, slats and flaps extended.

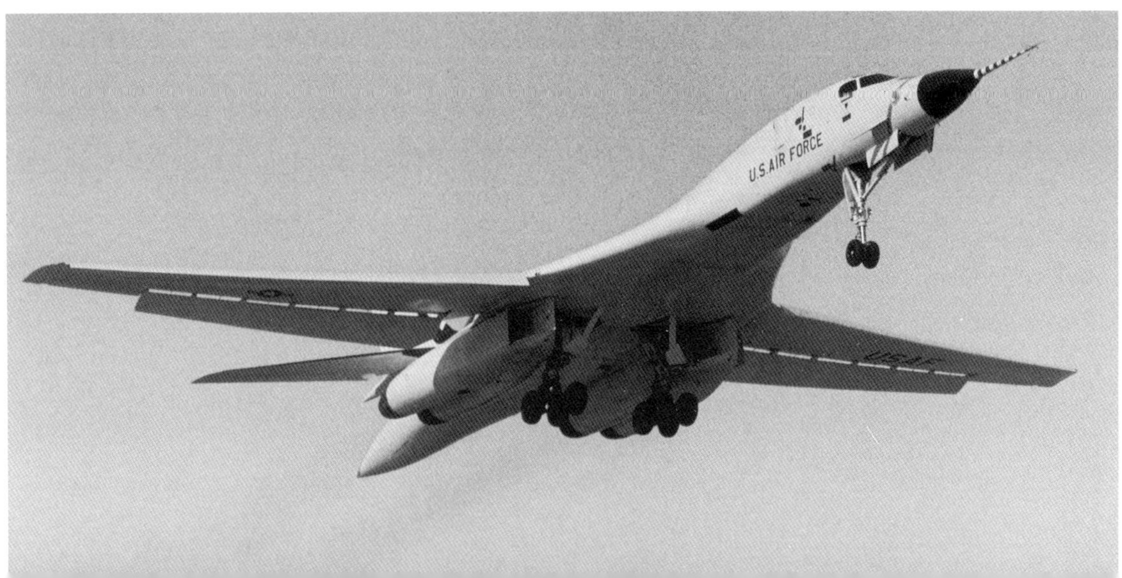

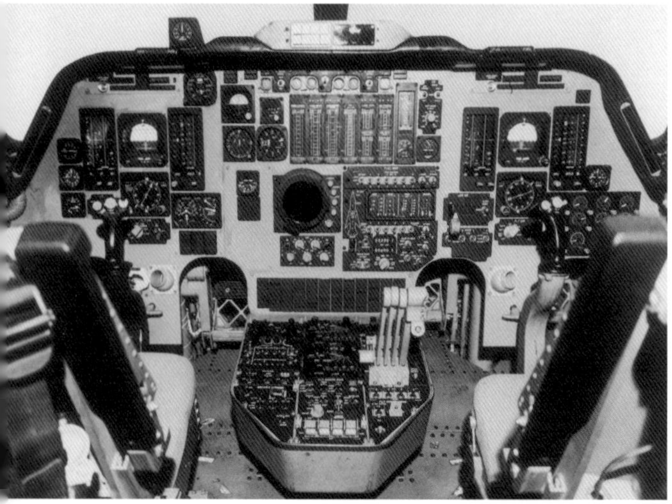

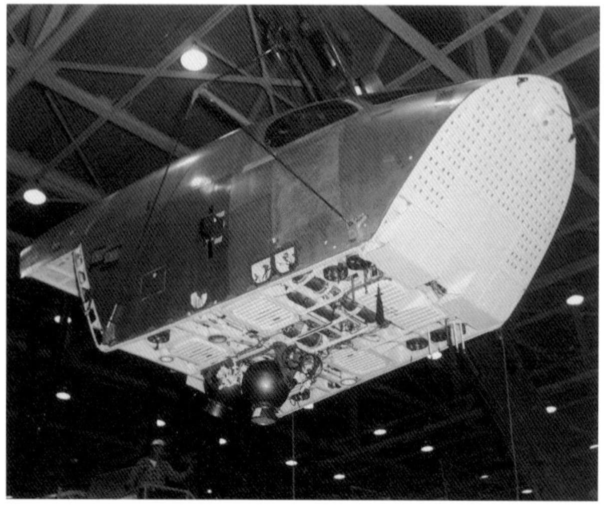

Above, the wide, almost luxurious B-1A flight deck, and right, the crew escape module. Below, the wings in fully swept position.

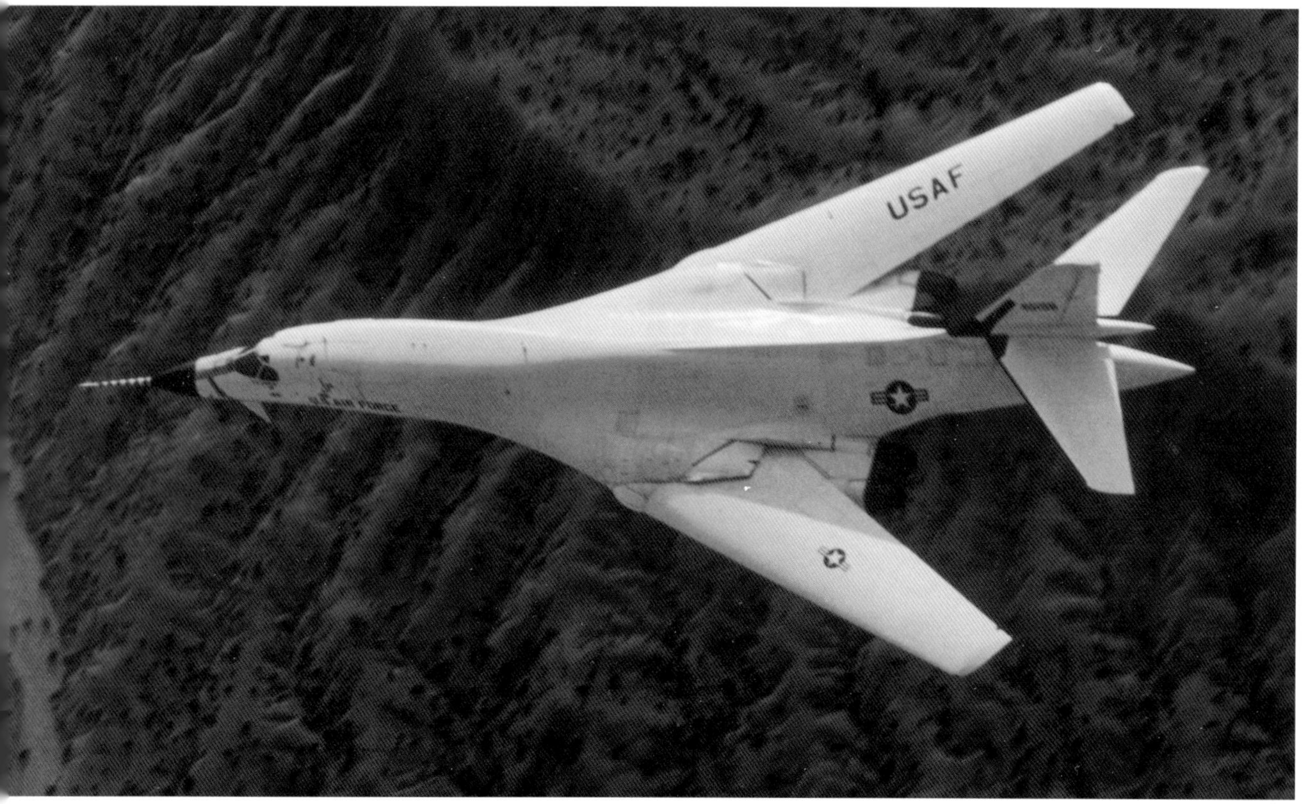

tion to a Radar Cross-Section evaluation utilizing a 36-foot model tested at Holloman AFB, New Mexico. Because the B-1A was intended to penetrate unfriendly airspace at low altitudes, a minimal RCS was incorporated into the design, construction and electronics of the aircraft. The RCS of the B-1A was approximately 1/100 that of the B-52.

The layout of the B-1A airframe was fairly conventional, with the exception of the variable-geometry swept wing that blended into the fuselage via large fairings. The wing sweep angle was 15 degrees fully forward, and 67.5 degrees fully swept. This variable geometry allowed the B-1A to be extremely versatile in speed, weapons loading and range. The four F101 engines were placed outboard and aft of the wing center section, two per side, and the horizontal stabilizers were located at approximately one-third of the vertical fin height. The main landing gear consisted of two sets of four wheels per side, with two wheels for the nose gear. In the lower part of the fuselage, weapons bays were located in the forward center section and between the engine nacelles.

The airframe was constructed of several different materials. Forty-one per cent of the structure was aluminum, with fiberglass, polyamide quartz and other materials accounting for more than 30 percent and titanium 21 percent. Steel and composites made up the remainder of the structure. These various hardened exotic materials were intended to withstand nuclear blasts and their use also helped reduce the RCS.

The assembly of the first prototype progressed through 1973 and most of 1974. On October 26, 1974, the first B-1A, AF 74-0158, was rolled out at Palmdale before a crowd of approximately 10,000. A series of ground tests was undertaken during November and the aircraft made its first flight on December 23, 1974. The crew was pilot Charles Bock Jr., co-pilot Colonel Emil Sturmthal and flight test engineer Richard Abrams. One month later, on January 23, 1975, the second flight was made, during which the landing gear was retracted and the wing-sweep mechanism activated.

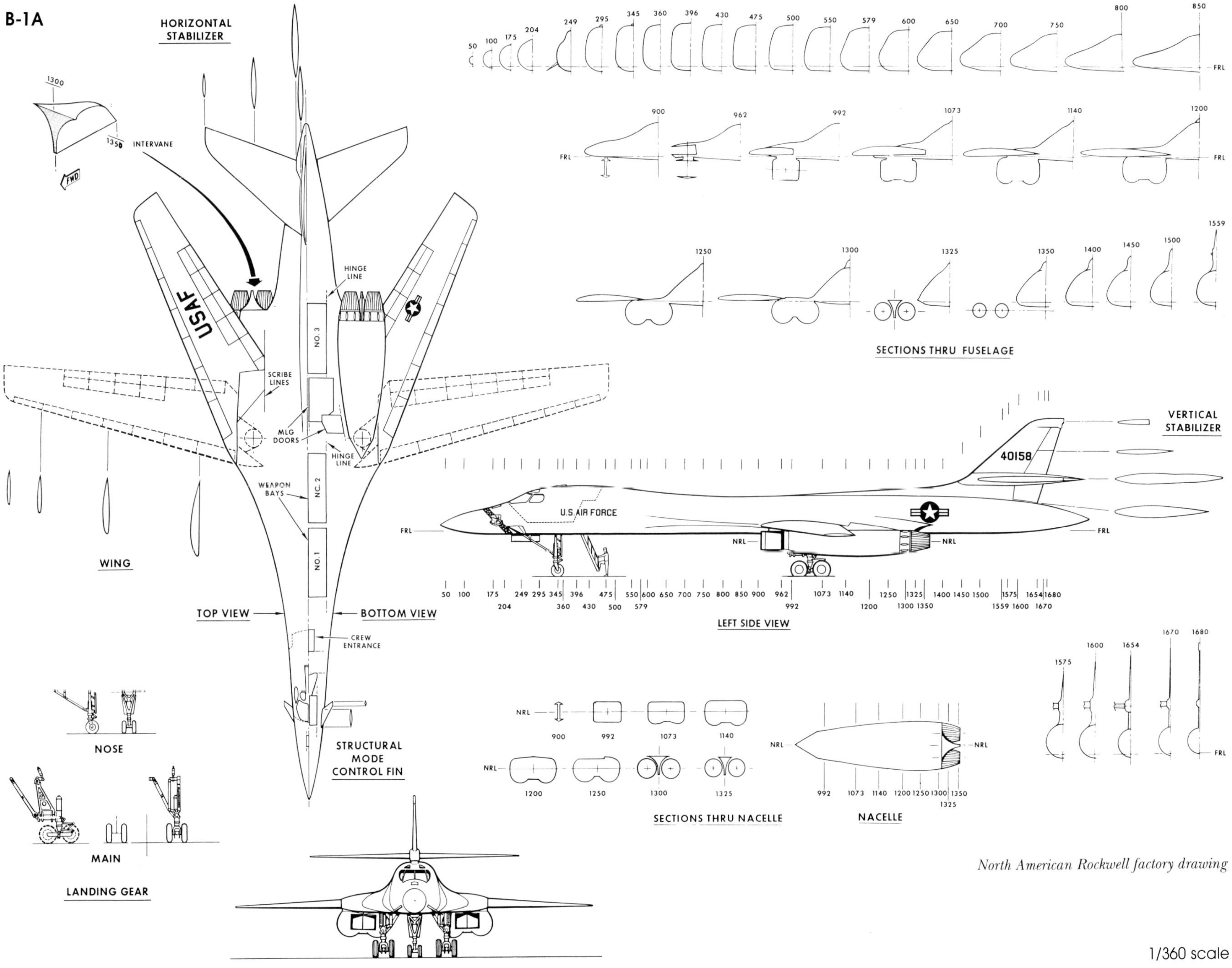

North American Rockwell factory drawing

1/360 scale

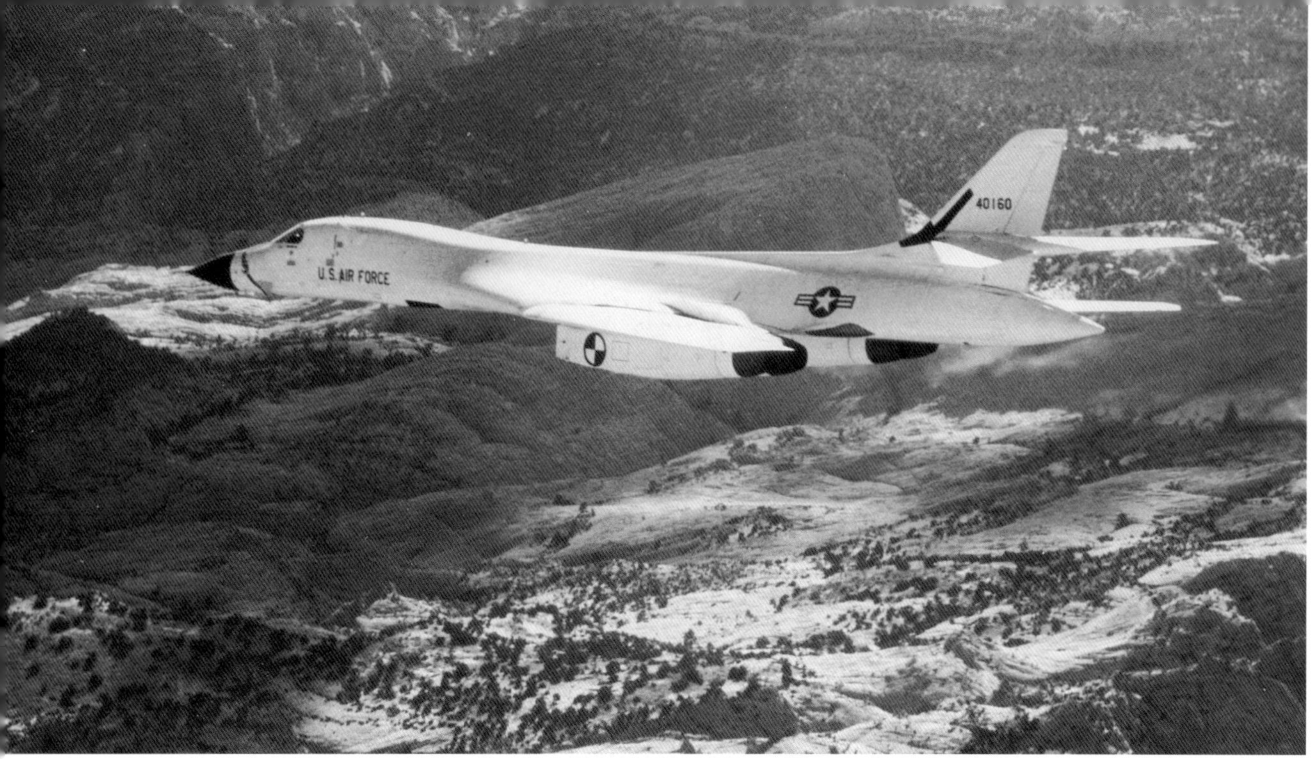

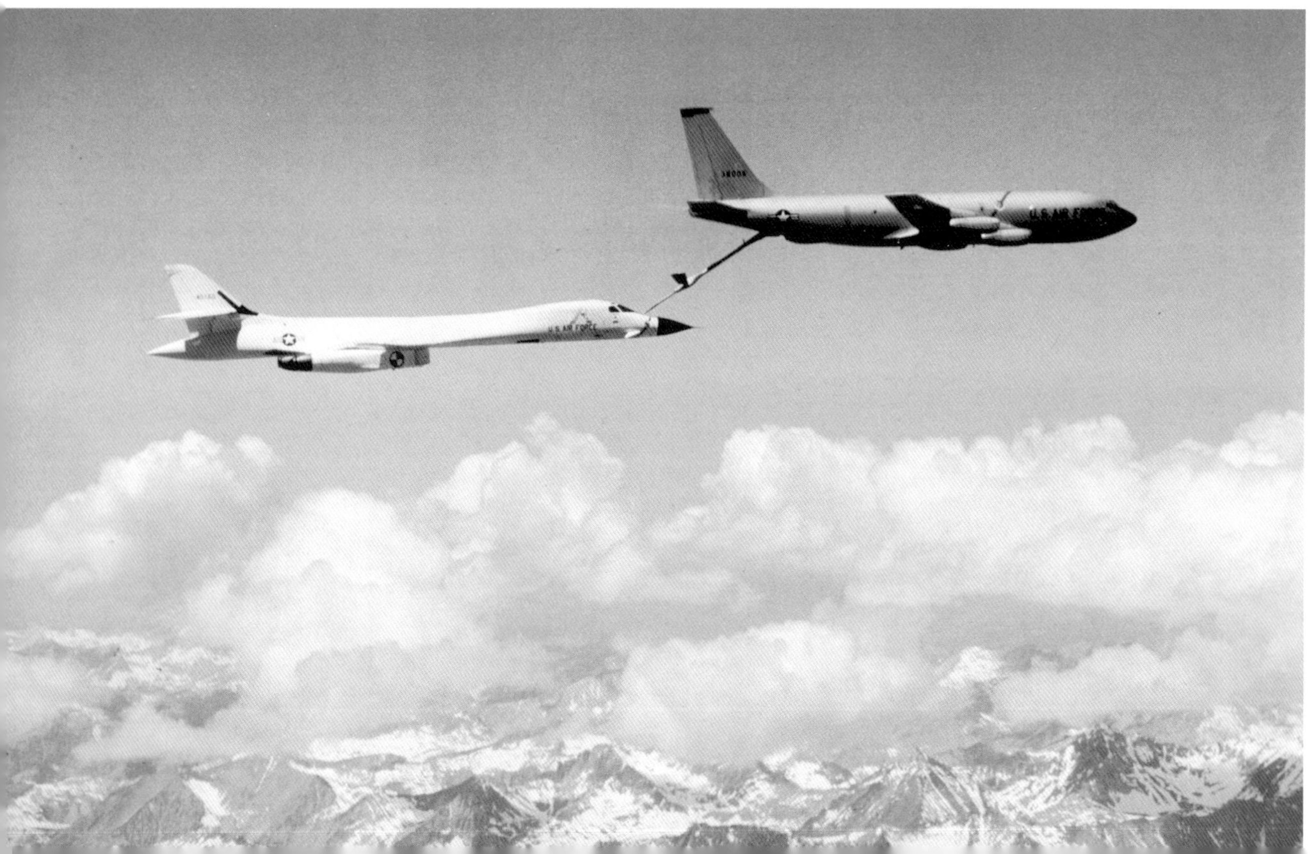

The third prototype, 74-0160, made its first flight on April 1, 1976, becoming the second B-1A to fly. The second prototype, 74-0159, was being utilized as a structural airframe at Lockheed's Palmdale facility, and did not fly until May 11, 1976. The fourth aircraft, 76-0174, flew on February 14, 1979.

Each of the four prototypes underwent a different flight program. The first logged 79 flights by April 1976 and was the first B-1A to achieve speeds of Mach 1.5 and later Mach 2. The second, used to evaluate structural loading, amassed 60 flights during its test program. It also achieved the highest speed flown by any B-1, Mach 2.22 on October 5, 1978. Aircraft numbers three and four served as offensive and defensive weapons testbeds, completing 138 and 70 flights respectively by late-April 1981. By the end of that month the four aircraft had flown a total of 1895 hours.

Although the B-1A program had been allocated funding under the FY 1978 budget in 1976—the final year of the Ford Administration—the cost of the program, the economic situation and the Air Force's doubts about the need for such an aircraft led to the cancellation of the program by President Carter in June 1977. Even though the program was terminated, the third and fourth aircraft continued testing with advanced electronic countermeasures equipment and Doppler beam-sharpening modifications to the attack radar.

In 1978 the Department of Defense called for an aircraft to replace the B-52 as a cruise-missile transport, and Rockwell submitted a detailed proposal that included the use of the third B-1A as a flight-test demonstrator. Both Rockwell and the USAF continued to study potential developments of the B-1, including Long Range Combat Aircraft. While Congress and the DoD both considered revival of the program in early 1981, the final flights of the B-1A prototypes were completed at Edwards AFB on April 30. The aircraft were subsequently

Left, two in-flight views of the third B-1A, 74-0160. The refuelling aircraft is a Boeing KC-135A Stratotanker, 63-8006.

put in storage to wait for further developments.

On October 2, after following the USAF recommendation that the LRCA version of the B-1 be selected over the General Dynamics FB-111H, newly elected President Reagan announced that the Air Force was authorized to order 100 B-1 derivatives under the designation B-1B. In January 1982, Rockwell and the USAF signed contracts for full-scale development, flight test and production. The full-scale development contract, which included modification of the four B-1A prototypes to the B-1B specification, totaled $1,317,000,000; the contract to produce the first B-1B and procure long-lead items for production lots totaled $886,000,000. General Electric was also awarded a contract to the manufacture the F100-GE-102 engines for the new aircraft, which was later given the name Lancer.

The B-1B configuration brought about a change in the aircraft's mission profile, turning it into a dedicated low-level penetration aircraft. With the lower-altitude mission came a reduction in speed. Because the original B-1A was intended to be capable of Mach 2, a variable inlet system was engineered into the intakes. The B-1B was not fitted with this system; the design changes to the engine nacelles reduced the RCS considerably and limited the top speed to Mach 1.25 (792 mph).

On August 29, 1984, the second B-1A prototype, fitted with B-1B subsystems, crashed after manual fuel management difficulties were encountered at low altitude, killing Rockwell test pilot Doug Benefield. The accident led to changes in the production B-1B fuel management system.

The aircraft's electronic defensive equipment consists of four sub-systems: a radio frequency surveillence/electronic countermeasures system, a tail warning function, a defensive man-

An interesting late-afternoon view of the first three B-1As, 74-0158 to -0160 left to right, at Edwards AFB in 1976. They appear almost identical, except for the much shorter nose probe of 74-0160.

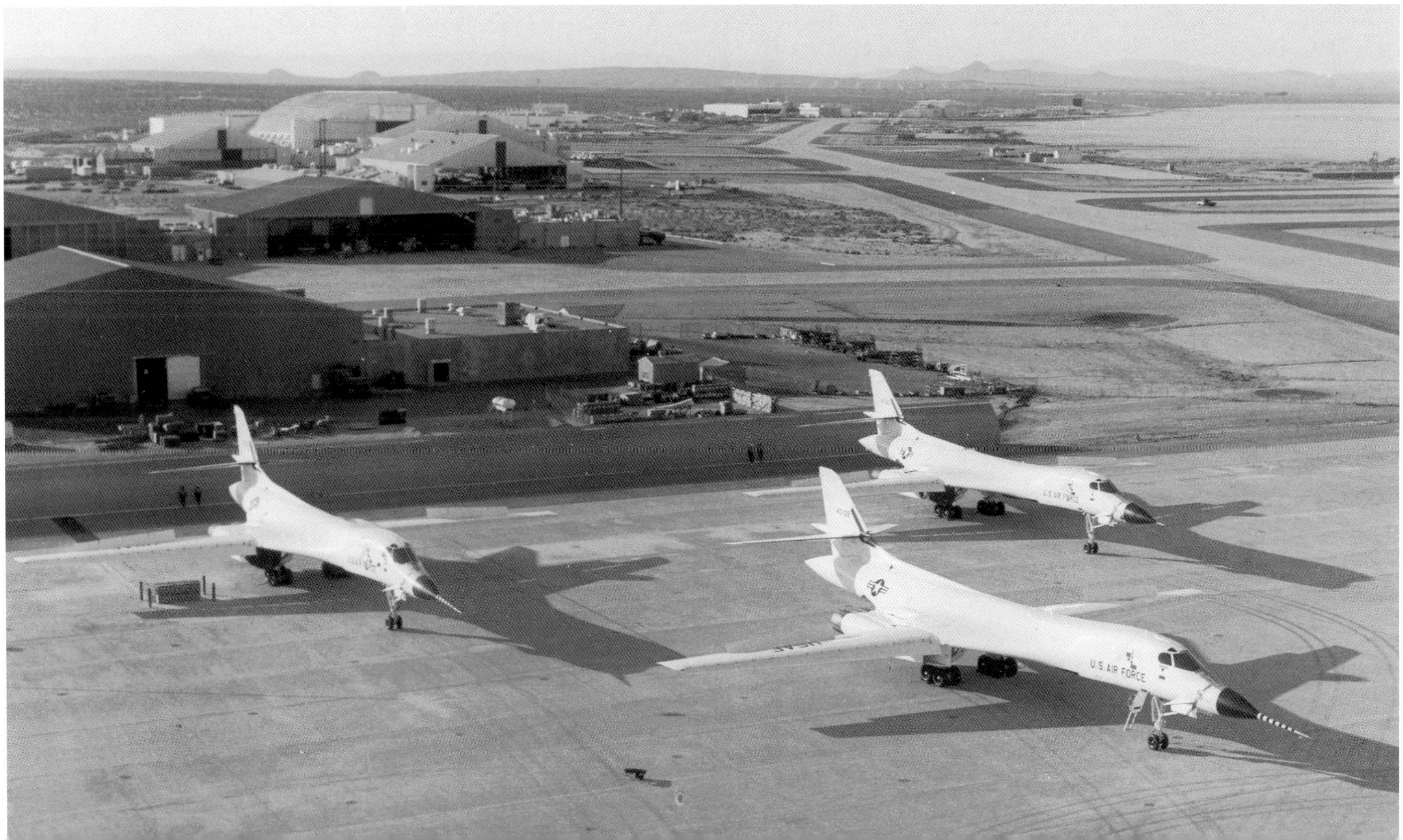

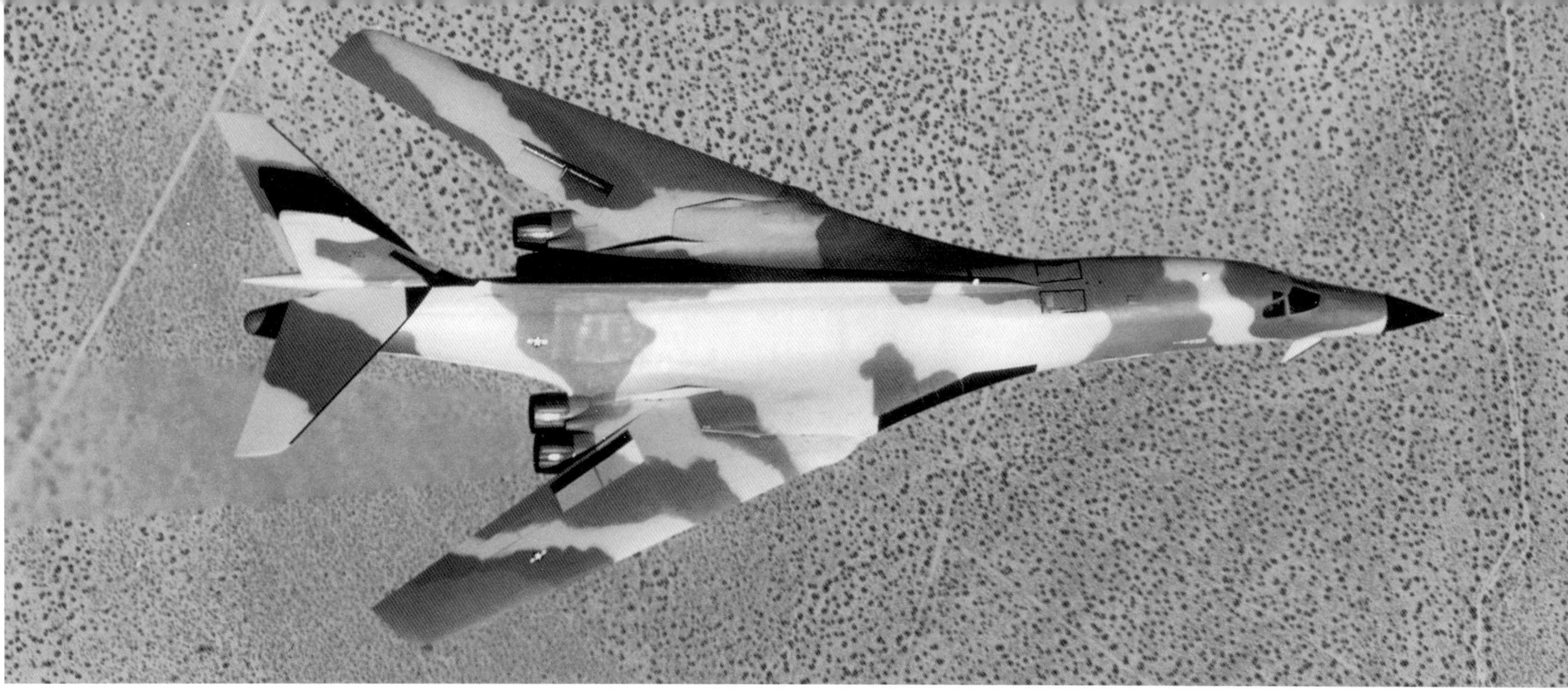

Above, tan and brown B-1A-3, 74-0160, with its prominent dorsal spine, a modification unique to this aircraft. Below, seven green and brown camouflaged B-1Bs with KC-135A tanker, 62-3561.

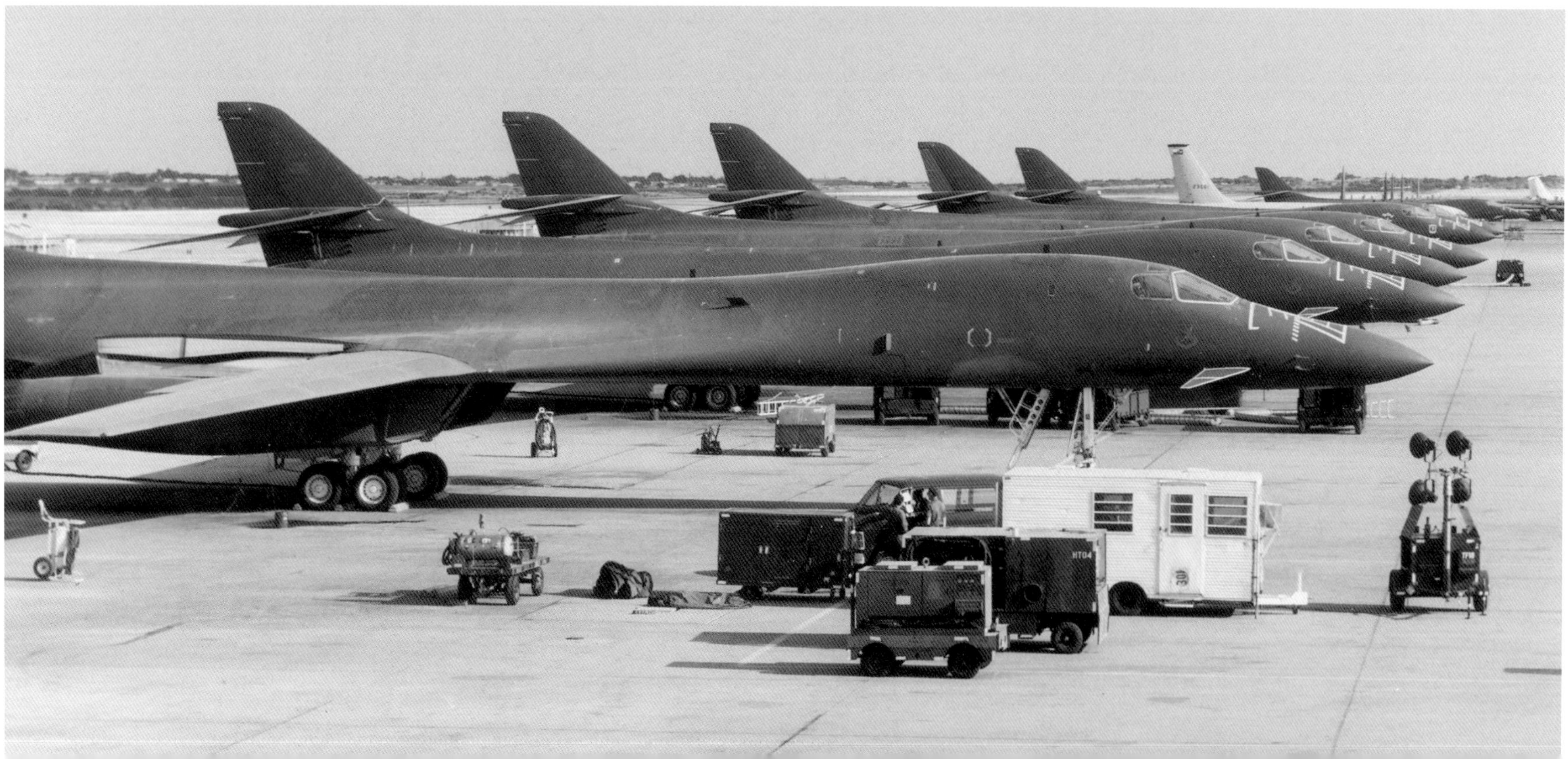

agement system and an expendable countermeasures system that utilizes chaff cartridges or flares.

The B-1B has been configured to deliver weapons in a variety of missions. Nuclear penetration with short-range attack missiles is possible, with the ability to carry up to 38 of these weapons. Thirty-two B-83 free-fall nuclear bombs can be carried for free-fall nuclear penetration missions. As a cruise-missile platform, the B-1B has the capacity to carry 22 AGM/86B or C cruise missiles externally with two additional fuel tanks carried in the weapons bays. Up to 172 Mk82 free-fall conventional bombs can be carried internally and externally. With such a wide variety of mission capabilities, the B-1B has shown great versatility as a weapons platform.

On September 4, 1984, the first production B-1B, AF 82-0001, was rolled out five months ahead of schedule and flew the following month. Deliveries began in early 1986 and concluded in August 1988. The 337th and 338th Bombardment Squadrons of the 96th Bombardment Wing at Dyess AFB, Texas, were the first Air Combat Command recipients of the type, followed by the 319th Wing at Grand Forks AFB, North Dakota; the 384th Wing at McConnell AFB, Kansas; and the 28th Wing at Ellsworth AFB, South Dakota. The 3400th Technical Training Wing at Lowry AFB, Colorado, is the Air Training Command squadron equipped with the B-1B, and the 6510th Test Wing of the Air Force Materiel Command at Edwards AFB constantly tests and upgrades the aircraft.

During the Gulf War in 1990, B-1Bs were not used against Iraq because they were equipped only for nuclear strikes. The aircraft are now equipped to carry a variety of conventional weapons and several B-1Bs have been deployed in the Persian Gulf as a deterrent to future aggression.

It is interesting to note that despite the versatility and performance increase the B-1B has over the B-52, the Stratofortress remains in the USAF inventory after 47 years of service. The B-1B may end up serving even longer, perhaps to the middle of the 21st century.

Specifications: Rockwell International / Boeing North American B-1B Lancer

Description: U.S. Air Force strategic and tactical heavy bomber. Low-wing, variable-geometry four-jet aircraft of metal and composite construction with retractable landing gear. Crew 4.

Engines: Four 30,750-lb/thrust General Electric F101-GE-102 afterburning turbofans.

Dimensions: Span 136 ft 8 in. unswept, 78 ft 2 in. swept, length 147 ft, height 33 ft 7 in, wing area 1950 sq ft.

Weights: 192,000 lb empty, 477,000 lb maximum takeoff. Fuel 30,084 gal.

Armament: 38 AGM-69A SRAM short-range attack missiles, 32 B-83 free-fall nuclear bombs, 22 AGM-86B/CALCM cruise missiles, or 128 Mk 82 500-lb free-fall conventional bombs, carried in three internal weapons bays and on eight external hardpoints.

Performance: Cruising speed 699 mph (Mach .92) at 500 ft, maximum 825 mph (Mach 1.25) at 50,000 ft, service ceiling 60,000 ft, maximum unrefueled range 7455 miles.

Production: Rockwell B-1A / Boeing North American B-1B Lancer

D481	B-1A	AF74-0158/0160, 76-0174	4
—	B-1B	AF82-0001/ 83-0065/0071, 84-0049/0058, 85-0059/0092, 86-0093/0140	100
Total B-1 bombers, 1969-1999			104

A B-1B making a low-speed pass for the public at the Reno Air Races in 1986. It later made a much quicker swept-wing pass.

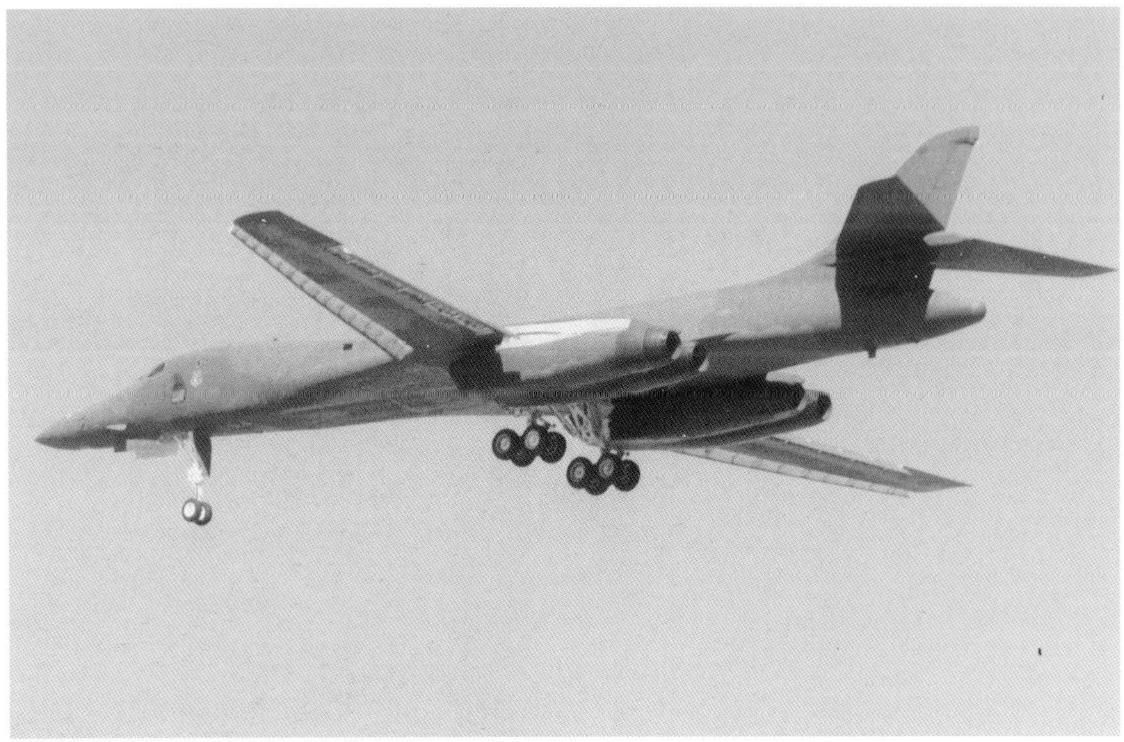

North American Rockwell NA-356 XFV-12A

IN NOVEMBER 1971 the Naval Air Systems Command invited U.S. aircraft manufacturers to submit design proposals for a shipboard, vertical/short take-off and landing (V/STOL) fighter/attack plane. This aircraft would serve on the proposed 12,500-ton Sea Control "flat top" ships that were scheduled to begin service in the late 1970s. Eight designs were considered, including tail-sitting proposals from Boeing and Lockheed and rotating-lift designs from Grumman, General Dynamics, Fairchild Republic and LTV Aerospace. McDonnell Douglas proposed an "Advanced Harrier" design with supersonic capabilities.

North American Rockwell's design was different from the others in that it used the thrust augmentation principle by which engine air is drawn through the top of the fuselage, then fed via ducts through the wings. The wings had three augmentor flaps, movable panels through which the thrust is expelled via augmentor exit nozzles. The lift is varied by changing the exit area between the augmentor flaps. During VTOL or hover the augmented flaps are vertical, then gradually rotated as the aircraft transitions into horizontal flight. This novel approach was claimed by Rockwell engineers to be 150 percent more efficient than tilt-lift engine designs, such as the Harrier.

The Rockwell design was a canard aircraft with some components utilized from Navy aircraft already in service. To be powered by the Pratt & Whitney F401 engine with 16,000 pounds of thrust (increased to approximately 21,000 pounds of lift thrust through the augmentor nozzles), Rockwell's V/STOL claimed an estimated top speed of over Mach 2 in level flight. The Navy was impressed with the projected performance and selected this design in May 1972. Two prototypes were to be built and evaluated under the designation XFV-12A.

Engineering work began at the Columbus Division in mid-1973. The layout of the XFV-12A was a central fuselage housing the single engine, small forward thrust-augmented canards, and larger thrust-augmented low-aspect wings at the

Two XFV-12A mock-ups, painted standard Navy gray and white and the one in the foreground wearing non-existent nose and tail numbers, at Rockwell's Columbus factory in 1974.

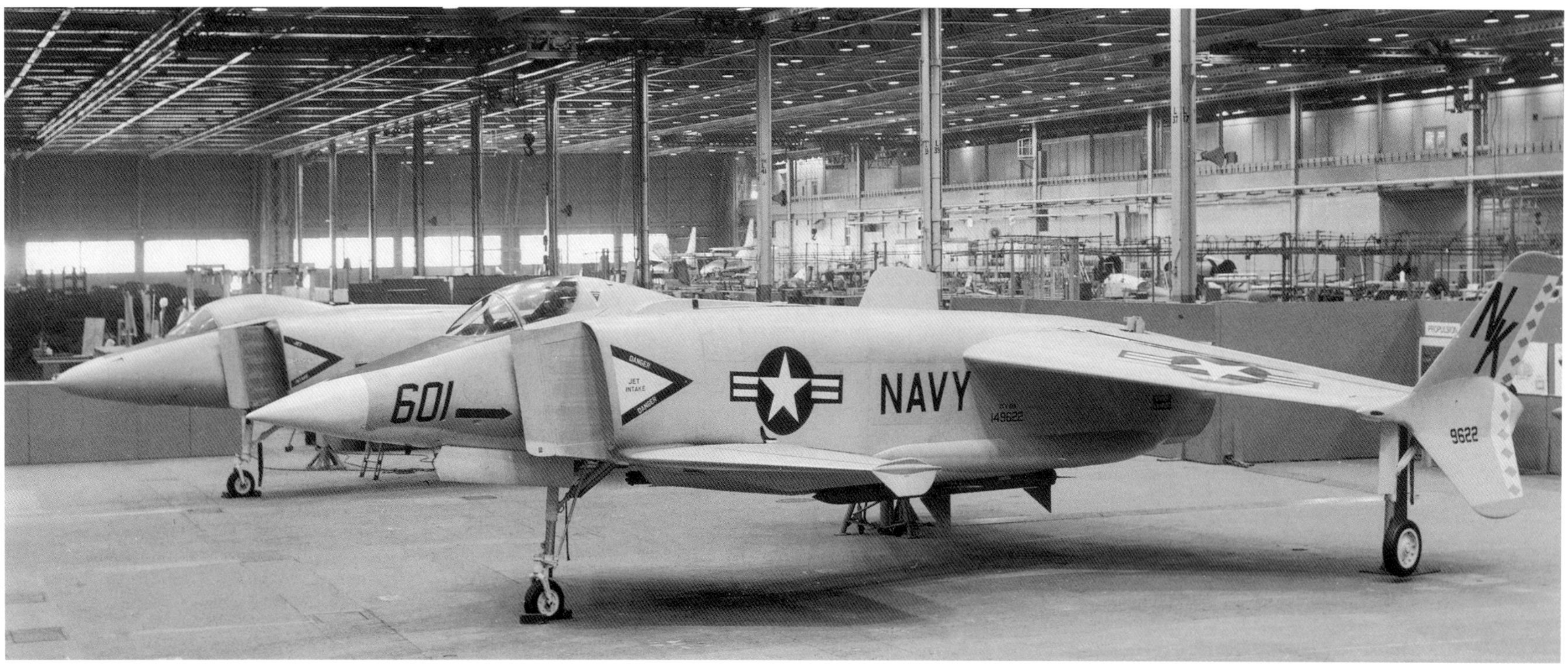

rear. The vertical stabilizers were positioned at the wingtips and mounted on streamlined pods housing the upward-retracting main gear. The landing gear and canopy were taken from the McDonnell Douglas A-4 Skyhawk, while certain portions of the main wing, fuel tanks and forward air intakes came from the F-4 Phantom. Beneath the fuselage were two hardpoints for proposed air-to-air missiles.

A full-scale mock-up was constructed with the finalized layout, but the actual prototype was very time-consuming to produce. The complexity of the propulsion system details and other engineering setbacks delayed the roll-out ceremony of the first prototype, BuAer 161080, until August 6, 1977. Within the first two weeks of September most of the ground tests were completed.

Tethered flight was the next step, and the prototype was transported in an Aero Spacelines Super Guppy to Langley AFB, Virginia, in November 1977. NASA lent its former lunar landing training gantry at Langley to the Navy for the purpose of testing the V/STOL. The gantry's massive box-like frame was well suited to the task of tethering the XFV-12A during its initial tests. Tethering was a must because the thrust augmentation systems would be difficult to synchronize, and free-flight operation of such new systems could be catastrophic.

Both restrained and unrestrained tethered tests were carried out in 1978. In the unrestrained mode the aircraft was suspended only from above and did not "feel" the tether, which allowed more-accurate flight simulation. Although the results of these tests were encouraging and the aircraft exhibited good handling characteristics, funding for the program was eventually terminated and the XFV-12A never achieved free flight.

Specifications: North American/Rockwell NA-356 XFV-12A

Description: V/STOL technology fighter prototype. Mid-wing canard single-jet aircraft of all-metal construction with retractable landing gear. Crew 1.

Engine: One Pratt & Whitney F401-PW-400 turbofan with 21,800 pounds of lift thrust.

Dimensions: Span 28 feet 6 in, length 43 feet 10 in, height 9 feet 5 in.

Weights: 13,800 lb empty, 19,500 lb gross VTOL, 24,250 lb gross STOL.

Armament: (Proposed) Two air-to-air or air-to-ground weapons mounted externally under the fuselage.

Performance: (Estimated) Maximum speed Mach 2.2-2.4, operational radius 575 miles.

Production: XFV-12A V/STOL fighter, BuAer 161080 1

Actual construction of two XFV-12A prototypes in 1976. The second, at the rear, was not completed. Right, a close-up of the complex forward thrust-augmented canard layout in April 1977.

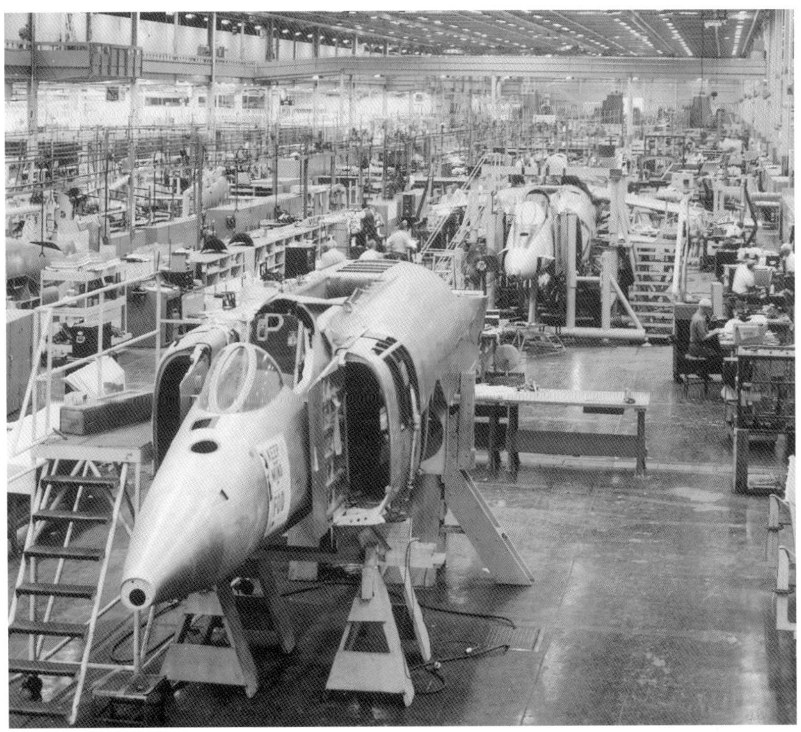

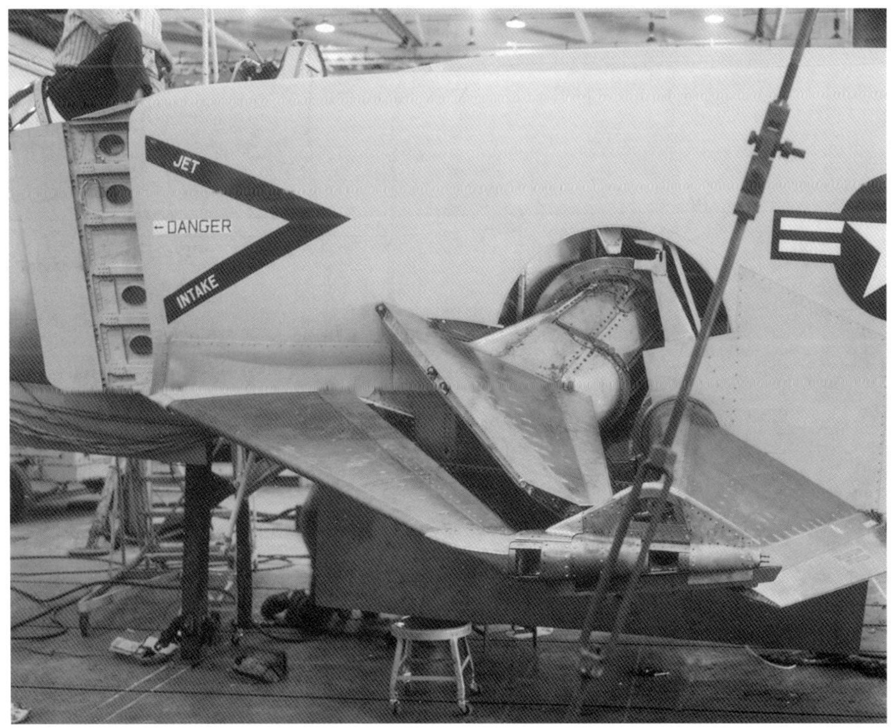

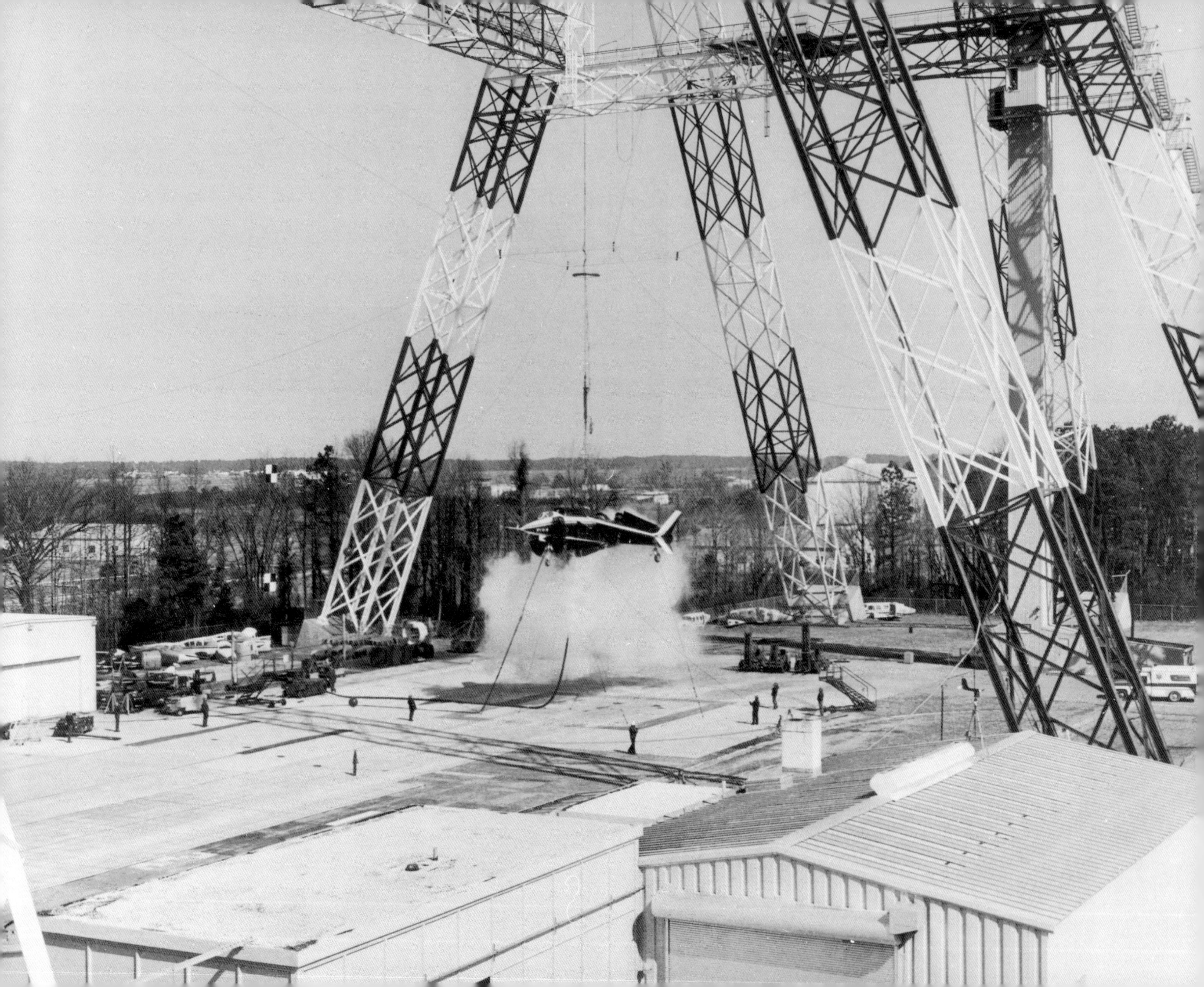

XFV-12A

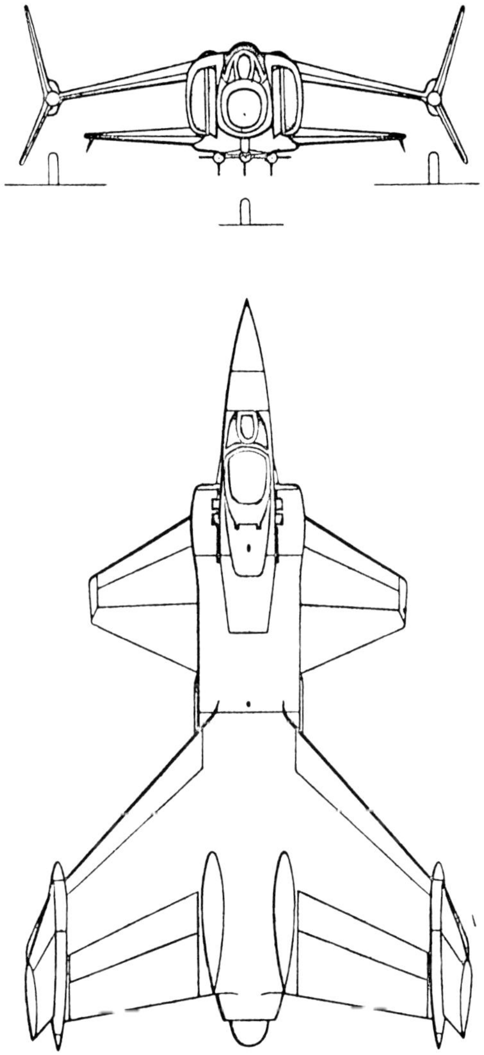

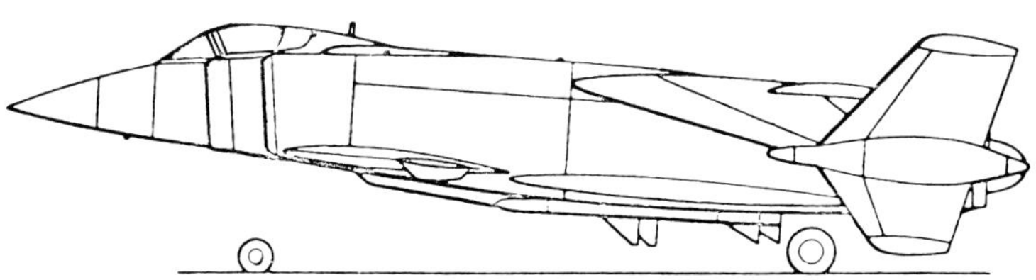

1/144 scale

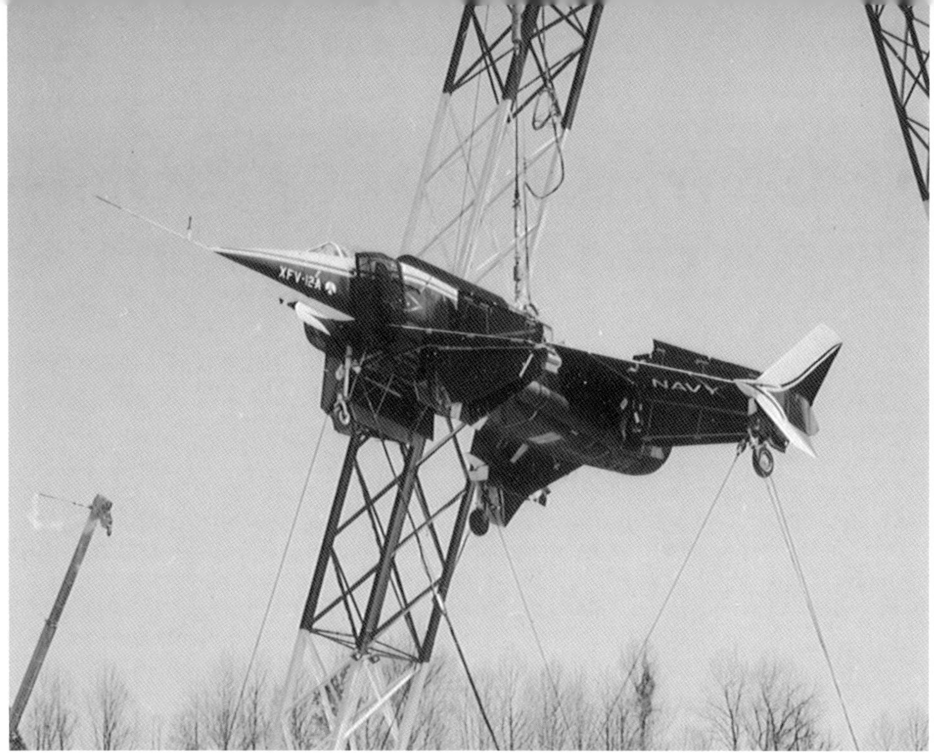

The XFV-12A, painted glossy blue and white, was tested on the former lunar-landing gantry at Langley, Virginia. The actual tethered operations, as shown opposite, produced satisfactory behavior under power.

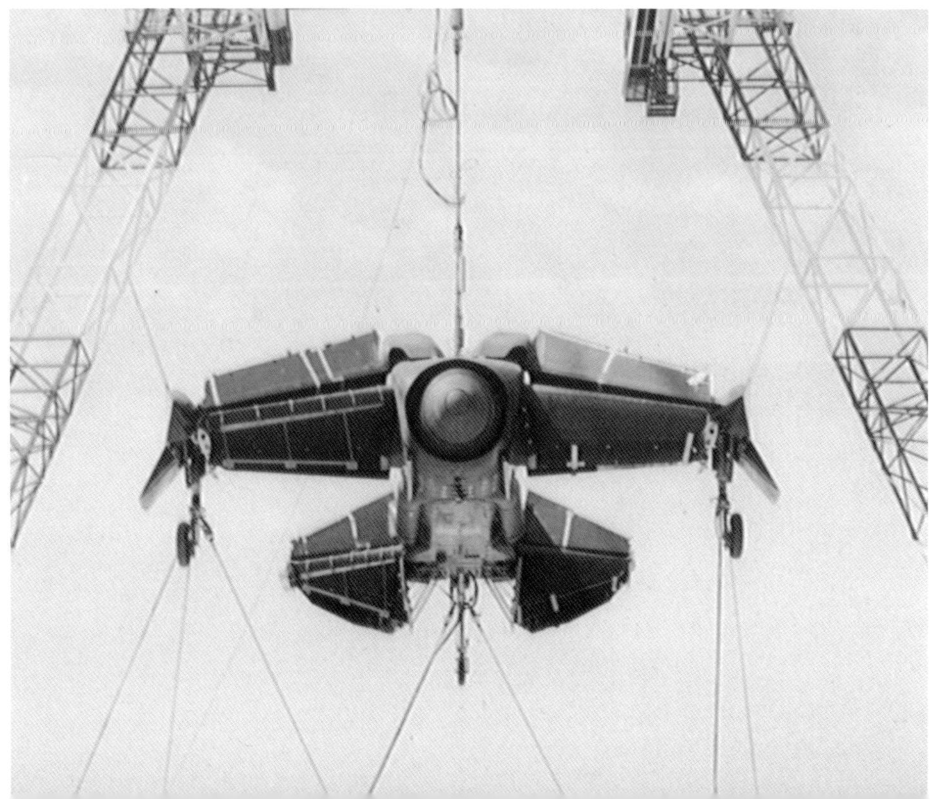

ROCKWELL/MBB / BOEING NORTH AMERICAN/ DAIMLER-BENZ AEROSPACE X-31A EFM

THE UNITED STATES Defense Advanced Research Projects Agency (DARPA) and the Federal Republic of Germany began a joint venture in 1977 to study the feasibility of developing a combat aircraft capable of high turning rates and low turning radii. The extraordinary maneuverability needed for better close-in combat performance led to the project designation EFM (Enhanced Fighter Maneuverability).

Rockwell joined the EFM program in 1983, establishing its own independent research and development program to evaluate the requirements for a demonstration vehicle. As IR&D studies progressed, it was concluded that thrust vectoring was required to achieve the projected performance. The U.S. Navy had previous success with thrust vectoring through the use of aft-mounted paddles for the exhaust of a Grumman F-14 spin demonstrator, and Rockwell intended to use a similar system on the EFM demonstrator. The General Electric GE F404 engine was selected for its high thrust-to-weight ratio, its tolerance of the inlet flow distortion to be experienced during high-maneuverability flight, and its excellent reliability. Rockwell and the German aerospace company Messerschmitt-Bölkow-Blohm intensified their cooperative studies, and on February 23, 1987, the U.S. Department of Defense assigned the designation X-31A to the joint-venture EFM design, of which two evaluation prototypes would be built. This was the first international X-plane, and as the primary funding was later shifted to the U.S. Naval Air Systems

The X-31 EFM test aircraft were the result of an international project but were funded through the U.S. Navy with BuAer numbers 164584/164585. The first is shown at Palmdale in 1990.

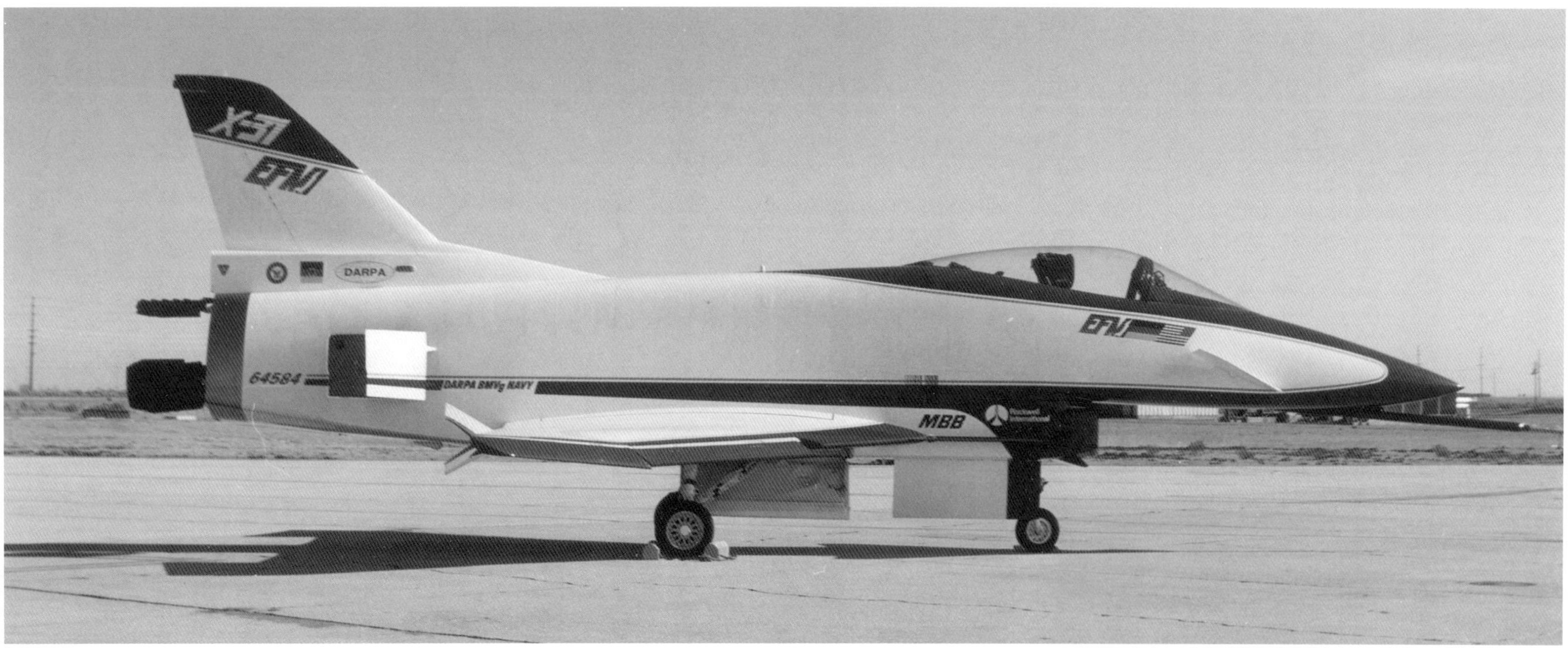

Command, it became the first Navy X-plane in more than twenty years.

Eventually, the X-31 program was developed under the cooperative direction of DARPA, which included the U.S. Air Force, U.S. Navy and NASA, as well as the Federal Republic of Germany and Deutsche Aerospace (formerly MBB, and later becoming Daimler-Benz Aerospace AG). Concurrently, an International Test Organization was formed with key representatives from all the participants. The cooperation of these companies led to the design of a relatively small, canard-layout aircraft utilizing many production components from existing aircraft. These off-the-shelf items, such as the canopy, landing gear, wheels and brakes as well as many electronic and hydraulic components, comprised 43 percent of total equipment and greatly contributed to the cost reduction of the prototypes.

The structures of the aircraft consisted of seven key materials. Although aluminum comprised 51 percent of the structural weight, graphite/epoxy was used extensively on the outer skin of the wings and forward part of the fuselage. The thrust-vectoring paddles were of the same material, which is able to withstand temperatures in excess of 1500 degrees Celsius. Fiberglass was used on the nosecone, the leading edge of the lower vertical stabilizer and the canopy frame. Titanium, steel, carbon and various other composite materials made up 25 per cent of the total structural weight.

The X-31 used an advanced fly-by-wire system with computers developed by Honeywell/Sperry. The forward canards were all-moving surfaces, operating as "weathervanes" with prevailing airflow, but also programmed for aerodynamic recovery from high angles of attack in case of thrust-vectoring system failure. Fixed strakes were mounted at the rear of the fuselage to supply additional nose-down pitch control from high angles of attack, and small nose strakes were fitted to control sideslip.

The flight demonstration program outlined for the two aircraft put great emphasis on agile flight in the post-stall regime, aerodynamic data,

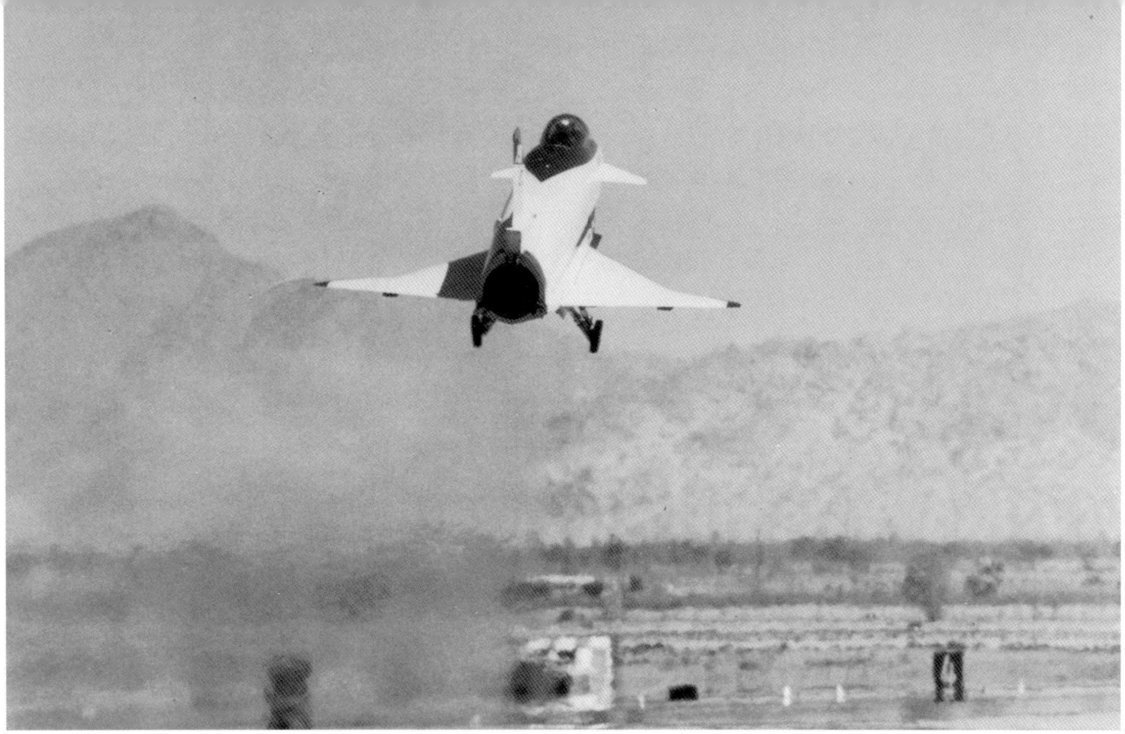

X-31A, BuAer 164584, making its first flight at Edwards AFB on October 11, 1990, clearly showing the canard layout.

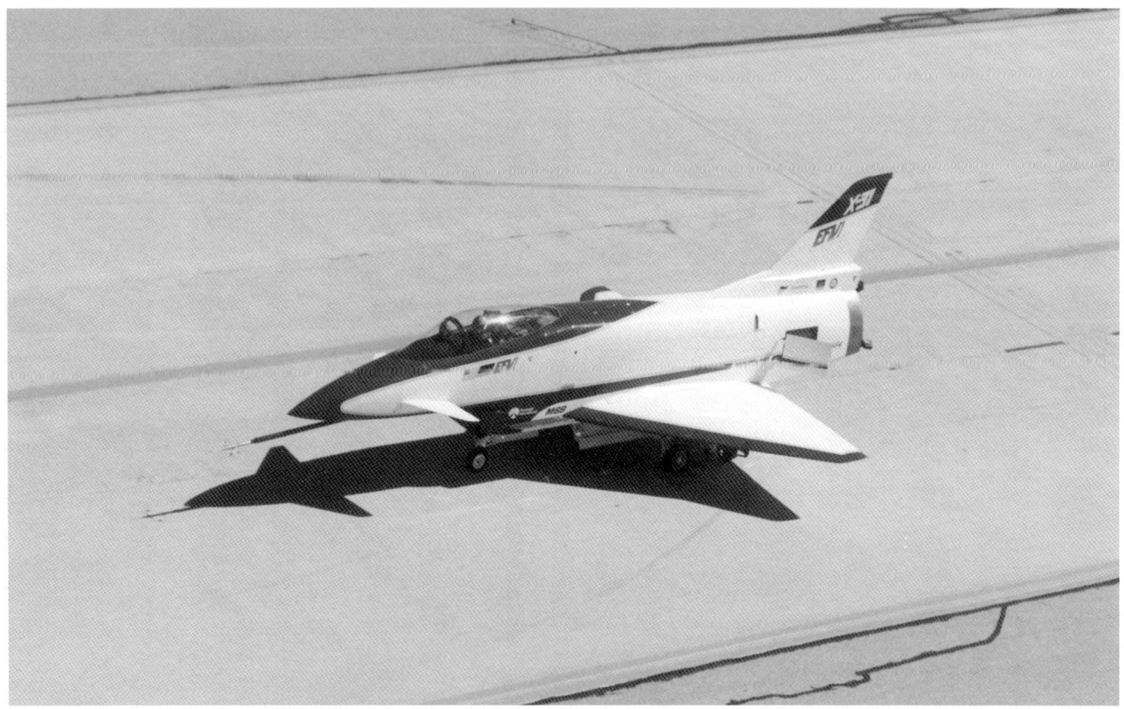

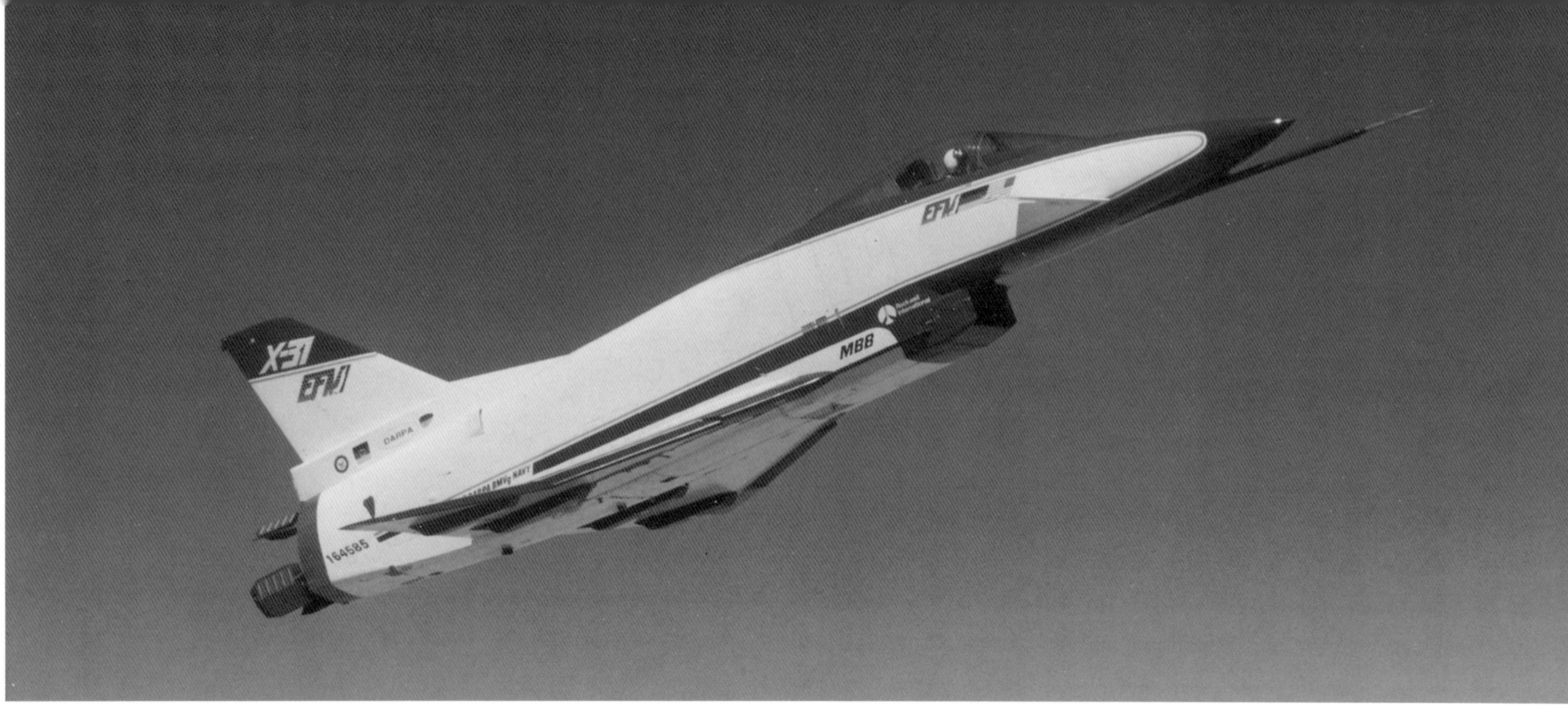

The second X-31A, BuAer 164585, in flight and using its single drag chute after landing. Note the X-31 program insigne on the fuselage. This aircraft, kept in storage at Edwards, is displayed each year.

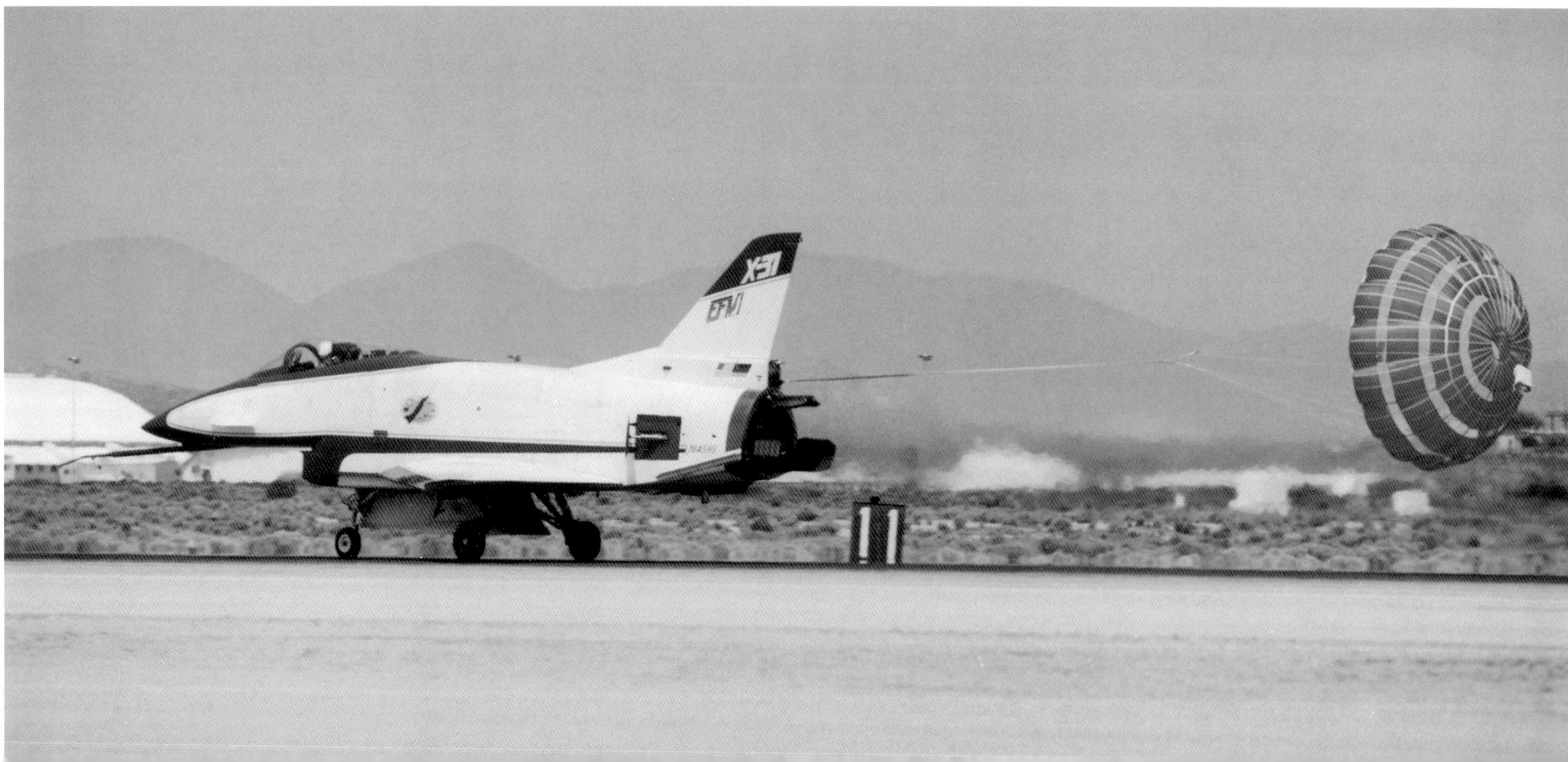

flight controls and a better understanding of airflow at high angles of attack. In addition to making future aircraft more maneuverable, the data gathered during these flights were intended to improve the safety of their operation.

The first X-31A, BuAer 164584, was rolled out at Air Force Plant 42, Palmdale, California, on March 1, 1990. The first flight occurred on October 11 with Rockwell chief test pilot Ken Dyson at the controls. The flight, described by Dyson as "picture perfect," lasted 38 minutes; the new aircraft was found to handle well and be extremely stable. It was flown at 340 miles per hour, reached 10,000 feet and the landing gear was successfully retracted. The second and third flights took place on October 17 and November 3, again with Dyson piloting. MBB test pilot Dietrich Seeck made the next three flights and was the first to use afterburner, though limiting speed to Mach .65. Before the end of November, Rockwell's Fred Knox flew the X-31A twice, and by this time all three pilots had put the aircraft through its paces at up to 3g.

On January 19, 1991, the second prototype, BuAer 164585, flew with Seeck at the controls. During 1991 the two X-31s logged 108 test flights, successfully achieving thrust vectoring as well as expanding the post-stall envelope to 40 degrees angle of attack. In February 1992 the flight test program moved from Palmdale to NASA's Dryden Flight Test Center at nearby Edwards AFB. This change was made at the request of DARPA and the flight envelope was further expanded.

The tests at Dryden focused on military utility evaluations, by pitting the X-31 against similarly equipped fighter types to assess its maneuverability in combat simulations. These flights were known as Close-In-Combat evaluations. The X-31 achieved controlled flight at 70 degrees angle of attack in November 1992. On April 29, 1993, the second prototype executed a minimum-radius 180-degree turn in post-stall configuration, an outstanding achievement well beyond the limits of conventional aircraft. By late 1993 the X-31 was flying simulated combat against F/A-18 adversaries, logged its 300th flight, and achieved a speed of Mach 1.28. In addition to these milestones, the two aircraft set a new annual record for experimental aircraft with 160 flights in 1993.

The CIC tests were concluded in March 1994, although the X-31's evaluation as a fighter maneuverability demonstrator continued into the following year. Later that year, a quasi-tailless configuration was tried on the X-31 to demonstrate the ability to stabilize a tailless aircraft at supersonic speeds utilizing thrust vectoring. All of the tests produced data that would enable future combat aircraft to be lighter, with reduced drag and radar cross-section.

On January 19, 1995, the first X-31 was destroyed in an accident north of Edwards, although pilot Karl Lang ejected safely. The second aircraft is currently stored at Edwards and can be seen as a static display during the annual October airshows. The X-31 program logged a record 524 flights in 52 months, with fourteen civilian and military pilots from the United States and Germany having flown the aircraft.

X-31A

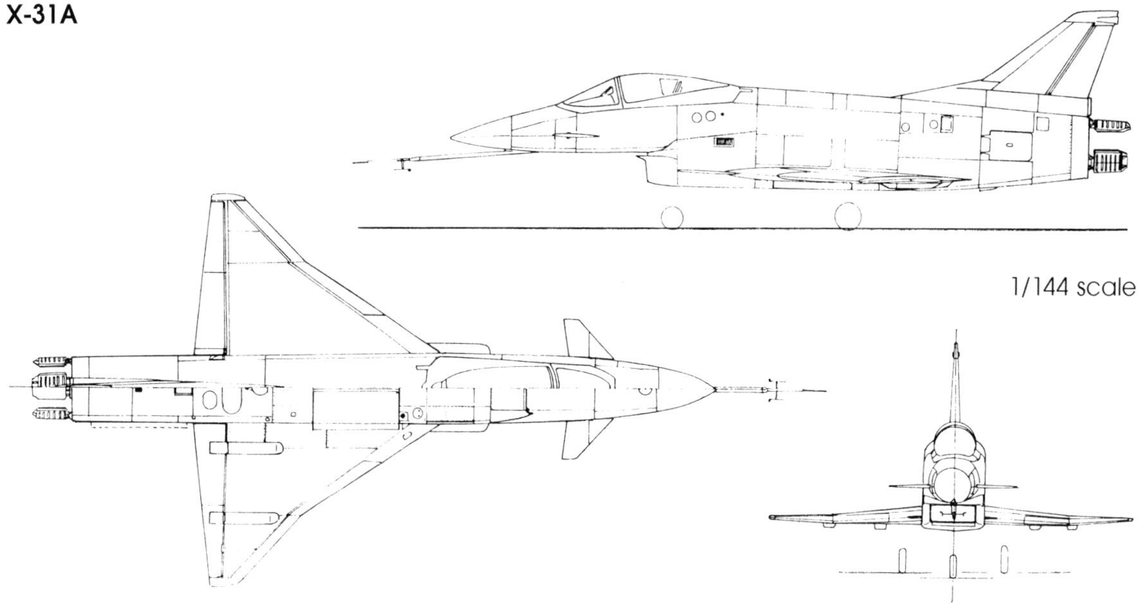

1/144 scale

Specifications: Rockwell/MBB / Boeing North American/ Daimler-Benz Aerospace X-31A EFM

Description: Experimental single-jet, low-wing canard aircraft of metal and composite construction with retractable landing gear. Crew 1.

Engine: One 16,000-lb/thrust General Electric P404-GE-400 turbofan with afterburner.

Dimensions: Span 23 ft 10 in, length 43 ft 4 in, height 14 ft 6 in, wing area 226.3 sq ft.

Weights: 11,410 lb empty, 16,100 lb maximum take-off.

Performance: Maximum speed Mach 1.28, typical envelope Mach 0.9. Service ceiling 40,000 ft.

Production: Rockwell/MBB / Boeing North American/ Daimler-Benz Aerospace X-31A EFM

—	X-31A	BuAer 164584/164585	2
Total, X-31 Enhanced Fighter Maneuverability research aircraft, 1987-1999			2

NORTH AMERICAN / ROCKWELL INTERNATIONAL / BOEING NORTH AMERICAN Missiles / Unpiloted Aircraft / Spacecraft / Subcontracts

AS THE RICH LEGACY of U.S. aircraft manufacturers has boiled down in merger after merger to a few huge corporations, the name North American has managed to survive as a component of The Boeing Company. Announcing its acquisition of Rockwell aerospace programs in full-page newspaper ads on December 5, 1996, Boeing presented its own aircraft but gave North American and Rockwell products—the P-51, F-86, B-1B and Space Shuttle—surprising emphasis. The newly acquired facilities were designated as Boeing North American and the Rockwell name for all intents and purposes disappeared from the international aerospace scene.

Predominantly a military contractor during its 65 years, North American Aviation manufactured huge quantities of extremely successful types during its first thirty years, followed by much smaller numbers of production aircraft and a few vastly more expensive experimental models during the Rockwell years. Traditional manned aircraft became fewer as missiles and spacecraft, especially the NASA Space Shuttle Orbiter program, became priorities.

During the 1950s NAA built the unpiloted radio-controlled X-10 cruise missile research aircraft, which led to the XSM-64 flight test model and the XSM-64A Navaho missile system. The AGM-28 Hound Dog was an air-breathing jet missile carried by B-52 bombers, capable of delivering a nuclear warhead from a range of 600 miles.

Among the North American and Rockwell unpiloted aircraft projected and produced since 1960 have been the NA-273/289 Redhead/Roadrunner MQM-42/-42A target missile systems for the Army, the NA-299/309/350/357/368/383/384/387/401 Condor AGM-53A missile systems for the Navy, the NA-314 AGM-65A missiles for the Air Force, the NA-319/322/326/334/337/341/347/353/359/361/362/364/366/388/389/393 Hobo missile testing, logistics and production for the Air Force, the NA-363 Modular Guided Glide Bomb-QRC system for the Air Force, the NA-394/395 Hellfire modular missile systems for the Army and the NA-406 Pavestrike GBU-15 Modular Glide

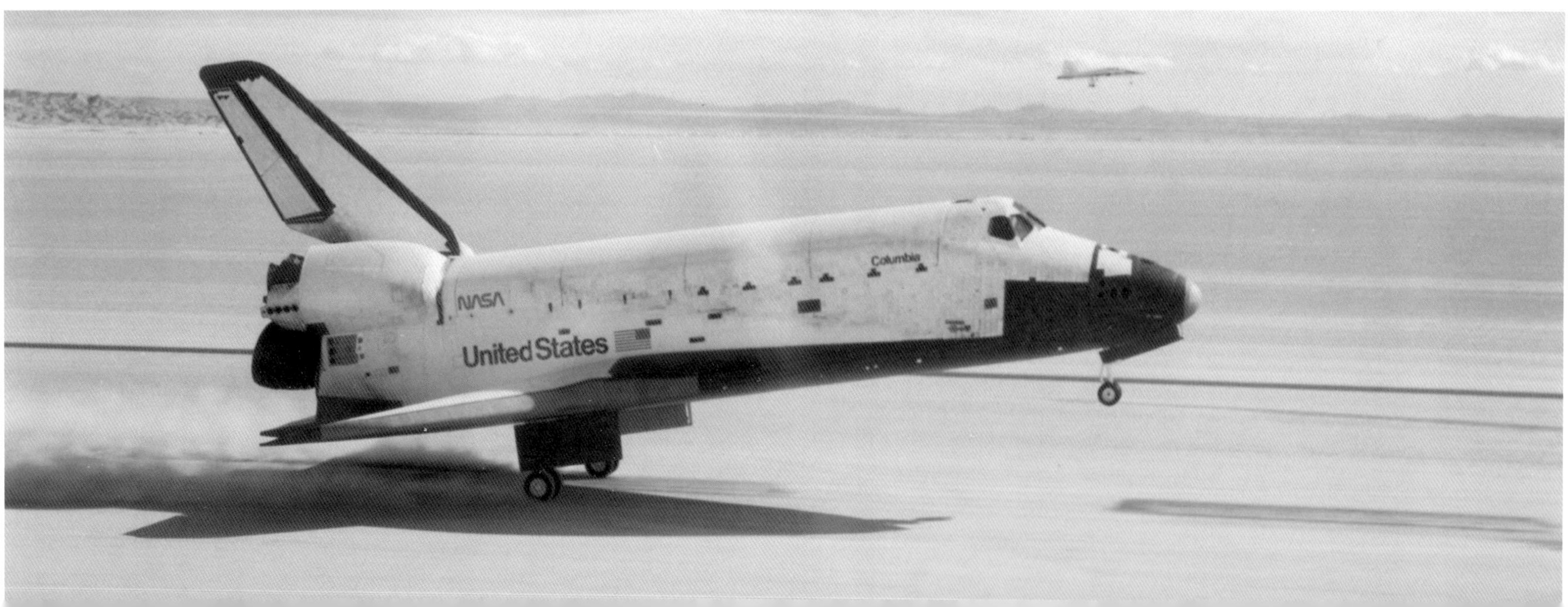

With military aircraft procurement drastically reduced, North American Rockwell shifted most of its resources to the space program. This is the famous STS Orbiter Columbia, *landing at Edwards in 1981.*

weapons system for the Air Force. In March 1978 Rockwell rolled out the HiMAT, an 0.44-scale Highly-Maneuverable Aircraft Technology demonstrator, controlled by a pilot on the ground to explore combat maneuvers up to 8g in the transonic speed range.

In 1966 North American produced the Hoverbuggy, a VTOL flying simulator, and during the 1960s and '70s Rockwell marketed several ranges of civil aircraft that it acquired from other companies, independently of North American—the Aero Commander (later Rockwell Commander) executive transports, lightplanes and agricultural aircraft.

The USAF/DARPA Rockwell FSW Sabrebat Forward-Swept Wing Technology Demonstrator, generally similar to the Grumman X-29A, was displayed in mock-up form in 1979. In 1989 Rockwell announced its participation in the X-30 National AeroSpace Program (NASP) for a single-stage-to-orbit ramjet/scramjet vehicle. During the 1990s Rockwell sought contracts to manufacture the German MBB/Deutsche Aerospace Fanranger and Ranger 2000 trainers, without any success.

Projects that did not go beyond the study or design stages included the NA-236 LRI Long Range Interceptor and NA-237 FBX Fighter-Bomber System for the Air Force, the NA-295 VAL Aircraft, the NA-323/324/328/329/330/333 VFX/F-14 and VSX weapon systems programs for the Navy, the NA-335 FX/F-15 development for the Air Force, the NA-339 AX program for the Navy, the NA-349 ABC Interceptor for the Air Force, and the NA-355 Multi-Purpose V/STOL Aircraft, NA-365 Carrier Onboard Delivery aircraft, NA-381 Long Endurance Sensor Carrier Aircraft, and NA-382 Medium V/STOL Transport, all for the Navy. The identities of the NA-250, NA-275 and NA-399 projects are not recorded in North American's Airframe Contract Record, Report "O."

As a subcontractor, North American was involved with Boeing in 1966 on NA-304, a project to fabricate wing pivot test assemblies for the unrealized SST supersonic transport.

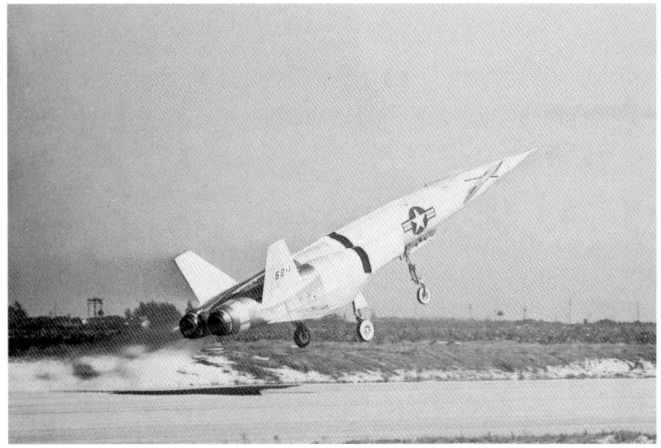

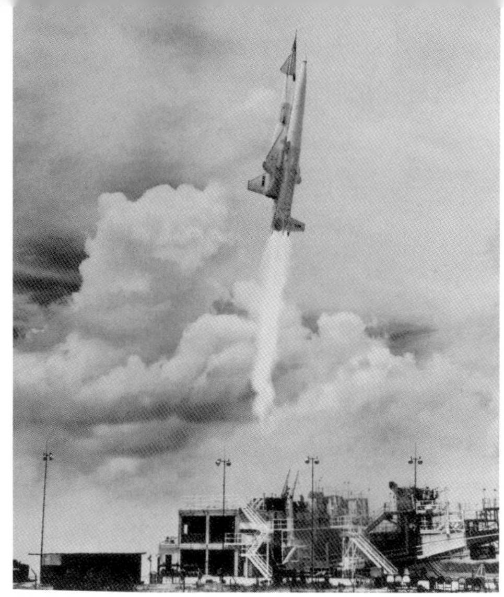

Above, one of thirteen X-10 unmanned test planes for the second stage of the SM-64 Navaho missile, being launched at right.

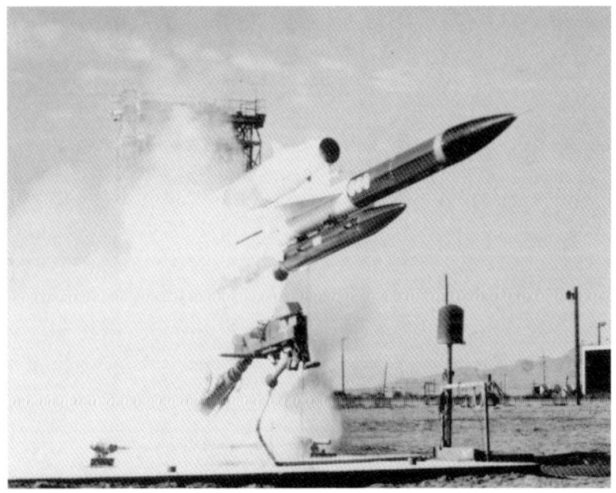

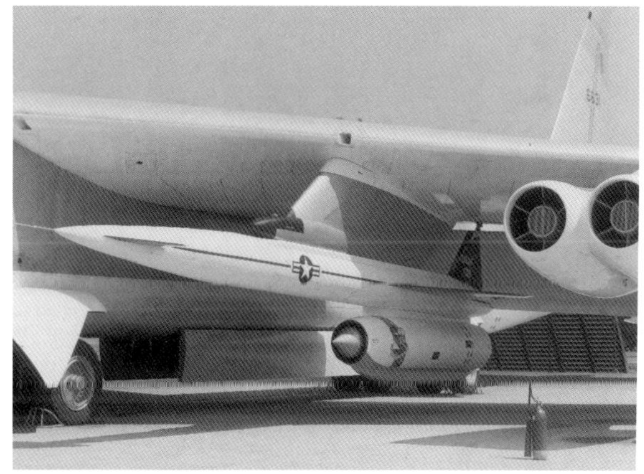

Above, launch of an MQM-42 Redhead/Roadrunner target drone. Right, a Hound Dog missile under wing of B-52G. Below left, an Army Hellfire missile fired from Hummer. Below right, the HiMAT unmanned combat test aircraft, "flown" by pilot on ground.

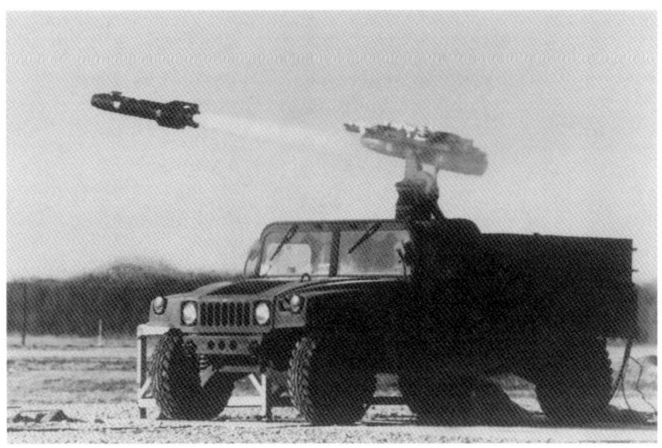

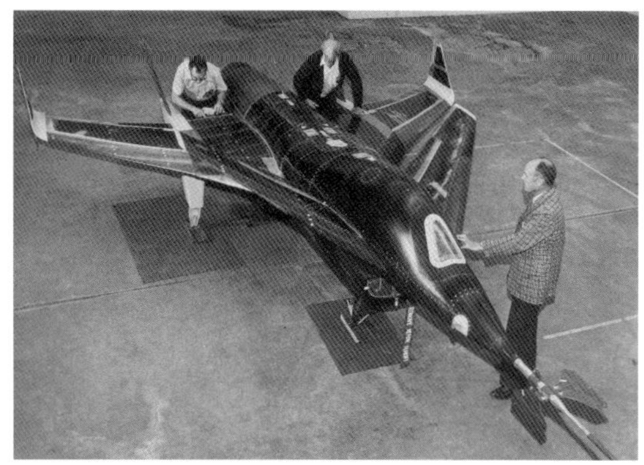

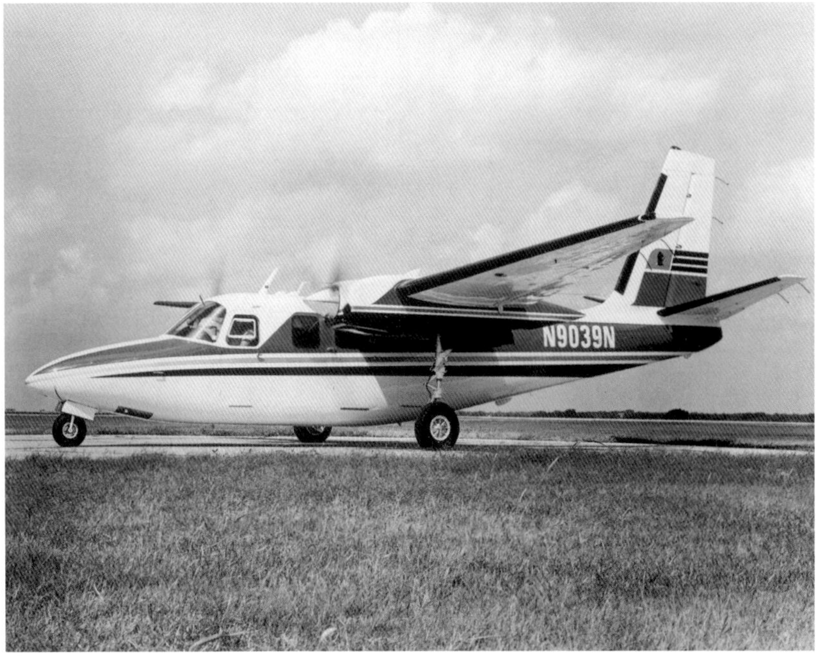

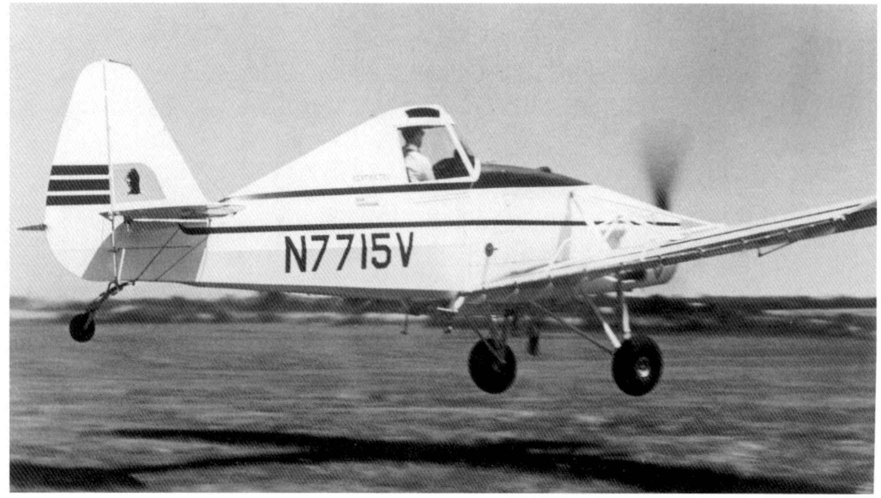

Clockwise from lower left: Rockwell Lark Commander, Shrike Commander and Quail Commander, and below, mock-up of Forward-Swept Wing Technology Demonstrator. Opposite page above, artists' renderings of VFX F-14 for the Navy and FX F-15 for the Air Force, and below, a concept for the X-30 National AeroSpace Plane, capable of taking off from runways and accelerating to Mach 25 at low Earth orbit, as proposed by Rockwell International in the 1980s for operations in the 21st Century.

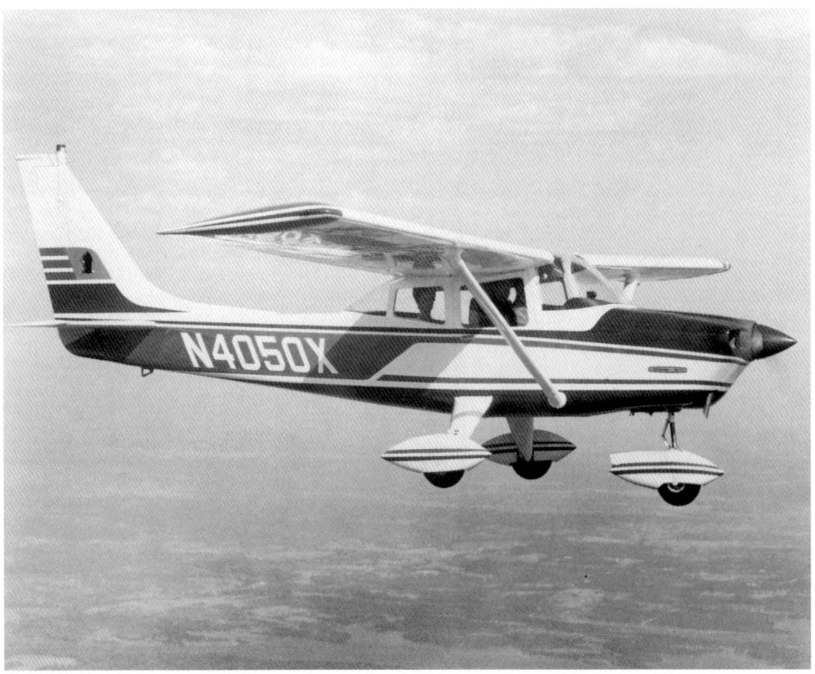

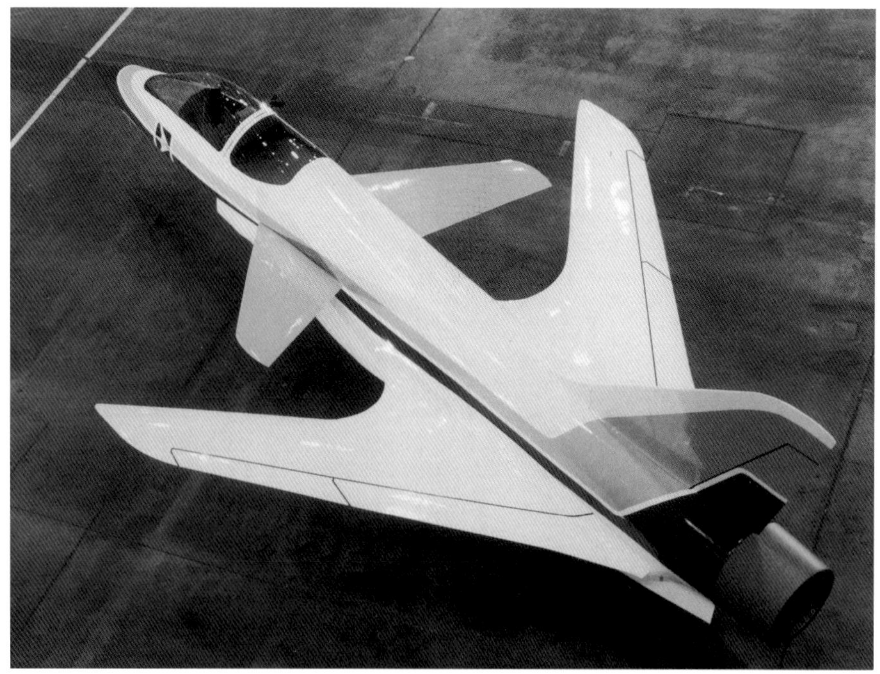

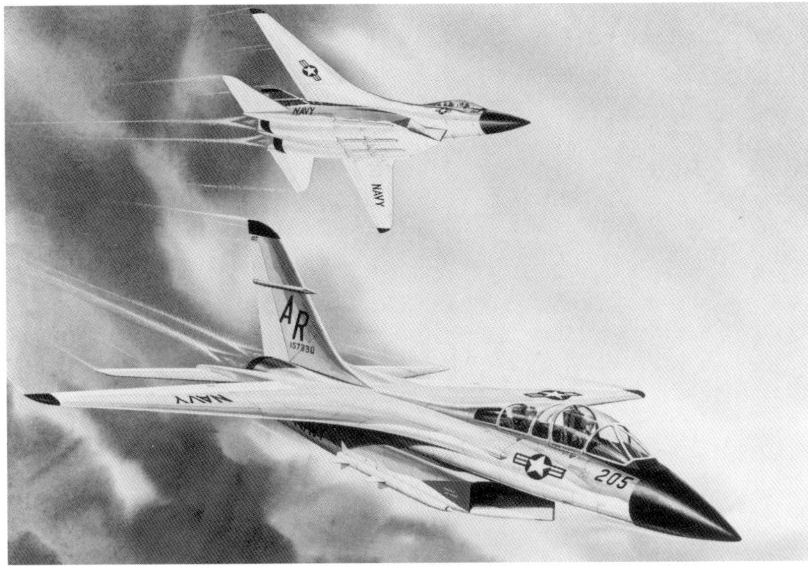
NAVY
AR
205

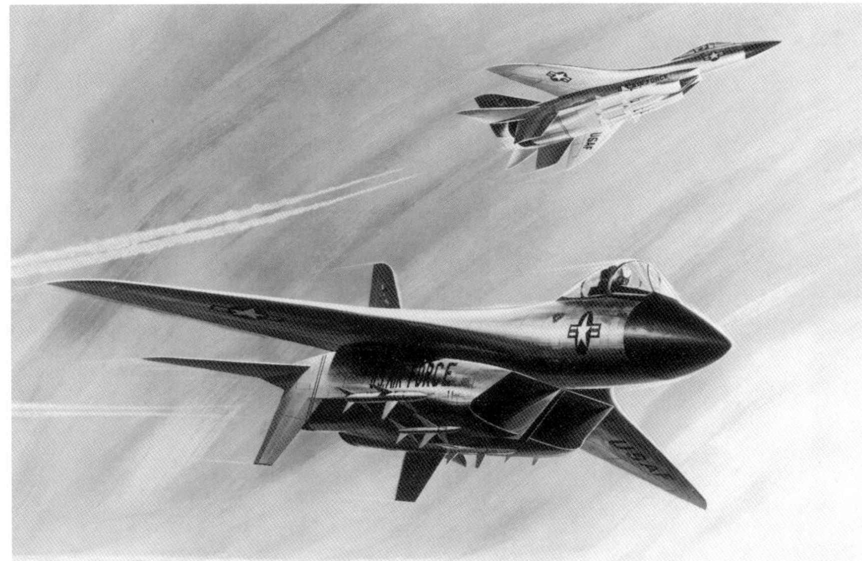

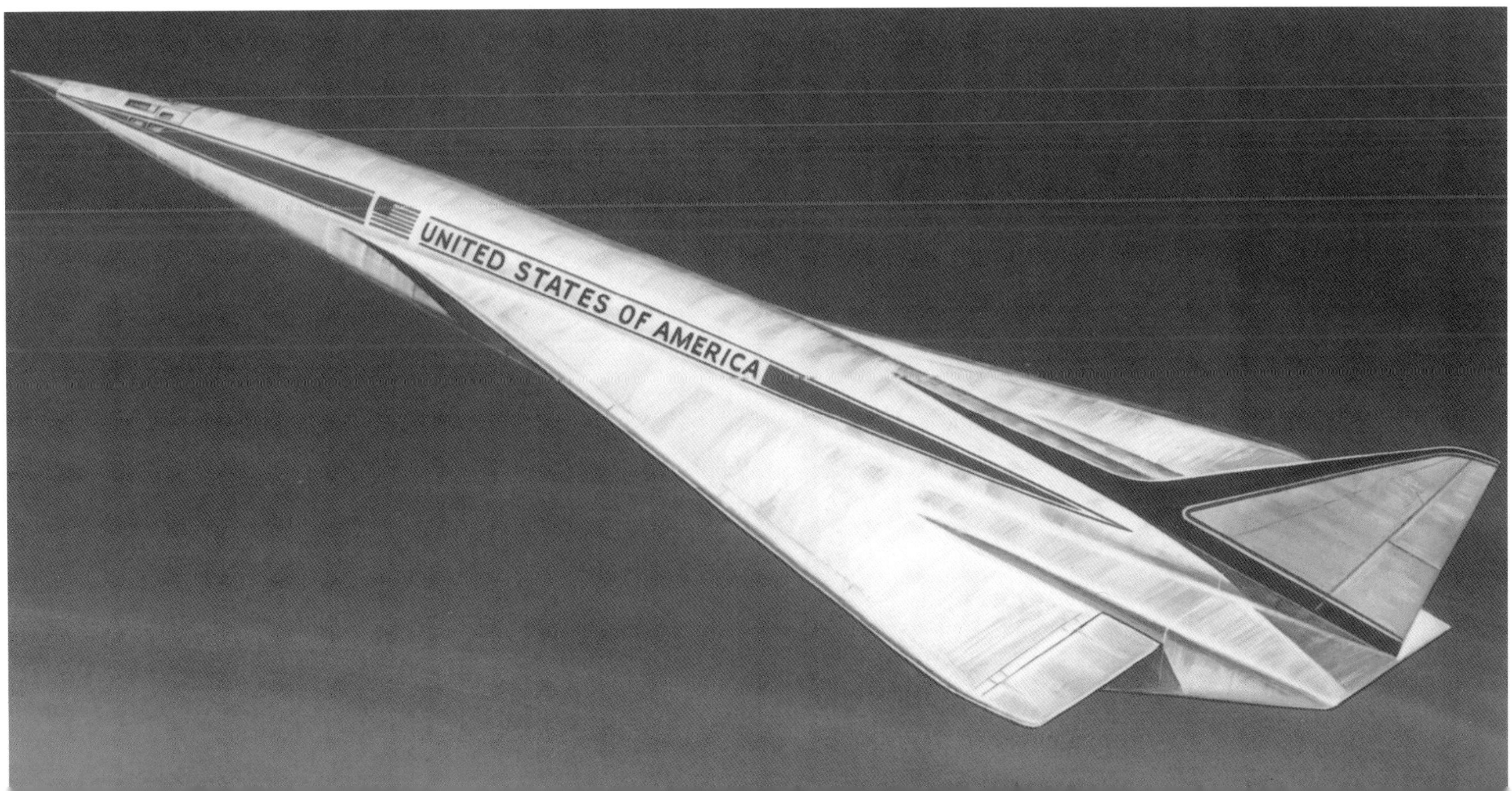
UNITED STATES OF AMERICA

NORTH AMERICAN Model Numbers and Production Figures

NORTH AMERICAN AVIATION kept one of the most thorough and accessible designation and production records of any aircraft manufacturer. Known as the Airframe Contract Record, Report "O," and produced from 1934 through the North American Rockwell and Rockwell International projects in 1976, it has provided a consistent framework for the efforts of historians and writers. The first North American charge (model) number was GA-15 (the numbers 1-14 having been assigned to previous General Aviation projects) and subsequent numbers were prefixed NA-16, etc. The following list, based on Report "O," contains a few corrections, clarifications and probable explanations regarding the actual aircraft built, as distinct from those originally specified under the charge numbers. It covers all NAA aircraft, including those in Volume 1 as well as in this book. In the following list, the number of individual aircraft actually constructed—not counting modified, remanufactured and renumbered aircraft—totals 66,750, a prodigious achievement approached by few manufacturers.

Initially the factory serial numbers started with NA, followed by the specific charge number and then the overall NAA production numbers in a continuing, essentially chronological sequence. Note that the production number -1 was used five times (GA-15-1, NA-16-1 twice, NA-18-1 and NA-19-1) before the system was firmly established. Theoretically the serial numbers should have indicated the total of NAA aircraft built up to any given point, but numerous blocks went unused when the aircraft for which they had been allocated were canceled. The serial prefix NA was dropped beginning with 25-203. In December 1947, to reduce the number of digits, the serial numbers for each charge number again started from 1 and ran to the total of that model produced, i.e. 157-1 and 157-2 (YF-93A), 159-1 to 159-268 (XT-28/T-28A). Even this system was modified when production that continued on new charge numbers still used the original charge number for the serials, i.e. NA-282 (Sabreliner) serials 282-1/35, then NA-290, -292, -293 and -297 Sabreliners with serials 282-36/123.

Gaps in the NA- charge number series indicate projects enumerated separately under Missiles/Unpiloted Aircraft/Spacecraft on pages 174-177.

NA Model	Designation	Factory Serials	Customer Serials	Number built
NA-15	XO-47	GA-15-1	Civil X-2079 / AC36-145	1
NA-16	NA-16-1	NA-16-1	Civil X-2080	1
NA-17	(161 sets of main and wingtip floats for Curtiss SOC-1)			—
NA-18	NA-18	NA-18-1 (NA-16-1 modified/renumbered)	Argentina (X-2080)	(1)
NA-19	BT-9	NA-19-1,-3,-5/11, -20/34, -50/67	AC36-28/69	42
NA-19A	BT-9A	NA-19-4, -12/19, -35/49, -68/83	AC36-88/127	40
NA-20	NA-16-2H	NA-16-2	Demonstrator/Honduras	1
NA-21	NA-21	NA-21-84	USAAC (no serial)	1
NA-22	NA-22	NA-16-1 (NA-19-11 modified/renumbered)	(AC36-36)	(1)
NA-23	BT-9B	NA-23-85/201	AC37-115/231	117
NA-24	(48 sets of main and wingtip floats for Curtiss SOC-2)			—
NA-25	O-47A	25-203/266	AC37-260/323	64
	O-47A	25-267/311	NG37-324/368	45
	O-47A	25-541/547	AC38-319/325	7
	O-47A	25-548/595	NG38-271/318	48
NA-26	BC-1	26-202	USAAC test/later Canada	1
NA-27	NA-16-2H	27-312	Fokker/Netherlands	1
NA-28	NJ-1	28-313/352	BuAer 0910/0949	40
NA-29	BT-9C	29-353/384	ACR37-384/415	32
	Y1BT-10	29-385	ACR37-383	1
	BT-9C	29-505/538	ACR38-224/257	34
NA-30	(Y1BT-10 drawings only)		USAAC	—
NA-31	NA-16-4M	31-386	Sweden	1
NA-32	NA-16-1A	32-387	Australia	1
NA-33	NA-16-2K	33-388	Australia	1
NA-34	NA-16-4P	34-389/418	Argentina	30
NA-35	NA-35	35-419	Civil NX14299	1
	NA-35	---	Civil NX21760	1
	(The second NA-35 was completed without an NA serial number because of the transfer of the project to Vega, which built another four)			(4)
NA-36	BC-1	36-420/504	AC37-372/456	85
	BC-1	36-596/687	AC38-356/447	92
NA-37	NA-16-4R	37-539	Japan	1
NA-38	NA-16-4M	38-540	Sweden	1
NA-39	XB-21 (NA-21development)		AC38-485	(1)
NA-40	NA-40	40-1052	USAAC (no serial)	1
NA-41	NA-16-4	41-697/731	China	35

NA Model	Designation	Factory Serials	Customer Serials	Number built
NA-42	NA-16-2A	42-691/692	Honduras	2
NA-43	NA-16-1G	(not built)	Brazil	—
NA-44	NA-44	44-747	Demonstrator/to Canada	1
NA-45	NA-16-1GV	45-693/695	Venezuela	3
NA-46	NA-16-4	46-972/977,		
		46-1991/1996	Brazil	12
NA-47	NA-16-4RW	47-696	Japan	1
NA-48	NA-16-3C	48-732/746	China	15
NA-49	NA-16-1E	49-748/947	RAF N7000/7199	200
		49-1053/1252	RAF P5783/5982	200
NA-50	NA-50	50-948/954	Peru	7
NA-51	O-47B	51-978/1001	AC39-65/88	24
	O-47B	51-1002/1051	NG39-89/138	50
NA-52	SNJ-1	52-956/971	BuAer 1552/1567	16
NA-53	(Single-seat fighter—canceled)		---	—
NA-54	BC-2	54-688/690	AC38-488/490	3
NA-55	BC-1A	55-1548/1566	NG39-798/816	
		55-1557/1606	ACR39-817/856	
		55-1607/1616	NG40-707/716	
		55-1617/1630	ACR40-726/739	83
NA-56	NA-16-4	56-1453/1502	China	50
NA-57	NA-57	57-1253/1452,		
		57-1518/1547	France 1-230	230
NA-58	BT-14	58-1655/1905	AC40-1110/1360	251
NA-59	AT-6	59-1631/1639	AC40-717/725,	
		59-1906/1990	AC40-2080/2164	94
NA-60	O-47A	(25-595 modified)	(NG38-318)	(1)
NA-61	NA-16-1E	61-1503/1517	RCAF 1321/1335	
		61-1640/1654	RCAF 1336/1350	30
NA-62	B-25/A/B	62-2834/3017	AC40-2165/2348	184
NA-63	XB-28	63-2233	AC40-3056	1
NA-64	NA-64	64-2033/2232,		
		64-3018/3047	France 1-230	230
NA-65	SNJ-2	65-1997/2032	BuAer 2008/2043	36
NA-66	Harvard II	66-2234/2833	RCAF 2501/3013, RAF	
			AH185/204, NZ901/967	600
NA-67	XB-28A	67-3417	AC40-3058	1
NA-68	NA-50A/P-64	68-3058/3063	(Siam) AC41-19082/19087	6
NA-69	NA-44/A-27	69-3064/3073	(Siam) AC41-18890/18899	10
NA-70	(NA-35 primary trainer production—canceled)			—
NA-71	NA-16-3	71-3074/3076	Venezuela 1-3	3
NA-72	NA-44	72-3077/3096,		
		72 -4757/4766	Brazil	30
NA-73X	NA-73X	73-3097	Civil NX 19998	1
NA-73	Mustang I	73-3098/3100,		
		73-3102/3106,		
		73-3108/3416,		

NA Model	Designation	Factory Serials	Customer Serials	Number built
NA-73	Mustang I	73-4767/4768,		
		73-7812	RAF AG345/664	320
	XP-51	73-3101, -3107	AC41-38/39	2
NA-74	NA-44	74-4745/4756	Chile	12
NA-75	Harvard II	75-3048/3057,		
		75-3418/3507	RCAF 3134/3233	100
NA-76	Harvard II	76-3508/3957	RAF AJ538/987	450
NA-77	AT-6A	77-3958/3982	AC41-149/173	
	AT-6A	77-4133/4282	AC41-174/323	
	SNJ-3	77-4283/4292	BuAer 6755/6764	
	AT-6A	77-4293/4312	AC41-324/343	
	SNJ-3	77-4313/4322	BuAer 6765/6774	
	AT-6A	77-4323/4352	AC41-344/373	
	SNJ-3	77-4353/4362	BuAer 6775/6784	
	AT-6A	77-4363/4382	AC41-374/393	
	SNJ-3	77-4383/4392	BuAer 6785/6794	
	AT-6A	77-4393/4422	AC41-394/423	
	SNJ-3	77-4423/4432	BuAer 6795/6804	
	AT-6A	77-4433/4452	AC41-424/443	
	SNJ-3	77-4453/4462	BuAer 6805/6814	
	AT-6A	77-4463/4492	AC41-444/473	
	SNJ-3	77-4493/4502	BuAer 6815/6824	
	AT-6A	77-4503/4532	AC41-474/503	
	SNJ-3	77-4533/4542	BuAer 6825/6834	
	AT-6A	77-4543/4582	AC41-504/543	
	SNJ-3	77-4583/4592	BuAer 6835/6844	
	AT-6A	77-4593/4632	AC41-544/583	
	SNJ-3	77-4633/4642	BuAer 6845/6854	
	AT-6A	77-4643/4682	AC41-584/623	
	SNJ-3	77-4683/4692	BuAer 6855/6864	
	AT-6A	77-4693/4722	AC410-624/653	
	SNJ-3	77-4723/4732	BuAer 6865/6874	
	AT-6A	77-4733/4744	AC41-654/665	637
NA-78	AT-6A	78-5932/6001	AC41-666/735	
	SNJ-3	78-6002/6011	BuAer 6875/6884	
	AT-6A	78-6012/6031	AC41-736/755	
	SNJ-3	78-6032/6041	BuAer 6885/6894	
	AT-6A	78-6042/6061	AC41-756/775	
	SNJ-3	78-6062/6071	BuAer 6895/6904	
	AT-6A	78-6072/6081	AC41-776/785	
	AT-6A	78-6082/6091	AC41-15824/15833	
	SNJ-3	78-6092/6101	BuAer 6905/6914	
	AT-6A	78-6102/6121	AC41-15834/15853	
	SNJ-3	78-6122/6131	BuAer 6915/6924	
	AT-6A	78-6132/6161	AC41-15854/15883	
	SNJ-3	78-6162/6171	BuAer 6925/6934	
	AT-6A	78-6172/6201	AC41-15884/15913	

NA Model	Designation	Factory Serials	Customer Serials	Number Built
NA-78	SNJ-3	78-6202-6211	BuAer 6935/6944	
	AT-6A	78-6212/6251	AC41-15914/15953	
	SNJ-3	78-6252/6261	BuAer 6945/6954	
	AT-6A	78-6262/6301	AC41-15954/15993	
	SNJ-3	78-6302/6311	BuAer 6955/6964	
	AT-6A	78-6312/6361	AC41-15994/16043	
	SNJ-3	78-6362/6371	BuAer 6965/6974	
	AT-6A	78-6372/6421	AC41-16044/16093	
	SNJ-3	78-6422/6431	BuAer 6975/6984	
	AT-6A	78-6432/6456	AC41-16094/16118	
	SNJ-3	78-6457/6496	BuAer 6985/7024	
	AT-6A	78-6497/6606	AC41-16119/16228	
	SNJ-3	78-6607/6636	BuAer 01771/01800	
	AT-6A	78-6637/6781	AC41-16259/16403	
	SNJ-3	78-6782/6816	BuAer 01801/01835	
	AT-6A	78-6817/6835	AC41-16439/16457	
	SNJ-3	78-6836/6851	BuAer 01836/01851	
	AT-6A	78-6852/6956	AC41-16474/16578	
	SNJ-3	78-6957/6993	BuAer 01852/01888	
	AT-6A	78-6994/7031	AC41-16616/16653	
	SNJ-3	78-7032/7070	BuAer 01889/01927	
	AT-6A	78-7071/7156	AC41-16693/16778	
	SNJ-3	78-7157/7198	BuAer 01928/01969	
	AT-6A	78-7199/7256	AC41-16821/16878	
	SNJ-3	78-7257/7301	BuAer 01970/01976, BuAer 05435/05472	
	AT-6A	78-7302/7357	AC41-16924/16979	
	SNJ-3	78-7358/7411	BuAer 05473/05526	1480
NA-79	SNJ-2	*78-3983/4007	BuAer 2548/2572	25
	*(Note the use of the earlier charge number in the serial)			
NA-80	(Project canceled)		—	—
NA-81	Harvard II	81-4008/4132	RCAF 3014/3033, 3761/3841; RAF BW184/207	125
NA-82	B-25C	82-5069/5931	AC41-12434/13296	863
NA-83	Mustang I	83-4769/5068	RAF AL958/999, AM100/257, AP164/263	300
NA-84	AT-6B	84-7412/7811	AC41-17034/17433	400
NA-85	(SNJ-3 production—transfered to NA-78)			—
NA-86	(Project canceled)		—	—
NA-87	B-25D-1/-5/-10/-15	87-7813/9012	AC41-29648/30847	1200
NA-88	AT-6C	88-9013/9033	AC41-32073/32093	
	SNJ-4	88-9034/9163	BuAer 05627/05656	
	AT-6C	88-9164/9178	AC41-32094/32108	
	Harvard IIa	88-9179/9313	EX 100/234	
	SNJ-4	88-9314/9331	BuAer 05657/05674	
	SNJ-4	88-9332/9478	BuAer 09817/09963	
NA-88	AT-6C	88-9479/9522	AC41-32109/32152	
	SNJ-4	88-9523/9612	BuAer 09964/10053	
	AT-6C	88-9613/9620	AC41-32153/32160	
	Harvard IIa	88-9621/9630	EX 235/244	
	AT-6C	88-9631/9640	AC41-32161/32170	
	Harvard IIa	88-9641/9650	EX 245/254	
	AT-6C	88-9651/9660	AC41-32171/32180	
	Harvard IIa	88-9661/9670	EX 255/264	
	AT-6C	88-9671/9680	AC41-32181/32190	
	Harvard IIa	88-9681/9690	EX 265/274	
	AT-6C	88-9691/9700	AC41-32191/32200	
	Harvard IIa	88-9701/9710	EX 275/284	
	AT-6C	88-9711/9720	AC41-32201/32210	
	Harvard IIa	88-9721/9730	EX 285/294	
	AT-6C	88-9731/9737	AC41-32211/32217	
	SNJ-4	88-9738/9747	BuAer 10054/10063	
	Harvard IIa	88-9748/9757	EX 295/304	
	SNJ-4	88-9758/9777	BuAer 10064/10083	
	Harvard IIa	88-9778/9787	EX 305/314	
	SNJ-4	88-9788/9807	BuAer 10084/10103	
	Harvard IIa	88-9808/9817	EX 315/324	
	SNJ-4	88-9818/9837	BuAer 10104/10123	
	Harvard IIa	88-9838/9847	EX 325/334	
	SNJ-4	88-9848/9867	BuAer 10124/10143	
	Harvard IIa	88-9868/9877	EX 335/344	
	AT-6C	88-9878/9912	AC41-32218/32252	
	Harvard IIa	88-9913/9922	EX 345/354	
	AT-6C	88-9923/9957	AC41-32253/32287	
	Harvard IIa	88-9958/9967	EX 355/364	
	AT-6C	88-9968/10007	AC41-32288/32327	
	Harvard IIa	88-10008/10017	EX 365/374	
	AT-6C	88-10018/10057	AC41-32328/32367	
	Harvard IIa	88-10058/10067	EX 375/384	
	AT-6C	88-10068/10100	AC41-32368/32400	
	Harvard IIa	88-10101/10112	EX 385/396	
	SNJ-4	88-10113/10152	BuAer 10144/10183	
	Harvard IIa	88-10153/10162	EX 397/406	
	SNJ-4	88-10163/10202	BuAer 10184/10223	
	Harvard IIa	88-10203/10212	EX 407/416	
	SNJ-4	88-10213/10246	BuAer 10224/10257	
	Harvard IIa	88-10247/10262	EX 417/432	
	SNJ-4	88-10263/10321	BuAer 10258/10316	
	SNJ-4	88-10322/10337	BuAer 26427/26442	
	AT-6C	88-10338/10369	AC41-32401/32432	
	SNJ-4	88-10370/10440	BuAer 26443/26513	
	AT-6C	88-10441/10525	AC41-32433/32517	
	Harvard IIa	88-10526/10677	EX 433/584	

NA Model	Designation	Factory Serials	Customer Serials	Number Built
NA-88	SNJ-4	88-10678/10752	BuAer 26514/26588	
	AT-6C	88-10753/10882	AC41-32518/32647	
	SNJ-4	88-10883/10922	BuAer 26589/26628	
	AT-6C	88-10923/11002	AC41-32648/32727	
	SNJ-4	88-11003/11042	BuAer 26629/26668	
	AT-6C	88-11043/11082	AC41-32728/32767	
	SNJ-4	88-11083/11122	BuAer 26669/26708	
	AT-6C	88-11123/11192	AC41-32768/32837	
	SNJ-4	88-11193/11232	BuAer 26709/26748	
	AT-6C	88-11233/11312	AC41-32838/32917	
	SNJ-4	88-11313/11352	BuAer 26749/26788	
	AT-6C	88-11353/11432	AC41-32918/32997	
	SNJ-4	88-11433/11472	BuAer 26789/26828	
	AT-6C	88-11473/11512	AC41-32998/33037	
	SNJ-4	88-11513/11552	BuAer 26829/26868	
	AT-6C	88-11553/11587	AC41-33038/33072	
	AT-6C	88-11588/11687	AC42-3884/3983	
	SNJ-4	88-11688/11757	BuAer 26869/26938	
	AT-6C	88-11758/11797	AC42-3984/4023	
	SNJ-4	88-11798/11867	BuAer 26939/27008	
	AT-6C	88-11868/11907	AC42-4024/4063	
	SNJ-4	88-11908/12028	BuAer 27009/27129	
	Harvard IIa	88-12029/12072	EX 585/628	
	SNJ-4	88-12073/12122	BuAer 27130/27179	
	Harvard IIa	88-12123/12142	EX 629/648	
	AT-6C	88-12143/12174	AC42-4064/4095	
	SNJ-4	88-12175/12224	BuAer 27180/27229	
	AT-6C	88-12225/12265	AC42-4096/4136	
	SNJ-4	88-12266/12314	BuAer 27230/27278	
	Harvard IIa	88-12315/12334	EX 649/668	
	SNJ-4	88-12335/12356	BuAer 27279/27300	
	AT-6C	88-12357/12400	AC42-4137/4180	
	SNJ-4	88-12401/12428	BuAer 27301/27328	
	AT-6C	88-12429/12484	AC42-4181/4236	
	SNJ-4	88-12485/12534	BuAer 27329/27378	
	Harvard IIa	88-12535/12554	EX 669/688	
	AT-6C	88-12555/12561	AC42-4237/4243	
	AT-6C	88-12562/12614	AC42-48772/48824	
	SNJ-4	88-12615/12664	BuAer 27379/27428	
	AT-6C	88-12665/12704	AC42-48825/48864	
	SNJ-4	88-12705/12754	BuAer 27429/27478	
	Harvard IIa	88-12755/12774	EX 689/708	
	AT-6C	88-12775/12814	AC42-48865/48904	
	SNJ-4	88-12815/12864	BuAer 27479/27528	
	AT-6C	88-12865/12924	AC42-48905/48964	
	SNJ-4	88-12925/12974	BuAer 27529/27578	
	Harvard IIa	88-12975/12994	EX 709/728	

NA Model	Designation	Factory Serials	Customer Serials	Number Built
NA-88	AT-6C	88-12995/13034	AC42-48965/49004	
	SNJ-4	88-13035/13084	BuAer 27579/27628	
	AT-6C	88-13085/13124	AC42-49005/49044	
	SNJ-4	88-13125/13174	BuAer 27629/27678	
	Harvard IIa	88-13175/13194	EX 729/748	
	AT-6C	88-13195/13219	AC42-49045/49069	
	AT-6C	88-13220/13234	AC42-43847/43861	
	SNJ-4	88-13235/13284	BuAer 27679/27728	
	AT-6C	88-13285/13344	AC42-43862/43921	
	SNJ-4	88-13345/13394	BuAer 27729/27778	
	Harvard IIa	88-13395/13414	EX 749/768	
	AT-6C	88-13415/13474	AC42-43922/43981	
	SNJ-4	88-13475/13524	BuAer 27779/27828	
	AT-6C	88-13525/13564	AC42-43982/44021	
	SNJ-4	88-13565/13587	BuAer 27829/27851	
	Harvard IIa	88-13588/13607	EX 769/788	
	AT-6C	88-13608/13616	AC42-44022/44030	
	SNJ-4	88-13617/13643	BuAer 51350/51376	
	AT-6C	88-13644/13758	AC42-44031/44145	
	SNJ-4	88-13759/13808	BuAer 51377/51426	
	AT-6C	88-13809/13888	AC42-44146/44225	
	SNJ-4	88-13889/13902	BuAer 51427/51440	
	Harvard IIa	88-13903/13922	EX 789/808	
	SNJ-4	88-13923/13958	BuAer 51441/51476	
	AT-6C	88-13959/14018	AC42-44226/44285	
	SNJ-4	88-14019/14068	BuAer 51477/51526	
	AT-6C	88-14069/14108	AC42-44286/44325	
	SNJ-4	88-14109/14158	BuAer 51527/51576	
	Harvard IIa	88-14159/14178	EX 809/828	
	AT-6C	88-14179/14218	AC42-44326/44365	
	SNJ-4	88-14219/14268	BuAer 51577/51626	
	AT-6C	88-14269/14314	AC42-44366/44411	
	SNJ-4	88-14315/14364	BuAer 51627/51676	
	Harvard IIa	88-14365/14382	EX 829/846	
	Harvard III	88-14383/14384	EX 847/848	
	AT-6D	88-14385/14424	AC42-44412/44451	
	SNJ-5	88-14425/14474	BuAer 51677/51726	
	Harvard III	88-14475/14494	EX 849/868	
	AT-6D	88-14495/14514	AC42-44452/44471	
	SNJ-5	88-14515/14539	BuAer 51727/51751	
	Harvard III	88-14540/14559	EX 869/888	
	AT-6D	88-14560/14605	AC42-44472/44517	
	SNJ-5	88-14606/14655	BuAer 51652/51801	
	Harvard III	88-14656/14675	EX 889/908	
	AT-6D	88-14676/14695	AC42-44518/44537	
	SNJ-5	88-14696/14715	BuAer 51802/51821	
	Harvard III	88-14716/14731	EX 909/924	

NA Model	Designation	Factory Serials	Customer Serials	Number Built
NA-88	AT-6D	88-14732/14799	AC42-44538/44605	
	SNJ-5	88-14800/14824	BuAer 51822/51846	
	AT-6D	88-14825/14844	AC42-44606/44625	
	SNJ-5	88-14845/14869	BuAer 51847/51871	
	Harvard III	88-14870/14889	EX 925/944	
	AT-6D	88-14890/14909	AC42-44626/44645	
	SNJ-5	88-14910/14934	BuAer 51872/51896	
	Harvard III	88-14935/14959	EX 945/969	
	AT-6D	88-14960/14999	AC42-44646/44685	
	SNJ-5	88-15000/15049	BuAer 51897/51946	
	Harvard III	88-15050/15069	EX 970/989	
	AT-6D	88-15070/15089	AC42-44686/44705	
	SNJ-5	88-15090/15114	BuAer 51947/51971	
	Harvard III	88-15115/15124	EX 990/999	
	Harvard III	88-15125/15139	EZ 100/114	
	AT-6D	88-15140/15159	AC42-44706/44725	
	SNJ-5	88-15160/15209	BuAer 51972/52021	
	AT-6D	88-15210/15230	AC42-44726/44746	
	AT-6D	88-15231/15234	AC41-34123/34126	
	SNJ-5	88-15235/15259	BuAer 52022/52046	
	Harvard III	88-15260/15279	EZ 115/134	
	AT-6D	88-15280/15299	AC41-34127/34146	
	SNJ-5	88-15300/15302	BuAer 52047/52049	
	SNJ-5	88-15303/15324	BuAer 43638/43659	
			(AC41-34373/34394)	
	Harvard III	88-15325/15348	EZ 135/158	
	AT-6D	88-15349/15360	AC41-34147/34158	
	SNJ-5	88-15361/15392	BuAer 43660/43691	
			(AC41-34395/34426)	
	AT-6D	88-15393/15572	AC41-34159/34338	
	SNJ-5	88-15573/15592	BuAer 43692/43711	
			(AC41-34427/34446)	
	Harvard III	88-15593/15612	EZ 159/178	
	AT-6D	88-15613/15646	AC41-34339/34372	
	AT-6D	88-15647/15652	AC41-34447/34452	
	SNJ-5	88-15653/15682	BuAer 43712/43741	
			(AC41-34453/34482)	
	Harvard III	88-15683/15702	EZ 179/198	
	AT-6D	88-15703/15722	AC41-34483/34502	
	SNJ-5	88-15723/15742	BuAer 43742/43761	
			(AC41-34503/34522)	
	Harvard III	88-15743/15752	EZ 199/208	
	SNJ-5	88-15753/15772	BuAer 43762/43781	
			(AC41-34523/34542)	
	Harvard III	88-15773/15792	EZ 209/228	
	AT-6D	88-15793/15825	AC41-34543/34575	
	SNJ-5	88-15826/15855	BuAer 43782/43811	
			(AC41-34576/34605)	

NA Model	Designation	Factory Serials	Customer Serials	Number Built
NA-88	Harvard III	88-15856/15876	EZ 229/249	
	AT-6D	88-15877/15896	AC41-34606/34625	
	SNJ-5	88-15897/15925	BuAer 43812/43840	
			(AC41-34626/34654)	
	AT-6D	88-15926/15943	AC41-34655/34672	
		88-15944/15952	EZ 250/258	
			(AC42-84163/84171)	
	SNJ-5	88-15953/15962	BuAer 43841/43850	
			(AC42-84172/84181)	
	Harvard III	88-15963/15982	EZ 259/278	
			(AC42-84182/84201)	
	AT-6D	88-15983/16022	AC42-84202/84241	
	SNJ-5	88-16023/16062	BuAer 43851/43890	
			(AC42-84242/84281)	
	Harvard III	88-16063/16082	EZ 279/298	
			(AC42-84282/84301)	
	AT-6D	88-16083/16122	AC42-84302/84341	
	SNJ-5	88-16123/16142	BuAer 43891/43910	
			(AC42-84342/84361)	
	Harvard III	88-16143/16152	EZ 299/308	
			(AC 42-84362/84371	
	AT-6D	88-16153/16172	AC42-84372/84391	
	SNJ-5	88-16173/16183	BuAer 43911/43921	
			(AC42-84392/84402)	
	AT-6D	88-16184/16203	AC42-84403/84422	
	SNJ-5	88-16204/16233	BuAer 43922/43951	
			(AC42-84423/84452)	
	Harvard III	88-16234/16253	EZ 309/328	
			(AC42-84453/84472)	
	AT-6D	88-16254/16293	AC42-84473/84512	
	SNJ-5	88-16294/16323	BuAer 43952/43981	
			(AC42-84513/84542)	
	Harvard III	88-16324/16343	EZ 329/348	
			(AC42-84543/84562)	
	AT-6D	88-16344/16383	AC42-84563/84602	
	SNJ-5	88-16384/16413	BuAer 43982/44011	
			(AC42-84603/84632)	
	Harvard III	88-16414/16423	EZ 349/358	
			(AC42-84643/84642)	
	AT-6D	88-16424/16463	AC42-84643/84682	
	SNJ-5	88-16464/16489	BuAer 44012/44037	
			(AC42-84683/84708	
		88-16490/16503	BuAer 84819/84832	
			(AC 42-84709/84722	
	Harvard III	88-16504/16523	EZ 359/378	
			(AC42-84723/84742)	
	AT-6D	88-16524/16563	AC42-84743/84782	
	SNJ-5	88-16564/16583	BuAer 84833/84852	
			(AC42-84783/84802)	

NA Model	Designation	Factory Serials	Customer Serials	Number Built
NA-88	Harvard III	88-16584/16603	EZ 379/398 (AC42-84803/84822)	
	AT-6D	88-16604/16663	AC42-84823/84882	
	SNJ-5	88-16664/16703	BuAer 84853/84892 (AC42-84883/84922)	
	Harvard III	88-16704/16713	EZ 399/408 (AC42-84923/84932)	
	AT-6D	88-16714/16753	AC42-84933/84972	
	SNJ-5	88-16754/16793	BuAer 84893/84932 (AC42-84973/85012)	
	Harvard III	88-16794/16813	EZ 409/428 (AC42-85013/85032)	
	AT-6D	88-16814/16853	AC42-85033/85072	
	SNJ-5	88-16854/16873	BuAer 84933/84952 (AC42-85073/85092)	
	AT-6D	88-16874/16883	AC42-85093/85102	
	Harvard III	88-16884/16903	EZ 429/448 (AC42-85103/85122)	
	AT-6D	88-16904/16963	AC42-85123/85182	
	SNJ-5	88-16964/17003	BuAer 84953/84992 (AC42-85183/85222)	
	Harvard III	88-17004/17013	EZ 449/458 (AC42-85223/85232)	
	AT 6D	88-17014/17073	AC42-85233/85292	
	SNJ-5	88-17074/17113	BuAer 84993/85032 (AC42-85293/85332)	
	AT-6D	88-17114/17193	AC42-85333/85412	
	SNJ-5	88-17194/17213	BuAer 85033/85052 (AC42-85413/85432)	
	AT-6D	88-17214/17273	AC42-85433/85492	
	SNJ-5	88-17274/17313	BuAer 85053/85092 (AC42-85493/85532)	
	AT-6D	88-17314/17393	AC42-85533/85612	
	SNJ-5	88-17394	BuAer 85093 (AC42 85613)	
	SNJ-5	88-17395/17433	BuAer 90582/90620 (AC42-85614/85652)	
	AT-6D	88-17434/17548	AC42-85653/85767	
	SNJ-5	88-17549/17568	BuAer 90621/90640 (AC42-85768/85787)	
	AT-6D	88-17569/17638	AC42-85788/85857	
	SNJ-5	88-17639/17678	BuAer 90641/90680 (AC42-85858/85897)	
	AT-6D	88-17679/17748	AC42-85898/85967	
	SNJ-5	88-17749/17788	BuAer 90681/90720 (AC42-85968/86007)	
	AT-6D	88-17789/17868	AC42-86008/86087	
	SNJ-5	88-17869/17888	BuAer 90721/90740 (AC42-86088/86107)	
	AT-6D	88-17889/17948	AC42-86108/86167	
	SNJ-5	88-17949/17988	BuAer 90741/90780 (AC42-86168/86207)	
	AT-6D	88-17989/18058	AC42-86208/86277	
	SNJ-5	88-18059/18098	BuAer 90781/90820 (AC42-86278/86317)	
	AT-6D	88-18099/18168	AC42-86318/86387	
	SNJ-5	88-18169/18198	BuAer 90821/90850 (AC42-86388/86417)	
	AT-6D	88-18199/18268	AC42-86418/86487	
	SNJ-5	88-18269/18308	BuAer 90851/90890 (AC42-86488/86527)	
	AT-6D	88-18309/18343	AC42-86528/86562	9331
NA-89	(B-28 production—canceled)		USAAC	—
NA-90	B-25C-5	90-11819/11980	AC42-53332/53493 (Netherlands N5-122/283)	162
NA-91	P-51	91-11981/12130	AC41-37320/37469	150
NA-92	(Boeing B-29 production—canceled)		USAAF	—
NA-93	B-25C-15	93-12491/12640	AC42-32383/32532	150
NA-94	B-25C-10	94-12641/12790	AC42-32233/32382	150
	B-25E	(94-12789)	(AC42-32281)	(1)
NA-95	(Consolidated B-24 production)			
	B-24G	95-15131/15560	AAF42-78045/78474	430
	B-24J	95-15561/15880	AAF42-78474/78794	
		95-36536/36751	AAF44-28061/28276	536
NA-96	B-25C-20,-25	96-16381/16680	AAF42-64502/64801	300
	B-25G-5,-10	96-16681/16780	AAF42-64802/64901	
		96-20806/21105	AAF42-64902/65201	400
NA-97	A-36	97-15881/16380	AAF42-83663/84162	500
NA-98	B-25H	98-21106/22105	AAF43-4105/5104	1000
NA-99	P-51A	99-22106/22415	AAF43-6003/6312	310
NA-100	B-25D-20,	100-20606/20805,	AAF42-87113/87312,	
	-25,-30,	-23306/23605,	AAF42-87313/87612,	
	-35	-23606/24195	AAF43-3280/3869	1090
NA-101	XP-51B	(91-12013,-12082)	(AC/AAF41-37352,-37421)	(2)
NA-102	P-51B-1	102-24541/24940	AAF43-12093/12492	400
NA-103	P-51C-1	103-22416/22815,	AAF42-102979/103378,	
	-5, -10	-25933/26882	AAF42-103379/103978,	
			AAF43-24902/25251	1350
NA-104	P-51B-5,	104-22816/23305,	AAF42-6713/7202	
	-10, -15	-24431/24540,	AAF42-106429/106538,	
		-24941/25340,	AAF43-6313/6712,	
		-25343/25930	AAF42-106541/106978,	
			AAF43-24752/24901	1588
NA-105	XP-51F	105-26883/26885	AAF43-43332/43334	3

NA Model	Designation	Factory Serials	Customer Serials	Number Built
NA-105	XP-51G	105-25931/25932	AAF43-43335/43336	2
	XP-51J	105-47446/47447	AAF43-76027/76028	2
NA-106	P-51D	106-25341/25342	AAF106539/106540	2
NA-107	(950 P-51C transfered to NA-103)		USAAF	--
NA-108	B-25J-1,	108-24196/24395	AAF43-3870/4069	
	-5, -10	-24396/24430	AAF43-4070/4104	
	-15, -20	-34486/35235	AAF43-27473/28222	
	-25, -30	-35236/35535	AAF43-35946/36245	
	-35	-31986/38385	AAF44-28711/31110	
		-37186/37585	AAF44-31111/31510	
		-47446/47651	AAF44-86692/86897	
		-47652/47669	AAF45-8801/8818	
		-47671/47674	AAF45-8820/8823	
		-47676/47679	AAF45-8825/8828	
		-47683	AAF45-8832	4318
(72 incomplete but flyable B-25J were included in the inventory:)				
		108-47670, -47675	AAF45-8819, -8824	
		-47680/47682	AAF45-8829/8831	
		-47684/47750	AAF45-8833/8899	72
NA-109	P-51D-5,	109-26886/28885,		
	-10, -15	-35536/36035	AAF44-13253/15752	2500
NA-110	P-51D-1	110-34386/34485	(Australia, unassembled)	100
NA-111	P-51C-10	111-28886/29285	AAF44-10753/11152	400
	P-51D-5	111-29286/29485	AAF44-11153/11352	200
	P-51K	111-29486/30885,		
		111-36036/36135	AAF44-11353/12852	1500
	P-51D-20	111-36136/36535	AAF44-12853/13252	400
NA-112	(P-51D-5 production--2000 transfered to NA-109)			--
NA-113	(B-25H production--canceled)		USAAF	--
NA-114	(B-25J production--1050 transfered to NA-108)			--
NA-115	(B-25J production--2400 transfered to NA-108)			--
NA-116	(4-engine twin-boom heavy bomber--canceled)			--
NA-117	(P-51H production--1945 canceled, 555 transfered to NA-126)			--
NA-118	(B-24G production--220 canceled, 430 transfered to NA-95)			--
NA-119	AT-6D	119-40086/40166	(Brazil, unassembled)	81
NA-120	XP-82	120-43742/43743	AAF44-83886/43887	2
NA-121	AT-6D	121-41567/42366	AAF44-80845/81644	800
	AT-6F	121-42367/43322	AAF44-81645/82600	956
NA-122	P-51D	122-30886/31885	AAF63160/64159	
		122-31886/31985	AAF44-72027/72126	
		122-38586/40085	AAF44-72127/73626	
		122-40167/41566	AAF44-73627/75026	4000
NA-123	P-82B	123-43746/43763	AAF44-65160/65177	18
	P-82C	123-43764	AAF44-65178	1
	P-82D	123-123-43765	AAF44-65179	1
NA-124	P-51D	124-44246/44845	AAF/AF44-84390/84989	
		124-48096/48496	AAF/AF45-11343/11743	1001

NA Model	Designation	Factory Serials	Customer Serials	Number Built
NA-125	(B-24G production—300 canceled)		USAAF	—
NA-126	P-51H	126-37586/38140	AAF/AF44-64160/64714	555
NA-127	(P-51D production—1400 transfered to NA-126/later canceled)			—
NA-128	(AT-6D production—1200 canceled)		USAAF	—
NA-129	(USAAF secret project—canceled)		USAAF	—
NA-130	XB-45	130-38391/38393	AAF/AF45-59479/59481	3
NA-131	(USAAF secret project—canceled)		USAAF	—
NA-132	(B-29 tail assemblies—canceled)		USAAF	—
NA-133	(Navy secret project—canceled)		USN	—
NA-134	XFJ-1	134-55996/55998	BuAer 39053/39055	3
NA-135	(Fairchild C-82 production—reduced from 792)			
	C-82N	135-49496/49498	AAF/AF45-25436/25438	3
NA-136	(USAAF/AF experimental project—canceled)			—
NA-137	(Lockheed P-80N production—1000 canceled)			—
NA-138	(P-51D production—629 canceled)		USAAF	—
NA-139	(P-51H production—2500 canceled)		USAAF	—
NA-140	XP-86	140-38424/38426	AAF/AF45-59597/59599	3
NA-141	FJ-1	141-38394/38423	BuAer 120342/120371	30
NA-142	XSN2J-1	142-38427/38428	BuAer 121449/121450	2
NA-143	Navion	143-1/2 (NAV-4-1/2)	Civil NX18928/18929	2
NA-144	F-82E	144-38141/38240	AAF/AF 46-255/354	100
NA-145	Navion	NAV-4-1/2 (143-1/2)	(Civil NX18928/18929)	(2)
		NAV-4-3/1027 (examples of a few known registrations)	Civil NX91100/91101, NC91102/91103, 91106/91135/ 91136/91170/91455/91528/ 91565/91785/91786,etc.	1025
NA-146	XAJ-1	146-38429/38431	BuAer 121460/121462	3
NA-147	B-45A	147-43401/43497	AAF/AF47-1/97	97
NA-148	(Commercial transport project—canceled)			—
NA-149	F-82F	149-38291/38381	AAF/AF46-405/495	91
NA-150	F-82G	150-38241/38269	AAF/AF46-355/383	29
	F-82H	150-38270/38274	AAF/AF46-384/388	5
	F-82G	150-38275/38290	AAF/AF46-389/404	16
	F-82H	150-38382/38390	AAF/AF46-496/504	9
NA-151	F-86A-1	151-33432/38464	AF47-605/637	33
	F-86A-5	151-43498/43685	AF48-129/316	188
NA-152	(F-86B production—188 of 190 transfered to NA-151 as A-5s)			—
NA-153	B-45C	153-38477/38486	AF48-1/10	10
	RB-45C	153-38487/38519	AF48-11/43	33
NA-154	L-17A	NAV-4-1028/-4-1110	AF47-1297/1379	83
NA-155	(Development of XAJ-1—mock-up only)		USN	—
NA-156	AJ-1	156-38465/38476	BuAer 122590/122601	12
Revised serial numbering system beginning with NA-157:				
NA-157	YF-93A	157-1/2	AF48-317/318	2
NA-158	(Development of XA2J-1—engineering only)			—
NA-159	XT-28	159-1 (static test)	AF49-1491	1

NA Model	Designation	Factory Serials	Customer Serials	Number Built
NA-159	XT-28	159-2/-3	AF48-1371/1372	2
	T-28A	159-4/268	AF49-1492/1756	265
NA-160	AJ-1	160-1/28	BuAer 124157/124184	28
NA-161	F-86A-5	161-1/333	AF1007/1339	333
NA-162	(B-45 production—2 B-45C and 49 RB-45C canceled)			—
NA-163	XA2J-1	163-1/ 2	BuAer 124439/124440	2
NA-164	YF-86D	164-1/2	AF50-577/578	2
NA-165	F-86D-1	164-1/37	AF50-455/491	37
	F-86D-5	164-38/63	AF50-492/517	26
	F-86D-10	164-64/99	AF50-518/553	36
	F-86D-15	164-100/122	AF50-554/576	33
	F-86D-15	164-123/153	AF50-704/734	31
NA-166	(F-93 production—1 all-weather F-93A canceled)			—
NA-167	F-86J	(161-63 modified)	(AF49-1069)	(1)
NA-168	T-6G	168-1/371	AF49-2897/3267	
		168-372/381	AF50-1317/1326 (ANG)	
		168-382/440	AF49-3278/3336	
		168-441/450	AF50-1277/1286 (ANG)	
		168-451/500	AF49-3337/3386	
		168-501/510	AF50-1287/1296 (ANG)	
		168-511/560	AF49-3387/3436	
		168-561/570	AF50-1297/1306 (ANG)	
		168-571/620	AF49-3437/3486	
		168-621/630	AF50-1307/1316 (ANG)	
		168-631/680	AF49-3487/3536	
		168-681/690	AF49-3268/3277	
		168-691	AF49-3537	691
	LT-6G	168-692/750	AF49-3538/3596	59
NA-169	AJ-1	169-1/15	BuAer 124850/124864	15
NA-170	F-86E-1	170-1/60	AF50-579/638	60
	F-86E-5	170-61/111	AF50-639/689	51
NA-171	T-28A	171-1/125	AF50-195/319	125
NA-172	F-86E-10	172-1/132	AF51-2718/2849	132
	F-86F-1	172-133/210	AF51-2850/2927	78
	F-86F-5	172-211/226	AF51-2928/2943	16
	F-86F-10	172-227/260	AF51-12936/12969	34
	F-86F-15	172-261/360	AF51-12970/13069	100
NA-173	F-86D-25	173-1/88	AF51-5857/5944	88
	F-86D-30	173-89/288	AF51-5945/6144	200
	F-86D-35	173-289/638	AF51-6145/6262, AF51-8274/8505	350
NA-174	T-28A	174-1/334	AF51-3463/3796	
		174-335/744	AF51-7482/7891	744
NA-175	AJ-2P	175-1/12	BuAer 128043/128054	
		175-13/23	BuAer 128185/129195	23
NA-176	F-86F-20	176-1/100	AF51-13070/13169	100
	F-86F-25	176-101/441	AF51-13170/13510	341
NA-177	F-86D-20	177-1/188	AF51-2944/3131	188
NA-178	(F-86 production—184 F-86E canceled)			—
NA-179	XFJ-2	179-1/2	BuAer 133754/133755	2
NA-180	YF-100A	180-1/2	AF52-5754/5755	2
NA-181	FJ-2	181-1/200	BuAer 131927/132126	200
NA-182	T-6G	182-1/824	AF51-14314/15137	824
NA-183	AJ-2P	183-1/7	BuAer 130422/130425, BuAer 134073/134075	7
NA-184	AJ-2	184-1/55	BuAer 130405/130421, BuAer 134035/134072	55
NA-185	XFJ-2B	185-1	BuAer 133756	1
NA-186	(T-6J design data for production by Canadian Car & Foundry)			—
NA-187	F-86H-1	187-1/115	AF52-1975/2089	115
	F-86H-5	187-116/175	AF52-2090/2124, AF52-5729/5753	60
NA-188	T-6G	188-1/107	AF51-15138/15237, AF51-16071/16077	107
NA-189	T-28A	189-1/59	AF52-1186/1242, AF52-3497/3498	59
NA-190	F-86D-40	190-1/450	AF52-3598/4047	450
	F-86D-45	190-451/600	AF52-4048/4197	150
	F-86D-50	190-601/707	AF52-4198/4304	107
	F-86D-55	190-708/901	AF52-9983/10176	194
NA-191	F-86F-30	191-1/859	AF52-4305/5163	859
	F-86F-35	191-860/967	AF52-5164/5271	108
NA-192	F-100A-1	192-1/10	AF52-5756/5765	10
	F-100A-5	192-11/23	AF52-5766/5778	13
	F-100A-10	192-24/63	AF53-1529/1568	40
	F-100A-15	192-64/103	AF53-1569/1608	40
	F-100A-20	192-104/203	AF53-1609/1708	100
NA-193	F-86F-25	193-1/259	AF52-5272/5530	259
NA-194	FJ-3	194-1/389	BuAer 135774/136162	389
NA-195	T-6G	195-1/11	AF51-17354/17364	11
NA-196	FJ-3	(FJ-2 modified with J65-W-2 engine)		(1)
NA-197	T-6G	197-1/50	AF52-8197/8246	50
		8002-1/60	AF53-4555/4614	60
NA-198	(SNJ-8 production—240 canceled)		USN	—
NA-199	T-28B	199-1/2	BuAer 137636/137637	2
NA-200	T-28B	200-1/173	BuAer 137638/137810	173
	T-28B	200-174/438	BuAer 138103/138367	265
NA-201	F-86D-55	201-1/225	AF53-557/781	225
	F-86D-60	201-226/515	AF53-782/1071	
		201-516/551	AF53-3675/3710	
		201-552/624	AF53-4018/4090	399
NA-202	F-86F-35	202-1/157	AF53-1072/1228	157
NA-203	F-86H-5	203-1/300	AF53-1229/1528	300
NA-204	TF-86F (F-86F-30, 191-712, AF52-5016, modified as 2-seater)			(1)

NA Model	Designation	Factory Serials	Customer Serials	Number Built
NA-205	YF-86K	(Two F-86D modified for tests of NATO version)		(2)
NA-206	F-86F	(Long-lead-time material)	USAF	(240)
NA-207	F-86K	(Airframe assemblies and parts for NATO aircraft)		50
NA-208	FJ-4	208-1/2	BuAer 139279/139280	2
NA-209	FJ-4	209-1/43	BuAer 139281/139323	
		209-44/175	BuAer 139424/139555	
		209-176/221	BuAer 141444/141489	221
NA-210	(Two F-86F modified as rocket-firing prototypes)			(2)
NA-211	(F-100 design and production engineering, including mock-up)			—
NA-212	F-107A	212-1/3	AF55-5118/5120	3
NA-213	F-86K	213-1/120	AF54-1231/1350	120
NA-214	F-100C-1	214-1/70	AF53-1709/1778	70
NA-215	FJ-3	215-1/69	BuAer 139210/138278	
		215-70/149	BuAer 141364/141443	149
NA-216	TF-86F	216-1 (F-86F 202-157, modified as 2-seater)	(AF53-1228)	(1)
NA-217	F-100C-1	217-1/30	AF54-1750/1769	30
	F-100C-5	217-31/75	AF54-1770/1814	45
	F-100C-15	217-76/120	AF54-1815/1859	45
	F-100C-20	217-121/230	AF54-1860/1969	110
	F-100C-25	217-231/381	AF54-1970/2120	151
NA-218	NA-218	218-1 (T-28B)	Japan	1
NA-219	T-28B	219-1/51	BuAer 140002/140052	51
NA-220	(FJ-4 production—45 transfered to NA-209)			—
NA-221	F-86K	(Airframe assemblies and parts for NATO aircraft)		70
NA-222	F-100C-10	222-1/25	AF55-2709/2733	25
NA-223	F-100D-1	223-1/12	AF54-2121/2132	12
	F-100D-5	223-13/31	AF54-2133/2151	19
	F-100D-10	223-32/101	AF54-2152/2221	70
	F-100D-15	223-102/183	AF54-2222/2303	82
	F-100D-20	223-184/283	AF54-3502/3601	100
	F-100D-25	223-284/383	AF54-3602-3701	100
	F-100D-30	223-384/496	AF54-3702/3814	113
NA-224	F-100D-35	224-1/10	AF55-2734/2743	10
	F-100D-40	224-11/50	AF55-2744/2783	40
	F-100D-45	224-51/130	AF55-2784/2863	80
	F-100D-50	224-131/175	AF55-2864/2908	45
	F-100D-55	224-176/221	AF55-2909/2954	46
NA-225	T-28C	(T-28Bs BuAer 138185 and 138187 modified with Hamilton Standard propellers and arrestor gear)		(2)
NA-226	T-28C	226-1/25	BuAer 140053/140077	25
	T-28C	226-26/243	BuAer 140449/140666	218
NA-227	F-86F-40	227-1/215	AF55-3816/4030	
		227-216/280	AF55-4983/5047	280
NA-228	(FJ-3 production—80 transfered to NA-215)			—
NA-229	(FJ-4 production—46 transfered to NA-209)			—
NA-230	TF-100C	(F-100C, AF54-1966, modified as a 2-seat trainer)		(1)

NA Model	Designation	Factory Serials	Customer Serials	Number Built
NA-231	F-86F	(Sets of airframe assemblies, no factory serials)	AF55-5048/5117	70
NA-232	F-86K	(Sets of airframe assemblies, no factory serials)	AF55-4881/4936	56
NA-233	(A3J-1 design, engineering and mock-up, Phase I)			—
NA-234	FJ-4F	(FJ-4s modified for mono-propellant rocket tests)		(2)
NA-235	F-100D-60	235-1	AF56-2903	1
	F-100D-90	235-2	AF56-2904	1
	F-100D-60	235-3/44	AF56-2905/2946	42
	F-100D-90	235-45	AF56-2947	1
	F-100D-60	235-46/60	AF56-2948/2962	15
	F-100D-65	235-61/120	AF56-2963/3022	60
	F-100D-70	235-121/240	AF56-3023/3142	120
	F-100D-75	235-241/296	AF56-3143/3198	56
	F-100D-90	235-297/444	AF56-3199/3346	148
NA-236	LRI	(Engineering for USAF long-range interceptor)		—
NA-237	FBX	(Development of USAF Fighter-Bomber system)		—
NA-238	F-86F	(Sets of airframe assemblies, no factory serials)	AF56-2773/2882	100
NA-239	(SAC bomber design, engineering and mock-up)			—
NA-240	X-15	240-1/3	AF 56-6670/6672	3
NA-241	(T2J-1 design and engineering)		USN	—
NA-242	F-86K	(Sets of airframe assemblies, no factory serials)	AF56-4116/4160	45
NA-243	F-100F-1	243-1/15	AF56-3725/3739	15
	F-100F-5	243-16/45	AF56-3740/3769	30
	F-100F-10	243-46/195	AF56-3770-3919	150
	F-100F-15	243-196/295	AF56-3920-4019	100
NA-244	FJ-4B	244-1/151	BuAer 143493/143643	151
NA-245	F-100D-80	245-1/28	AF56-3351/3378	28
	F-100D-85	245-29/113	AF56-3379/3463	113
NA-246	UTX	246-1	Civil N4060K	1
NA-247	YA3J-1	247-1/2	BuAer 145157/145158	2
	A3J-1	247-3/11	BuAer 146694/146702	9
NA-248	FJ-4F	(FJ-4s modified for bi-propellant rocket tests)		(2)
NA-249	T2J-1	249-1/6	BuAer 144217/144222	6
NA-250	(No designation or contract number released)			—
NA-251	FJ-4F	(FJ-4s modified for variable-thrust rocket tests)		(2)
NA-252	T-28C	252-1/56	BuAer 146238/146293	56
NA-253	T2J-1	253-1/121	BuAer 145996/146015, BuAer 147430/147530	121
NA-254	(F-100D production—13 canceled)		USAF	—
NA-255	F-100F-20	255-1/29	AF58-1205/1233	29
NA-256	F-86F	(Sets of airframe assemblies, no factory serials)	AF57-6338/6457	120
NA-257	F-108	(Design for long-range interceptor)		—
NA-258	(A3J-1 engineering, Phase II--transfered to NA-247)			---

NA Model	Desig- nation	Factory Serials	Customer Serials	Number Built
NA-259	(B-70 engineering, Phase I) (		USAF	—
NA-260	T-28A	(Modification to T-28A)	USAF	(1)
NA-261	F-100F-15	261-1/9	AF58-6975/6983	9
NA-262	F-100F-15	262-1/6	AF59-2558/2563	6
NA-263	A3J-1	263-1/14	BuAer 147850/147863	14
A-264	(B-70 engineering, Phase II)		USAF	—
NA-265	T-39A	265-1/5	AF59-2868/2872	5
	T-39A	265-6/36	AF60-3478/3508	31
	T-39A	265-37/88	AF61-634/685	52
NA-266	T2J-1	266-1/90	BuAer148150/148239	90
NA-267	(B-70 engineering, Phase II, Part 2)		USAF	—
NA-268	(F-108 engineering, Phase II)		USAF	—
NA-269	A3J-1	269-1/10	BuAer 148924/148933	10
	A3J-2	269-11/34	BuAer 149276/149299	24
	YA-5C	269-35/40	BuAer 149300/149305	6
	A3J-2(A-5B)	269-41/52	BuAer 149306/149317	12
NA-270	T-39B	270-1/6	AF59-2873/2874, AF60-3474/3477	6
NA-271	(T-39A for FAA Type Certificate)		Commercial	—
NA-272	(A3J-1 long-lead-time material—canceled)			—

(The following NA charge numbers were assigned to missile projects: NA-273, -275, -289, -299, -309, -314, -319, -322, -325, -326, -334, -337, -339, -341, -347, -350, -351, -353, -357, -359, -361, -362, -363, -364, -366, -379, -383, -384, -387, -388, -389, -393, -394, -395, -399, -401, -406, -407)

NA Model	Desig- nation	Factory Serials	Customer Serials	Number Built
NA-274	YB-70	(Weapon system development)	USAF	—
NA-276	T-39A	276-1/55	AF62-4448/4502	55
NA-277	T-39D	277-1/10	AF62-150542/150551	10
NA-278	XB-70	62-001, 62-207	AF62-001, 62-207	2
NA-279	RA-5C	279-1/20	BuAer 150823/150842	20
NA-280	YT-2B	(Modified T2J-1)	(BuAer 144218, 145997)	(2)
NA-281	XB-70	(Flight test program)	(AF62-001)	(1)
NA-282	Sabreliner	282-1/35	(Remmert-Werner)	35
NA-283	RA-5C	279-21/43	BuAer 151615/151634, BuAer 151726/151728	23
NA-284	YAT-28E	(Modified T-28A)	(AF51-3786, 52-1242)	(2)
NA-285	T-39D	285-1/32	BuAer 150969/150992, 151336/151343	32
NA-286	XB-70	(Termination, ship #3)	---	---
NA-287	Sabreliner	(Autonetics test bed)	---	---
NA-288	T-2B	288-1/10	BuAer 152382/152391	10
NA-290	Sabreliner	282-36-49	Series 40 (Remmert-Werner)	14
NA-291	T-2B	288-11/46	BuAer 152440/152475	36
NA-292	Sabreliner	282-50/65	Series 40 (Remmert-Werner)	16
NA-293	Sabreliner	282-66/98	Series 40 (Remmert-Werner)	33
NA-294	T-2B	294-1/18	BuAer 153538/153555	18
NA-295	(Proposed VAL aircraft--canceled)			---

NA Model	Desig- nation	Factory Serials	Customer Serials	Number Built
NA-296	RA-5C	279-44/70 (Conversion of A-5A)	BuAer 145157, 146695, 146696, 146698, 146701, 146702, 147850, 147858, 148925, 148932, 148933, 149276/149281, 149283/149289, 149291, 148293, 149294	(27)
NA-297	Sabreliner	282-99/123	Series 40 (Remmert-Werner)	25
NA-298	RA-5C	263-3/5, 7, 8, 10/12, 279-3, 5, 6, 30/34, (Conversion of A-5A)	BuAer 147852/147854, 147856, 147857, 147859/147861, 148926. 148928, 148929, 149295/149299	(16)
NA-300	OV-10A	300-1/7	BuAer 152879/152885	7
NA-301	(OV-10 transport--canceled)		---	---
NA-302	(OV-10A reconnaissance--canceled)			---
NA-303	XB-70	(Flight test of 281-1)	(AF62-001)	(1)
NA-304	(Fabrication of Boeing SST wing pivot test assembly)			---
NA-305	OV-10A	305-1/223 (114 to USMC; 109 to USAF)	BuAer 155390/155503 AF66-13552/13562, AF67-14604/14701	223
NA-306	Sabreliner	306-1/2	Series 60 (Remmert-Werner)	2
NA-307	(T-28C production--canceled)		---	---
NA-308	Sabreliner	306-3/37	Series 60 (Remmert-Werner)	35
NA-310	T-2B	310-1/36	BuAer 155206/155241	36
NA-311	(OV-10A training, technical services, publications)			---
NA-312	(OV-10A spares and repair parts)			---
NA-313	(OV-10A special support equipment)			---
NA-315	XB-70	(Flight test of 281-1)	(AF62-001)	(1)
NA-316	RA-5C	316-1/46 (last 10 canceled)	BuAer 156608/156653	36
NA-317	(OV-10A night forward control program)			(20)
NA-318	T-2C	318-1/48	BuAer 156686/156733	48
NA-320	Sabreliner	306-38/55	Series 60	18
NA-321	OV-10A	321-224A-110/ 271A-157	AF68-3784/3831	48
NA-323	(VFX/F-14 program)		USN	---
NA-324	(VSX/F-14 program)		USN	---
NA-327	Sabreliner	327-1/26	Series 60	26
NA-328	(VFX/F-14 contract definition)		USN	---
NA-329	(VFX/F-14 proposal sustaining effort)			---
NA-330	(VFX/F-14 development--canceled)			---
NA-331	(VFX/F-14 production--canceled)			---
NA-332	T-2C	332-1/36	BuAer 157030/157065	36
NA-333	(VFX/F-14 proposal sustaining effort)			---
NA-335	(F-15 development)		USAF	---
NA-336	Sabreliner	---	Series 70	25

NA Model	Designation	Factory Serials	Customer Serials	Number Built
NA-338	OV-10B	--- (Germany)	BuAer 158292/158297	6
	OV-10B(Z)	--- (Germany)	BuAer 158298/158309	12
NA-339	(AX program)	---	USN	---
NA-340	T-2C	---	BuAer 158310/158333	24
NA-342	OV-10C	--- (Thailand)	BuAer 158396/158411	16
NA-343	Sabreliner (NA-327 Series 60 redesignated)			(26)
NA-344	Sabreliner (NA-336 Series 70 redesignated)			(25)
NA-345	Sabreliner (NA-320 Series 60 redesignated)			(18)
NA-346	T-2C	---	BuAer 158575/158610	36
NA-348	(OV-10 fixed-wing ship aircraft)		USN	---
NA-349	(ABC interceptor)		USAF	---
NA-352	T-2C	---	BuAer 158876/158911	36
NA-354	OV-10E	--- (Venezuela)	BuAer 159057/159064	8
NA-355				
NA-356	XFV-12A	356-1(-2 unfinished)	BuAer 161080(161081)	1
NA-358	T-2D	--- (Venezuela)	BuAer 159330/159341	12
NA-360	OV-10C	--- (Thailand)	BuAer 159134/159149	16
NA-365	(Carrier onboard delivery aircraft)		USN	--
NA-367	T-2C	---	BuAer 159150/159173, BuAer 159704/159727	48
NA-369	Sabreliner	---	Series 40	16
NA-370	Sabreliner	---	Series 70	9
			Series 80	2
NA-371	Sabreliner	---	Series 60	18
NA-372	CT-39G	---	BuAer 158843/158844, BuAer 159361/159365, BuAer 160053/160058	13
NA-373	Sabreliner	(USN/Sierra TACANs sets)		(6)
NA-374	Sabreliner 80	(Sea Sabre proposal for USN/USCG)		(40)
NA-375	Sabreliner	---	Series 80	25
NA-376	(Sabre 75A conversion drawings from Sabre 75)			---
NA-377	Sabreliner	---	Series 60	10
NA-378	Sabre	---	Series 75A	24
NA-380	(T-2C long-lead time materials)		USN	—
NA-381	(Long Endurance Sensor Carrier)		USN	—
NA-382	(Medium V/STOL Transport)		USN	—
NA-385	(Sabre Series 85—canceled)		—	—
NA-386	Sabreliner	---	Series 60	18
NA-390	(Sabreliner 40A/60/80 material, parts and assemblies)			—
NA-391	(Sabreliner 40A/60/80 material, parts and assemblies)			—
NA-392	(Sabreliner 40A/60/80 material, parts and assemblies)			—
NA-396	T-2E	--- (Greece)	BuAer 1600059/160098	40
NA-397	OV-10F	--- (Indonesia)	BuAer 160216/160227, BuAer 160295/160298	16
NA-398	T-2D	--- (Venezuela)	BuAer 160228/160239	12
NA-400	(VFX Strike Attack aircraft)		USN	—
NA-402	Sabreliner	---	Series 40A	26
NA-403	Sabreliner	---	Series 40A	15
NA-404	OV-10E	--- (Venezuela)	BuAer 159065/159072	8
NA-405	(T-2 production for Turkey--canceled)			—
Rockwell	Sabreliner Series 40R converted from 40A)			(6)
Rockwell	D481	B-1A	AF74-0158/0160, AF76-0174	4
		B-1B	AF82-0001, 83-0065/0071 AF84-0049/0058, AF85-0059/0092, AF86-0093/0140	100
Rockwell/MBB		X-31A	BuAer 164584/164585	2

Total, North American and North American Rockwell aircraft completed, 1934-1999, including license-built models 66,750

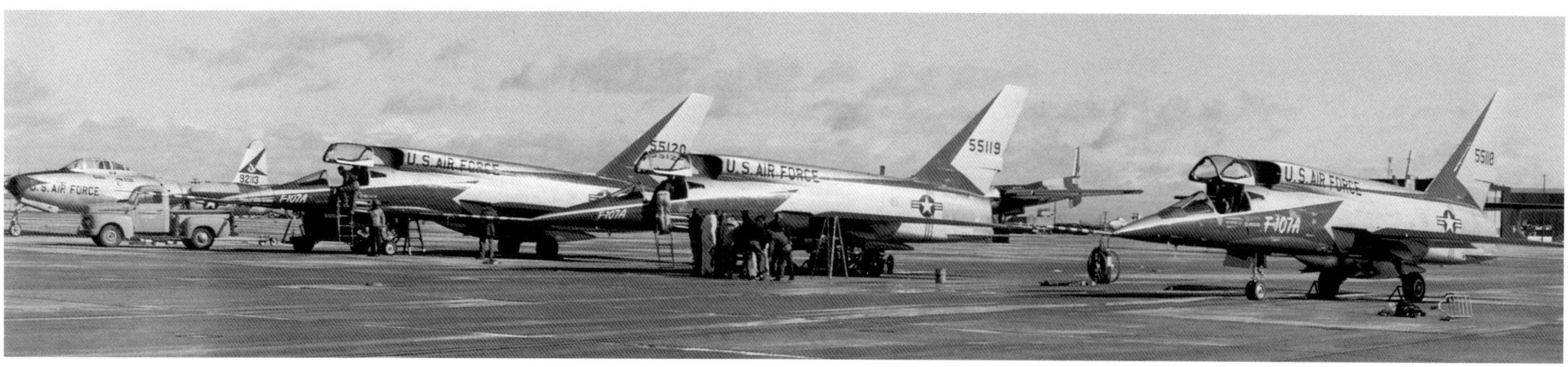

Bibliography

This is a selected list of the more significant published material on the North American and Rockwell aircraft presented in this volume, supplementing the bibliography in Volume 1. It does not include the many standard aviation reference works with North American sections, nor the countless magazine articles relating to NAA and Rockwell aircraft, except for those that have provided more than ordinary information. In addition to the books the author used in his research, the publisher has added other titles of particular interest.

John Leland Atwood and company history:

Russ Murray, *Lee Atwood...Dean of Aerospace,* Rockwell International Corp., Los Angeles, California, 1980.

North American and Rockwell aircraft, all models:

Gordon Swanborough, *North American, An Aircraft Album,* Ian Allan, Shepperton, Surrey, 1973.

Ray Wagner, *American Combat Planes,* Third Enlarged Edition, Doubleday, Garden City, New York, 1982.

Bill Yenne, *Rockwell, The Heritage of North American,* Crescent Books, New York, 1989.

René Francillon, *The Air Guard,* Aerofax Inc., Austin, Texas, 1983.

1960 United States Aircraft, Missiles and Spacecraft, National Aviation Education Council, Washington, D.C., 1960.

North American Aviation Inc., A Brief History of Operations, NAA Inc., June 30, 1945.

Rockwell International brochure, *Lineage of Rockwell International Military Aircraft, 1934-1997.*

North American Aviation Airplanes, folder, c.1948.

History and Highlights, the Columbus Division of North American Rockwell Corporation, May 14, 1970.

Lloyd S. Jones, *U.S. Fighters,* Aero Publishers, Fallbrook, California, 1975.

Jay Miller, "The X-Series," *Aerophile,* January/February and March/April 1977.

Steve Pace, *X-Planes at Edwards,* Motorbooks International, Osceola, Wisconsin, 1995.

X-Planes, Famous Aircraft of the World No. 67, Bunrindo Co. Ltd., Japan, November 1997.

Boeing News, Vol. 57 & 58, various issues, 1998-1999.

NA-142 / XSN2J-1:

Hal Andrews, "SN2J," *Naval Aviation News,* March-April 1995.

NA-159 / T-28 Trojan:

Frank G. Compton, "November Seven Nine Zulu, Story of the North American Nomad," *Sport Aviation,* June 1983.

Steve Ginter, *North American T-28 Trojan,* Naval Fighters Number Five, Simi Valley, California, 1981.

Al Adcock, *T-28 Trojan in Action,* Squadron Signal Publications, Carrollton, Texas, 1989.

Lindsay Peacock, "North American T-28 Trojan," *Scale Aircraft Modelling,* February 1982.

David Brazelton, *The North American T-28,* Profile Publications, Leatherhead, Surrey, 1967.

M.L. Shettle, Jr., *United States Naval Air Stations of World War II,* Volumes 1 and 2, Schaertel Publishing Co., Bowersville, Georgia, 1995 and 1997.

NA-143 / Navion:

C. C. King, "The Navion Airplane, A North American Detour," *North American Retiree News,* September 1997.

Joseph P. Juptner, *U.S. Civil Aircraft,* Volume 8, Aero Publishers Inc., Fallbrook, California, 1980.

Norm Peterson, "Four Generations! Hale Andrews' Ryan Navion 260," *Vintage Airplane,* April 1994.

Budd Davisson, "It's a Grand Old Plane," *Private Pilot,* January 1971.

Dorr B. Carpenter and Mitch Mayborn, *Ryan Guidebook,* Flying Enterprise Publications, Dallas, Texas, 1975.

Ed Cassagneres, *The Spirit of Ryan,* Tab Books, Blue Ridge Summit, Pennsylvania, 1982.

NA-146 / AJ Savage:

Steve Ginter, *North American AJ-1 Savage,* Naval Fighters Number Twenty-Two, Simi Valley, California, 1992.

All-yellow carrier-based T-28C, BuAer 140474.

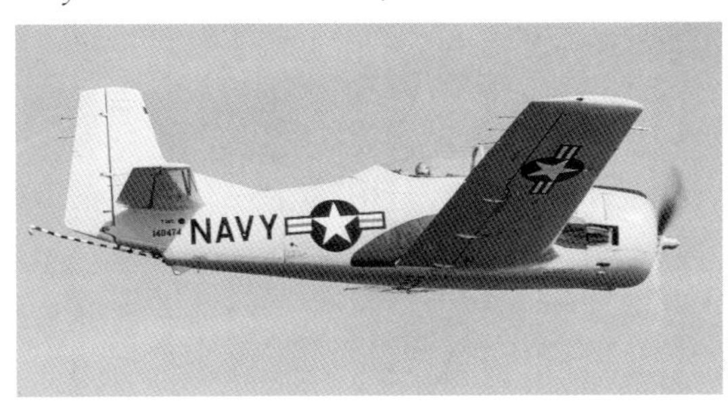

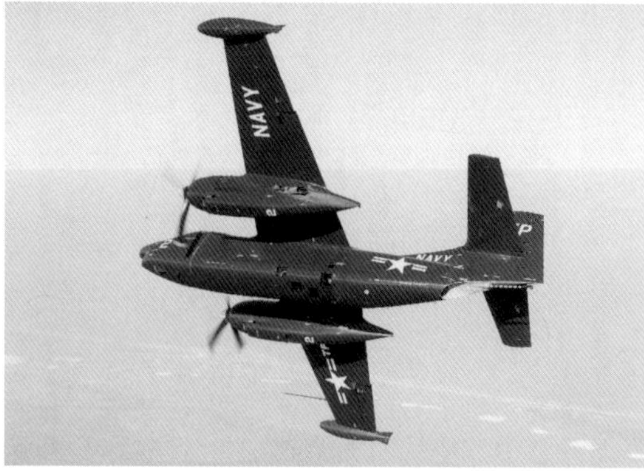

AJ-2P Savage photo reconnaissance, BuAer 128046.

CAPT William E. Scarborough, USN (Ret), "The North American AJ Savage," Parts One and Two, *The Hook,* Fall and Winter 1989.

Steve Pace, "Shortlived Savage," *Airpower,* July 1987.

NA-180 / F-100 Super Sabre:

Ned Barnett, MSGT Dave W. Menard, Phil Friddell and Jim Wogstad, "The Hun, North American F-100 Super Sabre," *Replica in Scale,* Volume 3, Number 2, 1975.

David A. Anderton, *North American F-100 Super Sabre,* Osprey Publishing, London, 1987.

Bert Kinzey, *F-100 Super Sabre in Detail & Scale,* Tab Books, Blue Ridge Summit, Pennsylvania, 1989.

Lou Drendel, *F-100 Super Sabre in Action,* Squadron/Signal Publications, Warren, Michigan, 1973.

Lindsay Peacock, *North American F-100 Super Sabre,* Warpaint No. 4, Hall Park Books Ltd., Husborne Crawley, England.

Robert Robinson and David Menard, *F-100 Super Sabre in Color,* Squadron/Signal Publications, Carrollton, Texas, 1992.

David W. Menard, *USAF Plus Fifteen,* Schiffer Military/Aviation History, Atglen, Pennsylvania, 1993.

"North American Super Sabre," *Illustrated Encyclopedia of Aircraft,* Issue 57, Orbis Publishing Ltd., London, 1982.

NA-212 / F-107A and NA-257 / F-108 Rapier:

William J. Simone, "One Oh Seven, North American's F-107A," *Aerophile,* Volume 1 Number 5, August 1978.

Steve Pace, "Supersonic Cavaliers," *Airpower,* November 1986.

Standard Aircraft Characteristics, F-108A, Secretary of the Air Force, 57WC 4984, Pre-Mockup, 2 May 58 and 1 Oct 58, and Mockup, 12 Jun 59.

North American Aviation, Inc., Press Release, *U.S. Air Force F-108 Rapier,* News Releases, May 30, 1958-April 14, 1959.

Steve Pace, *X-Fighters,* Motorbooks International, Osceola, Wisconsin, 1991.

NA-223 / A3J Vigilante:

Michael Grove and Jay Miller, *North American Rockwell A3J/A-5 Vigilante,* Aerofax Minigraph 9, Aerofax Inc., Arlington, Texas, 1989.

"North American Vigilante," *The Illustrated Encyclopedia of Aircraft,* Issue 197, Orbis Publishing Ltd., 1985.

"Vigilante, Tactical Strike and Reconnaissance," *Flight International,* 13 August 1964.

"Vigilante, The Eyes of the Fleet," *Air International,* November 1975.

NA-240 / X-15:

Ben Guenther, Jay Miller and Terry Panopalis, *North American X-15/X-15A-2,* Aerofax Datagraph 2, Aerofax Inc., Arlington, Texas, 1985.

The X-15 Research Airplane, National Aeronautics and Space Administration, U.S. Government Printing Office, Washington, D.C., 1963.

"Special Report, X-15 Ready for Man's Conquest of Space," *Los Angeles Skywriter,* September 1959.

Why Manned Aircraft, X-15 Manned Research Vehicle, North American Aviation, Inc. Los Angeles Division, c.1960.

"North American X-15: The Black Bullet," *The Illustrated Encyclopedia of Aircraft,* Issue 187, Orbis Publishing Ltd., 1985.

"North American X-15," *Airplane,* Part 179, Orbis Publishing Ltd., 1993.

NA-241 / T2J Buckeye:

Steve Ginter, *North American Rockwell T-2 Buckeye,* Naval Fighters Number Fifteen, Simi Valley, California, 1987.

T-2 Jet Trainer, The Trainer/Attack Aircraft for Complete Jet Training, NR74H-146, Columbus Aircraft Division, Rockwell International, c.1974.

"Rockwell Buckeye, Success the Second Time Round for the T-2," *Air Enthusiast,* October 1973.

NA-246 / T-39 Sabreliner:

"North American T-39 Sabreliner," *Replica in Scale,* Volume 3, Number 3, 1976.

"Special Report, Twin-Jet Trainer Joins Sabre Family," *Los Angeles Skywriter,* September 1958.

"Versatile Twin," *Royal Air Force Flying Review,* March 1962.

Kenneth Munson, *U.S. Commercial Aircraft,* Janes

Publishing Co. Ltd., London, 1982.

NA-259 / B-70 Valkyrie:

XB-70-2-1, Development Manual (General Airplane) USAF Series XB-70A Aircraft, Air Force NAA, Los Angeles, California, 30 July 1966, Changed 31 March 1967.

B-70 "Valkyrie" Manned Mach 3 Intercontinental Weapon System, brochure, Los Angeles Division, North American Aviation, Inc., 1961.

Thomas G. Foxworth, "North American XB-70, Half Airplane—Half Spacecraft," Parts 1 and 2, *Historical Aviation Album,* Volumes VII and VIII, 1969-1970.

Iain Pike, "B-70 The State of the Art Improver," *Flight International,* 25 June and 2 July, 1964.

"Valkyrie—the Bomber that Never Was," *Flying Review International,* January 1965.

"Flight Diary of the Valkyrie," *Flying Review International,* May 1966.

"XB-70 Valkyrie: the Bomber with the Mostest," *Aeroplane Monthly,* April 1975.

"Valkyrie: Doomed Dream," *The Illustrated Encylopedia of Aircraft,* Issue 214, Orbis Publishing Ltd., London, 1985.

Steve Pace, "Triplesonic Twosome," *Wings,* February 1986.

Mike Machat, "Valkyrie Flight," *Airpower,* November 1992.

NA-300 / OV-10 Bronco:

Jim Mesko, *OV-10 Bronco in Action,* Squadron/Signal Publications, Carrollton, Texas, 1995.

Robert Mills, "North American OV-10 Bronco," *Replica in Scale,* August 1972.

Donald E. Fink, "North American Accelerates COIN Work," *Aviation Week & Spacec Technology,* February 8, 1965.

"OV-10: Bucking Bronco," *Airplane,* Part 63, Orbis Publishing Ltd., 1990.

Rockwell B-1 Lancer:

Don Logan and Jay Miller, *Rockwell International B-1A/B,* Aerofax Minigraph 24, Aerofax Inc., Arlington, Texas, 1986.

Steve Pace, *Boeing North American B-1 Lancer,* Warbird Tech Volume 19, Specialty Press, North Branch, Minnesota, 1998.

David W.H. Godfrey, "B-1, Airborne Stategic Deterrent," *Air International,* February 1975.

James L. H. Peck, "Bomber One," *Airpower,* March 1986.

"Rockwell B-1, the Fast and Low Bomber," *The Illustrated Encyclopedia of Aircraft,* Issue 203, Orbis Publishing Ltd., 1985.

"Rockwell B-1," *Airplane,* Part 184, Orbis Publishing Ltd., 1993.

Rockwell XFV-12A:

Clarence A. Robinson, Jr., "XFV-12 May Spur Navy VTOL Family," *Aviation Week & Space Technology,* April 16, 1973.

CAPT Gerald O'Rourke, "A New Vertical Approach," *Air Enthusiast International,* February 1974.

Craig Covault, "XFV-12A Ground Tests 90% Completed," *Aviation Week & Space Technology,* September 12, 1977.

"XFV-12A Undergoes Tethered Tests," *Aviation Week & Space Technology,* July 17, 1978.

Rockwell/MBB X-31:

Press Information, *Rockwell International/MBB X-31 EFM,* North American Aircraft, May 1990.

"X-31 First Flight is Picture Perfect!!!," *Rockwell News,* October 24, 1990.

"X-31 Resumes Flight Test at NASA," *Rockwell News,* May 7, 1992.

X-31 Enhanced Fighter Maneuverability Demonstrator Fact Sheet, NASA Facts On Line, Dryden Flight Research Center, April 2, 1999.

Missiles / Unpiloted Aircraft / Spacecraft:

Space Shuttle, For Down to Earth Benefits, brochure, Space Division, Rockwell International, July 1974.

North American Rockwell Skyline, Special Issue: Space Shuttle, Volume 30, Number 2, 1972.

HiMAT brochure NA-78-893, Rockwell International, Los Angeles, California, September 1978.

Clare E. Wise, "Fighter of the Future Poised for Flight," *Machine Design,* June 22, 1978.

HiMATs were B-52-launched, controlled from the ground.

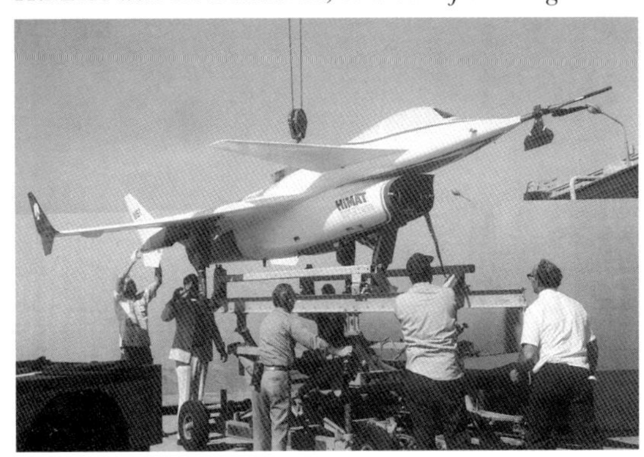

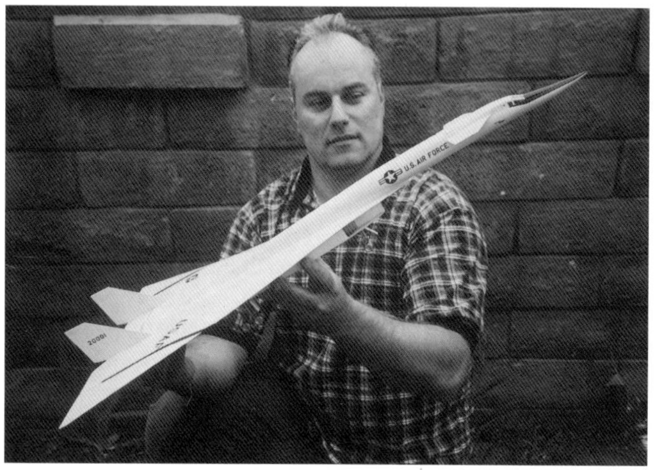

KEVIN THOMPSON, the author of *North American Aircraft 1934-1999,* Volume 2, is a long-time aviation enthusiast, an award-winning scale model builder and a member of the Planes of Fame Museum at Chino, California, where he has also been an airshow announcer. Professionally a skilled Mercedes-Benz, BMW and Ferrari technician, he is an expert on the history of Formula 1 Grand Prix racing and a contributor to *Forza* magazine. Kevin is married and has two sons participating in Civil Air Patrol and Marine Corps activities.

Rockwell International

BOEING